# HYDRAULIC FILL MANUAL

# Hydraulic Fill Manual

## For dredging and reclamation works

*Editors*

Jan van 't Hoff
*Van 't Hoff Consultancy B.V., Zeist, The Netherlands*

Art Nooy van der Kolff
*Royal Boskalis Westminster, Papendrecht, The Netherlands*

CRC Press
Taylor & Francis Group
Boca Raton   London   New York   Leiden

CRC Press is an imprint of the
Taylor & Francis Group, an **informa** business

A BALKEMA BOOK

CURNET publication: 244     ISBN: 978-0-415-69844-3

**Keywords**
Fill, fill area, reclamation area, dredgers, dredging methods, compaction, soil improvement, soil investigation, soil data, borrow area, fill design, fill strength, fill stiffness, soft soils, liquefaction, monitoring.

**Readers interest**
Infrastructural managers and engineers, consultants, planning and other consenting authorities, environmental advisors, Contractors, civil engineers, geotechnical engineers, hydraulic engineers.

**Classification**

| | |
|---|---|
| AVAILABILITY | Unrestricted |
| CONTENT | Advice/guidance |
| STATUS | Committee-guided |
| USERS | Infrastructural managers and engineers, consultants, consenting authorities, environmental regulators, Contractors, |

**Referencing this publication**

When referencing this publication in other written material please use the information below:

Title:      Hydraulic Fill Manual (1st edition)
Editors:    Jan van 't Hoff & Art Nooy van der Kolff
Date:       November 2012
Publisher:  CRC Press/Balkema Taylor & Francis Group

*CRC Press/Balkema is an imprint of the Taylor & Francis Group, an informa business*

© 2012 CURNET, Gouda, The Netherlands

Typeset by V Publishing Solutions (P) Ltd, Chennai, India
Printed and bound in India by Replika Press Pvt. Ltd.

Published by:  CRC Press/Balkema
               P.O. Box 447, 2300 AK Leiden, The Netherlands
               e-mail: Pub.NL@taylorandfrancis.com
               www.crcpress.com – www.taylorandfrancis.co.uk – www.balkema.nl

*Library of Congress Cataloging-in-Publication Data*
Hydraulic fill manual for dredging and reclamation works/editors, Jan van 't Hoff, Van 't Hoff Consultancy B.V., Zeist, The Netherlands, Aart Nooy van der Kolff, Royal Boskalis Westminster, Papendrecht, The Netherlands.
    pages cm
  Includes bibliographical references and index.
  ISBN 978-0-415-69844-3 (hardback : alk. paper)
  1. Hydraulic filling. I. Hoff, Jan van 't. II. Kolff, Aart Nooy van der.
  TA750.H85 2012
  627'.5--dc23

                                              2012032875

ISBN: 978-0-415-69844-3 (Hbk)
ISBN: 978-0-203-11998-3 (eBook)

# RECOMMENDATION

There are many reference books and sources of information on dredging techniques and dredging equipment but very little has been written solely on planning, design and construction of land reclamation using hydraulic fill. This manual, a first of its kind, is an ideal reference for all involved in the development of such infrastructure projects. Written and reviewed by expert practitioners who have been involved in many such projects around the world, this manual provides a useful and practical overview and reference guide for clients, developers, consultants and contractors who are engaged in planning, design and construction of reclamation works.

A lot of hard work has gone into the development and compilation of this manual. It is our pleasure to be able to recommend this document to all those involved in the civil engineering and dredging industries.

| | |
|---|---|
| Piet Besselink | Executive Board Royal Haskoning DHV |
| Payam Foroudi | Global Technology Director, Ports and Maritime – Halcrow Group Ltd |
| Jan de Nul | Managing Director, Jan de Nul N.V. |
| Ronald Paul | Chief Operation Officer, Port of Rotterdam Authority |
| Frank Verhoeven | Member of the Board of Management, Royal Boskalis Westminster N.V. |
| Wim Vlasblom | Professor, Emeritus Delft University of Technology |

# CONTENTS

PREFACE      xvii

ACKNOWLEDGEMENTS      xxi

NOTATION      xxv

ABBREVIATIONS      xxix

1   INTRODUCTION TO THE MANUAL      1
   1.1   Land reclamation by hydraulic filling      1
   1.2   History and prospects      1
   1.3   Context and objectives      3
   1.4   Design philosophy      5
   1.5   Structure, content and use      6

2   PROJECT INITIATION      11
   2.1   General      11
   2.2   Basic elements of a land reclamation project      13
     2.2.1   Conceptual design      14
     2.2.2   Availability of fill sources      14
     2.2.3   Data collection      14
     2.2.4   Environmental requirements      15
     2.2.5   Feasibility study      16
     2.2.6   Initial project planning      20
     2.2.7   Legal aspects      20
     2.2.8   Types of contracts      20
   2.3   Design      24
     2.3.1   Design phases      24
   2.4   Considerations for selecting construction method      25
   2.5   Systems Engineering      26

3   DATA COLLECTION      31
   3.1   Introduction      32
   3.2   Interpretation of data, contractual aspects      34
   3.3   Desk study      37
   3.4   Required data      38
     3.4.1   Bathymetrical or topographical data      38
     3.4.2   Geological and geotechnical information      39

| | | 3.4.2.1 | Geological and geotechnical information in the borrow area | 39 |
| | | 3.4.2.2 | Geological and geotechnical information of the subsoil in the reclamation area | 44 |
| | 3.4.3 | | Hydraulic, meteorological, morphological and environmental data | 44 |
| | | 3.4.3.1 | Hydraulic data | 45 |
| | | 3.4.3.2 | Meteorological data | 46 |
| | | 3.4.3.3 | Morphological and environmental data | 49 |
| | 3.4.4 | | Seabed obstructions | 50 |
| 3.5 | | | Typical sand search site investigation | 51 |
| 3.6 | | | Reporting | 54 |
| | 3.6.1 | | Soil and rock classification and description | 54 |
| | 3.6.2 | | Soil classification based on CPT measurements | 57 |
| 3.7 | | | Use of data during different project stages | 59 |
| 3.8 | | | Geostatistical methods | 61 |
| | 3.8.1 | | General | 61 |
| | 3.8.2 | | Methods | 61 |
| | 3.8.3 | | Geostatistical software | 63 |
| | | | | |
| 4 | | | DREDGING EQUIPMENT | 67 |
| 4.1 | | | Introduction | 69 |
| 4.2 | | | Dredging equipment | 71 |
| | 4.2.1 | | Suction dredging | 71 |
| | 4.2.2 | | Mechanical dredging | 74 |
| | 4.2.3 | | Other types of equipment | 77 |
| | 4.2.4 | | Combinations of equipment or dredge chains | 77 |
| 4.3 | | | Operational limitations | 79 |
| | 4.3.1 | | Waves and swell | 79 |
| | 4.3.2 | | Currents | 81 |
| | 4.3.3 | | Hindrance to shipping and other parties | 81 |
| | 4.3.4 | | Environmentally driven limitations | 82 |
| 4.4 | | | Dredging of fill material | 83 |
| | 4.4.1 | | Introduction | 83 |
| | 4.4.2 | | Volume and dimensions of borrow area | 84 |
| | 4.4.3 | | Minimum thickness of fill deposits | 84 |
| | 4.4.4 | | Dredgeability | 85 |
| 4.5 | | | Transport of fill | 87 |
| | 4.5.1 | | Introduction | 87 |
| | 4.5.2 | | Hydraulic transport through a pipeline | 87 |
| | 4.5.3 | | Transport by trailing suction hopper dredger or barge | 88 |
| 4.6 | | | Utilisation characteristics of dredging equipment | 89 |
| 4.7 | | | Basis of cost calculation for dredging | 96 |

5   SELECTION BORROW AREA                                                    101
    5.1   Considerations for the selection of a borrow area                  102
    5.2   Quality of the potential fill material                             103
          5.2.1   Change of the grading as a result of dredging             104
          5.2.2   Alternative fill materials                                104
    5.3   Data collection in the borrow area                                105
          5.3.1   Data collection for quality assessment                    105
          5.3.2   Data collection for quantity assessment                   106
          5.3.3   Data collection for dredgeability assessment              106
    5.4   Quantity of fill material available                               107
          5.4.1   Bulking                                                   107
          5.4.2   Losses                                                    108
          5.4.3   Slope stability                                          109
          5.4.4   Geo-statistical methods                                  111
    5.5   Boundary conditions                                               111

6   PLANNING AND CONSTRUCTION METHODS RECLAMATION                           113
    6.1   Planning of the works                                             115
          6.1.1   Introduction                                             115
          6.1.2   Work preparation                                         116
                  6.1.2.1   Establishment of project team                  117
                  6.1.2.2   Provision of housing and offices
                            for personnel                                  117
                  6.1.2.3   Execution of engineering works                 117
                  6.1.2.4   Create access to site and development
                            of lay-down areas                             118
                  6.1.2.5   Preparation and mobilization of equipment      118
          6.1.3   Construction and monitoring                              118
          6.1.4   Demobilisation, clean-up and maintenance                 119
          6.1.5   Example of a project schedule                            119
    6.2   Work plan for reclamation works                                   121
    6.3   Placement methods                                                 121
    6.4   Construction of containment bunds                                 122
          6.4.1   General                                                  122
          6.4.2   Methods of bund construction                             125
    6.5   Placement of fill material                                        129
          6.5.1   Underwater placement in bulk of fill material            129
          6.5.2   Placement of fill material using a discharge pipeline    131
          6.5.3   Rainbowing                                               134
          6.5.4   Spraying                                                 136
    6.6   Fill mass properties related to method of placement               137
    6.7   Management of poor quality materials                              139
          6.7.1   Use of cohesive or fine grained materials               139
          6.7.2   Settling ponds                                           139

7   GROUND IMPROVEMENT                                                    143
    7.1   Introduction                                                    145
    7.2   Benefits of ground improvement                                  145
    7.3   Overview of ground improvement techniques                       146
    7.4   Pre-loading with or without vertical drains                     149
          7.4.1   Purpose and principle of pre-loading                    149
          7.4.2   Vertical drains                                         153
    7.5   Compaction                                                      156
          7.5.1   Introduction                                            156
          7.5.2   Vibratory surface compaction                            156
          7.5.3   Deep vibratory compaction                               157
                  7.5.3.1   General                                       157
                  7.5.3.2   Vibratory probes without jets                 161
                  7.5.3.3   Vibroflotation                                163
          7.5.4   Dynamic compaction techniques                          167
          7.5.5   Explosive compaction                                    170
    7.6   Soil replacement                                                172
          7.6.1   Introduction                                            172
          7.6.2   Soil removal and replacement                            172
          7.6.3   Stone columns                                           173
                  7.6.3.1   Purpose and principle                         173
                  7.6.3.2   Execution of stone columns
                            by the vibro-replacement technique            173
          7.6.4   Sand compaction piles (closed end casing)               176
          7.6.5   Geotextile encased columns                              176
          7.6.6   Dynamic replacement                                     178
    7.7   Admixtures and in-situ soil mixing                              180

8   DESIGN OF RECLAMATION AREA                                            183
    8.1   Design philosophy                                               185
    8.2   Basic mass properties                                           187
          8.2.1   Strength of fill mass: Bearing capacity and slope stability   187
          8.2.2   Stiffness of fill mass: Settlements, horizontal
                  deformations and tolerances                             188
          8.2.3   Density of the fill mass and subsoil: Resistance
                  against liquefaction                                    189
          8.2.4   Permeability of fill mass: Drainage capacity            190
          8.2.5   Platform level: Safety against flooding and erosion     190
    8.3   Density                                                         190
          8.3.1   Definition of key parameters                            191
          8.3.2   Density ratios                                          192
          8.3.3   The use of densities or density ratios in specifications  194
          8.3.4   Effect of grain size distribution on the density
                  of a soil sample                                        196

|  | 8.3.5 | Density measurement | 198 |
|  |  | 8.3.5.1 Measurement of reference densities (minimum and maximum density) | 198 |
|  |  | 8.3.5.2 Direct measurement of in situ density | 199 |
|  |  | 8.3.5.3 Indirect measurement of relative density by cone penetration testing | 199 |
|  |  | 8.3.5.4 Indirect measurement of relative density by SPT testing | 201 |
|  |  | 8.3.5.5 Measurement of in situ state parameter $\psi$ by cone penetration testing | 201 |
|  | 8.3.6 | Typical relative density values of hydraulic fill before compaction | 202 |
| 8.4 | Strength of the fill mass and subsoil (bearing capacity and slope stability) | | 202 |
|  | 8.4.1 | Introduction | 202 |
|  | 8.4.2 | Shear strength | 204 |
|  |  | 8.4.2.1 High quality fill material | 204 |
|  |  | 8.4.2.2 Poor quality fill material | 210 |
|  |  | 8.4.2.3 Assessment of shear strength | 215 |
|  | 8.4.3 | Relevant failure modes | 221 |
|  |  | 8.4.3.1 Introduction | 221 |
|  |  | 8.4.3.2 Safety approach | 222 |
|  |  | 8.4.3.3 Analytical calculation models versus Finite Element Method (FEM) | 227 |
|  |  | 8.4.3.4 Bearing capacity | 228 |
|  |  | 8.4.3.5 Punch through | 232 |
|  |  | 8.4.3.6 Squeezing | 234 |
|  |  | 8.4.3.7 Slope stability of fill and subsoil | 236 |
|  |  | 8.4.3.7.1 Design methods | 236 |
|  |  | 8.4.3.7.2 Limit Equilibrium Methods | 238 |
|  |  | 8.4.3.7.3 Finite Element Method | 242 |
|  |  | 8.4.3.8 Construction of a slope on soft soil | 243 |
|  |  | 8.4.3.9 Effect of groundwater flow on slope stability | 244 |
|  |  | 8.4.3.10 Earthquakes and slope stability | 250 |
|  |  | 8.4.3.11 Stabilising measures for slope stability | 250 |
|  |  | 8.4.3.11.1 Optimizing the slope geometry by using counterweight berms | 251 |
|  |  | 8.4.3.11.2 Staged construction | 251 |
|  |  | 8.4.3.11.3 Soil replacement (sand key) | 252 |
|  |  | 8.4.3.11.4 Stone columns, sand compaction piles | 252 |
|  |  | 8.4.3.11.5 Geosynthetics | 253 |

|  |  |  |  |
|---|---|---|---|
| 8.5 | | Stiffness and deformation | 254 |
| | 8.5.1 | Introduction | 254 |
| | 8.5.2 | Stiffness | 254 |
| | | 8.5.2.1 General considerations | 254 |
| | | 8.5.2.2 Stiffness of subsoil | 256 |
| | | 8.5.2.3 Stiffness of fill material | 257 |
| | 8.5.3 | Deformations | 258 |
| | | 8.5.3.1 General considerations | 258 |
| | | 8.5.3.2 Settlement calculation methods | 260 |
| | | 8.5.3.3 Additional considerations | 262 |
| | | 8.5.3.4 Vertical deformation of a reclamation surface | 264 |
| | | 8.5.3.5 Vertical deformations of structures | 269 |
| | | 8.5.3.6 Horizontal deformations | 270 |
| | 8.5.4 | Techniques for limiting settlement | 271 |
| 8.6 | | Liquefaction and earthquakes | 272 |
| | 8.6.1 | Overview | 272 |
| | 8.6.2 | History of understanding | 275 |
| | 8.6.3 | Flow slides versus Cyclic softening | 280 |
| | 8.6.4 | Assessing liquefaction susceptibility | 282 |
| | | 8.6.4.1 Codes & Standards | 283 |
| | | 8.6.4.2 Loading: Estimating CSR by site response analysis | 285 |
| | | 8.6.4.3 Resistance, Step 1: Susceptibility to large deformations | 289 |
| | | 8.6.4.4 Resistance, Step 2: Evaluation of CRR | 292 |
| | 8.6.5 | Movements caused by liquefaction | 296 |
| | | 8.6.5.1 Slope deformations | 296 |
| | | 8.6.5.2 Lateral spreads | 299 |
| | | 8.6.5.3 Settlements | 300 |
| | 8.6.6 | Fill characterization for liquefaction assessment | 303 |
| | | 8.6.6.1 Necessity for in situ tests | 303 |
| | | 8.6.6.2 Required number of CPT soundings | 305 |
| | | 8.6.6.3 CPT calibration | 306 |
| | | 8.6.6.4 Supporting laboratory data | 307 |
| | 8.6.7 | Note on soil type (Calcareous and other non-standard sands) | 307 |
| **9** | | **SPECIAL FILL MATERIALS AND PROBLEMATIC SUBSOILS** | 309 |
| 9.1 | | Cohesive or fine-grained fill materials | 311 |
| | 9.1.1 | Introduction | 311 |
| | 9.1.2 | Segregation of fines | 313 |
| | 9.1.3 | Soft clay or soft silt | 315 |
| | | 9.1.3.1 Suitability of soft (organic) clay or silt as fill material | 316 |

|  |  | 9.1.3.2 | Workability of clay | 316 |
|  |  | 9.1.3.3 | Effects of winning method | 317 |
|  |  | 9.1.3.4 | Measures to improve the fill properties after disposal | 318 |
|  |  | 9.1.3.5 | Construction aspects of soft soils in case of application above the waterline | 334 |
|  |  | 9.1.3.6 | Construction aspects of soft soils in case of application below the waterline | 336 |
|  | 9.1.4 | Stiff clay or silt | | 336 |
| 9.2 | Carbonate sand fill material | | | 342 |
|  | 9.2.1 | Introduction | | 342 |
|  | 9.2.2 | Origin and composition of carbonate sands | | 343 |
|  | 9.2.3 | Typical properties of carbonate sands | | 344 |
|  | 9.2.4 | Mechanical behaviour of carbonate sands | | 349 |
|  | 9.2.5 | The use of carbonate sand as fill | | 357 |
|  |  | 9.2.5.1 | Typical behaviour during dredging and hydraulic transport | 357 |
|  |  | 9.2.5.2 | Cone Penetration and Standard Penetration testing in carbonate sands | 359 |
|  |  | 9.2.5.3 | Laboratory testing | 361 |
|  |  | 9.2.5.4 | Field compaction | 363 |
| 9.3 | Hydraulic rock fill | | | 364 |
|  | 9.3.1 | Introduction | | 364 |
|  | 9.3.2 | Lump size | | 364 |
|  | 9.3.3 | Compaction and measurement of compaction result | | 364 |
|  | 9.3.4 | Grading | | 367 |
|  | 9.3.5 | Fines | | 367 |
|  | 9.3.6 | Wear and tear | | 368 |
|  | 9.3.7 | Pumping distance during rock dredging | | 368 |
|  | 9.3.8 | Specifications rock fill | | 369 |
| 9.4 | Problematic subsoils | | | 370 |
|  | 9.4.1 | Sensitive clay | | 370 |
|  | 9.4.2 | Peat | | 372 |
|  | 9.4.3 | Glacial soils | | 374 |
|  | 9.4.4 | Sabkha | | 375 |
|  | 9.4.5 | Karst | | 379 |
|  | 9.4.6 | Laterite | | 383 |
| 10 | OTHER DESIGN ITEMS | | | 387 |
| 10.1 | Introduction | | | 388 |
| 10.2 | Drainage | | | 388 |
|  | 10.2.1 | Infiltration | | 390 |
|  | 10.2.2 | Surface runoff | | 390 |
|  | 10.2.3 | Artificial drainage systems | | 391 |

| | | |
|---|---|---|
| 10.3 | Wind erosion | 394 |
| 10.4 | Slope, bank and bed protection | 398 |
| 10.5 | Interaction between reclamation and civil works | 400 |
| | 10.5.1 General | 400 |
| | 10.5.2 Foundations | 400 |
| | 10.5.3 Construction sequence | 401 |
| | 10.5.4 Impact on existing structures | 402 |
| 10.6 | Earthquakes | 405 |
| 10.7 | Tsunamis | 409 |
| **11** | **MONITORING AND QUALITY CONTROL** | **413** |
| 11.1 | Introduction | 414 |
| 11.2 | Quality Control Plan | 417 |
| 11.3 | Monitoring and testing | 420 |
| | 11.3.1 Geometry | 420 |
| | 11.3.2 Fill material properties | 420 |
| | 11.3.2.1 Grain size distribution | 420 |
| | 11.3.2.2 Minimum and maximum dry densities | 422 |
| | 11.3.2.3 Mineralogy | 422 |
| | 11.3.3 Fill mass properties | 422 |
| | 11.3.3.1 Shear strength | 422 |
| | 11.3.3.2 Stiffness | 425 |
| | 11.3.3.3 Density, relative compaction and relative density | 428 |
| | 11.3.4 Environmental monitoring | 431 |
| **12** | **TECHNICAL SPECIFICATIONS** | **433** |
| 12.1 | Introduction | 434 |
| 12.2 | Roles and responsibilities | 434 |
| 12.3 | Checklist project requirements | 437 |
| 12.4 | Commented examples of technical specifications | 442 |
| | 12.4.1 Introduction | 442 |
| | 12.4.2 Description of the works | 443 |
| | 12.4.3 Standards | 444 |
| | 12.4.4 Data collection (see Chapter 3) | 444 |
| | 12.4.5 Dredging equipment and working method (see Chapter 4) | 445 |
| | 12.4.6 Selection borrow area—quality fill material (see Chapter 5) | 447 |
| | 12.4.7 Construction methods reclamation area (see Chapter 6) | 450 |
| | 12.4.8 Environmental impact | 452 |
| | 12.4.9 Design of a land reclamation (see Chapter 8) | 453 |
| | 12.4.10 Ground improvement (see Chapter 7) | 458 |

|  | 12.4.11 | Special fill materials (see Chapter 9) | 461 |
|  | 12.4.12 | Other design aspects (see Chapter 10) | 462 |
|  | 12.4.13 | Monitoring and quality control (see Chapter 11) | 463 |
|  |  | 12.4.13.1 Geometry | 464 |
|  |  | 12.4.13.2 Testing fill material properties (see Section 11.3.2) | 466 |
|  |  | 12.4.13.3 Testing fill mass properties (see Section 11.3.3) | 467 |
|  |  | 12.4.13.4 Settlement monitoring (see Appendix B.5.3) | 470 |
|  |  | 12.4.13.5 Performance testing | 470 |
|  |  | 12.4.13.6 Reporting | 472 |
|  |  | 12.4.13.7 Monitoring and Quality Control Program (see Section 11.2) | 472 |

APPENDICES

| A | Equipment | 477 |
| B | Field and Laboratory Tests | 529 |
| C | Correlations and Correction Methods | 593 |
| D | Geotechnical Principles | 609 |

REFERENCES 627

# PREFACE

Hydraulic fills are often used to reclaim land for large infrastructure projects such as airports, harbours, industrial and domestic areas and roads. The quality of the borrow material and construction methods are crucial for the quality of the end product. The end product or application will ask specific performance requirements and the characteristics of the fill mass will determine how well these performance criteria are met.

Given the fundamental importance of hydraulic fill to infrastructure projects, a need was felt for a single volume bundling the wide range of the design and construction aspects of hydraulic fills. The Hydraulic Fill Manual is the result.

The Manual represents the concerted effort of Clients, Consultants and Contractors to arrive at a rational and transparent process of project initiation, design, specification and construction of hydraulic fills. The aim of the book is to point the way for each particular project to realise an optimum design based on:

- the available quality and quantity of fill material;
- boundary conditions like the soil conditions, bathymetry, wave climate and tectonic setting of the proposed fill area;
- the selection of dredging equipment with its related construction methods;
- appropriate functional and performance requirements.

Such an optimum design is achieved by making the process from project initiation to construction a clear, iterative process. The Manual promotes this iterative process in which the results of each step are compared with the starting points and results of the previous step and/or with the functional requirements of the project.

This process follows the "System Engineering" approach, a method often applied to the realisation of engineering projects. The underlying idea of this approach is that process transparency and the implementation of sound engineering principles should lead the involved parties to suitable specifications for the construction of the hydraulic fill. Suitability of specifications implies that the functional requirements of the fill mass will be met within the wanted safety margins (and, hence, without excessive costs), but at the same time ensures that the hydraulic fill can be constructed in a feasible and economic manner. This will reduce excessive costs, unwanted disputes, arbitrations and lawsuits.

As it is the intention of the authors that the Manual can be used all over the world on land reclamation projects by hydraulic filling it will not necessarily adhere to (local) standards, norms and/or Codes of Practice. When considered to be relevant references to such documents will be made, but this will be limited to generally accepted and often used systems like the American Standards, the British Standards and/or the European Codes. It may nevertheless be important to be fully informed about the local codes and standards as they may form part of the jurisdiction of the country in which the project has to be realized.

For Clients the Manual presents the most important elements of a land reclamation project (planning, design, data collection, legal and contractual aspects) and explains how the land reclamation forms part of an overall cost-benefit analysis. Clients and Consultants will also learn that to make a project feasible, the fill material may not have to be restricted to sandy material but that

with certain technical measures and under certain conditions, cohesive and fine-grained materials (clay, silt) also may be used. The use of carbonate sands is also highlighted.

The Manual also advises about the various types of dredging equipment, fill material and soil improvement techniques and what geotechnical data are required for production estimates of dredging equipment and for analysing the suitability of fill material. Emphasis is placed on how to translate performance and functional requirements into a measurable properties of the fill mass, with special attention focussed on density, strength and stiffness characteristics and to liquefaction and breaching.

The Manual concludes with examples of practical geotechnical specifications for the construction of a fill mass.

Readers are warned that for proper understanding of design issues some background knowledge in geotechnical engineering is required. For specialist knowledge the reader is referred to handbooks on these subjects.

# Acknowledgements

This publication is the result of a joint project of Clients, Consultants and Contractors from Belgium, United Kingdom and the Netherlands. The overall management has been performed by CUR Building & Infrastructure. The manual has been reviewed by CIRIA/UK.

| Authors/ reviewers | Ken Been | Golder Associates |
| --- | --- | --- |
| | Rik Bisschop | Arcadis |
| | Erik Broos | Port of Rotterdam Authority |
| | Egon Bijlsma | Arcadis |
| | Henk Cloo | Royal Haskoning DHV |
| | Jurgen Cools | Royal Haskoning DHV |
| | David Dudok van Heel | Rotterdam Municipality Consultancy |
| | Arnoud van Gelder | Royal Haskoning DHV |
| | Reimer de Graaff | Arcadis |
| | Jarit de Gijt | Rotterdam Municipality/Delft University of Technology |
| | Robert de Heij | Witteveen + Bos |
| | Ilse Hergarden | Royal Haskoning DHV |
| | Jan van 't Hoff | Van 't Hoff Consultancy bv |
| | Richard de Jager | Royal Boskalis Westminster N.V. |
| | Dirk-Jan Jaspers Focks | Witteveen + Bos |
| | Mike Jefferies | Golder Associates |
| | Wouter Karreman | Van Oord Dredging and Marine Contractors |
| | Lieve De Kimpe | Van Oord Dredging and Marine Contractors |
| | Edwin Koeijers | Rotterdam Municipality Consultancy |
| | Rob Lohrmann | Witteveen + Bos |
| | Joop van der Meer | Van Oord Dredging and Marine Contractors |
| | Piet Meijers | Deltares |
| | Patrick Mengé | DEME Group/Dredging International |
| | Mario Niese | Royal Haskoning DHV |
| | Art Nooy van der Kolff | Royal Boskalis Westminster N.V. |
| | Cissy de Rooij | Royal Boskalis Westminster N.V. |
| | Rob Rozing | Arcadis |
| | Berten Vermeulen | Jan de Nul |

| | |
|---|---|
| Editors | Jan van 't Hoff & Art Nooy van der Kolff |

| | |
|---|---|
| Peer reviewers | In response to a request from CUR Bouw & Infra CIRIA has managed a peer review of this manual. This review was conducted by the following UK experts: |

| | |
|---|---|
| John Adrichem | Royal Haskoning DHV |
| Ken Been | Golder Associates |
| Nick Bray | HR Wallingford |
| Gijsbert Buitenhuis | Royal Haskoning DHV |
| Chris Capener | Royal Haskoning DHV |
| Chris Chiverrell | CIRIA |
| Scott Dunn | HR Wallingford |
| Payam Foroudi | Halcrow |
| Helge Frandsen | Royal Haskoning DHV |
| Prasad Gunawardena | Mott MacDonald |
| Greg Haigh | Arup |
| Mike Jefferies | Golder Associates |
| David Jordan | Scot Wilson/HR Wallingford |
| Roderick Nichols | Halcrow |
| Philip Smith | Royal Haskoning DHV |

| | |
|---|---|
| CIRIA Project Managers peer reviews | Kristina Gamst<br>Gillian Wadams |

| | |
|---|---|
| Review English translation | Marsha Cohen |

| | | |
|---|---|---|
| Technical review | Chris Chiverrell<br>Aad van den Thoorn | CIRIA<br>CURNET |

| | | |
|---|---|---|
| Executive steering board | Jurgen Cools | Royal Haskoning DHV |
| | Jarit de Gijt | Rotterdam Municipality/<br>Delft University of Technology |
| | Jan van 't Hoff | Van 't Hoff Consultancy bv |
| | René Kolman | IADC |
| | Joop Koenis (upto Dec. 2010) | CUR Bouw & Infra |
| | Dirk Luger | Deltares |
| | Art Nooy van der Kolff | Royal Boskalis Westminster N.V. |
| | Daan Rijks (upto Dec. 2009) | Royal Boskalis Westminster N.V. |
| | Ger Vergeer (from Jan. 2011) | CUR Bouw & Infra |
| | Wim Vlasblom (chair) | Delft University of Technology |

|  | The overall project management has been performed by CUR Building & Infrastructure: Joop Koenis (up to Dec. 2010) en Ger Vergeer (from Jan. 2011) |
|---|---|
| Project sponsors | Ministry of Infrastructure and the Environment, International Dredging Association (IADC) , China Communications Construction Company (CCCC), Foundation FCO-GWW, Port of Rotterdam Authority, Inros Lackner, PAO Delft University |

CUR wishes to thank the following organisations for providing photographs: DEME, Jan de Nul, Royal Boskalis Westminster, Van Oord Dredging and Marine Contractors, Port of Rotterdam Authority.

# NOTATION

| | | |
|---|---|---|
| $a_g$ | = Ground acceleration | (m/s²) |
| $B$ | = Bulk Modulus | (kPa) |
| $B_q$ | = CPT excess pore pressure ratio $B_q = (u_c - u_0)/(q_t - \sigma_{v0})$ | (–) |
| $c'$ | = Effective cohesion | (kPa) |
| $c_u$ | = Undrained shear strength | (kPa) |
| $c_v$ | = Vertical coefficient of consolidation | (m²/year) |
| $c_h$ | = Horizontal coefficient of consolidation | (m²/year) |
| $c'_k$ | = Characteristic effective cohesion | (kPa) |
| $c_u(t)$ | = Undrained shear strength at time $t$ after loading | (kPa) |
| $c_{u,0}$ | = Initial undrained shear strength | (kPa) |
| $c_{u,k}$ | = Characteristic undrained shear strength | (kPa) |
| $c_u^u$ | = Undrained shear strength at the upper side of a soft layer | (kPa) |
| $c_u^l$ | = Undrained shear strength at the lower side of a soft layer | (kPa) |
| $C_\alpha$ | = Peak ground acceleration | (m/s²) |
| $C_\alpha$ | = Secondary compression index | (–) |
| $C_c$ | = Compression index | (–) |
| $C_c$ | = Coefficient of curvature | (–) |
| $C_r$ | = Recompression index | (–) |
| $C_u$ | = Coefficient of uniformity | (–) |
| $d_e$ | = Equivalent diameter of the zone of influence of a drain | (m) |
| $d_w$ | = Equivalent diameter of a cylindrical drain column | (m) |
| $d_q, d_c$ | = Depth factors | (–) |
| $D_{50}$ | = Mean grain size | (mm) |
| $e$ | = Void ratio | (–) |
| $e_0$ | = Void ratio of layer with initial thickness $h_0$ | (–) |
| $e_p$ | = Void ratio of layer with thickness $h_p$ after primary settlement | (–) |
| $E$ | = Modulus of Elasticity | (kPa) |
| $E_{dyn}$ | = Dynamic Modulus | (kPa) |
| $E_y$ | = Young's Modulus | (kPa) |
| $E_{DMT}$ | = Dilatometer Modulus | (kPa) |
| $E_{PLT}$ | = Plate Load Test Modulus | (kPa) |
| $E_{PMT}$ | = Pressiometer Modulus | (kPa) |
| $E_s$ or $E_{oed}$ or $M$ | = Constrained Modulus | (kPa) |
| $E_{sec}$ | = Secant Modulus | (kPa) |

| | | |
|---|---|---|
| $E_{tan}$ | = Tangent Modulus | (kPa) |
| $E_u$ | = Undrained Modulus | (kPa) |
| $E_0$ | = Young's Modulus at very small deformations | (kPa) |
| $E_{50}$ | = Young's Modulus at 50% of the failure stress | (kPa) |
| $F_R$ | = Friction Ratio CPT test | (−) |
| $G$ | = Shear Modulus | (kPa) |
| $G_0$ | = Shear Modulus at very low strain | (kPa) |
| $G_{50}$ | = Shear Modulus at very low strain | (kPa) |
| $h_0$ | = Initial thickness of layer | (m) |
| $h_p$ | = Thickness of the considered layer after primary settlement | (m) |
| $H$ | = Layer thickness | (m) |
| $i_q, i_c, i_\gamma$ | = Inclination factors | (−) |
| $I_{S50}$ | = Point load strength | (MPa) |
| $I_P$ | = Plasticity index | (−) |
| $I_C$ | = Consistency index | (−) |
| $I_L$ | = Liquidity index | (−) |
| $K_o$ | = Coefficient of active earth pressure at rest | (−) |
| $k_h$ | = Horizontal seismic coefficient | (−) |
| $k_v$ | = Vertical seismic coefficient | (−) |
| $k_y$ | = Yield coefficient | (−) |
| $M$ | = Earthquake magnitude | (−) |
| $M, E_s$ or $E_{oed}$ | = Constrained Modulus | (kPa) |
| $M_L$ | = Local magnitude | (−) |
| $M_S$ | = Surface wave magnitude | (−) |
| $M_W$ | = Moment magnitude of earthquake | (·) |
| $n$ | = Porosity | (−) |
| $n_0$ | = Initial porosity | (−) |
| $N'$ | = Number of blows per per foot (300 mm) penetration of SPT | (−) |
| $N_k$ | = Empirical factor to correlate the undrained shear strength to the cone resistance | (−) |
| $N_q, N_c, N_\gamma$ | = Bearing capacity factors | (−) |
| $p'$ | = Mean effective stress | (kPa) |
| $p_a$ | = Atmospheric pressure | (kPa) |
| $q_{allow}$ | = Allowable load | (kPa) |
| $q_c$ | = Measured cone resistance | (MPa) |
| $q_{ck}$ | = Characteristic cone resistance for liquefaction assessment | (MPa) |
| $q_t$ | = Corrected cone resistance | (MPa) |
| $Q$ | = Dimensionless CPT resistance based on mean stress, $Q = (q_t − \sigma_{v0})/\sigma'_{v0}$ | (−) |
| $Q_u$ | = CPT resistance modified on pore pressure | (−) |
| $Q_u$ | = Ultimate bearing capacity | (kPa) |

| | | |
|---|---|---|
| $q_u$ | = Unconfined compressive strength | (kPa) |
| $r_d$ | = Response coefficient | (–) |
| $R_e$ | = Relative void ratio | (–) |
| $R_n$ | = Relative porosity | (–) |
| $R_c$ | = Degree of compaction | (–) |
| $s_r$ | = Residual undrained shear strength | (kPa) |
| $s_q, s_c, s_\gamma$ | = Shape factors | (–) |
| $S$ | = Degree of saturation | (–) |
| $S_{min}$ | = Minimum settlement to be reached at time of hand-over | (m) |
| $S_{total}$ | = Total primary settlement | (m) |
| $t$ | = Time | (year) |
| $t_p$ | = Time at end of primary settlement (full consolidation) | (year) |
| $t_f$ | = Time at which the secondary compression has to be calculated | (year) |
| $T_h$ | = Time factor for horizontal consolidation | (–) |
| $T_s$ | = Fundamental period | (s) |
| $T_v$ | = Time factor | (–) |
| $u_2$ | = Pore pressure measured behind the cone | (kPa) |
| $u_0$ | = In situ pore pressure | (kPa) |
| $U_v$ | = Average degree of consolidation due to vertical drainage | (–) |
| $UCS$ | = Unconfined compressive strength | (MPa) |
| $U_h$ | = Average degree of consolidation due to horizontal drainage | (–) |
| $U_{vh}$ | = Average degree of consolidation | (–) |
| $U_{(t)}$ | = Degree of consolidation at time t after loading | (–) |
| $w$ | = Water content | (–) |
| $wL$ | = Liquid limit | (–) |
| $wP$ | = Plastic limit | (–) |
| $\alpha$ | = Peak horizontal ground acceleration | (m/s²) |
| $\gamma$ | = Volumetric weight | (kN/m³) |
| $\gamma_{dry}$ | = Dry unit weight | (kN/m³) |
| $\gamma_{sat}$ | = Saturated unit weight | (kN/m³) |
| $\varepsilon_g$ | = Shear strain | (–) |
| $\Delta e$ | = Change in void ratio from a layer with initial void ratio $e_0$ | (–) |
| $\Delta h$ | = Compression of layer with initial thickness $h_0$ | (m) |
| $\Delta h_s$ | = Secondary compression of layer with thickness $h_p$ | (m) |
| $\Delta\sigma'$ | = Effective stress increment in the middle of the considered layer | (kPa) |
| $\Delta\sigma'_{ref}$ | = Reference stress (usually taken equal to 1 kPa) | (kPa) |

| | | |
|---|---|---|
| $\Delta S_{allow}$ | = Allowable residual settlement at time of hand-over | (m) |
| $\lambda$ | = Slope of CSL for semi-log idealization | (–) |
| $\rho$ | = Density | (kg/m³) |
| $\rho_b$ | = Bulk density | (kg/m³) |
| $\rho_d$ | = Dry density | (kg/m³) |
| $\rho_g$ | = Particle density | (kg/m³) |
| $\rho_s$ | = Density of solid particles | (kg/m³) |
| $\rho_{sat}$ | = Saturated density | (kg/m³) |
| $\sigma$ | = Normal stress | (kPa) |
| $\sigma'_0$ | = Initial effective stress in the middle of the considered layer with initial thickness $h_0$ | (kPa) |
| $\sigma'_m$ | – Mean effective stress | (kPa) |
| $\sigma'_n$ | = Effective normal stress | (kPa) |
| $\sigma'_p$ | = Pre-consolidation stress in the middle of the considered layer with initial thickness $h_0$ | (kPa) |
| $\sigma_{v0}$ | = Total stress | (kPa) |
| $\sigma'_v$ | = Effective vertical stress at foundation level next to foundation | (kPa) |
| $\sigma'_{v0}$ | = Effective vertical stress | (kPa) |
| $\Delta\sigma'_v$ | = Increase effective vertical stress due to loading after full consolidation | (kPa) |
| $\sigma'_{v,0}$ | = Effective vertical stress at foundation level | (kPa) |
| $\varphi'$ | = Effective friction angle | (°) |
| $\varphi'_{crit}$ | = Critical state friction angle | (°) |
| $\varphi'_k$ | = Characteristic effective friction angle | (°) |
| $\varphi'_{max}$ | = Peak effective friction angle | (°) |
| $\varphi'_s$ | = Secant friction angle | (°) |
| $\varphi_u$ | = Undrained friction angle | (°) |
| $\tau$ | = Shear strength | (kPa) |
| $\psi$ | = State parameter | (–) |

# ABBREVIATIONS

| | |
|---|---|
| ADCP | Acoustic Doppler Current Profilers |
| CBR | California Bearing Ratio |
| CIRIA | Construction Industry Research and Information Association |
| CPT | Cone Penetration Test |
| CRR | Cyclic Resistance Ratio |
| CSL | Critical State Locus |
| CSR | Cyclic Stress Ratio |
| CSWS | Continuous Surface Wave System |
| CTD | Conductivity, temperature, depth meter |
| CUR | Centre for Civil Engineering, Research and Codes |
| DC | Dynamic Compaction |
| DIN | German Institute for Standardization |
| DO | Dissolved Oxygen |
| DSM | Deep Soil Mixing |
| EAU | Recommendations of the Committee for Waterfront Structures Harbours and Waterways EAU 2004 |
| EC7 | Eurocode 7 |
| EC | Explosive Compaction |
| ECM | Electromagnetic Current Meter |
| EMS | European Macroseismic Scale |
| FEM | Finite Element Method |
| FS | Factor of Safety |
| FS | Safety Against Instability |
| $FS_L$ | CRR/CSR = Safety Against Failure by Liquefaction |
| GEC | Geotextile Encased Columns |
| GWL | Ground Water Level |
| HEIC | High Energy Impact Compaction |
| ISSMGE | International Society for Soil Mechanics and Geotechnical Engineering |
| LAT | Lowest Astronomical Tide |
| LEM | Limit Equilibrium Method |
| MBES | Multibeam Echo Sounding |
| MDD | Maximum Dry Density |
| MPM | Material Point Method |
| MSL | Mean Sea Level |
| NTU | Nephelometric Turbidity Units |
| NCEER | National Center for Earthquake Engineering Research |
| OCR | Over Consolidation Ratio |
| PGA | Peak Ground Acceleration |

| | |
|---|---|
| PIANC | Permanent International Commission for Navigation Congresses |
| PLT | Plate Load Test |
| PVD | Prefabricated Vertical Drain |
| RIC | Rapid Impact Compaction |
| RQD | Rock Quality Designation |
| SASW | Spectral Analysis of Surface Waves |
| SCR | Solid Core Recovery |
| SLS | Serviceability Limit State |
| SPT | Standard Penetration Test |
| SSM | Shallow Soil Mixing |
| SSS | Side Scan Sonar |
| TCR | Total Core Recovery |
| TSS | Total Suspended Solids |
| ULS | Ultimate Limit State |
| ZLT | Zone Load Test |

CHAPTER 1

## INTRODUCTION TO THE MANUAL

### 1.1  Land reclamation by hydraulic filling

Land reclamation is generally defined as the process of creating new land by raising the elevation of a seabed, riverbed or other low-lying land ('filling') or by pumping out the water from a watery area that is enclosed by dikes ('polder construction').

Land reclamation by filling may be undertaken by dry earth movement, but also by hydraulic filling. Hydraulic filling is defined as the creation of new land by the following consecutive activities:

1. dredging of fill material in a borrow area or dredging area by floating equipment (dredgers);
2. transport of fill material from the borrow area to the reclamation site by dredger, barge or pipeline;
3. placement of fill material as a mixture of fill material and (process) water in the reclamation area.

It is the hydraulic filling that forms the main subject of this Manual. For information on other reclamation methods like dry earth movement or the construction of polders reference is made to other publications like manuals, guides, state of the art reports and/or codes of practice that more specifically deal with these techniques.

In most cases land reclamation will be a part of a more comprehensive project such as the construction of a port, an airport, a housing project or an industrial complex. Whereas superstructures will not be discussed in this Manual, their presence will impose certain requirements on the quality of the reclaimed land, its response to external forces such as currents, waves, precipitation and wind and its ability to withstand hazards such as earthquakes and tsunamis.

### 1.2  History and prospects

Archaeological evidence indicates that land reclamation is not a recent invention, but has existed for thousands of years. Some 2000 years ago inhabitants of the swampy and tidal areas along the Wadden Sea in the north of The Netherlands and Germany lived on so-called 'terpen' or 'wierden', artificial dwelling mounds built to protect themselves against flooding in periods of high water levels. Further attempts to prevent their agricultural land from being flooded by the sea included the construction of dikes between those dwelling mounds.

Around 1500 A.D. a new method of land reclamation came into use: "Polders" were constructed by building a ring dike in shallow watery areas, after which the water was removed from the enclosed low-lying area by windmill-driven pumps. Once steam engines became available in the 19th century some of the windmills were replaced by pumping stations.

A transformational moment came with the development of the modern centrifugal pump that enabled the current large-scale reclamation projects by hydraulic filling. According to the International Association of Dredging Companies (*Terra et Aqua*, 2005) one of the first major reclamation works (Bay of Abidjan, Ivory Coast) was carried out in the 1960s.

As a result of the strong growth of the world's population and the subsequent urbanisation and economic development, in particular in densely populated coastal areas, the last decades have witnessed an ever-increasing demand for new land. This demand has resulted in a large number of reclamation projects ranging from numerous small-scale projects all over the world to well-known, large-scale projects such as the Palm Island Project in Dubai or the construction of Maasvlakte 2 in Rotterdam Europoort, The Netherlands (see Figure 1.1).

Demographic forecasts suggest that in the foreseeable future this demand for new land will remain or even increase, see Figure 1.2.

Figure 1.1 *Construction of Maasvlakte 2 in progress, Rotterdam Europoort, The Netherlands, October 2011.*

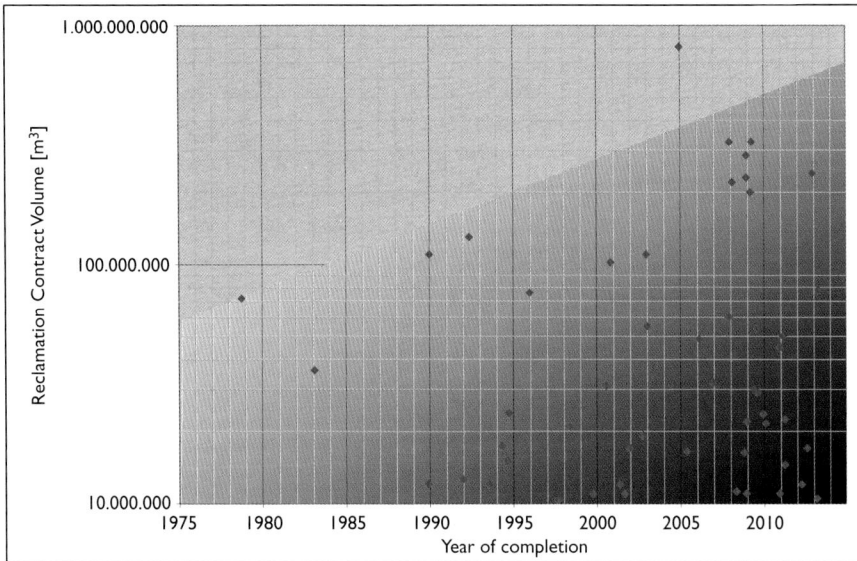

Figure 1.2   *Trend of reclamation contract volumes over the past 30 years.*

## 1.3   Context and objectives

This Hydraulic Fill Manual is written to supply the wants of the dredging industry to create a handbook that helps to improve the understanding between the various parties (i.e., Clients, Consultants and Contractors) involved in a hydraulic fill project. It contains the latest developments in the field of design and construction of hydraulic fills and presents clear guidelines for initiation, design and construction of a hydraulic fill project.

The design and construction of a hydraulic fill project requires specific knowledge of a wide variety of disciplines, such as hydraulic, geotechnical and environmental engineering in combination with practical know-how and experience in dredging and filling techniques.

Moreover, a new generation of dredging equipment, increasing awareness of the marine environment and the tendency to reduce construction time (i.e., return on investment period) will affect the design and construction methods requiring new standards.

Worldwide experience indicates that in recent years the technical specifications of reclamation projects have become more stringent. No rational basis for such a trend exists as the intended use of this newly created land (i.e., functional requirements) has not changed significantly nor has an increase of failures been reported. In a number of cases this trend has led to inadequate and conflicting specifications, to construction requirements that could not be met and/or to excessive costs for

3

fill treatment and testing. These developments frustrate the tender process, cause serious problems during construction and quality control and may lead to long-lasting, costly arbitration.

This Hydraulic Fill Manual has been written to avoid these problems. It includes theoretical and practical guidelines for the planning, design, construction and quality control of hydraulic fills.

The Manual covers the interfaces between the areas of interest of the contractual parties usually involved in reclamation projects (see Figure 1.3). It will:

- enable the Client to understand and properly plan a reclamation project;
- provide the Consultant with adequate guidelines for design and quality control;
- allow the Contractor to work within known and generally accepted guidelines and realistic specifications.

This Manual is believed to be the first handbook to date that covers all these aspects that are relevant to the construction of hydraulic fills.

The authors and reviewers have endeavoured to gather the most up-to-date knowledge regarding the design and construction of hydraulic fills with the goal that with time this Manual will be a standard for all parties involved in the implementation of hydraulic fill projects.

The structure of the Manual assumes that the design and construction of a hydraulic fill should be a rational process that ultimately results in the best and most economical match between the specified properties of the land reclamation, the requirements imposed by its future use and the environment in which it is located.

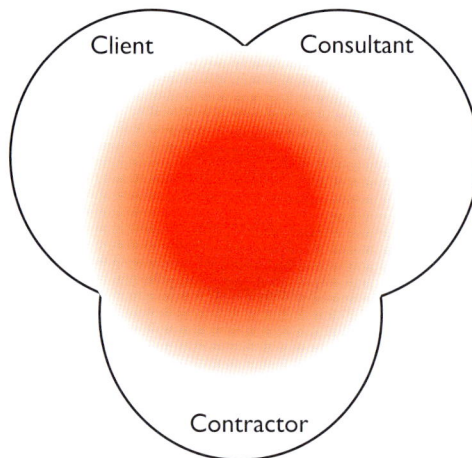

Figure 1.3   *Focus of the Manual: Interfaces between three contractual partners.*

## 1.4  Design philosophy

Land reclamation projects are undertaken for various purposes and under varying conditions. The performance requirements imposed on a fill depend on the future use of the reclamation and, hence, they vary for each individual project.

Boundary conditions are often site and project specific as well. Physical site conditions, such as wave climate, currents, water depth, subsoil properties and the vulnerability of the environment to dredging and reclamation activities will differ from one site to another. The quality and quantity of the fill available for construction will strongly depend on the location of the project. These conditions will not only affect the design of a reclamation, but they must also be taken into account when selecting the most suitable dredging equipment and construction method.

A rational design must integrate the functional and performance requirements considering the boundary conditions of the project in order to adequately specify the geometry and properties of the fill mass. The same rationality must be applied with respect to the construction of the reclamation requiring an appropriate selection of equipment and working method.

### Functional and performance requirements

A functional requirement defines what a system must do, while a performance requirement specifies something about the system itself and how well it performs its function. A fill mass (and its subsoil) can be regarded as a system with functional and performance requirements.

Starting point of a design must always be the future land use. The functional requirements of the fill mass follow directly from the intended use of the fill area. These functional requirements may be formulated in general terms (for instance: "*the reclamation area must accommodate an airfield with runways, aprons, a terminal building and a traffic control tower*"), but can also be more specific (for instance: "*the fill mass must support a structure founded on a strip footing having a width of 1.5 m, an embedment depth of 1.0 m and a bearing load of 80 kPa*") which may vary over the area depending on the lay-out of the future development.

The functional requirements and the design of the superstructures (i.e., their Ultimate Limit State and/or Serviceability Limit State, see section 8.4.1) lead to performance requirements of the fill mass such as maximum allowable settlement of the superstructures (buildings, roads, storage areas, runways, revetments, tunnels, etc.), and sufficient safety against slope failure or liquefaction. The required basic mass properties like strength, stiffness, density and permeability can be derived from these performance requirements.

The definition of functional requirements and their subsequent translation into performance criteria form the basis of System Engineering, see section 2.5, which may be used as a tool to control the development cycle of a reclamation project.

Following an approach in terms of functional and performance requirements, the design of a reclamation project becomes an iterative process. Functional requirements, dictated by structural criteria and other project-dependent boundary conditions, will not be discussed in this Manual.

## 1.5 Structure, content and use

Rather than following a chronological sequence of events (project initiation, design and construction), the structure chosen for this Manual intends to put the main emphasis on the design of a land reclamation. To that end the first chapters describe not only the collection of data required for the design but also present basic information on dredging equipment and construction methods before touching upon the design aspects.

The Manual concludes with a discussion of the technical specifications that result from a design. Additional information can often be found in the referenced Appendices. Figure 1.4 illustrates the set-up of this Manual.

Following the scheme of Figure 1.4 the contents of this Manual can briefly be summarised as follows:

Chapter 2: Project initiation, gives an overview of the most relevant elements in the procedure to realise a reclamation project and the way they are related to each other. It introduces the development cycle to realize a project, including the iterative nature of that cycle and concludes with an illustrative scheme of activities leading to the construction of a reclamation project.

Chapter 3: Data collection, presents the data required for the design of a hydraulic fill project. It deals not only with the type of information needed for the design, but also with the methods to collect the information, the reporting and the processing of data.

Commonly used dredging equipment and its use can be found in Chapter 4: Dredging equipment. Possibilities and limitations of the various types of dredgers and their vulnerability to the physical conditions of the project site are also included.

```
┌─────────────────────────────────────────────────────────┐
│ 2. Project initiation                                   │
└─────────────────────────────────────────────────────────┘
  ┌───────────────────────────────────────────────────────┐
  │  3. Data collection                                   │
  └───────────────────────────────────────────────────────┘
    ┌─────────────────────────────────────────────────────┐
    │   4. Dredging equipment                             │
    └─────────────────────────────────────────────────────┘
      ┌───────────────────────────────────────────────────┐
      │   5. Selection borrow area                        │
      └───────────────────────────────────────────────────┘
        ┌─────────────────────────────────────────────────┐
        │   6. Planning and construction methods reclamation │
        └─────────────────────────────────────────────────┘
          ┌───────────────────────────────────────────────┐
          │   7. Ground improvement                       │
          └───────────────────────────────────────────────┘
```

┌─────────────────────┐      ┌──────────────────────────────────────────┐
│                     │ ◁────│ 9. Special fill materials and problematic subsoils │
│    8. Design        │      └──────────────────────────────────────────┘
│                     │ ◁────┌──────────────────────────────────────────┐
└─────────────────────┘      │ 10. Other design items                   │
                             └──────────────────────────────────────────┘

┌─────────────────────┐
│ 11. Monitoring and  │
│  quality control    │
└─────────────────────┘

┌─────────────────────┐
│ 12. Specifications  │
└─────────────────────┘

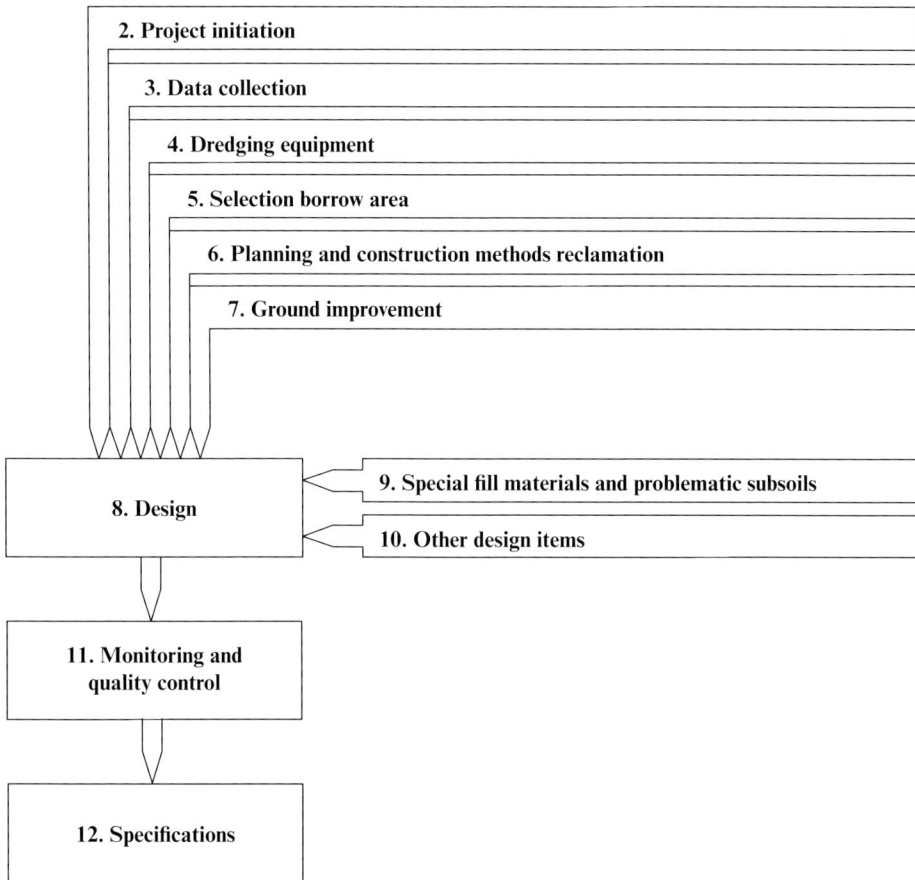

Figure 1.4   *Structure of the Manual.*

The feasibility of a project strongly depends on the availability of sufficient suitable fill material in the vicinity of the reclamation site. Chapter 5: Selection borrow area, describes the most important criteria for the selection of a borrow area.

Chapter 6: Planning and construction methods reclamation, deals with the construction methods of a reclamation area. This not only includes the deposition of the material, but also the planning, preparation and monitoring of the operations.

In the case where the existing subsoil and/or the fill behaviour do not meet the requirements, ground improvement may be required. Chapter 7: Ground improvement, gives an overview of the most relevant ground improvement techniques.

Chapter 8: Design, discusses the geotechnical design of a land reclamation. The main sections deal with density, (shear) strength, stiffness and deformations of the fill mass. A special section of this chapter is dedicated to the phenomena liquefaction and breaching.

In some areas of the world land reclamation projects have to be undertaken using cohesive materials or carbonate sands rather than with the more frequently encountered quartz sands. Furthermore, some subsoils may exhibit a different behaviour when loaded by fill. Chapter 9: Special fill materials and problematic subsoils, describes the behaviour of these special fill materials and problematic subsoils.

In addition to the geotechnical behaviour, a design should also take into account aspects like drainage of the reclaimed area, wind erosion, and slope, bed and bank protection. A short introduction to these subjects and some relevant references are presented in Chapter 10: Other design items.

Chapter 11: Monitoring and quality control, is about monitoring and quality control requirements during and after construction of the reclamation.

Finally, Chapter 12: Specifications, makes recommendations for specifying the construction of a hydraulic fill area which logically follow from the engineering philosophy adopted in this Manual.

# CHAPTER 2

# PROJECT INITIATION

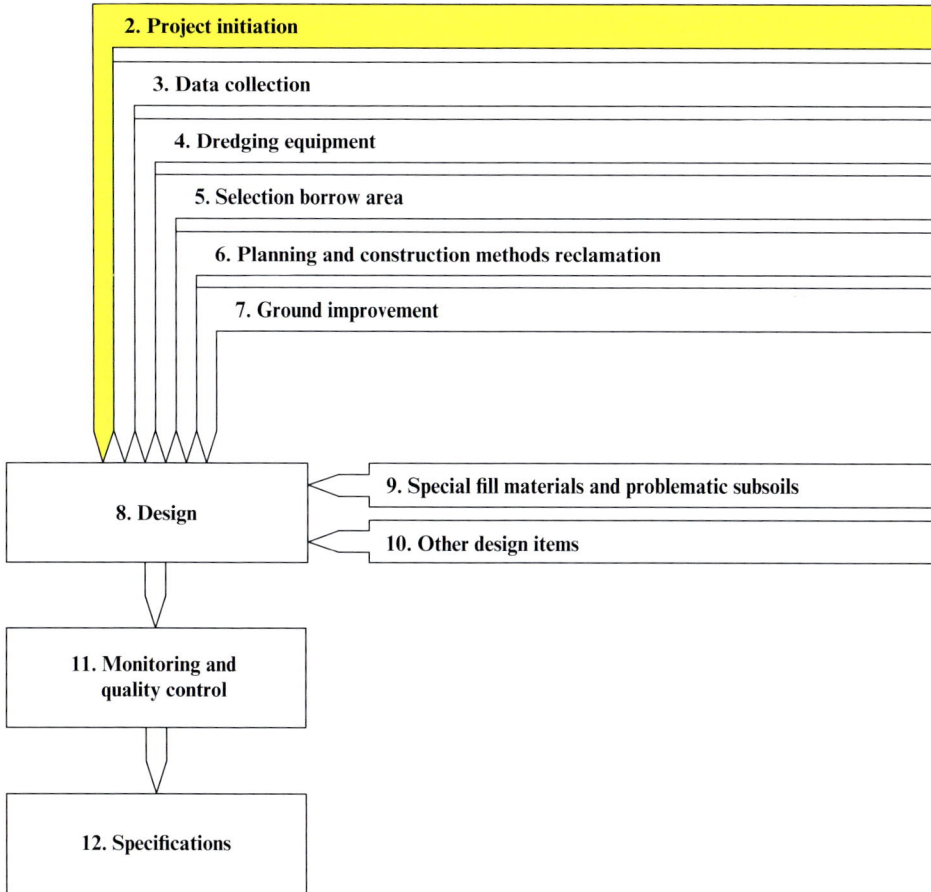

| 2. Project initiation |
|---|

| 3. Data collection |
|---|

| 4. Dredging equipment |
|---|

| 5. Selection borrow area |
|---|

| 6. Planning and construction methods reclamation |
|---|

| 7. Ground improvement |
|---|

| 8. Design |
|---|

| 9. Special fill materials and problematic subsoils |
|---|

| 10. Other design items |
|---|

| 11. Monitoring and quality control |
|---|

| 12. Specifications |
|---|

## 2.1  General

The steps to develop a land reclamation project are generally its initiation, the design and the construction:

– *Project initiation.* This phase is characterized by the Client defining his needs, securing the financing, identifying the resources, analysing costs and benefits and quantifying the risks in order to assess the feasibility of the project. A Client may call in the assistance of a Consultant to complete these tasks;
– *Project design.* Once the scope of a project has been defined and its feasibility assessed, the needs of the Client have to be translated into a detailed design.

Such a design can be divided into the design of the superstructures (e.g., buildings, pavements, tunnels) and the design of the reclamation area (fill mass) itself. The design of the reclamation area is usually prepared by a Consultant or, depending on the type of contract, a Contractor and the results of this are rendered in drawings and Technical Specifications;

– *Construction of the reclamation.* Construction of the reclamation area includes hydraulic filling to the lines and levels and Technical Specifications as set out in the Contract. Generally the construction is carried out by a specialised Dredging Contractor, while a Consultant is often in charge of monitoring and quality control.

The development of a project starts by defining the scope of the project according to the Client's needs. This will generally include the following steps:

– *Intended land use.* In most cases the Client has a specific reason to reclaim land, for instance, to extend an airport or a harbour or construct a refinery;
– *A first assessment of the capacity of the project.* For a container terminal, the needs may be determined by the size of vessels and required storage area in order to cope with the container throughput. For an airport this may include the throughput of passengers, which is reflected in the size of the terminal buildings, the types of planes and the number and dimensions of landing strips For a housing project this would mean, for instance, assessing the number and types of houses, required public facilities, roads, rainwater drainage systems, and so on. For a dike this may encompass the height of the water level against the outer embankment. If the dike has a traffic function, the intensity of the traffic has to be defined which leads to the dimension of a road.
– *Select location of the project.* The selection of the location of a project may be a rather complex decision. In a broad context, national or regional, this may be driven by economic considerations to develop a particular region or activities in that region, usually applicable to large scale projects (ports, airports, housing or industrial development). In a narrower context projects may form part of existing infrastructure. The eventual project position will be governed by considerations such as its access, available sand sources, subsoil conditions, possibilities for future extension and not in the least environmental opportunities and constrains.
– *Determine basic lay-out of the project plan.* This will be based on the anticipated capacity and purpose of the project and leads to the main project dimensions including its levels and a first impression of required fill volumes.
– A first impression is also obtained of the potential capacity and availability of a *sand borrow area*.
– At this stage *access to the project site* must be considered both from land and from water. Special access roads or access channels for shipping including Contractor's floating equipment may be required.

To realise the Client's needs the questions must be considered whether these needs are technically and economically feasible and whether the project can be realised regarding local environmental and planning regulations for which permits may be required. In this context certain compensation or mitigation measures may need to be incorporated.

## 2.2 Basic elements of a land reclamation project

The project initiation phase is generally characterised by many uncertainties and conceptual approaches as at this stage the information available is still limited. Nevertheless, important decisions have to be made on the basis of the results of studies undertaken as part of this phase. The most important outcome of this phase is:

- The Terms of Reference defining the purpose and structure of the project.
- The results of the business case capturing the reasoning for initiating the project and its financial and technical feasibility and the subsequent decision whether or not to proceed with the project (go/no go).
- A conceptual design presenting all ideas, needs and wishes of the Client translated into basic functional requirements and preliminary drawings.
- An initial planning with preliminary milestones.

Within the context of this Manual subjects such as business case and Terms of Reference will not be elaborated any further, while the feasibility study is only discussed briefly.

Other important tasks to be undertaken during the initiation phase of a project may be summarised as follows:

- A preliminary search for suitable fill sources (borrow area, capital dredging) in the vicinity of the envisaged reclamation site. This may require a desk study and initial geotechnical site investigations.
- A preliminary assessment of other boundary conditions such as subsoil characteristics, water depths and hydraulic, morphological and meteorological conditions (waves, currents, wind, ….) at the site.
- A first reconnaissance of the environmental vulnerability of the site and the envisaged borrow area.
- A study of the legal aspects of the proposed project (required permits, planning regulations, ….).
- A decision on the type of contract defining the obligations, responsibilities and relations between the parties involved in the project.

### 2.2.1 *Conceptual design*

A conceptual design of a land reclamation project defines the main outlines of the project. It describes the project in terms of a set of integrated ideas and concepts about what it should do, behave and look like. It includes the (general) functions of the fill mass performance and it should be sufficiently detailed as to allow a preliminary cost estimate and risk assessment of the project.

The final product of a conceptual design is generally a set of basic functional requirements and a number of conceptual or preliminary drawings based on the Client's ideas about the future land use. A conceptual design often forms the basis for the feasibility study of the project. The Client may involve a Consulting Engineer and/or a Contractor to assist him with his task.

### 2.2.2 *Availability of fill sources*

The availability of sufficient, good quality fill material is essential for realising a land reclamation project. Fill material may be obtained from a designated borrow area, from maintenance dredging or from capital dredging. The feasibility of the project depends to a large extent on the quality and quantity of the fill material available and its haulage distance to the reclamation site. Because of the common need for surcharge material it is possible that there will be a surplus of dredged material at the end of a project. This can have a significant impact on costs, environment and planning. Therefore it is advisable to make a proper cut and fill balance of the work during the planning stage and to update this regularly during the design process and during the execution phase.

Fine to medium quartz sands are to be preferred as fill material. If this is not available lesser quality material may have to be accepted such as carbonate sand, silt, clay or other alternative materials. This may, however, have an impact on the design, the work method which may possibly include ground improvement, equipment and therefore also on the cost.

### 2.2.3 *Data collection*

Already in an early stage of a project, starting to collect data of the physical (subsoil, bathymetry, waves, currents, wind) and environmental setting of the proposed reclamation site and borrow area is important. This information is not only required to support the feasibility study, but it will in later stages also be helpful to better plan successive site investigations. Some studies, like an Environmental Impact Assessment (EIA), may require much time to complete and should be started as soon as possible to avoid interference with the overall project planning.

This implies that activities such as the determination of the reference condition (existing situation before start of the reclamation works) have to commence in the project initiation phase.

Data collection includes desk studies, field work and laboratory testing. During the project initiation phase the main focus will usually be on desk studies and preliminary field and laboratory operations.

### 2.2.4  *Environmental requirements*

Each reclamation project will have its positive and negative effects on the environment in which it is executed: borrowing fill material will disturb and change the seabed, affect the water quality and its benthic habitat, dredging equipment creates noise and vibrations, changes the air quality, deposition of fill will cover the original benthic habitat, may increase the suspended solids in the water column and change the morphology, but may also create opportunities for the development of new habitats.

It is therefore important for stakeholders not only to focus on potential negative effects and the options to mitigate these but also to explore and incorporate the positive environmental impacts as a result of a project. This concept is promoted as 'Building with Nature' (www.ecoshape.nl) or similarly 'Working with Nature' (www.pianc.org/workingwithnature.php).

In most countries legislation requires that an Environmental Impact Assessment (EIA) be carried out before the necessary construction permits can be obtained. Such an EIA must be based on thorough, well-documented research and provides an evaluation of the positive and negative environmental consequences of the works. If the predicted impacts are considered to be unacceptable, the study may recommend mitigating measures ranging from prescribed dredging methods to compensation requirements.

Insight into environmental aspects is important for the following reasons:

- to show that no contaminants are present both in the borrow and the reclamation areas;
- to determine the impact of the project on flora and fauna and the opportunities of the project to enhance the sustainability of the altered environment;
- to determine the potential impact caused by the extraction of fill material, on coastal geomorphology and hydro dynamics with beach erosion or accretion as a result.

When large volumes of fill material are dredged, transported and placed, some of this material may be eroded during the construction or operational phase.

Determining whether or not this eroded material settles elsewhere and whether or not this may cause a problem (local siltation) is crucial. Should unacceptable sedimentation occur, measures must be defined to prevent this process.

The placement of fill material in a coastal zone or a riverbed can change the sea currents or river flow, which can cause changes in erosion patterns and siltation. This must be taken into consideration in the design stage as well as during construction.

Impact on the environment must be within the limits set out in permits and described in an Environmental Impact Report (EIR). If the effects are not within the prescribed limits, measures must be defined in order to minimise or compensate the environmental impact. Therefore, obtaining a first impression of the environmental impact at an early stage of the project is essential.

### 2.2.5 *Feasibility study*

An important activity during the project initiation phase is to analyse the project feasibility. This study assesses the viability of the (total) project and must support the decision to go ahead with the project or not. Although not really within the scope of this Manual, the feasibility study is briefly discussed in this section because the physical and environmental conditions at the envisaged borrow area and reclamation site, the equipment proposed for executing the reclamation works and the design (including technical specifications) may affect the outcome of this study.

The feasibility study should not only encompass the economics of the project, but should also address its technical viability, legal consequences, social and environmental issues. Although the land reclamation itself may only form a small part of the total project costs, failing to meet the planning could be a major risk factor as it may delay all subsequent construction activities which could affect the financing structure of the project.

**Financial and technical analysis**
Revenues of the Client can be a result of selling or leasing of land, but could also comprise the profits of land use such as the exploitation of an airport or a golf course, storage of containers or storage of LNG and petrochemicals, and so on. Potential project benefits may also be derived from port dues in the case of a harbour extension, economic benefits in the case of new roads or railways, safety against loss of land as well as loss of life and goods in the case of a dike structure.

The project development costs include the construction of the reclamation and its superstructures and facilities. A financial and technical analysis must be carried out at an early stage of the project definition.

The financial feasibility is determined by the costs and benefits of a project. Most important cost items are:

- engineering costs;
- costs related to environment;
- construction costs;
- costs for access roads or canals;
- costs of land acquisition, -exploitation costs;
- maintenance costs.

Dredging and reclamation projects are often of such a magnitude that long-term costs and benefits have to be considered.

The technical feasibility study includes an evaluation of the design, especially regarding soil conditions and material available in the borrow area. For that reason information on the volumes and quality of sand in the borrow area and an assessment regarding the dredgeability of the material is necessary. For instance, suitable sand may be overlain by unsuitable materials such as clay, peat or very fine sand, which have to be removed and disposed of. The borrow area may be sheltered from waves or positioned offshore exposed to wave action, which is of consequence for the dredging method. These physical circumstances require an accurate insight into the characteristics and possibilities of dredging equipment. A feasibility study further requires the evaluation of alternative designs in order to select the optimal solution.

The subsoil conditions in the reclamation area determine the additional fill volume needed to compensate for settlements. In the case of soft underlying clay layers, the fill may have to be placed in thin layers to avoid instabilities. Soft compressible clay layers may also have to be removed before the sand fill is placed. Vertical drains may be installed in the soft layers to accelerate consolidation and to strengthen the subsoil. In the case of removal, such material may be stored in depots on land or even removed from the site. These aspects have a consequence for the dredging method, the dredging cost and ultimately for the feasibility of the project.

In the initial stage of the project, the realisation costs may have to be assessed using rough estimates. As the project evolves more data become available and more accurate figures can be produced. The feasibility must be judged at a distinctive decision moment, such as after completion of the conceptual or preliminary design.

**Project risks**
An important feature of the feasibility analysis is an evaluation of the project risks inherent to the realisation of a project. The nature of these risks may be technical, financial, political and economic. Whether the Client is a governmental agency or a public or private company determines the legal position of the Client and

whether these risks can be insured. However, the primary approach should be to mitigate the risks whatever the case is.

Examples of technical risks are:

– uncertainty about the quality and volume of the fill material, required and available at the reclamation site and borrow area respectively;
– uncertainty about the nature of the subsoil at the reclamation site;
– uncertainty about the availability of equipment suitable to realise the project, especially for larger projects;
– uncertainty about settlements and slope stability;
– uncertainty about physical conditions (adverse weather, heavy seas and such.);
– delays in the planning.

The technical uncertainties may be eliminated by conducting an appropriate soil investigation both in the fill and borrow areas and by acquiring sufficient physical data, i.e., waves, current, wind etc. in a timely manner. Suitably qualified and experienced specialists should be employed for defining a sound (conceptual) design.

Examples of financial and economic risks are:

– unexpected cost increases caused by adverse soil conditions. This can result in other types of equipment to be deployed, additional equipment and/or a longer execution time, which involves considerable extra costs;
– additional costs caused by inflation (fuel, wages, materials, ….);
– risk of either party going bankrupt;
– risk of unstable currency;
– risk of less project income than envisaged;
– change in economic situation;
– too optimistic budgeting;
– increase of royalties for the material from borrow areas.

Examples of political risks are:

– change of government;
– political unrest, strikes;
– social/economic demands.

Political risks are usually difficult to forecast and can hardly be managed.

**Cost-benefit analysis**
The project planning forms the basis for the financial planning. An (on-going) cost estimate must be made as the project proceeds. These are the costs for project preparation (including Client's advisors), construction and maintenance costs, operational costs, costs of financing and the costs of risks. These costs must be offset against the expected project income. Such a financial analysis is an important tool to manage the project and is also often required to obtain

project financing from financial institutes, relevant governmental agencies, and investors.

Risks can be managed using an on-going cost-benefit analysis with all relevant parameters as input. A regular update of the input for such an analysis (as the project proceeds) enables the Client to take proper measures to cope with risks. On the benefit side the risks may include a change in the market demand which could be of consequence for the phasing of the project. For instance, less or more port users than originally predicted or less or more capacity required per port user would result in a change in the phasing of the construction schedule such as a delayed or earlier start of parts of the ports infrastructure (e.g., quay wall or container stacking areas).

Risks and benefits should be in balance. A scheme of such balance is the cost-benefit analysis presented in Figure 2.1.

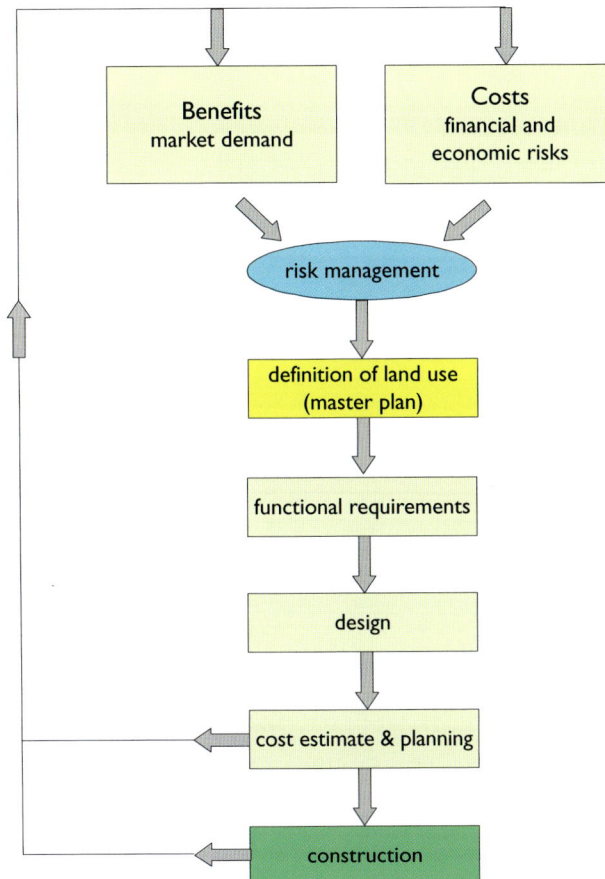

Figure 2.1  *Cost-benefit analysis for a hydraulic fill project.*

The cost-benefit analysis is not only critical during the project initiation stage but should also continue during the construction and operational stages using input relevant for each of these stages.

### 2.2.6 *Initial project planning*

In the initial stage the project is planned in accordance with the main milestones of the Master Plan which results from the feasibility analysis of the project. The planning indicates the dredging capacity required to produce sufficient sand volume within a certain anticipated time frame. Other important inputs for the planning are the time required for the design, to obtain permits and to acquire the ownership of the site where the project is to be realised. The outcome of the initial findings is important information in view of the judgement with respect to the feasibility of the project.

### 2.2.7 *Legal aspects*

Legal aspects related to various legislations and governmental directives, including natural and environmental zoning plans, as well as federal and local regulations must be accounted for. The permits required for the work should be checked and defined. The procedural period to cope with legal aspects is often very long which should be realised during the planning phase of a project. Legal aspects will not be elaborated further in this Manual.

### 2.2.8 *Types of contracts*

The type of construction contract is usually chosen in an early phase of a project. Important tasks that must be completed in every construction process relate to:

- Initiative
- Design
- Building processes
- Financing
- Long-term maintenance and management.

The main players in the construction process are the Client, the Contractor, the Advisor and/or Consultant, the End-user and the Financier. Depending on the chosen contract format, the role of each player will be different. The following main contract formats can be distinguished:

- traditional contracts, with split responsibilities for each party;
- integrated contracts, whereby the Contractor is involved in the design process;

– a Partnerships or an Alliance contract, whereby various tasks of the construction process are shared between Client and Contractor;

The selected contract format will definitely have a bearing on the risk profile of each party.

**Traditional "Client design" contract**

In the setup of a traditional contract, the Client is assisted by a specialist Consultant for the planning, design and construction of the project. Typically the Consultant:

– defines the soil investigation and survey programs;
– interprets the data and reports;
– makes an initial and detailed design of the project;
– provides a cost estimate;
– prepares the Tender Documents;
– assists in selecting the Contractor;
– is often also the Engineer during the construction.

Sometimes this last activity is contracted to another firm. This is not an ideal situation since, for the judgment of monitoring results, a direct line with the original and responsible designer is more convenient and efficient.

The tender includes detailed technical and contractual information upon which the Contractors base their bids. The tender procedure usually requires Contractors to prove their ability to construct the works (technical bid). Once the Contractors have gone through this selection procedure successfully, the project is usually favoured to the lowest bidder. The Tender Documents make up the basis of the contract between the Client and the Contractor.

Contractual risks are inherent to any contract and can be dealt with in a number of ways. The most ideal situation is a contract whereby risks are borne by the party who can handle specific risks the best.

Within a traditional contract, examples of risks normally born by the Client are:

– The risk of price increase of fuel.
– The risk of the accuracy of the required fill volume. The dredge and reclamation volume is usually paid based on a unit rate per $m^3$ and according to the actual measurement of this volume. In a traditional contract the risk of the actual volume is borne by the Client, unlike a lump- sum contract where the risk of the dredge and reclamation volume is for the Contractor.
– The risk of the correctness of the provided soil information upon which the Contractors base their bids.

Examples of Contractor risks are:

- The interpretation of the provided soil information both for dredging and reclamation works.
- The selected work method.
- The planning of the works within the specified milestones.

This risk allocation shows the responsibility of the Client to provide accurate soil information to enable Contractors to define their work method and to make their cost assessment. Failing this, "adverse soil conditions" may be encountered which could have a consequence for the work method such as deployment of additional equipment, change of equipment and extension of the construction period. This may have huge cost consequences leading to claims and even to arbitration.

A prerequisite for this "balance of risks" is that both parties, i.e., the Client and the Contractor, have professional knowledge/competencies in these specific areas.

An advantage of the traditional "Client's design contract" is that the Client has full control over the design of the project.

A disadvantage is that the Contractor's experience and know-how is not implemented in the design, the work method and the planning of the construction. This may lead to higher costs and possible changes in the design and the planning during the execution of the works. This disadvantage may be avoided if the Contractors are allowed to submit alternative bids which suit their equipment, planning and work methods, leading to lower project costs. Work methods may however already be defined in the permit for executing the works and this may not allow for alternative work methods or constructions. In that case permission for an alternative work method or construction may not be granted or may be subject to a lengthy process.

**Integrated contracts**

There are different ways of Contractor's involvement depending on which functions (design – construct – finance – maintenance – operate) are integrated in the Contract. Most commonly used is the "design and construct contract" which calls upon the Contractor's knowledge and experience at an early stage (planning) of a project and where the Contractor has design responsibility. Based on the definition of the use of the land reclamation, the Contractor formulates the performance requirements and prepares the design based on international standards and recommendations. In this process the Contractor may be assisted by a Consultant. The soil investigation and surveys may also be conducted by the Contractor. This would be the ideal situation because the risk of encountering adverse soil conditions is now fully borne by the Contractor. However, for practical reasons (since this can hardly be conducted by all potential bidders) often the soil investigation and survey is conducted under the responsibility of the Client.

The advantages of a "design and construct" contract are that:

– the Contractor is able to implement his work method and equipment into the work permit, the planning and the construction of the project which normally leads to reduction of costs;
– the Client draws on specialist knowledge of the Contractor regarding design disciplines in the preparation of the project.

During execution the position of the Client is restricted to check on "milestones" defined in the contract.

Disadvantages of a "design and construct" contract are:

– the detailing of the functional requirements into performance requirements by the Contractor may be a lengthy process;
– the Client is required to have specialist knowledge to check the Contractor;
– where more Contractors are involved during the bidding procedure, all Contractors have to go through this lengthy process with subsequently high tender preparation cost, which may not be fully retrieved from the Client;
– the Client is not fully in control of the resulting design. Once the design is conducted according to international standards and recommendations and the Contractor is convinced of the ability to construct his design, the Client can hardly reject the outcome of that process.

"Design and construct" contracts demand that the Client has complete confidence in the professional standards of the Contractor.

**Partnership contracts**
A joint venture between Client and Contractor, which can be applicable to both traditional and integrated contracts, is the partnership or Alliance contract. In this type of contract the Client and the Contractor share one or more tasks in the process based on a "best for project" philosophy. This means that costs and risks are shared according to a predetermined division. The advantage is both parties share an interest to minimise risks and to find ways for cost savings. Another advantage is that, due to early Contractor involvement, the total project implementation time can be shortened since a number of activities can be executed in parallel instead of serially. Furthermore, the Client can draw upon Contractor's know-how and experience to get the necessary environmental licenses in place.

Based on the recent success of various projects which have been executed as Alliance contracts in the Netherlands, this type of contract has lately become a favoured as well as successful tool for the implementation of complex projects.

## 2.3 Design

As a result of their nature and scale, hydraulic fill projects often require major investments, have long operational lifetimes and may have significant impacts on the environment in which they are constructed. These aspects ask for careful planning and design in order to obtain in a timely fashion the desired result in the most cost-effective way without unwanted environmental effects.

An adequate design must ensure the satisfactory behaviour of the land reclamation throughout the anticipated lifetime of the structure. Such satisfactory behaviour does not only include an adequate response of the fill mass and the existing underlying soils to loads of future superstructures, but also comprises sufficient resistance against natural forces such as waves, currents, earthquakes, and other events. To achieve this requires a wide variety of disciplines like geotechnical engineering, environmental engineering and morphology, hydraulic and coastal engineering in combination with practical knowledge and experience of dredging and reclamation techniques.

A design process can often be divided into phases, but the transitions between the various phases may not always be that distinct. The design usually starts with a conceptual design and ends with the detailed design (see below). A phased design approach allows for concurrent activities like a preliminary EIA and data collection and enables more detailed studies or even a revision of the design as more data become available.

As the Manual focuses on hydraulic fill, bearing loads, maximum allowable settlements and other requirements related to the superstructures are considered to be starting points for the design of the fill mass. The structural design of the superstructures is not part of this discussion.

### 2.3.1  *Design phases*

The following phases can generally be distinguished in the design process of a reclamation project regardless of the type of contract:

**Conceptual design, (pre-FEED = front end engineering design)**
In the conceptual design phase of a hydraulic fill project, the main outlines of the proposed reclamation area are determined. Also requirements, restrictions, assumptions and regulations are gathered during this phase so that the design can take shape in the subsequent stages.

**Preliminary design (basic engineering for FEED)**
The conceptual design is followed by the preliminary design. For this design more detailed data are needed. This includes mostly geotechnical and hydrodynamic studies which have a great effect on the development of the design.

In addition to the technical data research, also economic and environmental assessments are made and a tentative planning is produced to verify the feasibility of the design. Alternative designs are evaluated accounting for feasibility issues. The designs are weighed and compared by all stakeholders to reach agreement in an early stage of the project.

**Detailed design**

The activities undertaken during the detailed design phase will result in a final design including technical specifications, drawings, cost estimates and the planning. These activities may include further ground investigation and laboratory testing, geotechnical calculations and other studies such as the resistance of the structure to waves and current and, if necessary, morphological studies. This may also include physical modelling. From these studies, a more enhanced view of environmental impacts may result. If necessary, measures have to be studied and proposed to counteract possible negative environmental impacts during the construction and/or after completion of the project.

## 2.4   Considerations for selecting construction method

Construction methods, i.e., equipment and related working methods, must meet the planning of the project, its specifications and the lines and levels set out in the contract. Equipment comprises dredging plant, means of transport of the dredged materials and equipment required at the fill area itself.

The construction method will also be influenced by the prevailing conditions at the designated borrow area, along the route of transportation and at the proposed reclamation site, these include but are not limited to:

– the required properties and volume of the fill mass;
– bathymetry;
– quality of the fill available;
– environmental restrictions;
– nature of the existing subsoil;
– wave climate.

In addition, physical and environmental issues as well as the availability of the anticipated dredging plant may have a significant impact on the work method adopted. For instance:

– reclamation activities in areas with a sensitive marine environment may require construction methods that limit the turbidity caused by the process water discharged into that environment;

− soft subsoil conditions at the reclamation site may ask for a construction method with lifts of limited thickness and/or soil improvement in order to ensure geotechnical stability, and so on.

The possibilities and limitations of the dredging plant in combination with the boundary conditions may also affect the technical and economic feasibility of a project. For instance:

− the installed pumping capacity of a dredger in combination with a boundary condition like the grain size distribution of the fill will determine the maximum achievable pumping distance;
− the draught of dredgers in combination with a boundary condition like the bathymetry in the immediate vicinity of the site determines how close the dredger can approach the reclamation area and hence affect the pumping distance.

Technical and economic evaluations of the feasibility of a dredging project clearly require both a full inventory of the requirements of the Client (expressed as the desired properties of the fill mass) and the boundary conditions, and also a thorough knowledge of the characteristics of the available dredging plant and reclamation methods.

The suitability of a piece of equipment for a particular job will primarily depend on its technical characteristics. Note however that its availability at the time of construction and the mobilisation costs may also be decisive factors for deployment.

## 2.5   Systems Engineering

The design of a hydraulic fill should be a rational process that ultimately results in the best and most economical match between the properties of the reclaimed land and the requirements imposed by its future use, all within the boundary conditions of the project and limitations set by the EIA. The design process described in this Manual follows the "Systems Engineering" approach.

Systems Engineering is a worldwide accepted tool for the design of complex projects. It is used in many sectors of industry such as aero and space technology, telephone systems, defence industry and computer technology. Also for infrastructure works this method has proven its credibility.

The essence of Systems Engineering is that it is interdisciplinary, focuses on defining customer needs and required functionality early in the development cycle and is subject to a structured evaluation and feedback system as this cycle rolls out. The customer's needs and functionality requirements are input to the design and feasibility analysis of the project. Steps in the development cycle are evaluated at certain pre-defined moments and if needed adjusted. Such adjustments may include the design itself, but also the needs and functional requirements formulated for the project. Structured evaluation and feedback throughout the development cycle is aimed to lead to the most effective and economical solution of the Client's needs. As the development cycle proceeds the outline of the system becomes increasingly detailed.

In order to make use of the appropriate expertise, the integrated approach may already enlist Advisors and Contractor specialists at an early stage of the development cycle.

This development cycle approach is reflected in the following definition of Systems Engineering (Bahill *et al.* 1998): "Systems Engineering is an engineering discipline whose responsibility is creating and executing an interdisciplinary process to ensure that the customer and stakeholder's needs are satisfied in a high quality, trustworthy, cost-efficient and schedule-compliant manner throughout a systems' entire life cycle."

This process is usually comprised of the following seven tasks:

– State the problem
– Investigate alternatives
– Model the system
– Integrate
– Launch the system
– Assess performance, and
– Re-evaluate.

These functions can be summarised with the acronym SIMILAR: State, Investigate, Model, Integrate, Launch, Assess and Re-evaluate. The Systems Engineering Process is shown in Figure 2.2. It is important to note that the Systems Engineering Process is not sequential. The functions are performed in a parallel and iterative manner (www.incose.org).

The diagram above covers the total project cycle from its initiation to its use. In a comparable way the principle of System Engineering is adopted in this Manual for the realization of hydraulic fill projects from the definition of the Client's needs to monitoring and quality control of the fill after completion.

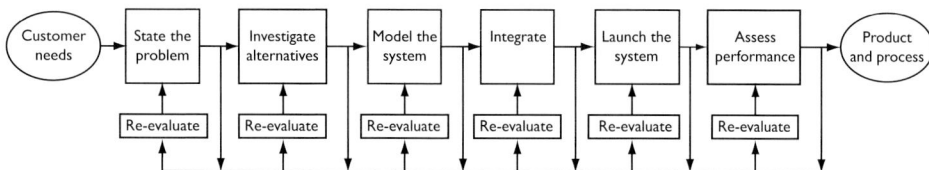

Figure 2.2   *The Systems Engineering Process from A.T. Bahill et al. 1998.*

For a reclamation project the development cycle is shown in Figure 2.3. At an early stage the use (or function) of the land reclamation is defined which is then reflected in functional requirements for the fill mass. This forms the basis for the design and the specifications which finally leads to the construction of the project. This is the main stream. The transformation of the functional requirements into a technical design is directed by the boundary conditions and the construction method. These boundary conditions are established by project data such as soil and environmental conditions.

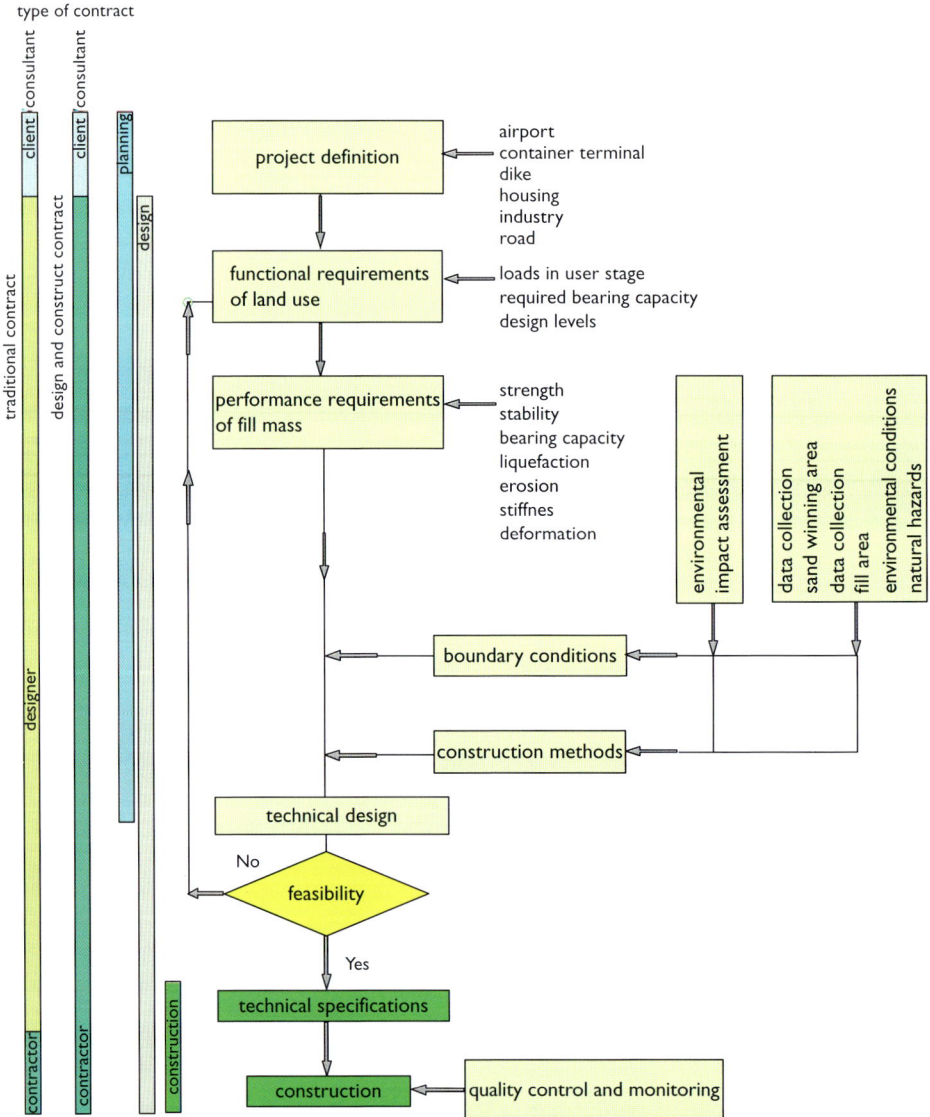

Figure 2.3    *Planning, design and construction phases of a hydraulic fill project.*

Given the project's boundary conditions and the construction methods, the feasibility of the design must be judged by the criteria set out for the land use and by the functional requirements of the fill mass. These may have to be re-considered should the criteria and requirements not be in agreement with each other.

The diagram in Figure 2.3 shows the involvement of the Client, Consultant and Contractor for different types of contract. In the classic contract, Contractor's involvement starts once the design is completed, specifications are written and tender documents are prepared. Modern design and construction contracts provide for early involvement of the Contractor in order to benefit from the Contractor's specialised knowledge. This early involvement helps avoid unnecessary risks and allows the design of the fill to take into account feasible and economical construction measures.

# DATA COLLECTION

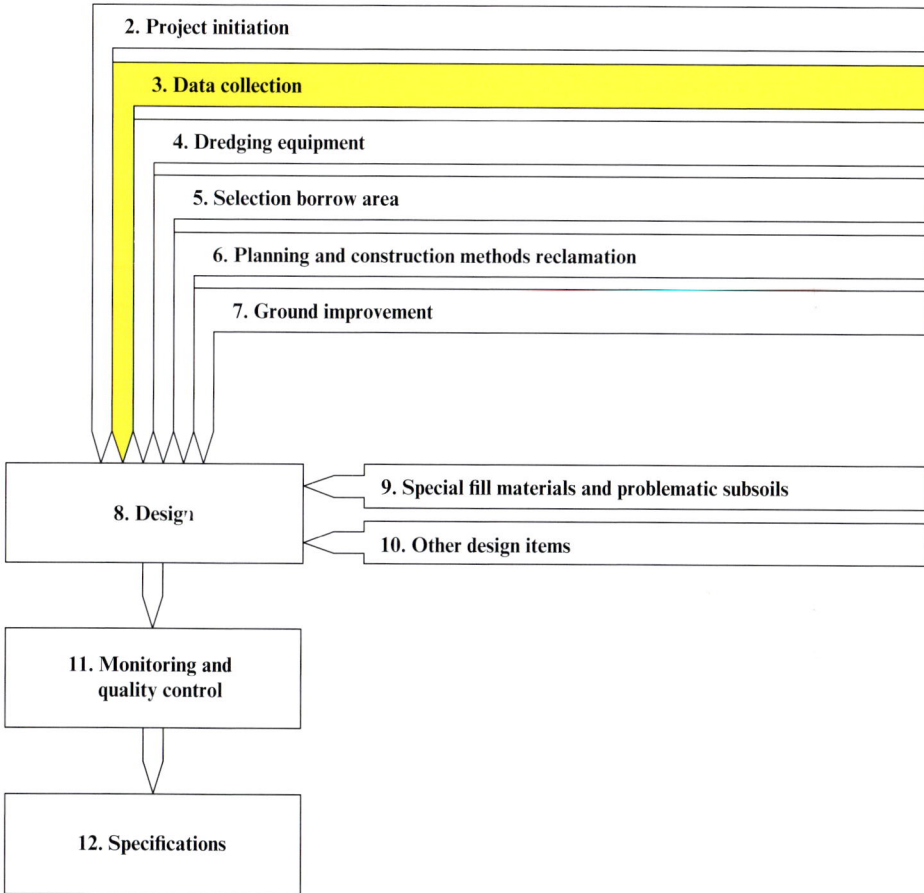

Chapter 3 describes which data are required for assessing a borrow area and designing a reclamation area. A brief description of the most common methods of data collection can be found in Appendix B, "Field and laboratory tests". Some of these data collection methods will also be discussed in other chapters of this manual, where appropriate. The flow chart shows the structure of this chapter and where to find information corresponding to various items.

Chapter 3 Data Collection

3.3 Desk study

| Dedicated borrow area | Dredging area (dredged material used as fill) | Reclamation area |

3.4 Required data

| Cohesive soils | Non cohesive soils | Rock |

3.4.1
Bathymetrical or topographical data

3.4.2
Geological and geotechnical information

3.4.3
Hydraulic, meteorological, morphological and environmental data

3.4.4
Seabed obstructions

Dedicated borrow area

Dredging area

Fill area

- Volumes
- Excavation methods & production
- Transport methods & production- abrasion, wear
- Bearing capacity
- Settlement
- Slope stability
- Liquefaction
- Environmental impact
- UXO / Wrecks

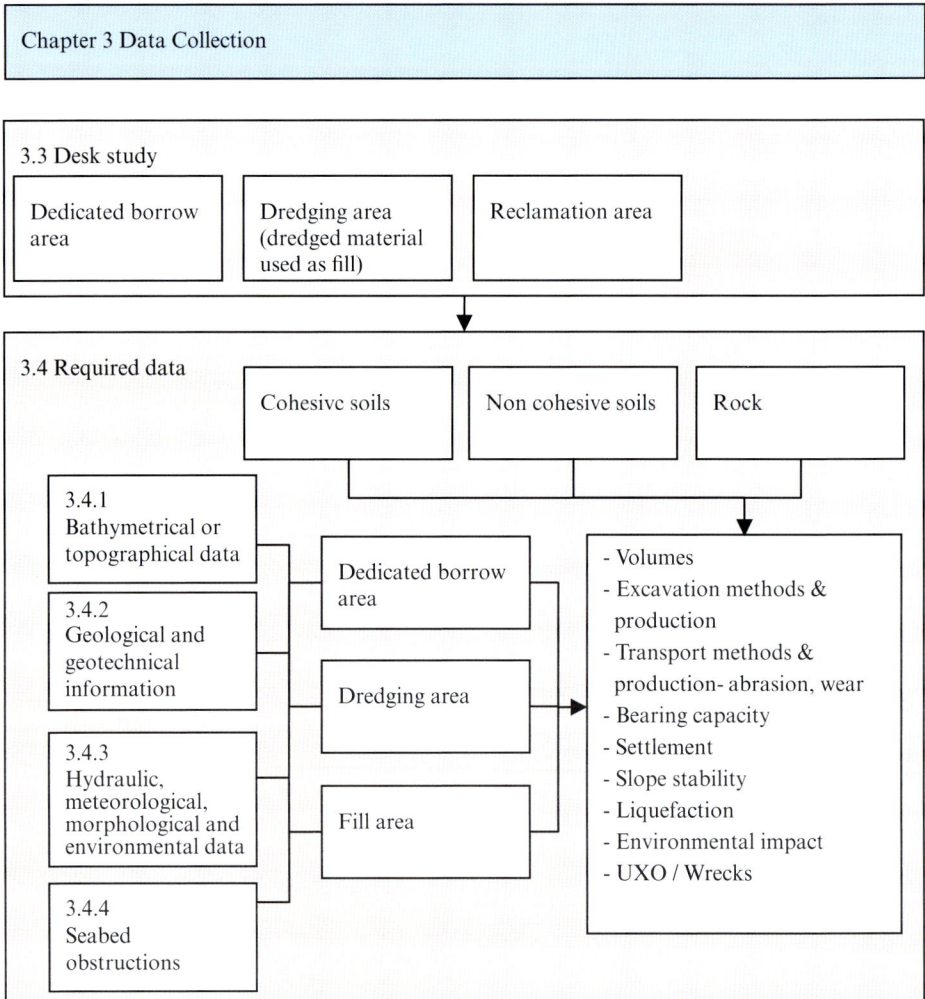

Figure 3.1   *Flowchart.*

## 3.1   Introduction

The required quality and quantity of data differs per project phase. Usually each successive step in the design process requires a higher level of detail in the data. However, even at an early stage of the project, adequate data should be available to assess the feasibility and location of both the reclamation area and the borrow area.

The borrow area can be:

– a dedicated borrow area purely used for soil extraction purposes;
– a dredging area like a harbour basin or navigation channel from which the resulting dredged soil or rock can be used as fill material.

When assessing the soil characteristics in a potential borrow area, the important aspect will be the suitability of the soil as fill material. It should be kept in mind that the dredging operations might alter the "in situ" properties of the soil (e.g. when dredging with a trailer suction hopper dredger, most of the silt/clay fraction will be washed out). This matter is discussed in more detail in Chapter 5, Selection borrow area.

The reclamation area can be situated onshore or offshore.

The data collection process is shown in Figure 3.1. The sections in this chapter corresponding to each subject are indicated on the flowchart.

A separate issue is data required for monitoring. For this information, the reader is referred to Chapter 11 monitoring. An overview of data required to monitor the impacts of dredging works on the surroundings is given in "Environmental Aspects of Dredging" (Bray, 2008).

A desk study is a cost-effective and logical first step of a site investigation carried out in an early stage of a project. As far as possible, the desk study should be completed and reported before commencement of surveys as well as conceptual design. The next step in the design process is to make an inventory of the required data for the design process.

The required data will differ depending on:

- type of investigated area (dedicated borrow area, dredge area or reclamation area);
- soil type: different parameters are required for cohesive soils, non-cohesive soils or rock;
- design issues which have to be investigated, such as volume calculations, slope stability, settlement; abrasion, possible dredging methods.

Based on the data gathered during the desk study, a site investigation can be set up to fill in the missing data. A site investigation campaign can be split up into 4 main topics:

- bathymetrical or topographical survey;
- geological and geotechnical investigations including geophysical investigations;
- metocean survey, environmental investigations;
- seabed scanning (side scan sonar, magnetometer survey).

Once the site investigation is completed, a report should be processed containing all gathered data and the interpretation thereof. It is important to include all factual data in order to allow all parties involved in the project to make their own interpretation. Often a party may not have been involved in the site investigation and has to rely solely on the report produced.

Ultimately, the total value of the site investigation is represented by the (quality of the) final report. This report will in most cases be part of the contract.

In reality, the data collection process may not be that straightforward. A preliminary survey, such as a preliminary soil investigation, may often be desirable in an early stage of the design process to determine the extent and nature of the detailed investigations. The data obtained in this preliminary survey may be used in the preliminary design. In any case, for a cost-effective design process, all soil properties and other data should be available before continuing with the detailed design.

## 3.2   Interpretation of data, contractual aspects

The responsibilities and liabilities of the Employer and the Contractor with regard to the interpretation and procurement of site data, depend strongly on the type of contract. Generally a FIDIC form of contract is adopted (see reference list at the end of this section). For typical dredge and reclamation works (not involving civil works) the FIDIC 2006 blue book is the most appropriate form of contract. In case civil works are involved for which the design has been done by the Employer, the FIDIC 1999 (new) red book can be adopted. In case of a design and build contract (design done by the Contractor), the FIDIC yellow book is more appropriate. Finally, for large EPC contracts, the FIDIC silver book is adopted.

In the blue, red and yellow book, the Contractor shall be responsible for the interpretation of the data provided by the Employer, but is not responsible for certain unforeseeable risks or conditions.

To further define what is reasonably foreseeable, particular conditions of contract are often added (for instance: The Contractor's price is based on the assumption that the SPT value of the sands to be dredged is lower than 30).

In the silver book, the Contractor is fully responsible, even for unforeseeable risks.

An overview of the relevant clauses can be found in the table below.

| FIDIC | Specific clauses on site data |
|---|---|
| Blue book | General conditions 2.3 site data |
| | The Employer shall have made available to the Contractor for his information prior to tendering, all data in the Employer's possession relevant to the execution of the Works, including hydrological, sub-water surface and sub-bottom conditions, and environmental aspects. The Contractor shall be responsible for interpreting all such data, and for inspecting |

the Site and making his own enquiries so far as is practicable (taking account of cost and time) before submitting his tender

10.4 Contractor's right to claim

If the Contractor incurs Cost as a result of any of the Defined Risks, the Contractor shall be entitled to the amount of such Cost, subject to any more specific provision in the Contract. If as a result of the Defined Risks, it is necessary to change the Works, this shall be dealt with as a Variation.

6.1 Defined risks

l)   Physical obstructions or physical conditions encountered on the Site during the performance of the Works, which obstructions or conditions were not reasonably foreseeable by an experienced Contractor and which the Contractor immediately notified to the Engineer.

m)  Climatic conditions more adverse than those specified in the Appendix.

Red book   General conditions 4.10 site data

The Employer shall have made available to the Contractor for his information, prior to the Base Date, all relevant data in the Employer's possession on sub-surface and hydrological conditions at the Site, including environmental aspects. The Employer shall similarly make available to the Contractor all such data which come into the Employer's possession after the Base Date. The Contractor shall be responsible for interpreting all such data.

To the extent which was practicable (taking account of cost and time), the Contractor shall be deemed to have obtained all necessary information as to risks, contingencies and other circumstances which may influence or affect the Tender or Works. To the same extent, the Contractor shall be deemed to have inspected and examined the Site, its surroundings, the above data and other available information, and to have been satisfied before submitting the Tender as to all relevant matters, including (without limitation):

(a)  the form and nature of the Site, including sub-surface conditions,

(b)  the hydrological and climatic conditions.

4.12 Unforeseeable physical conditions

In this Sub-Clause, "physical conditions" means natural physical conditions and manmade and other physical obstructions and pollutants, which the Contractor encounters at the Site when executing the Works, including sub-surface and hydrological conditions but excluding climatic conditions.

If and to the extent that the Contractor encounters physical conditions which are Unforeseeable, gives such a notice, and suffers delay and/or incurs Cost due to these conditions, the Contractor shall be entitled subject to Sub-Clause 20.1 [Contractor's Claims] to:

(a) an extension of time for any such delay, if completion is or will be delayed, under Sub-Clause 8.4 [Extension of Time for Completion], and

(b) payment of any such Cost, which shall be included in the Contract Price.

Yellow book    4.10 Site data: Identical to the red book;

4.12 Unforeseeable physical conditions: Identical to red book.

Silver book    General conditions 4.10 site data

The Employer shall have made available to the Contractor for his information, prior to the base date, all relevant data in the Employer's possession on subsurface and hydrological conditions at the site, including environmental aspects. The Employer shall similarly make available to the Contractor all such data which come into the Employer's possession after the base date.

The Contractor shall be responsible for verifying and interpreting all such data. The Employers shall have no responsibility for the accuracy, sufficiency or completeness of such data, except as stated in sub clause 5.1 (General design responsibilities).

4.12 Unforeseeable difficulties

Except as otherwise stated in the contract:

The Contractor shall be deemed to have obtained all necessary information as to risks, contingencies and other circumstances which may influence or affect the works.

> By signing the contract the Contractor accepts total responsibility for having foreseen all difficulties and costs of successfully completing the works; and
>
> The contract price shall not be adjusted to take account of any unforeseen difficulties or costs.

## 3.3  Desk study

A desk study is usually the first phase of a site investigation and is necessary to obtain optimal results of the subsequent phases. It includes the collection, review and verification of information already available about a site, and identifies potential areas of information conflict or deficiency. It is carried out at an early stage of site appraisal to guide the remainder of the site investigation. Depending on the location of the site (offshore, nearshore or onshore) desk studies may also include a visual inspection of the site and its surrounding area (usually called a walkover survey).

The desk study needs to consider a broad range of issues, all of which may affect a project both practically and logistically. The desk study should include a review of all sources of appropriate information including historical records, and collect and evaluate all available relevant data for the site including, for example:

- bathymetric information;
- admiralty charts and other local (more detailed) sea charts;
- geological information;
- information and records of seismic activities (earthquake risk and magnitude, tsunami activity);
- existing geotechnical and geohydrological data and information;
- previous experiences with comparable activities (such as dredging) or other activities of interest (such as foundations) in the area;
- meteorological, oceanographic information and hydrological information including water levels, tides, currents, wind and wave regimes;
- non-technical data such as the presence of submarine pipelines, restricted areas, shipping movements, fishing or military activity, sailing distance between borrow area and reclamation area, site accessibility, permits, housing and office facilities, etc.
- ecology: presence of marine reserves, indicators of the presence of protected species.

In any desk study of seabed conditions, a qualitative and quantitative assessment of existing data is mandatory. Potential sources for information and data sets include companies with proprietary data, national geological survey organisations, academia, other maritime operators and Contractors. Internet sites contain meteorological data for some offshore regions, but sources need to be checked and the data verified.

Invariably, the sources for this information cannot be held liable for the accuracy, or relevance, of their information and therefore it should be used with discretion at the engineer's (or user's) risk. Much third party information will have copyright or other restrictions applied to its use and these should be properly researched, and where necessary the owners' permission sought, before being cited.

Often a desk study is considered the most cost effective part of a site investigation. They provide early recognition of the characteristics of the site and potential geohazards by:

- facilitating appropriate scoping of the later stages of the site investigation;
- avoiding money being wasted on inappropriate intrusive ground investigations;
- enabling formulation of the preliminary geological ground model;
- leading to the formulation of efficient designs for subsequent works on the reclaimed area (e.g., foundations, retaining walls, slope stabilization works, etc.);
- providing early warning of possible delays to programme and/or budget implications.

Additionally they are:

- often required by authorities for planning approval;
- part of current good practice for phased site investigations;
- an essential tool for risk assessment and risk management.

A desk study alone would rarely be sufficient for detailed engineering purposes, but should be sufficient for conceptual engineering to move forward in a focussed manner and provide the basis upon which to design and plan subsequent site investigation work. Desk study reports, or linked documents, should include recommendations and requirements for the next phase of the investigation.

## 3.4 Required data

### 3.4.1 *Bathymetrical or topographical data*

Accurate bathymetrical information in the borrow area is required to calculate the available volume of fill in this area. In case the fill has to be borrowed from capital dredging works (i.e., deepening of harbour basin) bathymetric surveys are required to assess the total volume to be dredged.

Since these surveys often form a basis for payment, they are often carried out jointly by the Contractor and the client. Distinction can be made between:

- in survey (before any dredge activity);
- intermediate survey (during the dredging activities);
- out survey (after completion of the dredging activities).

In addition bathymetrical data may be needed to investigate the accessibility of the dredge area to the dredging equipment. The sailing route to and from the dredge area should also be mapped.

An accurate bathymetrical or topographic survey in the proposed reclamation area is required to assess the required fill quantities. The access routes and the immediate vicinity of the reclamation area need to be surveyed as well. For instance, when a trailer suction hopper dredger has to discharge its load through a floating pipeline, information about the water depths in the vicinity of the reclamation area is crucial to determine how close the vessel can approach the reclamation area.

Making the collected data available in a digital format is strongly advised. This will facilitate importing those data in specific software tools for further analysis.

### 3.4.2   Geological and geotechnical information

There is an important difference in geological and geotechnical required data for the analysis of a borrow area compared to the data required for the analysis of a reclamation area.

Tests done in the borrow area are mainly aimed at determining the quality, quantity and dredgeability of the soil. The soil properties (e.g. grading) might change during dredging and therefore not all tests are useful or at least need to be treated with care (e.g. Particle size distribution tests).

Test done in the reclamation area are mainly aimed at estimating settlements caused by the placement of the reclamation fill and at checking the stability, bearing capacity and liquefaction resistance of the reclamation area and its edges (revetments, retaining walls, bunds etc.).

In the sections 3.4.2.1 and 3.4.2.2, the geotechnical data requirements for borrow areas and reclamation areas are treated separately.

### 3.4.2.1   Geological and geotechnical information in the borrow area

The information gathered during the desk study should ideally be used to determine the geological history and processes of the region and to establish a hypothetical geological model for the borrow area. This geological understanding will help to define the required scope for the site investigations, but also plays an important role in the interpretation of the results of the investigations. Reference is made to "Site investigation requirements for dredging works. PIANC Report of Working Group No. 23, Brussels, 2000" (presently under review).

Table 3.1 *Summary of required geotechnical material properties for dredging works (modified from "Site investigation requirements for dredging works–Report of PIANC workgroup 23–2000").*

| | Data requirement | | | | |
|---|---|---|---|---|---|
| Properties of in situ material | Excavation methods & production | Transport methods & production | Abrasion (excavation and transport wear costs) | Dredged slope stability | Behaviour as fill material |
| **Cohesive soils** | | | | | |
| Particle size distribution | ✓ | ✓ | ✓ | | ✓ |
| Strength/Stress-strain | | ✓ | | ✓ | ✓ |
| Plasticity/water content | | ✓ | | ✓ | ✓ |
| In Situ Density | ✓ | ✓ | | ✓ | ✓ |
| Mineralogy | | ✓ | ✓ | | ✓ |
| Particle Specific Gravity | ✓ | ✓ | | | ✓ |
| Gas content | ✓ | ✓ (soft soils) | | | |
| Rheologic properties | ✓ (soft soils) | ✓ | | | ✓ |
| Organic content | ✓ | ✓ | | | ✓ |
| **Non-cohesive soils** | | | | | |
| Particle size distribution | ✓ | ✓ | ✓ | | ✓ |
| Strength/Stress-strain relationship | ✓ | | | ✓ | |
| Relative density | ✓ | | | ✓ | |
| In Situ density | ✓ | ✓ | | ✓ | |
| Mineralogy | | ✓ | ✓ | | ✓ |
| Particle Specific Gravity | | ✓ | ✓ | | ✓ |
| Angularity/roundness | ✓ | ✓ | ✓ | ✓ | ✓ |
| Permeability | ✓ | | | ✓ | ✓ |
| **Rock** | | | | | |
| Material strength | ✓ | ✓ | ✓ | ✓ | ✓ |
| Mass strength | ✓ | | | ✓ | ✓ |
| Elasticity | ✓ | | | | ✓ |
| Mineralogy | | ✓ | ✓ | | ✓ |
| Structure | ✓ | ✓ | ✓ | ✓ | ✓ |
| Density | ✓ | ✓ | ✓ | | ✓ |

The soil used for reclamation purposes can be split into 3 main categories:

– cohesive soils (clay, mud, silt);
– non cohesive soils (sand, gravel);
– rock.

Each of these soil categories requires its own approach. A first guideline when evaluating the required geological and geotechnical information for a dredge area can be found in Table 3.1.

When looking for a dedicated borrow area, the objective will most likely be to find an area with good quality non cohesive soil at a reasonable haulage distance. In case such an area cannot be found near the project, other soil types might be envisaged. In case soil or rock resulting from a dredging project is used as fill material, rock and cohesive soils will often be part of the dredged material.

There are no fixed rules concerning the extent of ground investigation, such as the number of boreholes required or the spacing of geophysical survey lines. The guiding principle should be one of undertaking sufficient investigation to gain a full understanding of the area in question, in terms of the factors which influence dredging and transport processes and, therefore, costs.

There are two approaches:

– intuitive approach based on detailed interpretation of the data by a qualified geologist or geotechnical engineer;
– the statistical approach using modern geostatistical methods and computer modelling.

The intensity of ground investigation will strongly depend on the heterogeneity and nature of the subsoil. Two main types of soil investigation can be defined:

a. Typical soil investigation campaign aimed at mapping a potential dedicated borrow area;
b. Soil investigation campaign for capital dredging. The reclaimed material is a by-product of the dredging works.

*ad a*

For a typical soil investigation campaign aimed at mapping a potential dedicated borrow area, the flowchart in Figure 3.2 can be used as a guideline (reference is also made to section 3.5).

As mentioned in Figure 3.2 the on-site geologist plays an important role in the determination of the required intensity of the soil investigation. An experienced geologist familiar with dredging techniques should be capable of determining whether the available data is sufficient to setup a sound geotechnical model for the borrow area.

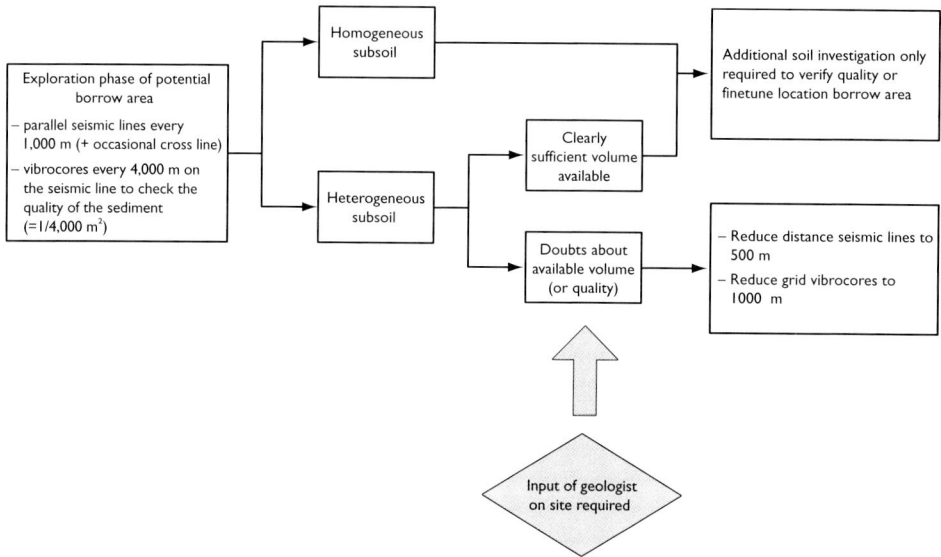

Figure 3.2  *Guidelines intensity of soil investigation for borrow area mapping.*

Another conclusion from Figure 3.2 is that it is almost impossible to exactly plan the required amount of soil investigation required prior to the execution thereof. Some provision and flexibility towards time, budget and equipment should be taken into account when planning a soil investigation.

Often a soil investigation campaign is set up in two phases. The results from the first phase can be used to plan the second phase.

An important aspect in the evaluation of the required amount of soil investigation is the risk and cost involved in case specific ground variations are not recognised. This risk is less pronounced when dealing with dedicated borrow areas. The unsuitable soil can simply be left in place. This risk can become very important when dealing with dredge areas such as harbour basins from which the dredged materials are to be re-used as fill material.

Although often neglected, information on the relative density in the borrow area is required for an accurate estimate of the dredging production. The in situ relative density can be determined by boreholes (with SPT testing) or CPT-testing.

*ad b*

For a soil investigation campaign for a capital dredging work, other considerations are involved. The reclaimed material is a by-product of the dredging works. All soils or rock above the design level will have to be removed.

In case of rock dredging, it is very important to verify that none of the rock above the dredge level is too hard to be dredged. The only alternative to remove this hard rock would be by hammering of by the use of explosives, which is a very costly solution.

Bates (1981) suggested using following formula to determine the number of boreholes that are required for a correct estimation and planning of a dredging project.

$$N = \frac{3 + (A^{0.5} D^{0.33})}{50}$$

In which:
$N$ = number of holes
$A$ = area of the dredging site (m$^2$)
$D$ = average thickness of material to be removed (m)

Applying this formula on 4 typical dredging areas ranging in size from 100 m × 100 m to 5000 m × 5000 m results in the graphs of Figure 3.3, which show the

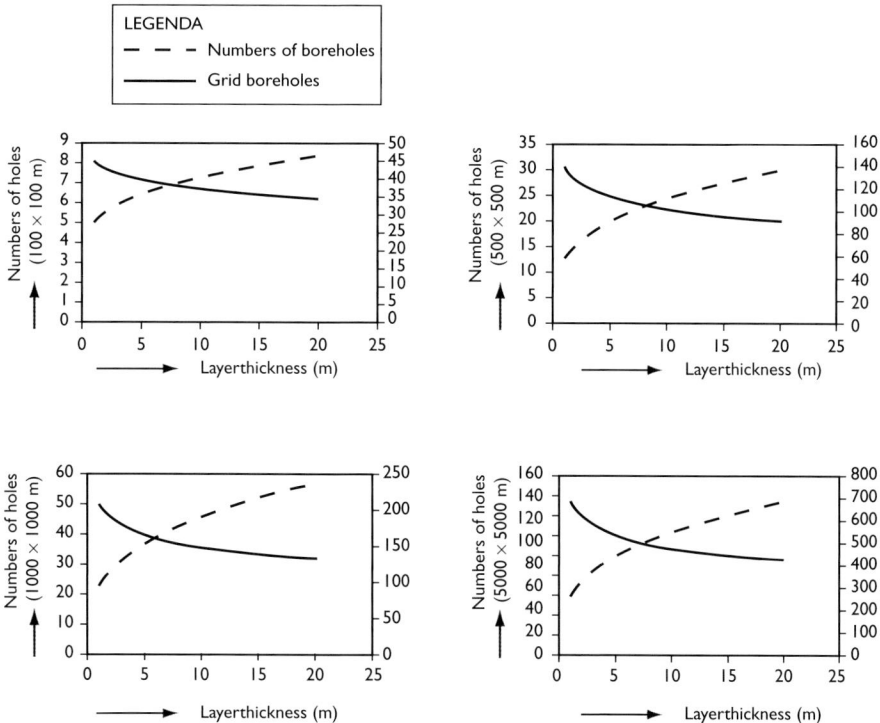

Figure 3.3  *Intensity of soil investigation, Bates 1981.*

required number of boreholes as a function of layer thickness and the corresponding grid spacing of the boreholes.

As a rule of thumb, the required number of boreholes can be reduced by half if a geophysical survey is undertaken, provided it is of good quality and can be correlated to the boreholes.

**Examples of sand search programmes**

Example 1 ("Maasvlakte 2" – Netherlands)
Offshore area dredged with a trailing suction hopper dredger:
Spacing between the seismic lines: 250 to 1000 m
sampling points (borehole or vibrocore) in a grid of 500 × 500 m.

Example 2 ("Saath Al Raaz Boot Field", Abu Dhabi, sand search 2010 for the creation of artificial islands) Irregular offshore dredge areas with local sand patches, to be dredged with a trailing suction hopper dredger:
Spacing between the seismic lines: 2000 m initially, refined to 500 m for interesting areas (total length of seismic lines = 900 km)
200 vibrocores.

3.4.2.2   Geological and geotechnical information of the subsoil in the
reclamation area

This section only deals with the soil investigation required to map the existing subsoil in a future reclamation area.

Once the reclamation area is filled, additional soil investigation will be required to check the properties of the fill and to monitor its behaviour. These issues are discussed in Chapter 11.

Data collection as possibly required for a fill area, corresponds to a large extent with the data collection for borrow areas. However, the soil properties of interest will be different. Consolidation properties of the subsoil for example can be of great importance for the fill area, but are of less importance for the borrow area. Therefore the geotechnical laboratory test programme of the fill area will differ from the test programme of the borrow area. Table 3.2 provides an overview of typical required geotechnical data of the subsoil of the fill areas. Depending on the future use of the subject area, specific additional data may be of interest as well.

3.4.3   *Hydraulic, meteorological, morphological and environmental data*

This section gives an overview of hydraulic, meteorological, morphological and environmental data that are relevant for dredging and reclamation projects. The

Table 3.2 *Summary of possible required geotechnical data at the fill area.*

| Data requirement | Cohesive soils | Non-cohesive soils | Rock |
|---|---|---|---|
| Particle size distribution | ✓ | ✓ | |
| Settlement & Consolidation characteristics | ✓ | ✓ | |
| Plasticity/water content | ✓ | | |
| In Situ Density | ✓ | ✓ | ✓ |
| Mineralogy | | | ✓ |
| Particle Specific Gravity | ✓ | ✓ | |
| Rheologic properties | ✓ | | |
| Organic content | ✓ | ✓ | |
| Relative density | | ✓ | |
| Compaction characteristics | | ✓ | |
| Permeability | ✓ | ✓ | |
| Strength | ✓ | ✓ | ✓ |
| Elasticity | ✓ | ✓ | ✓ |
| Seismic wave propagation [1] | ✓ | ✓ | ✓ |
| Structure (RQD, TCR, SCR) | | | ✓ |

[1] Required in areas prone to earthquakes

data are required as input for the planning and design of reclamation works, but also for monitoring the works in order to assess the short- and long-term impacts. A complete overview of relevant data for design purposes is given in the Rock Manual (2007).

The required data for a dredge area can be split up into 3 categories:

- hydraulic data;
- meteorological data;
- morphological data and environmental data.

### 3.4.3.1 Hydraulic data

Relevant hydraulic data may include:

**Water levels (tide, surge)**
Water levels can be considered as the most relevant data parameter for the design of reclamation works. The extreme water levels mainly determine the heights of the defence structures and the level of the reclamation itself.

**Currents**
Currents in rivers, estuaries and coastal areas are driven by gravity (rivers), density variations, tidal water level variations or wind shear during storms (surge currents). Where water flows around structures (e.g., groins or breakwaters),

horizontal circulation patterns will develop, so-called eddies. Current measurements at one or multiple locations in the area of interest will give insight in these conditions, and will provide data for use in numerical models or to monitoring the hydraulic conditions during dredging works. Defining the type, number, location and duration of current measurements as well as deploying current measurements is not straightforward and requires specialist services.

**Waves**
Wave data are not only important for the design of the surrounding structures of a land reclamation, but also are used for planning and monitoring of the dredging and reclamation works.

Extreme estimates of wave heights and associated parameters such as wave periods, directions and water levels are used in design formulae to assess:

– wave loadings for design of the armour type of the structure;
– wave-overtopping and run-up to determine the slope and crest height of the structure.

There is a major difference between the nature of required data for dredging works and for design studies. For instance, the design study for a breakwater or embankment is mainly guided by the highest expected wave with a certain return period. For the study of a dredging work, insight in the full wave statistics is necessary in order to determine the operational limitations of the vessels, see section 4.3. This difference in required data is also illustrated in Figure 3.4.

The reclamation works and offshore construction works can affect the wave climate. 3D modelling may be required to study and predict the effect of such works. This may include numeric and physical modelling.

Important aspects with respect to the wave climate are:

– the significant wave height;
– the wave period;
– the seasonal variation of the wave climate.

3.4.3.2   Meteorological data

Meteorological data which may be of interest for the design and the construction are:

**Wind**
Although not a direct input parameter for the design of reclamation works, wind is an important parameter describing the physical site conditions (i.e., seasonal variations, storm behaviour) and is the basic input information required in deriving

## Typical required data for analysis of dredging works

### Monthly distribution of wave height (m)

| lower | upper | Jan | Feb | Mar | Apr | May | Jun | Jul | Aug | Sep | Oct | Nov | Dec |
|---|---|---|---|---|---|---|---|---|---|---|---|---|---|
| 0 | 0.2 | 1.4 | 0.3 | 0 | 0 | 0 | 0 | 0 | 0 | 0 | 0 | 0 | 0.9 |
| 0.2 | 0.4 | 30.5 | 21.4 | 7 | 3.7 | 0.4 | 0 | 0 | 0 | 0 | 1.9 | 14.5 | 29.7 |
| 0.4 | 0.6 | 45.9 | 41.4 | 36.4 | 25.8 | 3.4 | 0 | 0 | 0 | 0.3 | 15.1 | 42 | 49.1 |
| 0.6 | 0.8 | 18.5 | 25.5 | 33.3 | 38.1 | 13.8 | 0.1 | 0 | 0 | 3.4 | 25 | 32.9 | 16 |
| 0.8 | 1 | 3.4 | 7.9 | 16.5 | 21.1 | 25 | 5 | 1 | 0.8 | 18.6 | 23.4 | 7 | 2.6 |
| 1 | 1.2 | 0.2 | 2.8 | 5.5 | 7.8 | 25.5 | 7.2 | 1 | 6.6 | 27.5 | 14.3 | 1.5 | 0.6 |
| 1.2 | 1.4 | 0 | 1.3 | 1.3 | 2.4 | 12.7 | 9.4 | 2 | 12.9 | 19.5 | 7.3 | 0.7 | 0.2 |
| 1.4 | 1.6 | 0 | 0.6 | 0.4 | 0.8 | 7.4 | 12.4 | 6.8 | 21.1 | 11.4 | 4.7 | 0.4 | 0.5 |
| 1.6 | 1.8 | 0 | 0.1 | 0.1 | 0 | 4.9 | 14.6 | 12.7 | 19.1 | 6.2 | 2.9 | 0.2 | 0.1 |
| 1.8 | 2 | 0 | 0 | 0 | 0.1 | 2.7 | 12.4 | 16 | 14.2 | 4.9 | 1.9 | 0.2 | 0.2 |
| 2 | 2.2 | 0 | 0 | 0 | 0.2 | 1.8 | 8.9 | 16.8 | 10.4 | 2.8 | 1.6 | 0.1 | 0 |
| 2.2 | 2.4 | 0 | 0 | 0 | 0.1 | 1 | 8.8 | 14.3 | 6.1 | 2.2 | 0.5 | 0.3 | 0 |
| 2.4 | 2.6 | 0 | 0 | 0 | 0 | 0.3 | 6.6 | 11.1 | 3.2 | 1.7 | 0.3 | 0 | 0 |
| 2.6 | 2.8 | 0 | 0 | 0 | 0 | 0.2 | 4.9 | 7.7 | 2.4 | 0.6 | 0.4 | 0 | 0 |
| 2.8 | 3 | 0 | 0 | 0 | 0 | 0.3 | 3.6 | 5.8 | 1.8 | 0.4 | 0.3 | 0 | 0 |
| 3 | 3.2 | 0 | 0 | 0 | 0 | 0.6 | 3.1 | 3.2 | 0.7 | 0.3 | 0 | 0 | 0 |
| 3.2 | 3.4 | 0 | 0 | 0 | 0 | 0.1 | 1.3 | 1.5 | 0.4 | 0.2 | 0.1 | 0 | 0 |
| 3.4 | 3.6 | 0 | 0 | 0 | 0 | 0 | 0.6 | 0.5 | 0.1 | 0 | 0 | 0 | 0 |
| 3.6 | 3.8 | 0 | 0 | 0 | 0 | 0 | 0.2 | 0.1 | 0.1 | 0 | 0 | 0 | 0 |
| 3.8 | 4 | 0 | 0 | 0 | 0 | 0 | 0.4 | 0 | 0 | 0 | 0 | 0 | 0 |
| 4 | 4.2 | 0 | 0 | 0 | 0 | 0 | 0 | 0 | 0 | 0 | 0 | 0 | 0 |
| 4.2 | 4.4 | 0 | 0 | 0 | 0 | 0 | 0 | 0 | 0 | 0 | 0 | 0 | 0 |
| 4.4 | 4.6 | 0 | 0 | 0 | 0 | 0 | 0 | 0 | 0 | 0 | 0 | 0 | 0 |
| 4.6 | 4.8 | 0 | 0 | 0 | 0 | 0 | 0 | 0 | 0 | 0 | 0 | 0 | 0 |
| total | | 100 | 100 | 100 | 100 | 100 | 100 | 100 | 100 | 100 | 100 | 100 | 100 |

### Percentage of occurrence of wave height (m) in rows versus mean wave period (s) in columns

| lower | upper | 3 | 4 | 5 | 6 | 7 | 8 | 9 | 10 | 11 | 12 | 13 | total |
| | | 4 | 5 | 6 | 7 | 8 | 9 | 10 | 11 | 12 | 13 | 14 | |
|---|---|---|---|---|---|---|---|---|---|---|---|---|---|
| 0 | 0.2 | 0 | 0 | 0 | 0 | 0 | 0 | 0 | 0 | 0 | 0 | 0 | 0 |
| 0.2 | 0.4 | 0 | 0 | 0.1 | 0.1 | 0 | 0 | 0 | 0 | 0 | 0 | 0 | 0.2 |
| 0.4 | 0.6 | 0 | 0 | 0.2 | 0.5 | 1 | 1.9 | 2.4 | 1.9 | 0.8 | 0.3 | 0.1 | 9.1 |
| 0.6 | 0.8 | 0 | 0.1 | 0.5 | 1.6 | 3 | 4.6 | 4.7 | 3.9 | 1.9 | 0.9 | 0.2 | 21.5 |
| 0.8 | 1 | 0 | 0 | 0.7 | 1.8 | 2.8 | 3.2 | 3.5 | 2.9 | 1.4 | 0.7 | 0.1 | 17.2 |
| 1 | 1.2 | 0 | 0.1 | 0.6 | 1.6 | 2.4 | 2.4 | 1.9 | 1.2 | 0.5 | 0.2 | 0.1 | 10.9 |
| 1.2 | 1.4 | 0 | 0 | 0.6 | 1.4 | 2.7 | 1.9 | 1.1 | 0.6 | 0.3 | 0 | 0 | 8.3 |
| 1.4 | 1.6 | 0 | 0 | 0.3 | 1.6 | 2.4 | 1.6 | 0.5 | 0.1 | 0.1 | 0 | 0 | 5.8 |
| 1.6 | 1.8 | 0 | 0 | 0.2 | 1.1 | 2.7 | 1.9 | 0.5 | 0.1 | 0 | 0 | 0 | 5.5 |
| 1.8 | 2 | 0 | 0 | 0.1 | 0.8 | 2.2 | 1.6 | 0.7 | 0.1 | 0 | 0 | 0 | 5.1 |
| 2 | 2.2 | 0 | 0 | 0 | 0.4 | 1.8 | 1.9 | 0.6 | 0.1 | 0 | 0 | 0 | 4.4 |
| 2.2 | 2.4 | 0 | 0 | 0.1 | 0.2 | 1.4 | 1.7 | 0.7 | 0.1 | 0 | 0 | 0 | 3.6 |
| 2.4 | 2.6 | 0 | 0 | 0 | 0.1 | 0.8 | 1.2 | 0.6 | 0.1 | 0 | 0 | 0 | 2.8 |
| 2.6 | 2.8 | 0 | 0 | 0 | 0 | 0.4 | 0.7 | 0.6 | 0 | 0 | 0 | 0 | 1.9 |
| 2.8 | 3 | 0 | 0 | 0 | 0 | 0.3 | 0.5 | 0.4 | 0.1 | 0 | 0 | 0 | 1.4 |
| 3 | 3.2 | 0 | 0 | 0 | 0 | 0.2 | 0.3 | 0.2 | 0.1 | 0 | 0 | 0 | 1 |
| 3.2 | 3.4 | 0 | 0 | 0 | 0 | 0.1 | 0.2 | 0.2 | 0 | 0 | 0 | 0 | 0.7 |
| 3.4 | 3.6 | 0 | 0 | 0 | 0 | 0 | 0.1 | 0.1 | 0 | 0 | 0 | 0 | 0.3 |
| 3.6 | 3.8 | 0 | 0 | 0 | 0 | 0 | 0 | 0.1 | 0 | 0 | 0 | 0 | 0.1 |
| 3.8 | 4 | 0 | 0 | 0 | 0 | 0 | 0 | 0 | 0 | 0 | 0 | 0 | 0.1 |
| 4 | 4.2 | 0 | 0 | 0 | 0 | 0 | 0 | 0 | 0 | 0 | 0 | 0 | 0 |
| 4.2 | 4.4 | 0 | 0 | 0 | 0 | 0 | 0 | 0 | 0 | 0 | 0 | 0 | 0 |
| 4.4 | 4.6 | 0 | 0 | 0 | 0 | 0 | 0 | 0 | 0 | 0 | 0 | 0 | 0 |
| 4.6 | 4.8 | 0 | 0 | 0 | 0 | 0 | 0 | 0 | 0 | 0 | 0 | 0 | 0 |
| total | | 0 | 0.2 | 2.7 | 10.2 | 22.5 | 26.6 | 18.7 | 11.1 | 5 | 2.3 | 0.6 | 100 |

## Typical required data for design studies

### Wave height (m) versus return period of 3 hour exceedance

| return period | wave height | lower limit | upper limit |
|---|---|---|---|
| month | 3.9 | 3.8 | 4.0 |
| 1 yr | 4.8 | 4.6 | 5.0 |
| 2 | 5.0 | 4.8 | 5.3 |
| 5 | 5.3 | 5.0 | 5.6 |
| 10 | 5.5 | 5.2 | 5.9 |
| 25 | 5.8 | 5.4 | 6.2 |
| 50 | 6.0 | 5.6 | 6.4 |
| 100 | 6.2 | 5.8 | 6.7 |
| 1000 | 6.9 | 6.3 | 7.4 |

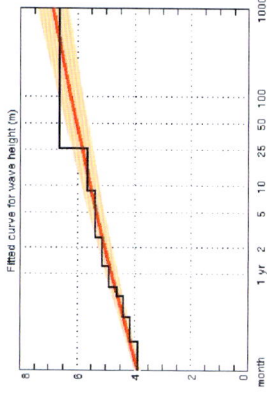

Fitted curve for wave height (m)

return period of 3 hour exceedance: month, 1 yr, 2, 5, 10, 25, 50, 100, 1000

Figure 3.4  Typical wave data required for analysis of dredging works (operational limitations) and for design studies.

design parameters such as extreme waves and water levels, which is primarily done by means of hydrodynamic modelling. Wind data may also be useful to evaluate the required dust control measures.

**Air pressure**

Time- and spatial varying air or barometric pressure data are used in combination with wind fields to derive extreme water levels by means of numerical storm surge modelling.

Sudden changes in barometric pressure readings observed during reclamation works are indicators for weather changes and may therefore be a relevant monitoring parameter during the works.

**Precipitation**

In the absence of river flow or discharge data, or to verify river discharge measurements, precipitation data can be used to estimate average or extreme river discharges by relating it with the river's catchment area.

Precipitation data (i.e., volume of precipitation per unit of time) are also required to enable the design of the drainage system of the reclamation area. For the design both peak precipitation intensity and average annual or seasonal values must be known.

**Air temperature**

Dredging and reclamation works in cold regions require specific design considerations. Information on air temperature will give insight into the duration and intensity of frost periods. General information can be derived from the literature and other general sources.

High air temperature may cause overheating of the engines of the dredging and reclamation equipment, which may affect the efficiency of this equipment.

**Water temperature and salinity**

Information on water temperature and/or salinity is relevant in areas with a large seasonal variation in water temperature or salinity and in areas influenced by thermal plumes or river outflows. Temperature and salinity gradients in the water will induce density currents and stratification and may lead to enhanced settlement of fines. For reclamation projects, it therefore is relevant as an input parameter for (three-dimensional) modelling to determine the potential impacts of the dredging works.

**Visibility**

The visibility is generally expressed in terms of distance. Low visibility can affect all forms of traffic including shipping traffic.

**Storm tracks**

When planning a dredging work, it is important to know if special measures or precautions need to be taken against heave storms or hurricanes.

In certain circumstances, information on ice formation and the groundwater regime (for inland borrow areas) may also be relevant.

3.4.3.3   Morphological and environmental data

The most important morphological and environmental data relate to:

**Sediment transport**

Dredging and reclamation works can trigger erosion or sedimentation of certain specific areas. To study these phenomena, a combination of bathymetrical, hydraulic, meteorological and geotechnical information is required. Since the impact zone can potentially be much larger than the actual dredge or reclamation area, the extent of the area for which data has to be collected can also be much larger. These data may be used to make a physical or numerical model of the vicinity of the fill area and or the borrow area in order to determine the impact of the future project on the present morphological regime.

These data, including the physical or numerical model, may also be used as input for an environmental impact assessment.

**Turbidity**

The turbidity of the water is defined as the degree to which water contains particles that cause cloudiness or backscattering and the extinction of light (Bray, 2008). High turbidity levels (in excess of the natural water turbidity) generally occur in the vicinity of dredging and reclamation projects caused by the release of organic and inorganic solids (i.e., fine sediment fractions) into the water column as a result of the dredging or placement activities.

One of the main areas of environmental concern is the effect of elevated levels of suspended sediment on the natural environment. Suspended sediments introduced in the marine environment limit the transparency of the water and prevent light from shining through. As such, suspended sediments will affect the primary processes in the water body. Furthermore, settling particles can potentially result in the smothering of sensitive habitats like seagrass and coral. Quite often, dredging and reclamation works have to be executed within stringent Total Suspended Solids (TSS) criteria.

As a result of the presence of suspended particles, water will lose its transparency. Increased turbidity levels will alter local primary photosynthesis processes. Consequently, the oxygen and temperature levels of a water body may be affected.

Because the amount of suspended particles corresponds to the intensity of scattered light, a turbidity measurement can quickly provide an estimate of the amount of suspended solids in a water body.

Background turbidity data relate to the measurement of the concentration levels of suspended material in the water column prior to dredging.

**Dissolved oxygen**
Organic matter released or stirred up during dredging and reclamation operations will reduce the levels of dissolved oxygen in a water body as a result of aerobic oxidation. Because of its importance to all life forms, a shortage of oxygen may have serious implications for the natural environment.

**Contaminants**
In case contaminated soils are to be dredged, detailed analysis of contaminants such as heavy metals, PCB's is required (Bray, 2008).

**Additional parameters**
The complexity with respect to water quality is reflected in the many types of physical, chemical and biological indicators. Therefore, depending on specific project requirements, information about a wide variation of additional parameters may be required in order to describe the state of and the effect on the water quality. Simple measurements such as temperature, pH, dissolved oxygen, conductivity, conductivity, alkalinity (pH) and ammonia can be made on-site in direct contact with the water source. More complex measurements such as toxic contaminants and the presence of micro organisms may require water sampling followed by an analysis in the laboratory.

### 3.4.4  Seabed obstructions

Should a desk study indicate the potential presence of wrecks, ordnance, sub-marine pipelines or other obstacles in the borrow area or the fill area, more detailed investigations are required.

Wrecks and other obstacles may hamper dredging operations and can cause significant damage to the dredging equipment, both during dredging and the filling process. In addition, if not removed, existing obstacles may obstruct the construction of future subsurface structures like piled foundations, tunnels etc.

Apart from damage to the dredging equipment, a wreck itself can sometimes be of archaeological importance and may not be disturbed. Therefore, all wrecks and obstacles should be localized with a wreck survey and if necessary removed from the seabed before dredging operations begin. Large wrecks or wrecks with a high archaeological importance can be a reason for relocation of the borrow area.

Obstructions can be detected either by side scan sonar or with a magnetometer. The side scan sonar cannot identify buried objects. A magnetometer can only detect metal objects which cause a magnetic anomaly. Therefore, one can never be sure whether the subsurface is free of obstacles. Shallow geophysical methods can be helpful for this.

The following types of hazards and obstacles may seriously hinder dredging operations:

- unexploded ordinance (UXO);
- pipelines;
- cables;
- large boulders;
- debris;
- wrecks.

Logical steps in a wreck survey are:

1. Desk study – Research into available archives (Wreck charts, Wreck database)

   Often, the local (harbour) authorities or hydrographical service has already done some survey work. This is always a good starting point for a wreck survey. Admiralty charts, special wreck charts or a wreck database can give valuable information about possible wrecks, especially when they are of archaeological interest. However, since these charts are not always up-to-date or accurate enough for dredging purposes an additional pre-dredging survey is required should wrecks be expected.

2. Additional pre-dredging survey – In order to accurately locate wrecks and obstacles in a pre-dredging survey, a number of techniques are being used.

   Localization is no easy task, especially when bearing in mind that many of the wrecks and other objects that have to be removed before dredging takes place are not visible to the naked eye or to standard hydrographical survey techniques (Dredging News Online, 2000). An overview of some common wreck survey techniques together with their advantages and drawbacks is given in appendix B/section 4 detection of seabed obstructions.

## 3.5  Typical sand search site investigation

Once the required information has been listed, an investigation campaign can be set up. A description of the most common methods of site investigation can be found in Appendix B. As an example, a typical sequence of soil investigations often adopted during a sand search is displayed. The term "sand search" implies the search for a dedicated sand borrow area. Typically following steps are taken during such an investigation:

Step 1. Desk study to map potential borrow areas (Figure 3.5).

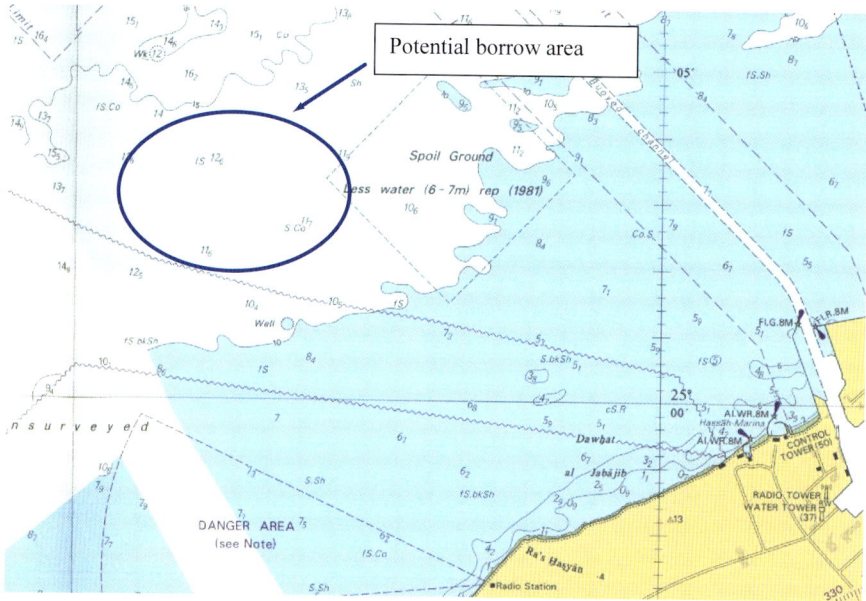

Figure 3.5   *Definition of a potential borrow area based on information on an admiralty chart (location: Dubai).*

Step 2. Reflection survey to map the thickness of the sediment (Figure 3.6).

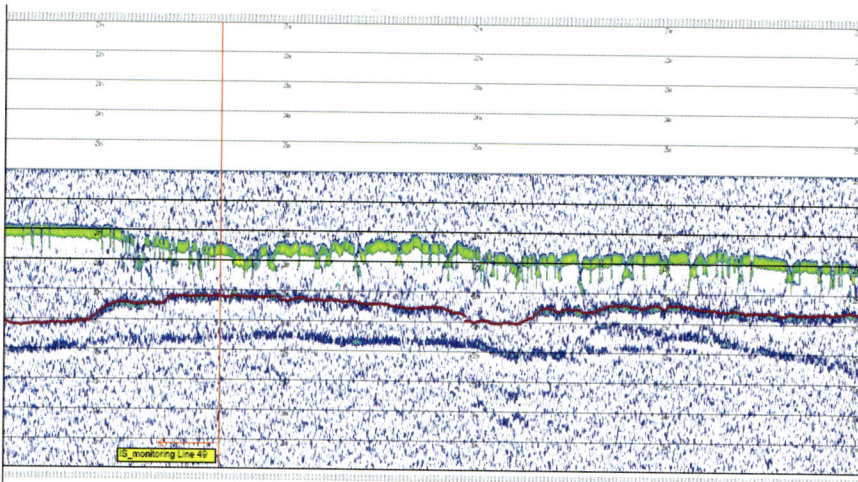

Figure 3.6   *Example of reflection survey in Dubai, showing 2 distinct seismic reflectors.*

Step 3. Vibrocore campaign to sample the sediment and verify the seismic survey (Figure 3.7).

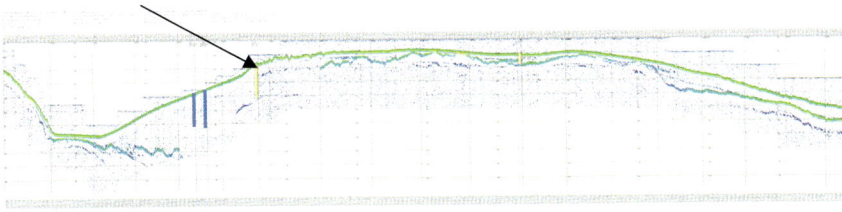

Figure 3.7    *Verification of seismic reflection by means of vibrocores.*

Step 4. Lab testing and situ testing on samples (Figure 3.8).

Figure 3.8    *Torvane test on vibrocore sample.*

Generally, a vibrocore will result in disturbed samples.

Step 5. Calculation of available suitable volume and setup of isopach maps indicating the sediment thickness in the areas suitable for sand extraction (Figure 3.9).

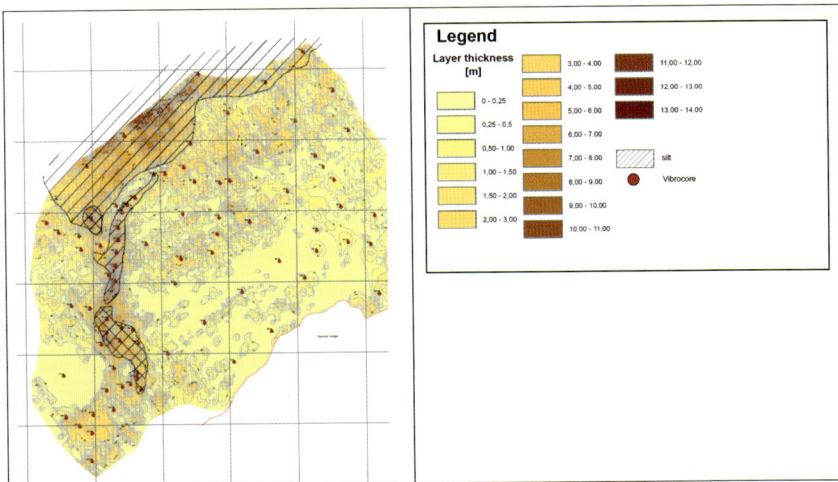

Figure 3.9    *Isopach chart mapping sediment thickness and suitable borrow areas, darker colours indicating a larger layer thickness.*

## 3.6  Reporting

Investigation reports fall into two categories:

–  Factual reports, which present the results of boreholes, geophysical surveys, field and laboratory tests, etc. without any attempt at interpretation or analysis.
–  Full interpretative reports which present the factual results of the investigation together with a detailed assessment of the ground conditions, including a geotechnical model supported by sections, plans and an analysis of the geotechnical properties of the materials which have been encountered and a quantitative and qualitative analysis of the potential resources.

The adopted standards and classification systems should be clearly referred to. (Ref section 3.6.1). In either case, the manner in which the report is presented is of vital importance to those who have to use the data. It is essential that the reports are written and presented in a clear and unambiguous manner and include full details concerning methodology and standards which have been used. In particular, it is stressed that the report should include all of the factual data, which may be required by the user to make his own basic interpretation from first principles.

It should also be stressed that this text often is the only way of communication between the investigation party and other parties involved in the project. In the end the final report is the only product of a site investigation and should be prepared with the utmost care as it represents the total costs of the field and laboratory investigations undertaken.

A proper recording and reporting of the investigations can also be important since test results might need to be compared with post dredging/reclamation tests.

### 3.6.1  *Soil and rock classification and description*

An important product of a soil investigation is the accurate and detailed description and classification of the soils and rocks, which have been encountered. It is a vital element in the proper understanding of the materials, which are to be dredged, and of the general geological and geotechnical characteristics of the dredge and reclamation area. The classification of the soil will give a first indication of the behaviour of the soil or rock. It is strongly advised to use internationally accepted standards.

The 3 major standards internationally used are:

–  European standards (Eurocode 7, EN ISO 14688 and EN ISO 14689);
–  British standards (BS 5930 and BS 6031);
–  ASTM (American Standard for Testing and Materials) (D 2487 and D5878).

An overview of the most commonly used tables in these standards can be found in Figure 3.10.

**Identification and classification of soils**

| application | usefull reference tables | description |
|---|---|---|
| **European Standard** | | |
| ISO 14688 Part 1<br>identification and description of soils | Figure 1 Page 3 | field description of soils (flow chart) |
| ISO 14688 Part 2<br>classification of soils | Table A1(annex)<br>Table B1 | principles for classification<br>informative example of classification based on grading alone |
| Earthworks (TC396)<br>standard classification for earthworks | in preparation | |
| **BS Code of Practice** | | |
| BS 5930 :1999 + A2:2010<br>Code of practice for site investigation | Table 12, page 113<br>Table 13, page 114<br>figure 18, page 122 | general identification and description of soils (flowchart)<br>Identification and description of soils<br>Plasticity chart |
| BS 6031: 2009<br>Code of practice for earthworks | Table 1a, page 7<br>Table 1c, page 8 | soil classification<br>comparison of soil definitions in different earthwork circumstances |
| **ASTM standard practice** | | |
| ASTM 2487 (UCSC)<br>standard practice for classification of soils for engineering purposes | Table 1 page 2<br>fig. 1 page 4<br>fig. 2 page 5<br>fig. 3, page 6<br>fig. 4 page 7 | soil classification chart<br>flow chart for classifying fine grained soil<br>flow chart for classifying organic fine grained soil<br>flow chart for classifying coarse grained soils<br>plasticity chart |

**Identification and classification of rock**

| application | usefull reference tables | description |
|---|---|---|
| **European Standard** | | |
| ISO 14689-1:2003<br>identification and classification of rock | table A.1 page 15 | Aid to rock identification for engineering purposes |
| **BS Code of Practice** | | |
| BS 5930-1999<br>Code of practice for site investigation | Table 14 page 128 | aid to identification of rock for engineering purposes |

Figure 3.10 *Overview of classification standards.*

It is noted that in Europe, gradually national standards (such as the BS) are being replaced or revised to get them in line with the European standards. The Eurocodes allow for national annexes with specific guidelines. BSI is issuing documents with non-contradictory complementary information in addition to the Eurocodes and its national annexes.

Some remarkable differences between the above-mentioned classification systems can be found. For example:

The definition of density index is not the same in ASTM and ISO, BS (see section 8.3.2).

The definition of fine grained soils (silt and clay) is different:

− BS: % of fraction passing 60 μm sieve >35%.
− ISO: % of fraction passing 63 μm sieve >40% (only mentioned as example of classification).
− ASTM: % of fraction passing 75 μm sieve >50%.

Obviously, this also results in a different definition of the term "SAND".

The descriptive relative density terms such as very loose, loose are directly correlated to SPT values in BS, while in ISO reference is made to the relative density $R_e$, see Table 3.3.

The undrained shear strength is described as soft, firm, stiff in BS; while in ISO the shear strength is described as low, medium, high. The corresponding undrained shear strength $c_u$ is identical, see Table 3.4.

ISO is describing the consistency of clay or silt as soft, firm, stiff according to the consistency index, see Table 3.5.

*Table 3.3 Sand and gravel: BS and ISO classification according to relative density.*

| | BS 5930 | | ISO 14688 | |
| | SPT | | $R_e$ (*) | |
| Description | From | To | From | To |
| --- | --- | --- | --- | --- |
| Very loose | 0 | 4 | 0 | 15 |
| Loose | 4 | 10 | 15 | 35 |
| Medium dense | 10 | 30 | 35 | 65 |
| Dense | 30 | 50 | 65 | 85 |
| Very dense | 50 | ... | 85 | 100 |

(*) $R_e$ is defined as $I_D$ in ISO 14688

Table 3.4   *Fine grained soil: BS and ISO classification according to undrained shear strength.*

| Description BS 5930 | Description ISO 14688 | $c_u$ [kPa] From | To | Observation |
|---|---|---|---|---|
| Very soft | Extremely low | 0 | 10 | |
| Very soft | Very low | 10 | 20 | Extrudes between fingers when squeezed |
| Soft | Low | 20 | 40 | Moulded by light finger pressure |
| Firm | Medium | 40 | 75 | Moulded by strong finger pressure |
| Stiff | High | 75 | 150 | Can be indented by thumb |
| Very stiff | Very high | 150 | 300 | Can be indented by thumb nail |
| Hard | Extremely high | 300 | … | |

Table 3.5   *Fine grained soil: ISO classification according to consistency.*

| Description | ISO 14688 $I_c(*)$ From | To |
|---|---|---|
| Very soft | – | 0.25 |
| Soft | 0.25 | 0.5 |
| Firm | 0.5 | 0.75 |
| Stiff | 0.75 | 1 |
| Very stiff | 1 | |
| $I_c$ = consistency index | | |

### 3.6.2   *Soil classification based on CPT measurements*

When CPT measurements are available, there is a wide variety of classification systems based on the measured cone resistance, friction ratio and pore pressure. Reference is made to specialized literature.

Some of the most frequently used classification systems are:

– Schmertmann (1978);
– Douglas & Olsen (1981);
– Robertson (1986, 1990).

Some of the most commonly used graphs for classification of the soil based on CPT data are plotted in Figure 3.11.

These classifications systems are based on:

– measured cone resistance $q_c$
– corrected cone resistance $q_t$

- friction ratio $FR$ (%)
- pore pressure parameter ratio $B_q = \dfrac{u_2 - u_0}{q_t - \sigma_{v0}}$

where $u_2$ = pore pressure measured behind the cone
$u_0$ = in situ pore pressure

It should be stressed that the classification systems displayed in Figure 3.11 are only an example. Many other classification systems exist, often linked to specific countries or soil conditions.

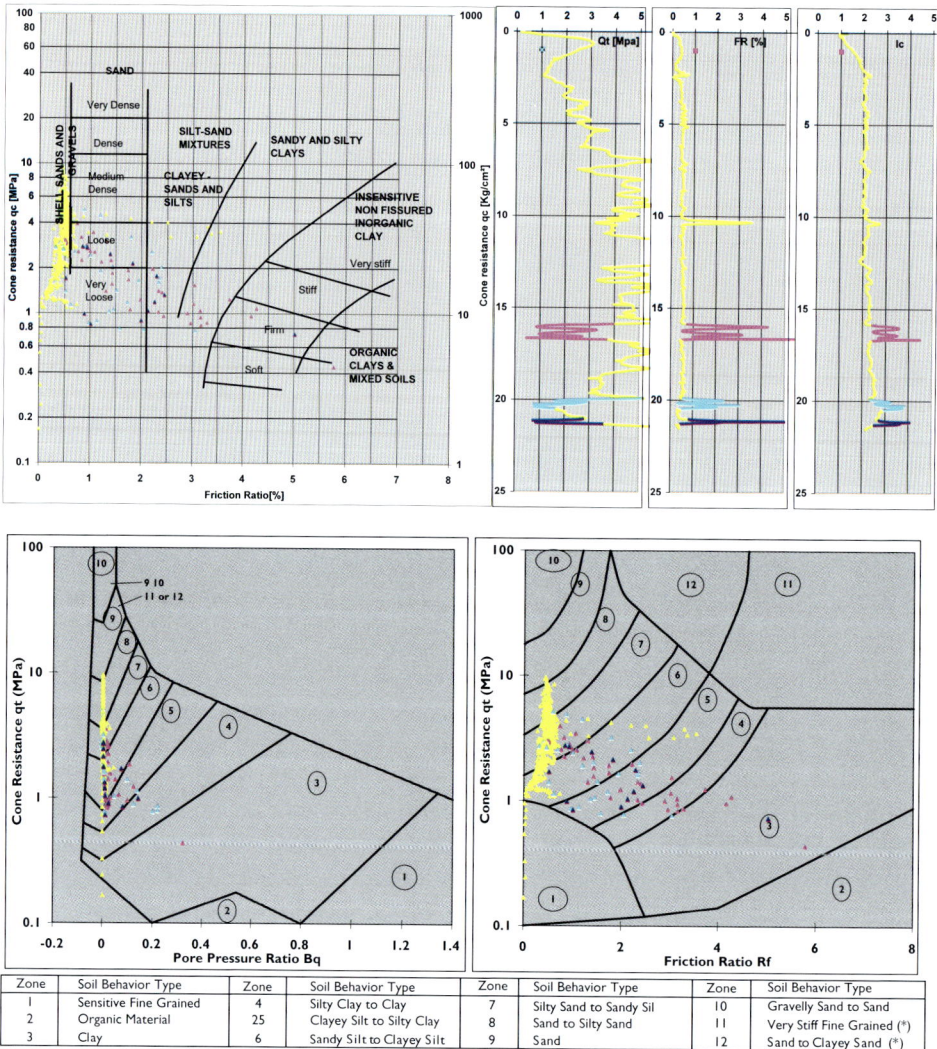

| Zone | Soil Behavior Type | Zone | Soil Behavior Type | Zone | Soil Behavior Type | Zone | Soil Behavior Type |
|------|-------------------|------|-------------------|------|-------------------|------|-------------------|
| 1 | Sensitive Fine Grained | 4 | Silty Clay to Clay | 7 | Silty Sand to Sandy Sil | 10 | Gravelly Sand to Sand |
| 2 | Organic Material | 25 | Clayey Silt to Silty Clay | 8 | Sand to Silty Sand | 11 | Very Stiff Fine Grained (*) |
| 3 | Clay | 6 | Sandy Silt to Clayey Silt | 9 | Sand | 12 | Sand to Clayey Sand (*) |

Figure 3.11   *Typical classification graphs based on CPT data (top: Schmertzmann 1978, bottom: Robertson 1986).*

**Thin layer effect**

Robertson, ref: www.cpt-robertson.com:

Quote

During penetration the cone tip senses ahead and behind due to the essentially spherical zone of influence. Hence, when pushing a cone through a sand layer, the cone tip will sense a softer clay layer before the cone actually reaches the clay and likewise when pushing in a clay layer the cone will sense a stiffer sand layer before reaching the sand layer. The distance over which the cone senses an interface between different soil types is a function of soil stiffness, with stiffer soils having a larger zone of influence. What appears on the CPT profile is a rapid variation of measured tip stress ($q_t$) through these transition zones (i.e. either sand to clay or visa-versa). The CPT data collected through these transition zones is not truly representative of the soil, since the data are 'in transition' from either a stiff to soft layer or visa-versa. When a thin sand layer is located within a softer clay deposit the cone data are influenced by the two transition zones (i.e. the top and bottom interface boundaries), resulting in the 'thin layer' effect.

Unquote

## 3.7 Use of data during different project stages

**Pre-construction stage**

Prior to the design and construction of the reclamation works, all data sources available in the project area need to be explored and collected. A review of the collected or available data will provide insight into the specific physical site conditions and will serve as input to planning and design studies of the dredging and reclamation works.

In the pre-construction stage data from similar and/or nearby projects can be used as desk study. But not only data, also the experience of these projects (what went well and what went wrong) can be used to identify project risks in an early stage.

Before the data can be applied in design formulae and/or numerical or physical modelling tools, the data will require proper quality control and dedicated analyses (e.g., to determine trends and extreme statistics). The data analysis may also show that the available data sources are unreliable, incomplete or insufficient. If so, it may be decided to collect additional data by means of field surveys and/ or data purchase in combination with numerical modelling. Numerical modelling can also be applied to determine the operational site conditions for use in planning the reclamation works.

**Construction stage**

During the dredging and reclamation works, in-situ measurements of for instance currents, wind and waves can be used to monitor the actual physical site conditions, including sedimentation and erosion. In case, for example, a storm builds up and the wind and wave conditions exceed certain limits, the dredging and/or reclamation activities may be stopped for safety and efficiency reasons. Nowadays, it usually is also a requirement to monitor the turbidity of the water in the area surrounding the on-going dredging and reclamation works. In this way, dredging activities can be planned and managed as not to exceed certain pre-defined levels at a distance from the works. The levels and monitoring locations can be derived from pre-construction impact assessments, i.e., by means of numerical modelling. See further Chapter 11 for environmental monitoring methods and procedures.

**Post-construction**

Short- and long-term impacts of the dredging or reclamation works on the hydraulics after construction are assessed by monitoring of, for instance, currents and waves in the vicinity of the project. In this way, the impacts predicted by (numerical) modelling studies in the pre-construction stage can be confirmed. The post-construction monitoring of seabed changes, silt deposition and/or water quality will be of most interest. A further description of these aspects is given in section 3.4.

**Data sets and requirements**

The required detail, duration and type of data that is relevant to dredging and reclamation projects depends on the location of the project and the stage the project is in. For instance, a coastal reclamation project in its design stage requires different data sets than long-term monitoring of the effects of a recent reclamation project in a river.

Requirements for the duration of data records differ per data type and application or analysis method. Relatively short data records are, for instance, sufficient to validate numerical models; long data records are required for design purposes, e.g., to determine the operational and extreme conditions.

An important aspect of the design of reclamations is the design of the surrounding structures and the minimum required height of the reclamation. The design levels are in most cases determined by the extreme conditions, e.g., during extreme storms or peak river discharges. To derive the extreme conditions analyses, multi-year data records are required of wind, water levels and waves. Analysis of the multi-year data will also provide the operational conditions including seasonal variations.

In general, the longer the data records, the more reliable the results of the analyses. In most cases multi-year data records at the reclamation site will not exist and data from nearby locations (or for instance from offshore databases) will need to be made applicable to the site by numerical modelling techniques.

## 3.8 Geostatistical methods

### 3.8.1 *General*

As mentioned before in section 3.4.2 geostatistical methods can be used to develop a geological model of the borrow area and reclamation area. In case of a borrow area, the geological model enables to estimate the quantity of the available fill material probabilistically including standard deviations of the best estimate. Geostatistical modelling of the geology of the reclamation site can be helpful to assess spatial variation in layer thicknesses and indirectly the best estimate and standard deviation of expected settlements and fill volumes needed. Furthermore geostatistics can assist in planning site investigation. Additional boreholes and sampling can best be planned at locations with highest spatial uncertainty.

In this section the use of geostatistics for design of hydraulic fills is outlined. Geostatistics is the application of different statistical methods to determine the spatial distribution of geological variables (Burnaman *et al.*, 1994). Variograms, kriging and conditional stochastic simulation are some statistical methods that can be applied to model spatial uncertainty. Together these methods are called 'Geostatistical methods'.

Geostatistical methods can be used for different reasons. Often the main objective is:

– to visualise the measured data (e.g. contouring);
– to quantify the spatial uncertainty of certain geological variables (e.g. layer thickness or lithology);
– to combine different types of data (e.g. borehole data and geophysical data).

Geostatistical methods are comprehensively described in literature by Chilès (1999), Sandilands & Swan (1995), Davis (1986), Srivastava (1989) and Goovaerts (1997).

### 3.8.2 *Methods*

Geostatistical methods are often used for modelling of the spatial uncertainty of a geological variable. Geological variables can be categorical (e.g. lithology) or continuous (e.g. height, thickness, grain size). For both type of variables histograms can be used to present the data. It shows how often the measured data fits within intervals or classes. From the histogram, statistical parameters can be derived, like mean value, variance, standard deviation and skewness.

In common, geological variables are not stationary, which means that the mean value and variance vary in space (Gorelick, 1996). By determining the trend within the data it is possible to take into account this spatial variation of the mean and variance. A trend analysis divides conditional (measured) data in two components: the regional trend and the deviation from the trend. An example of a method for trend analysis is polynominal regression. The theory of trend analysis is described by Davis (1986).

For almost all geostatistical methods, variograms are needed as input to predict values at unmeasured points. A variogram is a mathematical method to quantify the correlation or continuity of a geological variable. It describes how variation in values changes with distance. This is also called 'spatial variability'. Different variogram models are defined, like semi-variogram, cross-semi-variogram and correlogram. The theory is described by Deutsch & Journel (1992). Hosgit & Royle (1974) describe typical variogram models for sand deposits. A typical semi-variogram is shown in Figure 3.12. The semi-variogram, is a graph describing the expected correlation of a variable between pairs of samples with a given relative orientation.

The value at which a variogram levels of is called the sill value. The distance at which the variogram approaches the sill value is called the range.

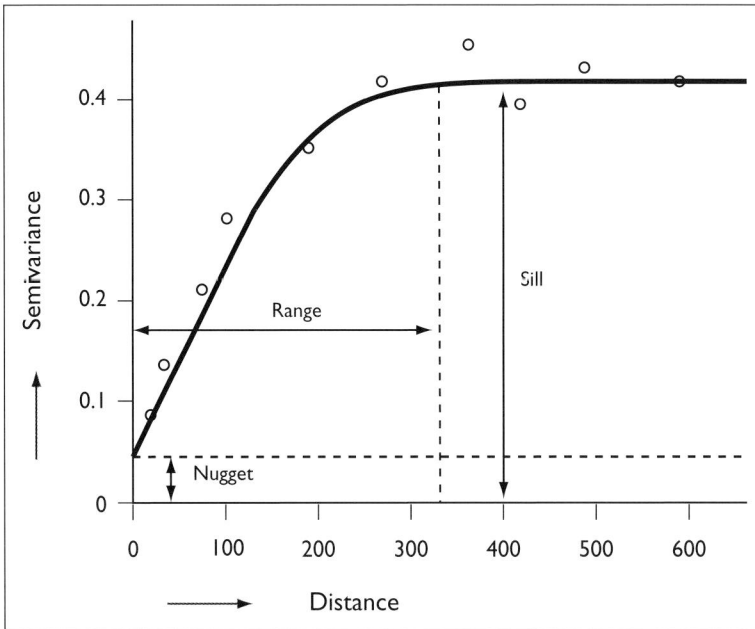

Figure 3.12   *Example semi-variogram model.*

The nugget-value reflects the differences found between neighbouring samples and/or sampling errors.

After deriving the histogram and vario-gram models, this data can be used for geological modelling. Basically the modelling methods comprise two elements: interpolation and stochastic simulation.

*Interpolation* means assessing a value for a certain variable at a certain point, based on the known values of neighbouring points. Interpolation can be done by methods like inverse distance and kriging. For probabilistic design kriging is the preferred method, because in addition to a best estimate also the minimum variance is determined. This variance expresses the uncertainty. For example, in planning site investigation additional boreholes can best be placed at locations with high variance.

Interpolation results in one representation or "map" of a variable. *Simulation* can be used to develop more than one representation. Each representation is consistent with its values at sampled locations and with its in situ spatial variability, as characterized by histogram and variogram models. The simulation method to be used mainly depends on the type of variable (categorical, continuous or objects) and the amount of data. Simulation methods are described by Srivastava (1994) and Gorelick & Kolterman (1996). The mathematical background is described by Goovaerts (1997).

Finally Monte Carlo Analysis is often used to quantify the uncertainty of the simulated geological model. The point values from different representations (maps) are used to create a probability density function. With this function the volumetric uncertainty of the geological model can be quantified.

### 3.8.3   *Geostatistical software*

Different geostatistical software packages are available. Often geostatistics tools are accessible within GIS or mapping programs (e.g. ArcView, Surfer, Datamine). These tools are user friendly; however input is needed from a specialist to prevent unrealistic outcome.

More advanced software package like ISATIS (Geovariances) or GS+ (Rockware) are developed. Often these types of programs can be linked to other resource modelling software packages, such as GOCAD (Paradigm), Gems (Gemcom), Petrel (Schlumberger) and RML (Beicip Franlab), allowing a combination of geostatistics and geology modelling. An example of 3D modelling is shown in Figure 3.13.

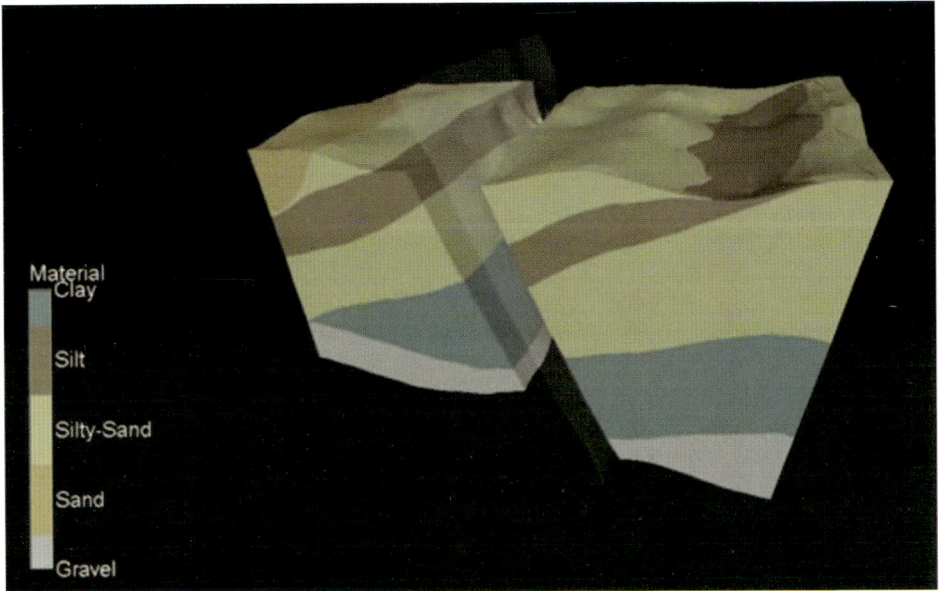

Figure 3.13    *Example of 3D geological modelling (Source: C Tech Development Corporation).*

# DREDGING EQUIPMENT

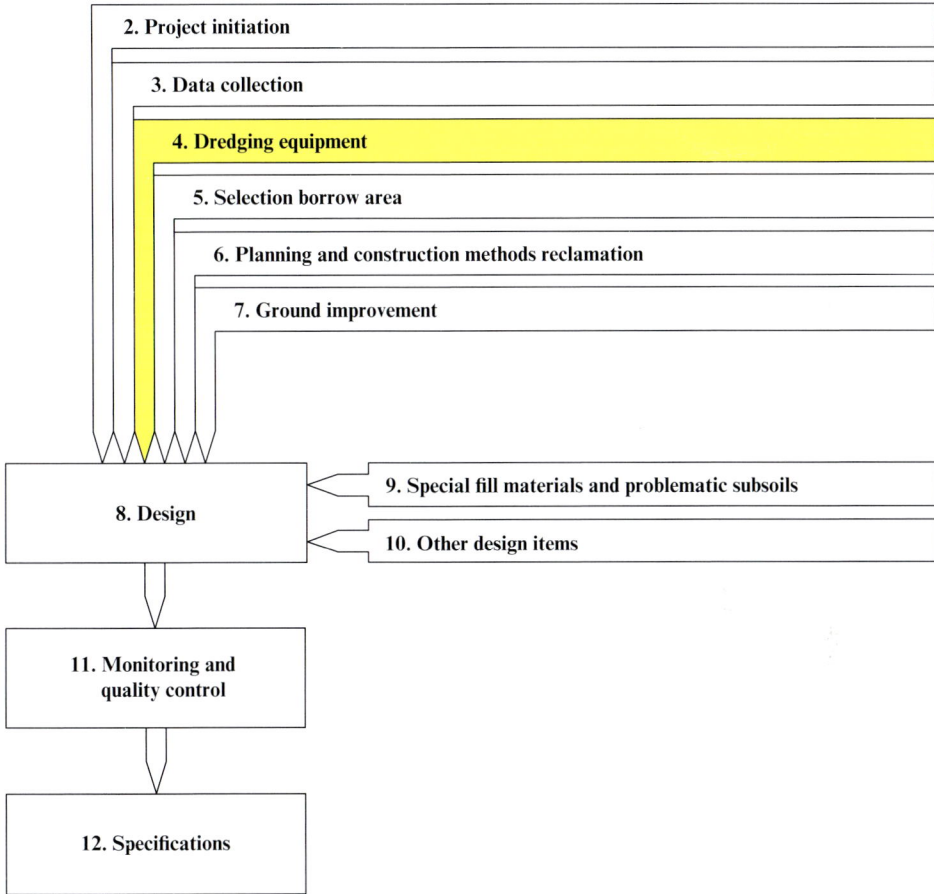

2. Project initiation

3. Data collection

4. Dredging equipment

5. Selection borrow area

6. Planning and construction methods reclamation

7. Ground improvement

8. Design

9. Special fill materials and problematic subsoils

10. Other design items

11. Monitoring and quality control

12. Specifications

## Chapter 4 Dredging Equipment

### 4.1 Introduction

### 4.2 Dredging equipment

| 4.2.1 Suction dredging | 4.2.2 Mechanical dredging | 4.2.3 Other types of equipment | 4.2.4 Combinations of equipment or dredge chains |

### 4.3 Operational limitations

| 4.3.1 Waves and swell | 4.3.2 Currents | 4.3.3 Hindrance to shipping and other parties | 4.3.4 Environmentally driven limitations |

### 4.4 Dredging of fill material

| 4.4.1 Introduction | 4.4.2 Volume and dimensions of borrow area | 4.4.3 Minimum thickness of fill deposits |

| 4.4.4 Dredgeability |

### 4.5 Transport of fill

| 4.5.1 Introduction | 4.5.2 Hydraulic transport through a pipeline | 4.5.3 Transport by trailing suction hopper dredger or barge |

### 4.6 Utilization characteristics of dredging equipment

### 4.7 Basis of cost calculation for dredging

## 4.1  Introduction

This chapter presents a brief description of dredging equipment including the soils which can be dredged, operational limitations and the combination of equipment for dredging, transport and placement. Considerations are given for:

- the selection of appropriate equipment in the borrow area;
- the manner dredged material is transported to the fill area either by discharge pipeline, barge or trailing suction hopper dredger;
- a summary of the operational characteristics of each type of equipment;
- the effects of process variables on the costs of dredging, such as transport distance, pipeline transport, barge transport and type of equipment.

Dredging can basically be divided in two main categories, namely, suction dredging and mechanical (cutting) dredging. These dredging techniques can also be combined, such as the cutter suction dredger (cutter head) and the trailing suction hopper dredger (draghead provided with a cutting edge).

The various types of suction dredgers are: the plain suction dredger, the cutter suction dredger and the trailing suction hopper dredger.

Mechanical dredgers are the (hydraulic) backhoe dredger and the grab dredger.

The dredged material may be transported:

- by pumping through a discharge pipeline directly into the fill area;
- by barges;
- by the dredger itself, i.e., the trailing suction hopper dredger.

The selection of dredging equipment for winning, transport and placement of dredged material depends on many factors such as:

- the location of the borrow area which can be at open sea exposed to waves or in sheltered waters;
- current velocity and wave heights in the borrow area;
- actual water depth;
- dimensions of dredge area;
- distance from the borrow area to the reclamation site;
- soil conditions in the borrow area;
- dredging in layers or not;
- required volume and quality of fill material;
- presence of shipping;
- restricted working hours;
- required dredging production.

All the above factors have to be accounted for in the selection of the type of equipment, to determine the capacity of that equipment and to decide how the dredged material will be transported to and placed in the fill area. The process resulting in the selection of the most appropriate equipment is schematised in Figure 4.1. This process finally leads to a cost estimate of the dredging works which may vary significantly per project, since all the above-mentioned differ per project.

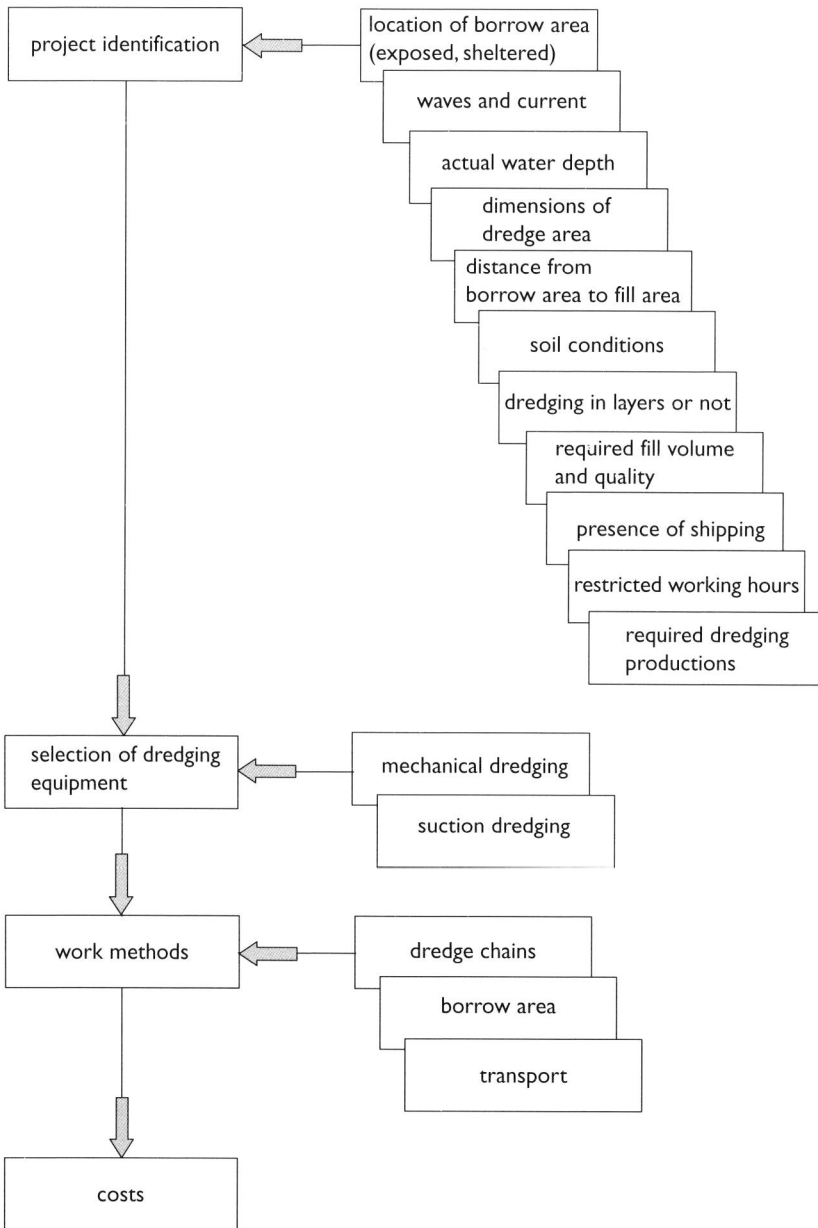

Figure 4.1    *Selection scheme for dredging equipment.*

In the following sections a brief description of types of dredging equipment and examples of combinations of equipment are given. For a more detailed description of the dredging equipment reference is made to Appendix A of this Manual.

## 4.2 Dredging equipment

In this section a brief overview of the main available dredging equipment is presented in order to provide the reader with some basic knowledge of these tools and their uses.

### 4.2.1 Suction dredging

**Plain Suction Dredger (SD)**
Suction dredging is based on the erosive action of water flow which enters the mouth of the suction pipe of the dredger as a result of the vessel's pumping action. Often jets attached to the suction mouth are used to loosen the material to be dredged. The eroded material is mixed with water and this sand-water mixture is pumped through the discharge pipeline towards the reclamation site or through a spreader discharge system into a barge alongside the dredger. The "plain suction dredger" and the "dustpan dredger" use this "plain" dredging principle.

The plain suction dredger (Figure 4.2) is generally used for sand winning and the dustpan dredger is popular in the United States for maintenance dredging. While dredging, the plain suction dredger's position is controlled with a wire anchor system.

Figure 4.2  *Plain suction dredger loading barges.*

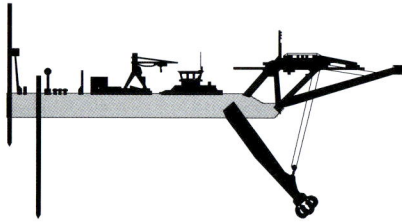

Figure 4.3    *A large cutter suction dredger with boom anchor system.*

### Cutter Suction Dredger (CSD)

The principle of the cutter suction dredger (Figure 4.3) is based on a combination of mechanical and suction dredging. The material to be dredged is cut with the dredger's cutter head and loosened with the erosive action of the water which flows towards the suction mouth by the vessel's pumping action. In addition, jets attached to the cutter head, or the lower end of the cutter ladder, may be used to loosen the material. The sand-water mixture is transported through the discharge pipeline into the reclamation area. Sometimes the material is pumped through a spreader system into a barge alongside the dredger.

CSDs are widely used in dredging and reclamation projects and can, depending on their installed power, dredge material ranging from silt and soft clay to moderately strong rock. While dredging, the cutter suction dredger swings sideways around a point fixed by the so-called working spud at the vessel's aft. The sideways movement is controlled with a wire anchor system attached to the lower end of the cutter ladder. Although the larger cutter suction dredgers are often self-propelled, most cutter suction dredgers are not.

### Trailing Suction Hopper Dredger (TSHD)

The trailing suction hopper dredge (Figure 4.4) also uses a combination of mechanical dredging (cutting) and suction power to loosen the material to be dredged.

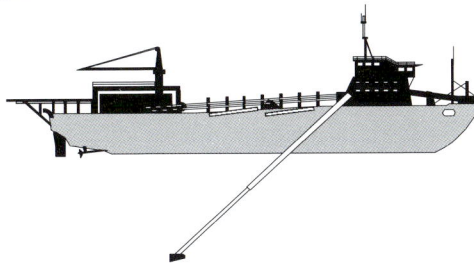

Figure 4.4    *A trailing suction hopper dredger pumping sand to the shore for a beach nourish-ment project.*

Unlike most other dredgers, this dredger is self-propelled. While operating, one or two suction pipes are lowered to the seabed for dredging the soil. The loosening of the material is a combination of cutting with the draghead attached to the lower end of the suction pipe, jetting, and the erosive action of water flowing towards the suction mouth. The material is pumped into the vessel's hopper and the process water is discharged overboard into the surrounding water via a controllable overflow system. Once the hopper is filled with dredged material, the vessel sails to a designated disposal area or reclamation site and discharges its load.

Discharging may be achieved by opening doors or valves in the bottom of the vessel's hopper, or by so-called rainbowing or pumping through a discharge pipeline. The selected discharge method depends on many factors like existing environmental constraints, distance to the reclamation site and water depth.

The trailing suction hopper dredger may be used for dredging sand, silt or clayey material. The larger dredgers are even capable of dredging (weak) rock. Dredging production will vary according to the strength of the material to be dredged.

### 4.2.2   Mechanical dredging

During mechanical dredging, material is cut and retrieved in a grab or a bucket. Dredged material may be placed in a barge alongside the dredger, in their own hopper or on land adjacent to the dredger. Typical mechanical dredgers are:

– the backhoe dredger;
– the grab dredger;
– the bucket dredger.

These type of dredgers are so called stationary dredgers and they are either fixed in the dredging position by spuds or by an anchor system.

**Backhoe dredger**
The backhoe dredger (Figure 4.5) is basically a hydraulic excavator installed on a pontoon. The pontoon uses spuds to secure its position when dredging. In the dredging position the spuds elevate the pontoon and in this way considerable excavation forces may be delivered to the seabed. Depending on the size and capacity of the backhoe pontoon, this type of dredger may dredge compacted sand, stiff clay and weak rock. Depending on the sophistication of the dredger, the backhoe can be more or less integrated with the pontoon.

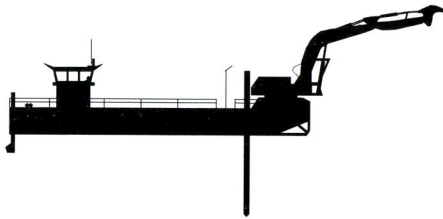

Figure 4.5   *A backhoe dredger secured on a pontoon.*

**Grab dredger**

The grab dredger (Figure 4.6) is a wire crane installed on a pontoon. The dredger may be anchored by wires or with spuds. Since the dredger uses a wire suspended grab, the maximum dredging force is determined by the weight and design of the grab only. Consequently, this dredger can operate in less strong material than the backhoe dredger.

The advantage of the wire suspended grab is that the dredging depth is not limited by the dimensions of the dredger, such as the length of suction pipes of (trailing) suction dredgers, or the length of the boom and stick of a backhoe on a backhoe dredger, since this depth is determined by the length of the wires.

A disadvantage is that with its wire suspended grab, a grab dredger cannot reach underneath structures, such as a jetty deck on piles, which to a certain extent a backhoe dredger can do.

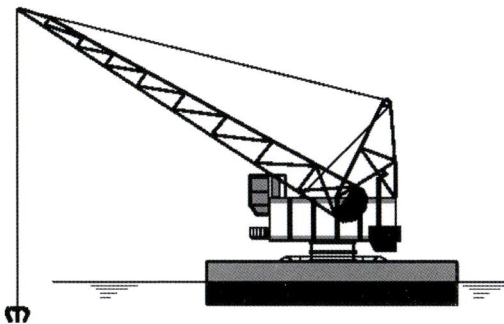

Figure 4.6    *Grab dredger with a wire suspended grab installed on a pontoon.*

**Bucket dredger**

The bucket dredger (Figure 4.7) is provided with a chain of buckets mounted on a ladder. When dredging, the ladder with the bucket chain is lowered to the seabed and material is cut by the individual buckets while the chain rotates around the top and lower end of the ladder. While dredging, the bucket dredger is anchored by means of a wire anchor system. The bucket dredger is able to dredge cohesive as well as non-cohesive material.

Figure 4.7    *Bucket dredger with its chain of buckets mounted on a ladder.*

### 4.2.3 Other types of equipment

Other dredging equipment are the Plough and the Water Injection Dredger (WID), of which the principle of removing material is based on hydronamic methods. These dredgers are, in general, not suitable for winning fill material but they can be used for special purposes. More information on these types of dredgers can be found in Appendix A.

### 4.2.4 Combinations of equipment or dredge chains

Construction of a hydraulic fill involves a number of consecutive phases:

– dredging/mining of fill material;
– transport of this material to the reclamation site;
– placement of the fill at the designated location;
– and, if necessary, improvement of the subsoil and/or the fill material.

Improvement of the existing subsoil may have to be undertaken prior to placement of the fill, but could also be carried out from the ground surface after (partial) placement of the fill above water level.

Depending on the prevailing boundary conditions these activities can be carried out by one piece of equipment or by more units. For dredging, transporting and placing of dredged material various combinations of equipment or dredge chains, can be thought of:

– For suction dredging, the dredged material may be pumped by the dredger through a discharge pipeline into the reclamation area. The material may also be pumped in a barge alongside the dredger (Figure 4.2).
– For mechanical dredging, the dredged material is placed in a barge alongside the dredger, in the hopper of the dredger itself (if available) or directly on land. Occasionally dredged material is placed in the feeder of a special pump and pumped through a discharge pipeline. When using barges for the transport of dredged material, these may be off loaded with a barge unloading dredger (see Figure 4.8).

In the following configurations (see Figure 4.9 and Figure 4.10) possible combinations of equipment are shown for dredging, transporting and placing of dredged material. In these configurations, a distinction is made between suction dredging and mechanical dredging. Fill material for a reclamation may be retrieved from a designated borrow area, but may also be obtained from dredging for new works, such as harbour basins or the deepening of existing navigation channels, and from maintenance dredging (Chapter 5).

Figure 4.8    *A barge unloading dredger which pumps dredged material from a barge alongside the dredger through a discharge pipeline to the shore.*

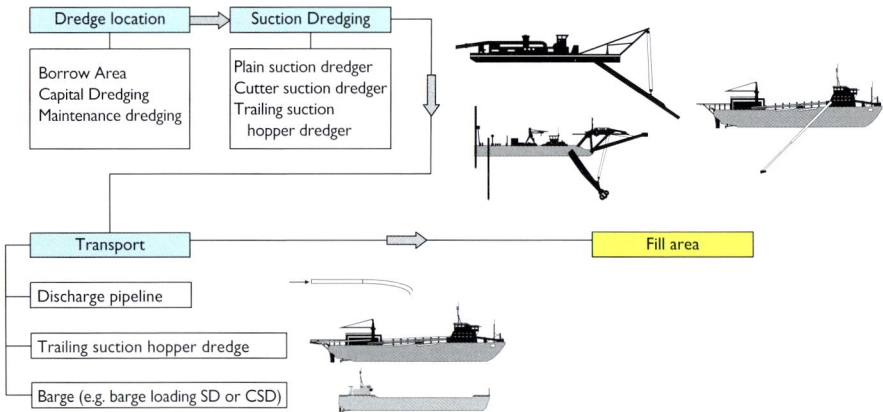

Figure 4.9    *Possible configurations of suction dredging and transport to the fill area.*

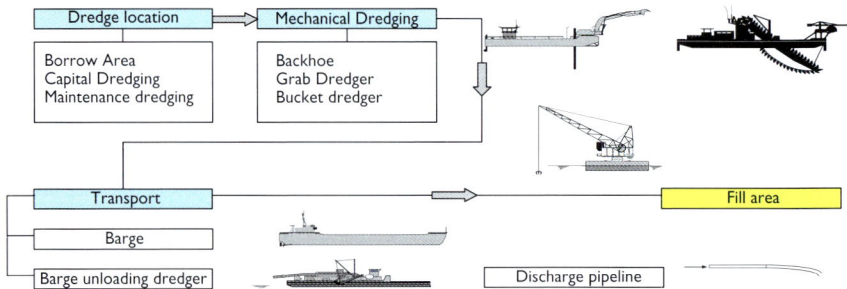

Figure 4.10    *Configurations of mechanical dredging and transport to the fill area.*

## 4.3   Operational limitations

Dredging projects are often undertaken in marine environments that can be hostile to people and equipment. Physical site conditions like waves, swell, currents, wind, temperature, rain, ice, and such may significantly affect the dredging operations. Not only can they reduce the dredging accuracy and efficiency or limit the operational hours of the plant, but they may even damage equipment and jeopardise safety of the crew. A thorough understanding of the possible effects of these conditions on the performance of the dredging equipment is therefore important.

Therefore the collection of information during the tender stage on the prevailing weather and sea conditions at the project site in the anticipated construction period is essential. These data may have a bearing on:

- the choice of the dredging equipment;
- selection of the most appropriate construction method;
- the predicted productions and, hence;
- the costs and planning of the project.

Accurate weather forecasts during the construction phase of a project are crucial to allow for a timely preparation of personnel and equipment to changing weather and sea conditions. Such preparations could not only include specific operational measures that may vary from one dredger to another, but may also contain evacuation plans for plant to more sheltered areas should this be required.

For an overview of methods to acquire data relevant to the physical environment of the project site reference is made to section 3.4. This section discusses the relationship between the physical site conditions and the dredging operations.

Apart from waves and currents dredging operations can be hindered by shipping. This applies to all plant but particularly to plant that is not self-propelled. Locating of borrow areas within shipping lanes should be avoided. However, this may not always be possible especially when the fill material has to be retrieved from the maintenance or deepening of a navigation channel.

Other operational limitations may be environmentally driven such as creating turbidity while dredging, making noise and working day and night.

### 4.3.1   *Waves and swell*

Waves and swell will not only induce excessive vertical and horizontal movements of the dredgers, barges, floating pipelines and other ancillary equipment, but may also affect the working conditions of the crew and the construction methods.

The response of dredging equipment to swell and waves depends on characteristics of the equipment such as:

– plant size;
– plant geometry;
– the plant's dynamic characteristics.

Although each individual vessel has its own specific response, in general, increasing the size and weight of plant will reduce its sensitivity to waves and swell.

Other important factors affecting the susceptibility of equipment to sea-state conditions are the connections between the floating parts of the dredger (the pontoon or hull) and the parts that are in direct contact with the seabed (mooring system, cutting device). A rigid mooring system like a spud will be more susceptible to waves and swell than a flexible anchoring system. The same applies to the connection between the cutting device and the dredger: a dredger with a rigid and heavy ladder accommodating a cutterhead or a bucket chain is much more sensitive to sea conditions than a trailer with a flexible suction pipe (often attached to a swell compensator).

As a result of their increased flexibility floating reinforced rubber pipelines are generally more able to resist swell, waves and movements of the dredger than steel pipelines. As it is placed on the seabed a sinker line is not or only minimally susceptible to sea conditions, but the installation of such a pipeline in adverse conditions could be very difficult.

Most relevant operational conditions that influence the response of dredging equipment to swell and waves include:

– the movement and orientation of the dredging equipment with respect to direction of the waves/swell and the water depths encountered on site;
– specific characteristics like wave length, height and return period.

If not made available in the information provided with the tender documents most dredging Contractors consult special databases to collect information on the characteristics of waves and swell near or at the future project site. In addition, during execution of their projects they gather information on the specific responses of their equipment when exposed to sea conditions. When tendering, they estimate the efficiency and operational days of their equipment based on these data.

In section 4.6 a range of limiting wave conditions is presented for various types of dredgers. The least susceptible to swell and waves are large trailing suction hopper dredgers, while stationary dredgers may prove to be very sensitive to sea state conditions.

### 4.3.2 *Currents*

Currents may have a strong impact on dredging and filling operations as they affect:

– the manoeuvrability and anchoring of dredgers and barges;
– the sailing speed of trailers and barges;
– the dredging accuracy of trailers and, in particular, grab dredgers;
– the handling of floating pipelines, the placement of sinker lines, and such;
– the environmental impacts;
– the accuracy and efficiency of fill placement under water.

Depending on their magnitude and direction, currents may hamper the dredging operations of cutter suction dredgers. As result of the lateral pressure on the dredger and floating pipeline two knots cross-current is generally considered to be the maximum allowable for large dredgers (Bray *et al.,* 1997). In general, cross-currents can induce large tensile forces in floating pipelines that may even cause them to break.

Manoeuvrability of trailing suction hopper dredgers and barges may be limited by the presence of currents. In particular dredging efficiency and accuracy may decrease considerably when operating perpendicular to the current. Sailing speed could be affected by currents and, in particular, tidal areas may require careful planning during the tender stage in order to put on a dredger or barge that has the most favourable cycle time compared to the tidal fluctuation in the project area.

According to Bray *et al.* (1997) backhoe dredgers may be able to operate in currents up to 3 knots. This figure may vary, depending on local conditions like water depth, orientation of the dredger, and the size of the dredger. Grab dredgers are more sensitive to currents as the positioning of the bucket becomes more difficult with increasing currents.

Strong currents during placement of fill in water will have adverse effects on the accuracy of this operation as they could transport significant volumes of material outside the boundaries of the reclamation area. The same transport capacity of currents could also cause unwanted siltation of environmentally sensitive areas.

As the presence of currents in the borrow -, transport - and reclamation areas can significantly affect the dredging operations, information about currents during the tender stage is essential for the Contractor.

### 4.3.3 *Hindrance to shipping and other parties*

Works carried out during the execution and the operational phase may influence activities in the direct surroundings and beyond. In order to coordinate these works and to avoid conflicting situations all activities should be listed and clear agreements must be made with all parties involved.

For instance, often a contract stipulates that dredging and reclamation works may not hinder the operations of third parties in the neighbourhood. However, dredging works often involve very large volumes of material and significant activities. Frequently, several dredging vessels are used to achieve the desired production level. This means a considerable increase in the traffic of dredgers sailing between borrow area and reclamation site, which may hinder existing shipping traffic. In such cases, extra nautical guidelines are required to ensure that shipping traffic will be efficient and safe during the dredging and reclamation activities. This is usually organised by the Port Authority in consultation with the parties involved.

### 4.3.4   *Environmentally driven limitations*

**Turbidity**
Dredging operations cause the suspension of fines in the water column, called turbidity (see also section 3.4.3.3, 11.3.4 and Appendix B.3.10). The cause of this turbidity is the stirring action by the dredge tool and the overflow of process water into the surrounding water by a Trailing Suction Hopper Dredger (TSHD) or the overflow of process water from a barge when hydraulically loaded.

In environmentally sensitive areas such overflow is not allowed especially when silty or clayey material is to be dredged. This restriction will lead to considerable cost increases.

Turbidity is inevitable in dredging operations. Its extent depends on the type of dredger, dredging method and soil to be dredged. Turbidity is the increase in concentration of solids expressed in mg/as an average over the water column at the perimeter of an area of $50 \times 50$ m around the centre of dredging activities (Manual Environmental Dredging Techniques Section 4, 2000 (Handboek Bodemsanerings-technieken, deel 4)).

Turbidity may be limited by taking special measures in the technical layout of the dredging equipment. For trailing suction hopper dredgers this may be the so-called green valve system. Backhoe dredgers and grab dredgers may be provided with visor grabs, which close using special valves when lifted through the water column.

Also special cutter suction dredgers are developed such as the bottom disk cutter dredger and the sweep head dredger. These measures and specially developed dredgers are used to dredge contaminated material for which it is essential to minimise turbidity in the water column (see Bray 2007).

For sand-winning, such measures and special dredgers are not economical and therefore unusual to employ. In section 4.6 ranges of turbidity are presented for different types of equipment and different types of soils.

**Working day and night and seasonal restrictions**
Usually dredging operations are carried out 24 hours a day, seven days a week (24/7), for which permission must be obtained from the authorities. In noise-sensitive areas the noise generation of dredgers may be a limiting factor and dredging may be restricted to day work only.

Seasonal restrictions may be imposed on the winning of fill material in the borrow area. Such restrictions may be related to bird life, sea habitat or tourism. Regulations are generally directed by the local authorities.

## 4.4  Dredging of fill material

### 4.4.1  *Introduction*

A project's borrow area may be a designated area known to contain sufficient fill material of adequate quality. However, a borrow area is also often an area that needs to be dredged as part of another project, without much consideration to the suitability of the dredged material as a construction material. For instance, a contract may stipulate that all materials dredged from an anticipated harbour basin will be used as fill for a nearby reclamation area – without considering the quality of the dredged material. Such a requirement may limit the quality of the land reclamation.

The maximum volume of fill that can be borrowed in an area depends on the dredging plant being used (e.g. the maximum achievable dredging depth of the dredger). Therefore, when selecting an area for borrowing a basic understanding of the dredging techniques and the possibilities and limitations of the various types of dredging equipment is important.

Other aspects that have to be considered when undertaking quantitative assessments of potential borrow areas are:

– the difference in density of the fill material in-situ and after placement and compaction within the reclamation area (bulking);
– material losses as a result of dredging activities (spill, overflow).

### 4.4.2   Volume and dimensions of borrow area

The volume of fill that can be mined in a borrow area depends, to a large extent, on the size and shape of such an area. However, the extractable volume also depends on the strength characteristics of the in-situ borrow materials and the ability of the dredging plant to excavate these. The size of the designated borrow area must be large enough to efficiently accommodate the dredging equipment. It must also produce the required volume of fill given boundary conditions like stable slopes and layer thickness of the in-situ materials.

The required size and shape of the borrow area to enable operations is different for all types of plant:

- Generally a trailing suction hopper dredger – as it is self-propelled and sailing while dredging – will need the largest area, preferably with an elongated shape.
- The area required by plain suction dredgers is dictated by slope development of the borrow area as a result of the dredging process. Economics usually restrict the volume which can be extracted from a borrow pit by a plain suction dredger to approximately 2/3 of the geometrical available volume.
- Operation of cutter suction dredgers demands sufficient area to allow for the width of the cuts, the positioning of the anchors and, possibly, the movement of barges.
- Grab, and backhoe dredgers need less space, generally determined only by the size of the dredging pontoon and the area needed to handle the transport barges.

### 4.4.3   Minimum thickness of fill deposits

The minimum thickness of fill deposits varies with the type of plant as follows:

- For an economical employment of a cutter suction dredger the minimum thickness of the fill deposits in the borrow area should typically not be less than the diameter of its cutter.
- A plain suction dredger requires a minimum layer thickness of in-situ borrow material of at least 6–8 m as its production is a function of the depth of the pit it creates.
- A trailing suction hopper dredger can operate economically in a minimum layer thickness of 1 to 2 m.
- For other types of dredgers, a minimum thickness of approximately 2–3 m may be sufficient. This all is subject to the availability of sufficient fill.
- For stationary dredgers, a limited thickness of the fill deposits will require more frequent relocation of the dredger. This will adversely affect the production rates and, hence, the economics of the dredging operations.

### 4.4.4  *Dredgeability*

The nature of a potential fill material may range from moderately weak rock, broken rock fragments after blasting to loose sand or even silt and clay. The properties of these materials will determine the ease of excavation and its behaviour during loading and transport. It will also determine the stability of the slopes at the borrow pit after dredging and the properties of the material after placement in the reclamation area. Hence, these properties must be known in an early stage of a project as they will define several important aspects that may have a significant impact on the feasibility of a reclamation project such as:

– the plant required to excavate the fill;
– the plant required to transport the fill;
– the anticipated soil improvement required after placement;
– the volume of fill that can be borrowed in an area.

**Rock**
Although dredging stronger rock is possible, economics generally dictate that rock with an unconfined compressive strength in excess of approximately 25 to 50 MPa for the large cutter suction dredgers, depending on the capacity of the dredger and on the rock mass characteristics, will have to be pre-treated by drilling and blasting in order to break this material prior to dredging. Note that this strength limit is strongly influenced by the tensile strength and the presence of joints and other discontinuities in the rock mass.

After blasting, the rock fragments can be excavated by backhoe, grab, bucket dredger, cutter suction dredger or (large) trailing suction hopper dredger depending on the maximum size of the broken fragments. Gravel and cobble-sized material can be dredged by all these dredgers; boulders can only be handled by large backhoe dredgers depending on the weight of the boulders and the size of the bucket.

The maximum size of the fragments that can be dredged by a trailing suction hopper dredger and a cutter suction dredger is defined by the diameter of the suction pipes, the minimum opening of the impellers of the pumps and – in the case of the CSD – the opening between the blades of the cutter. The maximum fragment size that can be handled is further determined by the installed pump capacity and is generally less than 30 cm.

Dredging of moderately strong rock with a typical maximum unconfined compressive strength of approximately 12.5 MPa will require a large cutter suction dredger. Recent developments have resulted in dragheads of trailing suction hopper dredgers equipped with ripper teeth. Although productions are modest, rock with a maximum unconfined compressive strength of approximately 5 MPa can now be dredged by the large (jumbo-sized) trailing suction hopper dredgers. Generally, dredging rock with a trailer suction hopper dredger should only be regarded as a possible economic solution when other equipment, such as cutter suctions dredgers, cannot be deployed (for instance, for reasons of adverse weather conditions or extreme water depth).

Wear and tear of the pickpoints of the cutter, the teeth of the bucket or the ripper of the draghead depend on the strength and mineralogy of the rock. The need to regularly replace these parts may seriously affect the production rate of the dredger. Siliceous rock will result in a higher wear compared to calcareous rock.

### Sand and silt
Sand and silt can be dredged by most types of plant. The plain suction dredger is usually only employed to dredge thick deposits of loose to moderately loose sand.

When dredging sand with a trailing suction hopper dredger, most of the fines will be removed from the sand by overflowing. Only the coarser sand particles will settle in the hopper. In the case of environmental restrictions, dredging can be done without overflowing. However, this means that only a fraction of the vessels carrying capacity will be used (generally less than 30%).

Production rates of most types of dredgers are affected by the in-situ density, the friction angle, the grain size, the fines content and the permeability of the sand/silt deposits.

### Clay
When dredging clay for reclamation purposes the work method is mostly aimed at limiting the disturbance of the material in order to preserve as much of the original strength as possible. The undrained shear strength, the Atterberg limits and the in-situ unit weight of the clay and/or silt deposits are the most important material properties.

Preserving the strength can best be achieved by using a backhoe or grab dredger.

However, dredging can also be carried out by a cutter suction dredger if the clay has a high undrained shear strength, a high plasticity and a low liquidity index – even though passage of the clay lumps through the pumps and subsequent hydraulic transport through a pipeline will generally erode the lumps to a certain extent. This will produce rounded clay lumps and a clay slurry that may adversely affect the quality of the fill mass after deposition. Also, dredging clay with a cutter may clog the cutter, interrupting operations and affecting the production.

Reference is made to section 9.1 for a more elaborate discussion on the suitability of clay as a fill material. Required data for judgement of the dredgeability of the various types of material see Table 3.1 in section 3.4.2.1.

## 4.5   Transport of fill

### 4.5.1   *Introduction*

Transport of fill material from the borrow area to the reclamation site can be carried out either by hydraulic transport through pipelines or by hopper dredgers or barges.

Detailed calculations are required to define the most economical transport method. Depending on the pump capacity of the dredger and boosters, the pumping distance for rock and gravel can be up to 4 to 5 km and for fine sands and silt the distance can be up to 15 km and more.

Booster pump stations can be employed to increase the pumping capacity. If the situation allows this, preference should be given to hopper dredgers or barges since the employment of booster pump stations usually has a significant impact on the dredging costs.

### 4.5.2   *Hydraulic transport through a pipeline*

Hydraulic transport through floating pipelines is the most common method of transport in reclamation operations. Part of the floating pipeline may be a sinker line to reduce hindrance to other shipping, to limit the pumping distance and to mitigate the effects of swell, waves and currents.

After the soil is mixed with water in the cutterhead of the Cutter Suction Dredger (CSD) or by the fluidisation system in the hopper well of the Trailing Suction Hopper Dredger (TSHD), the resulting slurry is pumped (commonly by centrifugal pumps) through a pipeline from the dredger to the required location. Water acts as a transport medium for the solid particles of the fill. In the reclamation area the solid particles will settle and the discharge water is then removed or drained off.

Cutter suction dredgers and plain suction dredgers may pump the fill directly from the borrow area (see Chapter 5) or re-handle pit (section 4.5.3), through a floating pipeline and, if required, an onshore pipeline to the point of discharge at the reclamation. The discharge point may be on land, but could also be a spreader pontoon, diffuser or another floating discharge device. TSHDs may also discharge their load after having connected to a floating or onshore pipeline.

Barges can be unloaded by a barge unloading dredger which pumps the material through a pipeline from the barge to the required discharge location.

### 4.5.3 *Transport by trailing suction hopper dredger or barge*

**Principles**
Once the hopper well of a THSD or barge has been filled with material, it will sail or be towed to the point of discharge. Several options exist for unloading the vessel:

- discharge the fill underwater using bottom doors, split hull or crane (for unloading the barge);
- discharge the fill through a pipeline using the vessels own dredge pumps;
- discharge the fill through a pipeline using a barge unloading dredger;
- rainbow the fill through a nozzle mounted at the bow of the vessel.

The point of discharge may be the designated fill area but can also be a re-handling pit from which the material is dredged again by, for instance, a cutter suction dredger and pumped to the reclamation.

The cycle production of a TSHD or barge is determined by:

- the volume and density of the fill material in the hopper;
- the sailing distance between the borrow area and the discharge location. Note that because the draught is different when the ship is loaded and after discharging, the vessel may follow two different routes when sailing to and from the borrow area;
- the sailing (or towing) speed of the TSHD or barge (loaded and unloaded);
- the loading and discharge time;
- the anchoring time (if the vessel requires an anchor to keep it in position during discharging);
- the time required to connect and disconnect to the discharge pipeline.

**Operational aspects**
The sailing route of the TSHD or barge may be an important factor in the overall production and, hence, the dredging cost. When choosing the sailing route the following aspects must be taken into account:

- the draught of the dredger in loaded and empty condition;
- the bathymetry;
- the tidal variations and the prevailing currents along the sailing route;
- the local navigational regulations/restrictions;
- other shipping traffic in the area, and so on.

Depending on the local situation (such as bathymetry, tide and currents), only partially loading the vessel and thus reducing the draught might be cost effective.

Sometimes the fill material is not directly placed in the reclamation area, but rather in a specially dredged pit to be re-handled by a cutter suction dredger. This has a number of advantages:

- It will reduce the discharge times of the TSHDs (direct discharge instead of pumping, so no coupling to a pipeline);
- It will prevent waiting times when more than one dredger wants to connect to the same floating pipeline;
- Use of a re-handle (interim) pit may create a more continuous process of fill supply to the reclamation area for the peak delivery of fill material. Especially with the large TSHDs the large peak discharge production might cause operational problems at the reclamation site. This can lead to interruptions in the discharge process or the need to reduce the discharge productions.

The use of a re-handling (interim) pit will optimise the use of the TSHDs, the pipeline, the dry earth moving equipment or the spreader pontoon. The obvious disadvantage is the need to create an additional pit and to employ a CSD as an extra vessel to the project, raising the operating costs. This method may, however, be preferable when haulage distances are long, hopper capacities are large and/or there is only one (long) discharge pipeline.

If the TSHD or barge can get close enough to the reclamation area the slurry could be discharged through a bow- or side-mounted nozzle ('rainbowing'). This process of placement is relatively inexpensive, but not very accurate and may, in some cases, not be allowed as rainbowing could produce a spray of (salt) water and fine mud over a certain area.

For the required data for selection of the transport methods, reference is made to Table 3.1 in section 3.4.2.1.

## 4.6   Utilisation characteristics of dredging equipment

This section presents a summary of utilisation characteristics for each type of equipment. In these tables the markings refer to the following codes:

| | |
|---|---|
| | *Applicable* |
| | *Possibly applicable* |
| | *Not applicable* |

The values of these characteristics must be considered as indicative and may differ from what can be achieved in practise. When special precautions are taken turbidity may be reduced, see section 4.3.4, this also applies to spillage.

Table 4.1   *Utilization characteristics of plain suction dredger.*

| plain suction dredger | | | | | | | |
|---|---|---|---|---|---|---|---|
| type of water | | small and large inland waters | | | | | |
| water depth [m] | | 2,5 | 5 | 10 | 15 | 20 | > 20 |
| soil type | | peat | clay | silt | sand | gravel | rock |
| project volume [situ m³ x 1.000] | | 50 | 100 | 500 | 1000 | 5000 | >10000 |
| limiting wave height [m], sea not swell | | 0.1 | 0.2 | 0.4 | 0.6 | 0.8 | >0.8 |
| presence of bulky refuse | | not sensitive | | moderately sensitive | | very sensitive | |
| horizontal dredging accuracy [±m] | | 2 | 4 | 4 | 6 | 8 | >10 |
| vertical dredging accuracy [±m] | | 1 | 2 | 4 | 6 | 8 | >10 |
| production (net m³/hour) | | 200 | 500 | 1000 | 1500 | 2000 | > 2500 |
| concentration [%] | | 20 | 40 | 60 | 80 | 100 | |
| | peat | | | | | | |
| | clay | | | | | | |
| | silt | | | | | | |
| | sand | | | | | | |
| | gravel | | | | | | |
| | rock | | | | | | |
| spillage [m] | | 0.05 | 0.1 | 0.15 | 0.2 | 0.25 | >0.25 |
| | peat | | | | | | |
| | clay | | | | | | |
| | silt | | | | | | |
| | sand | | | | | | |
| | gravel | | | | | | |
| | rock | | | | | | |
| turbidity [mg/l] | | 30 | 50 | 200 | > 200 | | |
| | peat | | | | | | |
| | clay | | | | | | |
| | silt | | | | | | |
| | sand | | | | | | |
| | gravel | | | | | | |
| | rock | | | | | | |

Table 4.2 *Utilization characteristics of cutter suction dredger.*

| cutter suction dredger | | | | | | |
|---|---|---|---|---|---|---|
| type of water | small and large inland waters | | | | | |
| water depth [m] | 2,5 | 5 | 10 | 15 | 20 | > 20 |
| soil type | peat | clay | silt | sand | grav | rock (1) |
| project volume [situ m³ x 1.000] | 50 | 100 | 500 | 100 | 500 | >10,000 |
| limiting wave height [m] | 0.1 | 0.2 | 0.4 | 0.6 | 0.8 | > 1.0 |
| presence of bulky refuse | Not sensitive | | moderately sensitive | | very sensitive | |
| horizontal dredging accuracy [±m] | 0.10 | 0.20 | 0.5 | 1.0 | 5 | > 5 |
| vertical dredging accuracy [±m] | 0.05 | 0.10 | 0.15 | 0.2 | 0.25 | >0.25 |
| production (net m³/hour) | 200 | 500 | 100 | 150 | 200 | > 2500 |
| concentration [%] | 20 | 40 | 60 | 80 | 100 | |
| | peat | | | | | |
| | clay | | | | | |
| | silt | | | | | |
| | sand | | | | | |
| | gravel | | | | | |
| | rock | | | | | |
| spillage [% of layer thickness] (2) | 10 | 20 | 30 | 40 | 50 | >50 |
| | peat | | | | | |
| | clay | | | | | |
| | silt | | | | | |
| | sand | | | | | |
| | gravel | | | | | |
| | rock | | | | | |
| turbidity [mg/l] | 30 | 50 | 200 | >200 | | |
| | peat | | | | | |
| | clay | | | | | |
| | silt | | | | | |
| | sand | | | | | |
| | gravel | | | | | |
| | rock | | | | | |

1 Dredging rock can only be done by the larger, heavy-duty cutter suction dredgers. The limiting rock strength depends very much on the extent of in-situ rock fractures (RQD) and the state of weathering. Depending on rock fractures, weathering and tensile strength, rock up to 25–50 MPa compressive strength may be dredged.
2 Spillage is expressed in terms of percentage per dredged layer thickness. Note that when dredging multiple layers, the spillage will be removed when dredging the next lower layer. If a certain depth must be achieved (capital dredging), spillage in the lowest layer – or when dredging only one layer – is generally removed by conducting a clean-up sweep.

Table 4.3   *Utilization characteristics of trailing suction hopper dredger.*

| trailing suction hopper dredger | | | | | | | |
|---|---|---|---|---|---|---|---|
| type of water | | large inland waters and coastal waters | | | | | |
| water depth [m] | | 2,5 | 5 | 10 | 15 | 20 | > 20 |
| soil type | | peat | clay | silt | sand | gravel | rock (1) |
| project volume [situ m³ x 1.000] | | 50 | 100 | 500 | 100 | 5000 | >10000 |
| limiting wave height [m] | | 1.0 | 1.5 | 2.0 | 2.5 | 3.0 | > 3.0 |
| presence of bulky refuse | | not sensitive | | moderately sensitive | | very sensitive | |
| horizontal dredging accuracy [±m] | | 1.0 | 5.0 | 10 | 15 | 20 | > 25 |
| vertical dredging accuracy [±m] | | 0.10 | 0.20 | 0.40 | 0.60 | 0.80 | >1.0 |
| production (net m³/hour) | | 200 | 500 | 100 | 150 | 2000 | > 2500 |
| concentration [%] | | 20 | 40 | 60 | 80 | 100 | |
| | peat | | | | | | |
| | clay | | | | | | |
| | silt | | | | | | |
| | sand | | | | | | |
| | gravel | | | | | | |
| | rock | | | | | | |
| spillage [m] | | 0.05 | 0.10 | 0.15 | 0.2 | 0.25 | >0.25 |
| | peat | | | | | | |
| | clay | | | | | | |
| | silt | | | | | | |
| | sand | | | | | | |
| | gravel | | | | | | |
| | rock | | | | | | |
| turbidity [mg/l] | | 30 | 50 | 200 | > 200 | | |
| | peat | | | | | | |
| | clay | | | | | | |
| | silt | | | | | | |
| | sand | | | | | | |
| | gravel | | | | | | |
| | rock | | | | | | |

1   Large (jumbo-sized) trailing suction hopper dredgers may be able to dredge rock with a maximum unconfined compressive strength of approximately 5 MPa, although productions are modest.

Table 4.4　*Utilization characteristics of backhoe dredger.*

| backhoe dredger | | | | | | | |
|---|---|---|---|---|---|---|---|
| type of water | | small and large inland waters | | | | | |
| water depth [m] | | 2,5 | 5 | 10 | 15 | 20 | > 20 |
| soil type | | peat | clay | silt | sand | gravel | rock |
| project volume [situ m³ x 1.000] | | 50 | 100 | 500 | 1000 | 5000 | >10000 |
| limiting wave height [m], sea not swell | | 0.1 | 0.2 | 0.4 | 0.6 | 0.8 | >0.8 |
| presence of bulky refuse | | not sensitive | | | moderately sensitive | | very sensitive |
| horizontal dredging accuracy [±m] | | 0.1 | 0.2 | 0.4 | 0.6 | 0.8 | >0..8 |
| vertical dredging accuracy [±m] | | 0.1 | 0.2 | 0.4 | 0.6 | 0.8 | >1.0 |
| production (net m³/hour) | | 200 | 500 | 1000 | 1500 | 2000 | >2500 |
| concentration [%] | | 20 | 40 | 60 | 80 | 100 | |
| | peat | | | | | | |
| | clay | | | | | | |
| | silt | | | | | | |
| | sand | | | | | | |
| | gravel | | | | | | |
| | rock | | | | | | |
| spillage [m] | | 0.05 | 0.1 | 0.15 | 0.2 | 0.25 | >0.25 |
| | peat | | | | | | |
| | clay | | | | | | |
| | silt | | | | | | |
| | sand | | | | | | |
| | gravel | | | | | | |
| | rock | | | | | | |
| turbidity [mg/l] | | 30 | 50 | 200 | > 200 | | |
| | peat | | | | | | |
| | clay | | | | | | |
| | silt | | | | | | |
| | sand | | | | | | |
| | gravel | | | | | | |
| | rock | | | | | | |

Table 4.5 *Utilization characteristics of grab dredger.*

| grab dredger | | | | | | |
|---|---|---|---|---|---|---|
| type of water | small and large inland waters | | | | | |
| water depth [m] | 2,5 | 5 | 10 | 15 | 20 | > 20 |
| soil type | peat | clay | silt | sand | gravel | rock |
| project volume [situ m³ x 1.000] | 50 | 100 | 500 | 1000 | 5000 | >10000 |
| limiting wave height [m], sea not swell | 0.1 | 0.2 | 0.4 | 0.6 | 0.8 | >0.8 |
| presence of bulky refuse | not sensitive | | moderately sensitive | | | very sensitive |
| horizontal dredging accuracy [±m] | 0.1 | 0.2 | 0.4 | 0.6 | 0.8 | >0..8 |
| vertical dredging accuracy [±m] | 0.1 | 0.2 | 0.4 | 0.6 | 0.8 | >1.0 |
| production (net m³/hour) | 200 | 500 | 1000 | 1500 | 2000 | > 2500 |
| concentration [%] | 20 | 40 | 60 | 80 | 100 | |
| | peat | | | | | |
| | clay | | | | | |
| | silt | | | | | |
| | sand | | | | | |
| | gravel | | | | | |
| | rock | | | | | |
| spillage [m] | 0.05 | 0.1 | 0.15 | 0.2 | 0.25 | >0.25 |
| | peat | | | | | |
| | clay | | | | | |
| | silt | | | | | |
| | sand | | | | | |
| | gravel | | | | | |
| | rock | | | | | |
| turbidity [mg/l] | 30 | 50 | 200 | > 200 | | |
| | peat | | | | | |
| | clay | | | | | |
| | silt | | | | | |
| | sand | | | | | |
| | gravel | | | | | |
| | rock | | | | | |

Table 4.6   *Utilization characteristics of bucket dredger.*

| bucket dredger | | | | | | | |
|---|---|---|---|---|---|---|---|
| type of water | | small and large inland waters | | | | | |
| water depth [m] | | 2.5 | 5 | 10 | 15 | 20 | > 25 |
| soil type | | peat | clay | silt | sand | gravel | rock |
| project volume [situ m³ x 1.000] | | 50 | 100 | 500 | 1000 | 5000 | >10000 |
| limiting wave height [m], sea not swell | | 0.1 | 0.2 | 0.4 | 0.6 | 0.8 | >0.8 |
| presence of bulky refuse | | not sensitive | | moderately sensitive | | | very sensitive |
| horizontal dredging accuracy [±m] | | 0.2 | 0.4 | 0.6 | 0.8 | 1.0 | >1.2 |
| vertical dredging accuracy [±m] | | 0.1 | 0.2 | 0.4 | 0.6 | 0.8 | >1.0 |
| production (net m³/hour) | | 200 | 500 | 1000 | 1500 | 2000 | > 2500 |
| concentration [%] | | 20 | 40 | 60 | 80 | 100 | |
| | peat | | | | | | |
| | clay | | | | | | |
| | silt | | | | | | |
| | sand | | | | | | |
| | gravel | | | | | | |
| | rock | | | | | | |
| spillage [m] | | 0.05 | 0.1 | 0.15 | 0.2 | 0.25 | >0.25 |
| | peat | | | | | | |
| | clay | | | | | | |
| | silt | | | | | | |
| | Sand) | | | | | | |
| | gravel | | | | | | |
| | rock | | | | | | |
| turbidity [mg/l] | | 30 | 50 | 200 | > 200 | | |
| | peat | | | | | | |
| | clay | | | | | | |
| | silt | | | | | | |
| | sand | | | | | | |
| | gravel | | | | | | |
| | rock | | | | | | |

## 4.7 Basis of cost calculation for dredging

One of the first questions raised when considering a reclamation project is "What will it cost?". The cost of a reclamation project is strongly related to the choice of the main dredging equipment. In this section, the cost of a reclamation project is briefly discussed. Although this is an important issue, it is not possible to give exact figures or even cost ranges. As demonstrated in this section, too many variables may influence the project cost. For that reason, contacting an experienced dredging Contractor or Consultant for advice on dredging and reclamation costs is highly recommended.

The cost of a dredging and reclamation project is composed of:

− non-operational costs such as mobilisation and demobilisation;
− operational costs, typically calculated by multiplying the number of weeks of the execution period with the weekly costs of the dredging and reclamation set.

The weekly costs of a dredging and reclamation set is composed of:

− Weekly costs of the dredger, its marine auxiliaries and pipelines, including:

Constant weekly costs such as depreciation and interest, maintenance, insurance premiums, which can be based on calculation guidelines such as the CIRIA publication "A guide to cost standards for dredging equipment 2009" by R.N. Bray.

Variable weekly costs such as:

Repairs (constant per operational hour)

Fuel and lubricants consumption (constant per operational hour)

Wear and tear (constant per m³ dredged) both on board (on board pipes and pumps, teeth consumption, cutter head, draghead, etc.) and in the discharge pipelines. The wear and tear increases with the coarseness and angularity of the dredged material and is also influenced by the mineralogy.
− Weekly general site costs, including, for instance, costs for project management, surveys, site offices, workshop, reclamation equipment (bulldozers, excavators, wheel-loaders, compaction equipment …) and labour.

The operational costs of a dredging and reclamation project are often expressed in a relevant currency unit per m³. This is called the unit rate. This unit rate can be expressed as a function of the dredging vessel's weekly production.

$$\text{Unit rate} = \frac{(\text{Weekly costs of dredger, marine auxiliary and pipeline} + \text{Weekly general site costs})}{(\text{Weekly production of dredger})}$$

As is clear from the formula above, the weekly production of the dredger will have a very direct impact on the unit rate. And a dredging vessel's weekly production can be influenced by a great number of parameters such as soil type, layer thickness, pumping distance, wave climate.

To demonstrate the impact of such parameters on the unit rate, a typical reclamation project by TSHD is selected as an example:

A TSHD with a hopper size of 12,000 m³ is dredging coarse sand at a borrow area located at a distance of 15 nm from the reclamation project. The sand is discharged through pipelines.

A typical example of the calculation of the vessel's weekly production is presented in Table 4.7.

In Table 4.8, various parameters affecting the vessel's production are discussed. For each parameter, an example is given indicating the possible impact on the unit rate as a percentage of the costs applicable to the case presented in Table 4.7. Note that these figures are only examples. The resulting figures can be very different under different circumstances.

Table 4.7   *Example of calculation of weekly production of a TSHD.*

| | | | |
|---|---|---|---|
| Hopper size = | 12,000 m³ | | |
| Vessel load = | 9,000 m³ | Sailing distance = | 15 nm |
| | | | |
| Dredging time = | 70 min | Dredge production = | 7,714 m³/h |
| Sailing time empty = | 56 min | Sailing speed empty = | 16 knots |
| Sailing time loaded = | 64 min | Sailing speed loaded = | 14 knots |
| Manoeuvring time = | 10 min | (speed on/off) | |
| Connection time = | 20 min | (connecting to floating line) | |
| Discharging time = | 90 min | Discharge production = | 6,000 m³/h |
| | | | |
| Cycle duration = | 310 min | | |
| | 5.17 h | | |
| | | | |
| Cycle production = | 1,742 m³/h | | |
| | | | |
| Operational hours/week = | 150 h/week (efficiency = 89%) | | |
| | | | |
| Weekly production = | 261,300 m³/week | | |

Table 4.8    *Influence change of factors on unit rates.*

| Influence factor | Base case | Description/ example | Calculation (example) | Influence on unit rate of Table 4.7 |
|---|---|---|---|---|
| Shipping delay | 150 operational hours/week | The number of shipping delays directly affects the unit rate. Example: 2 hours shipping delay per day. | Operational hours/ week = 136 h/week Week production = 237,000 $m^3$/week | +10% |
| Weather delay | 150 operational hours/week | The number of weather delays directly affects the unit rate. Example: Hs >1.5 m for 50 % of time. Unsheltered connection point | Operational hours / week = 80 h/week Week production = 139,000 $m^3$/week | +88% |
| Overflowing not allowed | Vessel load = 9,000 $m^3$ | If overflowing is not allowed for environmental reasons, the vessel load will be significantly lower | Vessel load = 4,000 $m^3$ Dredging time = 30 min cycle time = 7.2 h week production = 133,000 $m^3$/week | +96% |
| Increased sailing distance to borrow area | Cycle time = 5.2 h | Example: Sailing distance increase from 15 nm to 30 nm | Sailing time empty = 113 min Sailing time loaded = 129 min Cycle time = 7.2 h Week production = 187,500 $m^3$/week | +39% |
| Increased discharge distance to reclamation area | discharge production = 6,000 $m^3$/h | The effect of the discharge distance on the discharge production is demonstrated in figure 16 in Appendix A. Example: 600 μm sand, discharge distance is increased from 500 to 2000 m | Discharge production = 3,000 $m^3$/h Discharging time = 180 min Cycle time = 6.7 h Week production = 202,500 $m^3$/week | +29% |
| Reclaiming in small layer thicknesses in reclamation area | Cycle time = 5.2 h | If the reclamation schedule requires reclaiming in small layer thicknesses, this could result in additional delays (and/ or require additional dry equipment) | Delay time per trip = 30 min Cycle time = 5.7 h Week production = 238,000 $m^3$/week | +10% |
| Small layer thicknesses in borrow area | Dredge production = 7,714 $m^3$/h | The dredge production is significantly lower in borrow areas with a small layer thickness of sand. Example: layer thickness <0.5 m | Dredge production = 3,800 $m^3$/h Dredging time = 142 min Cycle time = 6.4 h Week production = 211,000 $m^3$/week | +24% |

(*continued*)

Table 4.8   (*Continued*).

| Influence factor | Base case | Description/ example | Calculation (example) | Influence on unit rate of Table 4.7 |
|---|---|---|---|---|
| Borrow area (too) small or unfavourably orientated | Cycle time = 5.2 h | The manoeuvring time of the vessel will increase because of frequent lifting of the pipe and turning. Example: borrow area of 500 m length | Manoeuvring time = 40 min Cycle time = 5.8 h Week production = 232,800 m³/week | +12% |
| Borrow areas with deeper water depth | Dredge production = 7,714 m³/h | The dredge production is reduced with increased dredging depth. Example: 50 m instead of 20 m water depth | Dredge production = 4,000 m³/h Dredging time = 135 min Cycle time = 6.3 h Week production = 214,000 m³/week | +22% |
| Very fine sand | Vessel load = 9,000 m³ | The vessel load is reduced because of difficult settling in the hopper; the dredge production is reduced because of increased overflow losses. Example: dredging 150 μm sand instead of 400 μm sand. | Vessel load = 7,500 m³ Dredge production = 4,500 m³/h Dredging time = 100 min Cycle time = 5.7 h Week production = 199,000 m³/week | +32% |
| Very coarse sand/gravel | Dredge production = 7,714 m³/h | The dredge production is reduced because of increased resistance in the pipelines. Example: 2 mm sand instead of 400 μm sand. | Dredge production = 5,500 m³/h Dredging time = 98 min Cycle time = 5.6 h Week production = 240,000 m³/week | +9% |
| Increased fine content | Dredge production = 7,714 m³/h | The dredge production is reduced because of increased overflow losses. Example: increase in fine content of 10% | Dredge production = 6,900 m³/h Dredging time = 78 min Cycle time = 5.3 h Week production = 254,700 m³/week | +3% |
| Increased relative density of sand in borrow area | Dredge production = 7,714 m³/h | The higher the relative density of the sand in the borrow area, the lower the dredge production. Example: relative density = 60% instead of 15% | Dredge production = 3,500 m³/h Dredging time = 154 min Cycle time = 6.6 h Week production = 205,000 m³/week | +27% |
| Presence of bulky refuse or unsuitable material in the borrow area | Dredge production = 7,714 m³/h | The presence of bulky refuse can lead to con-siderable delay for cleaning the dragheads and thus reduced dredge production. Example: assume 20 min delay every dredge cycle | Dredge production = 6,000 m³/h Dredging time = 90 min Delay = 20 min Cycle time = 5.8 h Week production = 232,759 m³/week | +12% |

# Chapter 5

# SELECTION BORROW AREA

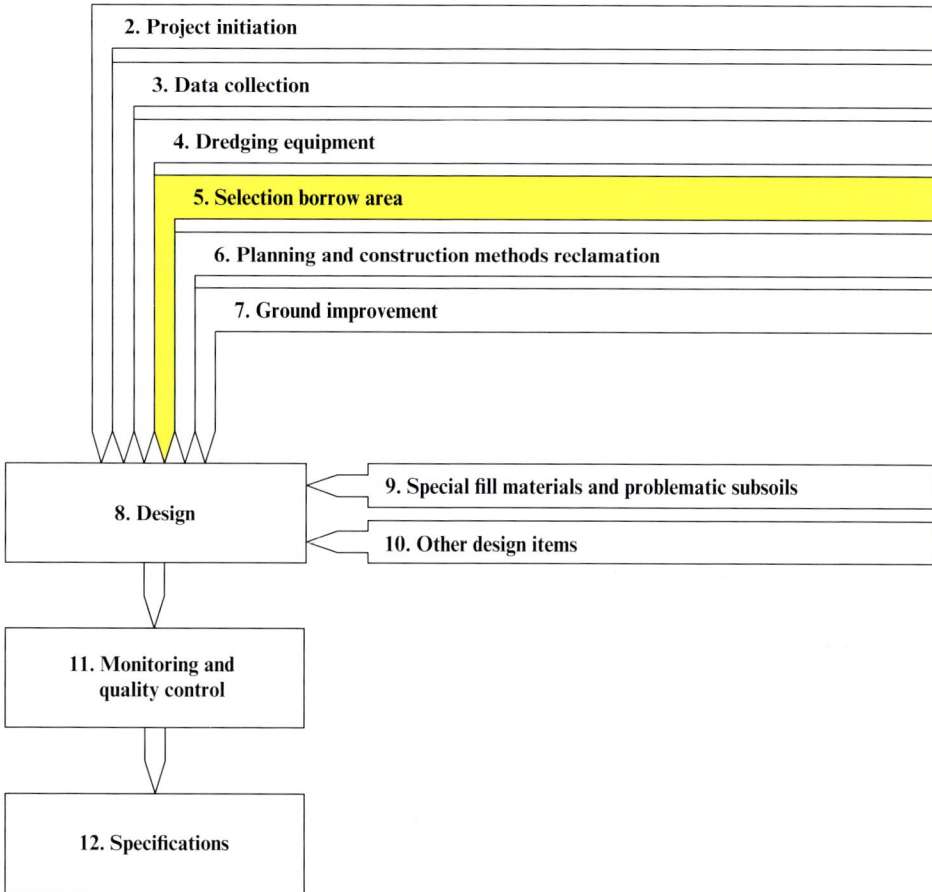

```
┌─────────────────────────────────────────────────────────────┐
│  2. Project initiation                                       │
  ┌─────────────────────────────────────────────────────────────┐
  │  3. Data collection                                        │
    ┌─────────────────────────────────────────────────────────────┐
    │  4. Dredging equipment                                    │
      ┌─────────────────────────────────────────────────────────────┐
      │  5. Selection borrow area                                 │
        ┌─────────────────────────────────────────────────────────────┐
        │  6. Planning and construction methods reclamation        │
          ┌─────────────────────────────────────────────────────────────┐
          │  7. Ground improvement                                  │
```

**8. Design**

**9. Special fill materials and problematic subsoils**

**10. Other design items**

**11. Monitoring and quality control**

**12. Specifications**

| 5 Selection borrow area |
| --- |

| 5.1 Considerations for the selection of a borrow area |
| --- |

| 5.2 Quality of the potential fill material | |
| --- | --- |
| 5.2.1 Change of the grading as a result of dredging | 5.2.2 Alternative fill materials |

| 5.3 Data collection in the borrow area | | |
| --- | --- | --- |
| 5.3.1 Data collection for quality assessment | 5.3.2 Data collection for quantity assessment | 5.3.3 Data collection for dredgeability assessment |

| 5.4 Quantity of fill material available | | | |
| --- | --- | --- | --- |
| 5.4.1 Bulking | 5.4.2 Losses | 5.4.3 Slope stability | 5.4.4 Geo-statistical methods |

| 5.5 Boundary conditions |
| --- |

## 5.1 Considerations for the selection of a borrow area

One of the most important conditions, having a significant impact on the feasibility of a land reclamation project, is the availability of sufficient suitable fill material within a reasonable haulage distance to the project location. In most cases special borrow areas are allocated for the supply of fill material, but fill may also have to be retrieved from capital or maintenance dredging. For instance: in harbour projects, deepening of basins or approach channels are often combined with reclamation projects.

The available quantity of fill material in the borrow area may affect the size of the reclamation. The quality of that material will have its bearing on the fill mass behaviour and the need for ground improvement. If the fill mass does not meet the functional requirements and ground improvement turns out to be too costly, then it will have consequences for the (foundation) design of the structures (see Figure 8.1).

It is also important to determine the in-situ characteristics of the materials within the borrow area as they have a significant impact on the dredging method, selection of the most suitable type of equipment and the planning.

In order to assess the quality, quantity and in-situ characteristics of the fill material available in the borrow area, soil investigations need to be undertaken. Such investigations generally consist of drilling boreholes, taking vibrocores, and performing in-situ testing and subsequent laboratory testing. In addition bathymetric and geophysical surveys are carried out, see section 3.5.

The dredging operations may have an impact on the environment of the area due to the generation of fines and the effects of exhaust fumes, noise and light. After completion of the project morphological changes of the seabed may have negative and/or positive effects on the hydrological, biological and ecological conditions of the area (Bray 2007).

Being such an important issue it is essential to identify potential borrow areas in an early stage of the project by collecting sufficient information for an adequate evaluation.

## 5.2 Quality of the potential fill material

The desired quality of potential fill material is dictated by its future behaviour after dredging, transport and deposition in the reclamation area. This implies that already during the design stage, samples retrieved from the borrow area must be tested. Based on these test results an assessment is made regarding the properties of a fill mass constructed with this material. This process may consequently also identify the need for ground improvement.

The quality of the fill material mainly depends on the stiffness and strength of the mass after deposition. These mass properties are affected by various material properties like mineralogy, grain shape, angularity, and particle size distribution. For a more detailed overview of mass properties, related testing and testing requirements reference is made to sections 8.3, and 8.5.

Generally, loose, medium grained, quartz sand is considered to be the most suitable fill material for reclamation purposes, but depending on project-specific conditions, other materials may sometimes be preferred instead. For instance, fine grained sand might be preferred in case of a very long pumping distance to the reclamation site.

It may often not be possible to select a borrow area based on required material properties only. Nowadays environmental, socio-economical (such as interests of local fishermen or the tourist industry) and other considerations may be important or even decisive selection criteria as well. When a project also includes capital or maintenance dredging, the materials produced by these activities may have to be used for reclamation purposes, irrespective of its properties. In such cases it is even more important to thoroughly investigate the borrow or dredge area because the nature of the fill strongly affects the choice of dredging equipment, the means of transport and the method of reclamation including the potential need for ground improvement techniques.

### 5.2.1 *Change of the grading as a result of dredging*

It is important to realise that the particle size distribution of fill material changes during the dredging process. During excavation fines are produced as a result of crushing. Degradation during hydraulic transport may further increase the fines content. Overflowing during the loading process of a TSHD or a barge will, on the other hand, reduce the quantity of fines. Segregation of fill material during deposition results in areas with little fines content (usually near the point of discharge) and other areas with high fines contents (at a larger distance of the point of discharge or near the weir box).

The increase in fines content is often significantly influenced by the mineralogy of the fill material: whereas strong minerals like quartz generally do not suffer from crushing, carbonates tend to be very susceptible to degradation as a result of crushing, see section 9.2.5.1. This change in grading must be taken into account when testing samples originating from the borrow area. It may imply that fines have to be added to or removed from the sample that is subjected to laboratory testing. The effect of fines on fill mass behaviour is dealt with in section 9.1.

### 5.2.2 *Alternative fill materials*

In case no suitable quartz sands are available from a borrow area within a reasonable distance to the site, or from maintenance dredging or capital dredging, other fill materials may have to be used such as clay, carbonate sand or rock. These materials have a different mineralogy, particle size distribution, grain shape, etc. which results in different and/or unexpected fill mass properties. However, by adequately addressing the typical properties of these materials, it may be possible to construct a reclamation that meets the specifications. The possibilities and restrictions of the use of these materials for fill is described in Chapter 9.

## 5.3    Data collection in the borrow area

Sufficient information on fill material must be collected within potential borrow areas. Chapter 3 presents the sources regarding the collection of such data.

Data collection in borrow areas is necessary in order to prepare:

- a quality assessment of the fill material;
- an estimate of the available quantity;
- and to assess the dredging characteristics of the material.

The information required for the quality assessment must focus on the material properties. The evaluation thereof should enable a prediction of the mass properties after deposition of the fill and the possible need for additional ground improvement.

A quantity estimate of a potential borrow area should rely on sufficient spatial information allowing for a delimitation of the fill deposits.

Results of the data collection must also include information on in-situ density and shear strength in order to enable an accurate selection of the type of dredging equipment to be employed within the borrow area.

### 5.3.1    *Data collection for quality assessment*

Soil investigations for the quality assessment of a potential borrow area include for instance drilling of boreholes, vibrocoring, and in-situ tests like Standard Penetration Tests and/or Cone Penetration Tests. The extent of the investigation depends on the spatial variability of the soil properties (see also section 3.4.2.1).

For granular fill material laboratory tests on samples retrieved from the boreholes include amongst others:

- sieve analysis;
- sedimentation test if say 20% < 60 μm;
- triaxial test/shear box test (to be undertaken at various densities);
- particle shape;
- angularity;
- minimum/maximum density;
- mineralogy (carbonate content, organic content, etc.).

When defining the laboratory test program, it must be taken into account that samples have to be reconstituted in order to represent the density of the fill material after dredging, transport, deposition and possibly also ground improvement. Uncertainties in this respect may require further sensitivity analyses.

Note that typical properties of some fill materials may affect the results of the laboratory tests. For instance, determination of the moisture-density relationship of carbonate sands could induce crushing of the particles which will change the particle size distribution of the sample. The resulting maximum dry density will therefore not be representative for the fill mass. Likewise for carbonate soils, gradings before and after compaction may also change because of the same crushing effect caused by compaction.

### 5.3.2 *Data collection for quantity assessment*

The information required to estimate the volume of suitable fill available generally consists of:

- a bathymetric survey;
- geotechnical data from the results of boreholes, vibrocores and subsequent laboratory tests and in-situ tests such as SPT's and CPT's;
- geophysical surveys.

This information can be used to identify and delimit the fill deposits that may be covered, intersected and/or underlain by unsuitable materials. The number of boreholes, vibrocores and/or Cone Penetration Tests can be reduced by carrying out a geophysical survey. Seismic reflection or refraction surveys and geo-electrical surveys are the most common techniques. Reference is made to Chapter 3.

### 5.3.3 *Data collection for dredgeability assessment*

Although the final choice with respect to the execution method also involves other aspects such as project planning, economics, hydrographic conditions, availability of plant, distance between borrow and reclamation area, local regulations and legislation, reliable geotechnical information on the proposed borrow area and fill material is crucial for a Contractor to select the most appropriate dredging tools.

The most important mass properties with respect to the dredegability of a material are summarized below, see also section 3.4.2.1 and Chapter 4:

| | |
|---|---|
| cohesionless, granular materials: | (relative) density, angle of internal friction, degree of cementation and permeability |
| cohesive materials: | undrained shear strength |
| rock and in-situ rock mass: | unconfined compressive strength, tensile strength, intensity, orientation of fracturing, degree of weathering |

## 5.4 Quantity of fill material available

The quantity of fill available is defined as the in-situ volume of extractable suitable material within the boundaries of the borrow area. Note that as a result of losses and/or bulking this volume may not be equal to the volume deposited at the reclamation site.

The extractable quantity is theoretically determined by the area of the concession, the difference between top and bottom of the suitable deposits and the allowable slope angles of the borrow pit. These slope angles may be restricted for stability reasons, but may also be subject to limitations dictated by the surroundings. See section 5.4.3.

### 5.4.1 *Bulking*

Bulking may affect the total volume of fill material required for a reclamation project. Bulking occurs when the in-situ density of the material within the borrow area is not equal to the density achieved within the reclamation. Positive bulking (volume increase) implies that the in-situ density of the material in the borrow area is larger than the density after deposition in the reclamation area. Negative bulking (volume decrease) is the opposite. Since it is often very difficult to determine the in-situ density of the potential fill material in the borrow area accurately, bulking can usually only be estimated by using correlations between the cone resistance (or SPT N-value), the effective stress and the relative density. Once both the minimum and maximum density of the fill material have been determined in the laboratory, an estimate of the in-situ density can be made. Further reference is made to section 8.3 of this Manual.

**Volume change**

During dredging, transport and placement of fill material a change in volume occurs. This is caused by a change of density (bulking) and by loss of fine material. This is schematically shown in Figure 5.1.

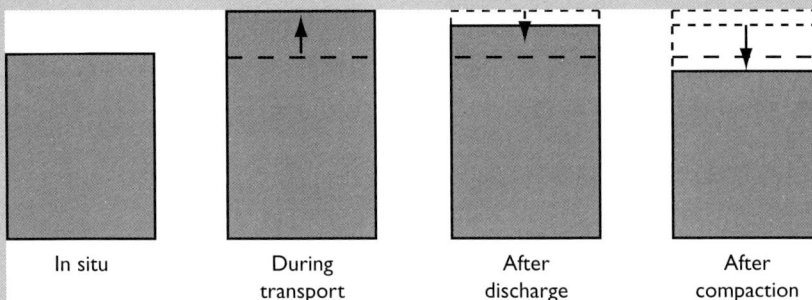

| In situ | During transport | After discharge | After compaction |

Figure 5.1   *Soil volume changes during different situations.*

### 5.4.2 *Losses*

Whereas bulking assumes a constant volume of solids with a varying porosity, a proper quantity assessment of a potential borrow area must also account for loss of material during the course of the dredging process. For instance, dredgers may not be able to dredge all material leaving behind a certain volume of spill. This may particularly become important when deposits of suitable fill occur as relatively thin strata.

Loss of suspended fines may occur as a result of:

– currents in the borrow area;
– overflow in TSHDs or barges (Figure 5.2);
– discharge through the weir box.

This may adversely affect the balance between the volume of fill required and the quantity available.

Figure 5.2 *Barge loading.*

5.4.3   *Slope stability*

In remote areas the stability of the slopes in borrow pits is not necessarily a big issue. The volume of sand deposits that can be borrowed in near-shore areas, lakes, rivers and other in-land waterways are, however, is often limited by minimum allowable slope angles. These minimum slope angles are not only a function of the properties of the soil mass, but they may also be dictated by environmental issues and/or the safety of adjacent structures (revetments, buildings, infrastructure like roads, pipelines, etc.). Slope stability analyses may be divided into:

– temporary (short term) slopes developed during dredging;
– final (long term) slopes.

When analysing the short term stability, it is important to examine the effects of the proposed dredging method: Dredging may induce or promote slope failure (the working principle of the stationary suction dredger is based on creating a local instability of the slope near the suction pipe!). This analysis should lead to a minimum allowable slope. Limitation of this slope angle could lead to the requirement to apply a dredging method that limits the risk of slope failure.

Analysis of the long term stability may require an assessment of loads that could cause failure in future. Such loads include both static loads like bearing loads of all types of structures and dynamic loads as a result of ship movements, earthquakes, etc. Special trimming of the slopes may be required to reduce the slope angles after dredging in order to ensure sufficient safety against failure in the long term.

Failures may occur as rotational slips (in homogeneous, medium dense to dense sands and clays), translational or compound slips (in case the failure surface is influenced by strata of different strength) or as flow slides (due to liquefaction, generally occurring in loose sands or uncontrolled breaching in denser deposits). For more details reference is made to sections 8.4.3 and 8.6. The result of uncontrolled breaching in a borrow pit is shown in Figure 5.3.

Although finite element methods may also be used, slope stability is generally analysed by conventional limiting equilibrium methods. These methods assume a known failure surface and consider the ratio between resisting forces (determined by the shear strength of the soil mass) and driving forces (weight of the soil mass and possible external loads) along this failure plane.

The analysis of flow slides is more complicated. Both the liquefaction and breaching phenomenon are discussed in section 8.6.

Determination of the shear strength of cohesionless material may require in-situ tests like the Standard Penetration Test or the Cone Penetration Test in

Figure 5.3   *Flow slide in inland borrow pit.*

combination with generally accepted correlations between the results of such tests and the angle of internal friction and effective stress, see section 8.4.2.

If the potential fill material has a cohesive nature, in-situ vane shear tests may be performed to determine the undrained shear strength (see section 3.4.2.1 and Appendix B.2.3.2). Alternatively, Cone Penetration Tests could be undertaken to indirectly assess the undrained shear strength using existing correlations (see Appendix C). Undisturbed samples of the cohesive strata could be recovered from boreholes which are subsequently tested in the laboratory. For more details on slope stability analysis reference is made to section 8.4.3.7 of this Manual.

Above water, the natural slope of a borrow pit may be as steep as 1(v):1.5(h) to 1(v):2(h). The natural slope under water shows more variation. It not only depends on the properties of the material and the dredging method, but also on the different mechanism of slope failure: rotational and translational failures usually result in slopes of 1(v):5(h) to 1(v):10(h) and are relatively regular in nature, while flow slides result in much more gentle slopes, usually between 1(v):10(h) to 1(v):25(h).

The volume of material involved in a flow slide is generally much larger than in a rotational or translational failure. As a result a flow slide may cause a significant regression of the top of the slope which may well extend beyond the boundaries of the designated borrow area.

### 5.4.4 *Geo-statistical methods*

By using geostatistical interpolation techniques it is possible to develop a geological model of the borrow area and to perform a probabilistic volume estimate of the quantity of suitable fill material available within its boundaries, including standard deviations for the best estimate, see section 3.8.2. In addition, it can assist in planning the site investigations by identifying the locations with the highest spatial uncertainty.

## 5.5 **Boundary conditions**

The suitability of a potential borrow area depends on more aspects than the quality, quantity and dredgeability of the fill material. Other important boundary conditions include:

– The exposure of the borrow area to waves, swell, wind, etc., that may hamper the dredging operations.
– The distance between the borrow area and the reclamation site may affect the economics of the operations.
– The volume of unsuitable overburden and intersecting layers or lenses that need to be removed prior to or during borrowing. These materials may have to be removed and disposed of at significant costs.
– The bathymetry at and near the borrow area which may prevent the employment of economic dredging equipment or may require dredging of an access channel.
– The environmental vulnerability that may hamper dredging or may ask for very costly measures to avoid unacceptable damage.
– The thickness of the suitable fill deposits or the area of the designated concession that may result in profitable or non-remunerative dredging operations.
– Governmental regulations that may prevent economic dredging operations (dredging permits, pipeline routing, anchoring locations, working at night, etc.).
– High costs of royalties for the fill that may favour the use of another borrow area.
– The presence of existing infrastructure (pipelines, cables, oil platforms, etc.), obstacles (wrecks, remnants of old foundations, etc.) and/or other potentially harmful and dangerous materials like unexploded ammunition within the potential borrow area.

Special investigations, surveys and studies may be required to identify such conditions. Consideration of all aspects could result in the need to use another borrow area containing a lesser quality fill material or may even require an adjustment of the initial design of the land reclamation and its superstructures.

# PLANNING AND CONSTRUCTION METHODS RECLAMATION

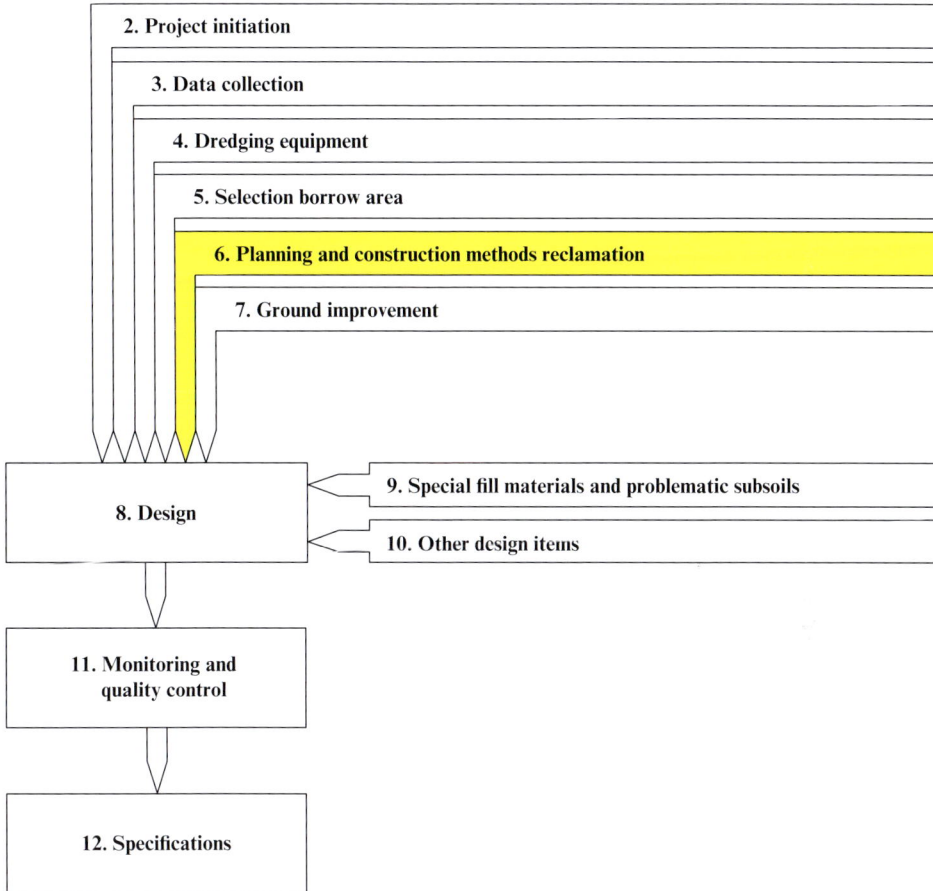

| 2. Project initiation |
|---|

| 3. Data collection |
|---|

| 4. Dredging equipment |
|---|

| 5. Selection borrow area |
|---|

| **6. Planning and construction methods reclamation** |
|---|

| 7. Ground improvement |
|---|

| 8. Design |
|---|

| 9. Special fill materials and problematic subsoils |
|---|

| 10. Other design items |
|---|

| 11. Monitoring and quality control |
|---|

| 12. Specifications |
|---|

**Chapter 6 Planning and construction methods reclamation**

## 6.1 Planning of the Works

| 6.1.1 Introduction | 6.1.2 Work preparation | 6.1.3 Construction and monitoring |

| 6.1.4 Demobilisation, clean-up and maintenance | 6.1.5 Example of project schedule |

| 6.2 Work plan for reclamation works | 6.3 Placement methods | 6.4 Construction of containment bunds |

## 6.5 Placement of fill material

| 6.5.1 Underwater placement in bulk of fill material | 6.5.2 Placement of fill material using a discharge pipeline | 6.5.3 Rainbowing |

| 6.5.4 Spraying |

## 6.6 Fill mass properties related to method of placement

## 6.7 Management of poor quality materials

| 6.7.1 Use of cohesive fine grained material assessment | 6.7.2 Settling ponds |

## 6.1   Planning of the works

### 6.1.1   *Introduction*

A project planning or (works programme) is the visualisation of the anticipated project progress on paper. By organizing and connecting activities in a logical way, the planning is a direct translation of the working method. Where the activities in a planning represent defined tasks that have to be executed under the scope of work, the relationships show the interdependency between them.

For proper project planning, which deals with future activities, it is essential to keep record of the progress of the works since change of production rates or scope of the Works will affect the planning.

The planning, together with the resources and costs that are required to execute the project, is an essential part of project management. This especially applies to complex projects. A schematic representation of the project management relationship is visualized in Figure 6.1. This relationship is characterized by a continuous and iterative process of planning, execution and monitoring. The project planning, the resources that are required to execute the activities in the planning and the costs of the resources are directly related to each other in this relationship. Changing one of them immediately influences the other two.

The aim of planning can vary from one situation and work phase to the other. The most important arguments for using a planning are:

- Progress control: to compare the actual progress of a project with the planned progress and to adjust/mitigate where and if required.
- Milestone(s): to show that (parts of) the project will finish on time.

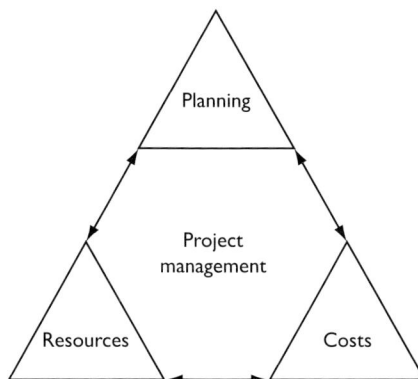

Figure 6.1   *Project management relationships.*

- Use of resources: to establish the most efficient contribution of resources (manpower, material and equipment) of the different work parts.
- Working method: to determine the optimum working method.
- Communication: to inform others (mainly the Client) about the way and the time frame in which the project is executed. For the contractual status of the planning, see FIDIC "Form of Contract" First edition 2006, clause 7.
- Contractual tool: to serve as a contractual tool for identifying and managing changes (variations, delays, etc.). For further reading, refer to Delay and disruption Protocol (2002).
- Cost control: to assist in the determination and control of the project costs.

To be able to fulfil the goals mentioned above, the planning will have to give an answer to the following questions:

- What has to be done (activities)?
- When is it done and how does this relate to the required completion (milestones, relationships)?
- Where is it done (location)?
- How an activity is executed (resources)?
- What is the duration of an activity (production)?
- Which quantities need to be handled during the activity (material)?

The planning of a realisation phase of a project discussed in the next sections can be divided into:

- the preparation phase;
- the construction phase;
- the maintenance period.

Note that the planning of a total project may include more phases: the initiation phase, the design phase and the tender phase.

The type of contract (Traditional, Design and Construct, etc.) will determine the activities undertaken as part of the design and realisation phases: if the project is to be built under a Design and Construct contract, the scope of the design phase preceding the realisation phase will be limited to a conceptual development (often referred to as a FEED or front end engineering design) and the detailed design is undertaken during the realisation phase. A Traditional contract, however, will require a full detailed design before the start of the realisation phase.

### 6.1.2  *Work preparation*

After award of a contract for a reclamation work, the preparation works begin. Preparation is the key to a successful execution of the project and is often under-estimated in dredging and reclamation works. Typical activities that have to be done in the preparation stage are:

- the establishment of a project team and mobilization of this team;
- provision of housing and offices for personnel;
- execution of Engineering works (design, permits, HSE management, environmental management, method statements, etc.);
- create access to site;
- furnish lay-down/working areas;
- preparation and mobilization of equipment.

All these activities require (a significant amount of) time and need to be accounted for in the planning of the works.

### 6.1.2.1 Establishment of project team

The first thing that has to be arranged before any work can be done is the establishment of a project team and mobilization of this team to, preferably, a location near the works. Whereas the project team is often established during the tender or negotiation phase before contract award, mobilization of this team to the country/ location where the job is executed is required for the other preparation activities that have to be done.

### 6.1.2.2 Provision of housing and offices for personnel

Except for projects that are located close to the main offices or near large cities where housing can be arranged by renting existing accommodation, housing must be specially provided. In remote areas, these accommodation camps are often specifically designed and built by the Contractor. Depending on the location, these camps can be made onshore, or can be developed as house boats. It is obvious that accommodation is a key activity in the preparation of the works. Without proper accommodation, the execution will not commence. It should be realised that the establishment of these camps can take considerable time and effort and needs to be accounted for in the project planning.

### 6.1.2.3 Execution of engineering works

Increasingly important in modern dredging and reclamation projects is adequate administration and documentation. The dredging Contractor has to provide documentation proving the Contractor's ability to work safely, according to high quality standards and within strict environmental restrictions. On top of that, in design and construct projects, the design responsibility lies with the Contractor as well. Application of and obtaining permits are more and more tasks that the dredging Contractor has to take on-board his Scope of Work.

In the planning, this implies that a significant amount of time shall be reserved for the development (and approval by the Client) of these engineering works. Even more important is to properly show the dependency of these activities with the activities in the execution phase. For example, the Contractor will not be allowed to start working without a proper safety management plan in place.

### 6.1.2.4   Create access to site and development of lay-down areas

On a reclamation project, floating equipment must have access to both the borrow area and the (designated) fill area. This may require a specially dredged channel for dredgers to approach the disposal location. Depending on the situation, access may stop short from the reclamation area and the remainder may be bridged using a discharge pipe line through which the material is pumped to the fill area. Details of these working methods are described in Chapter 4.

For storage and maintenance of equipment and stockpiling of construction materials, such as rock armour protection for the reclamation perimeter, lay-down areas have to be created. Temporary unloading facilities may have to be constructed to allow for the delivery of construction materials and land-based equipment transported over water.

### 6.1.2.5   Preparation and mobilization of equipment

Finally, all equipment that is required to execute the works has to be mobilized to site. This is not limited to the dredging vessels, but also includes all other main and auxiliary equipment, such as:

- booster stations;
- pipelines, bends, Y-pieces, sinker lines etc.;
- survey boat, multicat, fuel barge etc.;
- land based equipment; wheel loaders, excavators, bulldozers, fuel truck etc.;
- cars, survey equipment, safety gear, environmental monitoring equipment, etc.

For most projects, the equipment has to fulfil the specific requirements in terms of HSE and environment.

### 6.1.3   *Construction and monitoring*

The planning of the dredging and land fill works is laid down in a construction program. As already highlighted in the introduction, this program is a visualisation of the working methodology and sequence on paper. As examples of work methods are described in Chapter 4, it is not further elaborated in this section. The construction program should clearly show the relationships (dependencies) between the various stages of construction and between activities.

A normative factor, largely determining the duration of the works or the amount of equipment used is the sailing distance between the borrow area and the location of the actual fill. This distance, when bridged by trailing suction hopper dredgers or barges, largely determines the size, number and type of equipment used.

The monitoring during construction may include verification of the quality of the fill material (particle size distribution, settlement measurements, in-situ testing such as density determination, CPT, SPT, etc.) and intermediate surveys, see Chapter 11. The planning must allow for these activities. However, they should be planned in alignment with the progress of the filling operations and delays caused by monitoring should be avoided.

### 6.1.4 Demobilisation, clean-up and maintenance

When the actual construction activities are completed, the final phase of the project starts: demobilization of the equipment and personnel, clean-up of the site and (if applicable) a maintenance period during which the Contractor is responsible for keeping the works as per Specifications. Note that maintenance of dredged areas (basins/channels) is often required during the construction as well.

During this final phase, the Contractor will still require his lay-down area, housing and offices until all operations on site have been completed. This needs to be included in the planning.

### 6.1.5 Example of a project schedule

A typical example of a project schedule is provided in Figure 6.2. This schedule represents a dredging project in 'Sandy Harbour' and shows the realization of the following project scope of work:

– dredging of a trench for the construction (by others) of a quay wall;
– dredging of a berth pocket and access channel;
– construction of a reclamation using dredged material.

As can be seen in Figure 6.2, the project schedule shows all main stages of the project and by using the activity relations described above, the activities in the stages are interlinked.

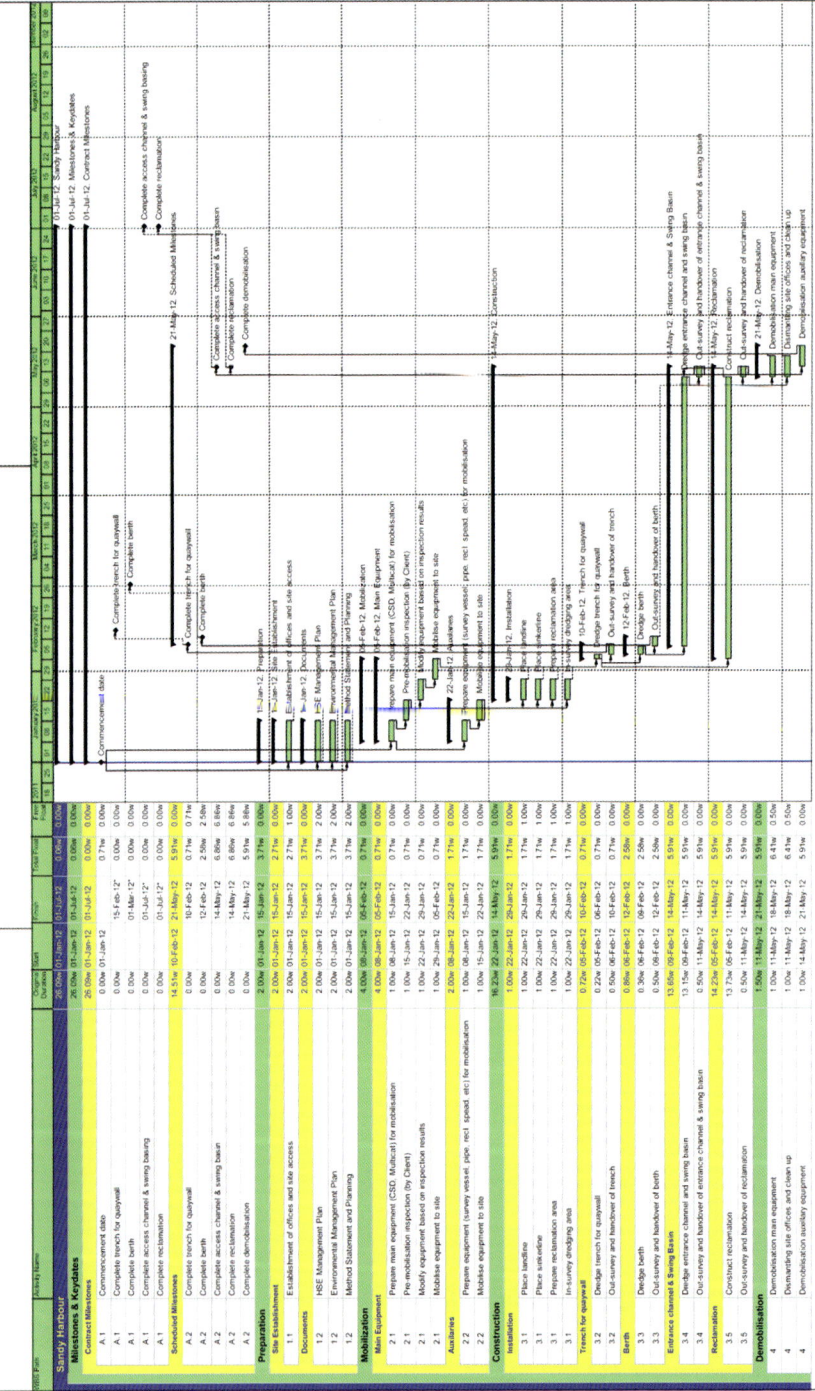

Figure 6.2 *Project planning Sandy Harbour project.*

## 6.2 Work plan for reclamation works

Before starting the filling operations, a plan should be made of the dredging, transport and reclamation activities. Such a plan could include:

- an inventory of all relevant contractual and technical specifications including the lay-out of the reclamation area, the required fill (mass) properties, testing and monitoring requirements, possible milestones, etc.;
- an inventory of all relevant boundary conditions like
  - o bathymetry (of borrow area, the sailing route of the dredging vessels, and of the reclamation area and its surroundings)
  - o nature of the existing subsoil at borrow area and reclamation site
  - o environmental regulations
  - o navigational regulations
- required permits;
- availability of plant;
- a detailed project planning;
- a dredging plan of the borrow area (including dredging equipment);
- a fill transport plan (pipeline, booster pump stations, TSHD, barges, etc.);
- the results of geotechnical analyses (settlement and stability of bunds and fill);
- a construction method of the bunds;
- a filling plan (including the plant and equipment required, the number and thickness of lifts, consolidation periods, testing schedules;
- a soil/fill ground improvement plan (vertical drains, compaction), if necessary;
- a monitoring and quality control plan;
- a risk assessment including possible mitigating measures.

Fill may be placed under water or above water with land based equipment or with water borne equipment. Working methods and equipment required for these operations differ considerably and resulting densities, natural slopes, segregation, etc., will be different as well.

In the following sections various construction methods are discussed. The most appropriate method depends usually heavily on the local conditions, availability of plant, contractual requirements, etc. and can only be selected after a careful inventory of all relevant aspects.

## 6.3 Placement methods

The following placement methods may be distinguished:

Under water:

- dumping by TSHDs and barges (bottom doors or valves, split hull);
- pumping through a spreader or diffuser at the end of a pipeline;

- dumping/placing by grab crane;
- rainbowing from TSHDs or specially constructed pontoons.

Above water:

- discharging via a pressure pipeline from cutter suction dredgers, TSHDs, plain suction dredgers or barge unloading dredgers;
- rainbowing from TSHDs or specially constructed pontoons.

Often the first layers of fill have to be placed under water, while in a later stage filling can continue above water.

The preferred placement method may depend on:

- the water depth at and within the vicinity of the proposed reclamation area;
- the nature of the existing subsoil;
- the nature of the fill material;
- the geometry of the proposed reclamation area;
- the required placement accuracy;
- the required fill mass properties;
- the environmental requirements.

To ensure that the fill is placed within the required boundaries without lateral spreading outside its footprint, the reclamation area is often partly or fully bounded by containment bunds.

Fines in the fill material tend to segregate during the deposition process. If the fines content is not too high, fill must be placed in such a way that the fines are well distributed within the reclamation.

Should the fill contain too many fines and/or other deleterious materials, measures may have to be taken. The surplus of these materials should either be removed from the reclamation area or redirected to locations where they can do no harm (for instance settling ponds). In both cases a working method must be adopted that promotes segregation of the unsuitable and suitable fractions of the fill (see section 6.7.2).

## 6.4    Construction of containment bunds

### 6.4.1    *General*

Containment bunds are constructed to:

- contain the fill and – sometimes – the suspended solids within the footprint of the reclamation area or settling pond;
- control the slope angle of the edge of the reclamation area;

- improve stability of fill placed on soft existing subsoil;
- control the water table within the reclamation area;
- control the flow of the discharge water in the fill area;
- accommodate pipelines and to allow access to the site.

Stability analyses have to be undertaken to investigate the integrity of the bunds during the filling operations. These analyses may include critical slip circle calculations, the verification of the bearing capacity and horizontal equilibrium. If required, more complicated finite element analyses should be done. Should the site be located in a seismic active region, it may also be required to analyse the response of the bunds to earthquake loading. Reference is made to section 8.4 for more details about these geotechnical analyses.

Natural slope angles of fill after hydraulic placement are usually (very) gentle. These slopes will depend on the placement method, the particle size distribution of the fill and the exposure to waves and currents, see Table 6.1. Should the fill not be contained by bunds, large volumes of fill may be deposited outside the actual footprint of the reclamation area.

By constructing bunds and profiling their slopes to their required stable slope angle, it is possible to limit the volume of fill. In general it can be stated that with increasing coarseness of the bund material the slopes can be constructed at steeper angles.

If the subsoil of the reclamation area consists of soft material having a low shear strength, a stable bund around the perimeter of the area may help to improve the stability of the fill. This bund can be constructed on top of a sand filled trench cut down to more competent strata. This construction method can prevent the need for entire removal of the soft strata or the construction of very gentle slopes.

Table 6.1  *Indicative natural slopes of granular materials hydraulically placed by pipeline discharge from above water (from Athmer & Pycroft 1986).*

| Grain size [mm] | Indicative range of slopes | | |
| | Above water | Below water Calm seas | Below water Rough seas |
| --- | --- | --- | --- |
| 0.060–0.200 | 1:50–1:100 | 1:6–1:8 | 1:15–1:30 |
| 0.200–0.600 | 1:25–1:50 | 1:5–1:8 | 1:10–1:15 |
| 0.600–2.000 | 1:10–1:25 | 1:3–1:4 | 1:4–1:10 |
| > 2.000 | 1:5–1:10 | 1:2 | 1:3–1:6 |

Figure 6.3   *Weir box.*

An area fully enclosed by bunds that reach above water is generally called a closed fill area or confined disposal facility (CDF). To allow the discharge water to drain, adjustable weir boxes are usually placed in the bunds. Figure 6.3 shows such a weir box.

At a partly closed or open fill area, which lacks bunds at one or more sides, the discharge water runoff freely through the open section.

If the bunds arc sufficiently watertight and stable it may be possible to raise and control the water level within the contained area by adjusting the level of the over flow of the weir boxes. This may affect the settling regime of the fill and in particular of the fines, see also section 6.7.2. It might also be a solution to create sufficient water depth to allow for floating equipment like a spreader pontoon in the reclamation area.

Bunds may be constructed just to guide the flow of the discharge water and, hence, the current velocity of the discharge water within the fill area. By adjusting this velocity it is possible to control the settling regime.

Sometimes bunds are required for operational reasons: to support pipelines, to allow access for equipment, to accommodate anchors, to compartmentalize the reclamation area, etc.

## 6.4.2   *Methods of bund construction*

Depending on its function, its retaining height, the availability of construction materials, etc. a bund may be constructed of dredged or imported rock fill or quarry run, borrowed granular or cohesive fill material or just material available in the immediate vicinity of the bund. Very soft to soft clay or silt, however, is not a suitable construction material. Sometimes these materials are used for filling geotubes which may be used to bund a reclamation area.

### Bund construction above water level

Bunds may be constructed in one or more layers to their final level. In case the bund is constructed in several layers, the bund is generally raised just above the level of the next fill phase. This method may provide a better control of the final slope, see Figure 6.4. Usually these bunds are compacted by bulldozers.

If the bund has to be constructed on soft subsoil, stability considerations may require a staged construction with a limited height for each lift and a gentle overall slope. Alternatively, it may be possible to construct the bund on top of a trench excavated into the soft soils see Figure 6.5. Another solution is to increase the shear strength of the subsoil prior to or during construction of the bund by soil improvement techniques as discussed in Chapter 7.

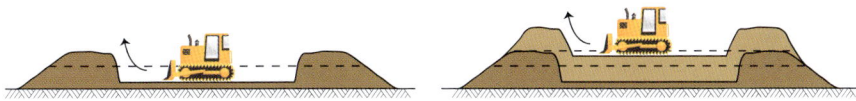

Figure 6.4   *Construction of bund in lifts.*

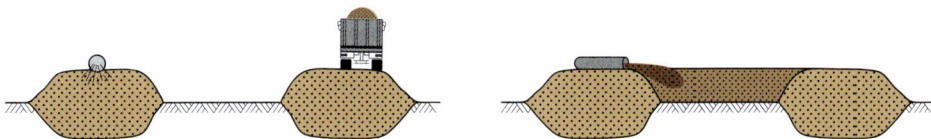

Figure 6.5   *Bund in soft soil.*

There are many ways to construct bunds. A distinction can be made between bund construction below water level and above water level.

Bunds that will be raised in one lift from below water level to above water level can be constructed by:

– reclaiming a strip at the inner side of the reclamation area by discharging through a pipeline or rainbowing. Bulldozers, excavators or backhoe dredgers can be used to profile the outer slope to the required slope angle;

– truck dumping ('end tipping') and subsequent profiling with an excavator or a backhoe dredger;
– a backhoe or a grab dredger side-casting material dredged in the immediate vicinity of the bund (usually limited to relatively shallow waters).

Figure 6.6 to Figure 6.11 show the principles of the various working methods.

Figure 6.6    *Rainbowing and subsequent shaping of the sand pancake.*

**Bund construction below water level**
Bunds built below water level (offshore) in various lifts will require water-borne equipment:

– when using quarry run, originating from onshore sources, a dumping vessel, a grab hopper dredger or barge with a grab dredger can be used to construct the bunds;
– when using sand fill, a TSHD can be used to raise the bunds by controlled dumping through bottom doors or rainbowing (it may be required to further profile the bund afterwards). Some TSHDs have the capability to dump sand by pumping it through their suction pipe ('reverse pumping'). Also a grab dredge may be used or controlled placement of hydraulic fill material using a diffuser.

Examples of these working methods are illustrated in Figure 6.8 and Figure 6.9.

Once the fill has reached a level well above water level, further construction of the bunds is carried out by bulldozers and excavators, see Figure 6.11. Whether the

Figure 6.7    *Truck dumping (end tipping) and profiling with excavators.*

Figure 6.8    *Side stone dumping vessel at work.*

bund is immediately raised to its final height or construction is continued using more subsequent lifts depends on various factors such as:

- the geotechnical stability of the slopes;
- the management of the discharge water in the fill area;
- the soil improvement requirements;
- other operational considerations.

Figure 6.9    *Bund construction using rainbowing.*

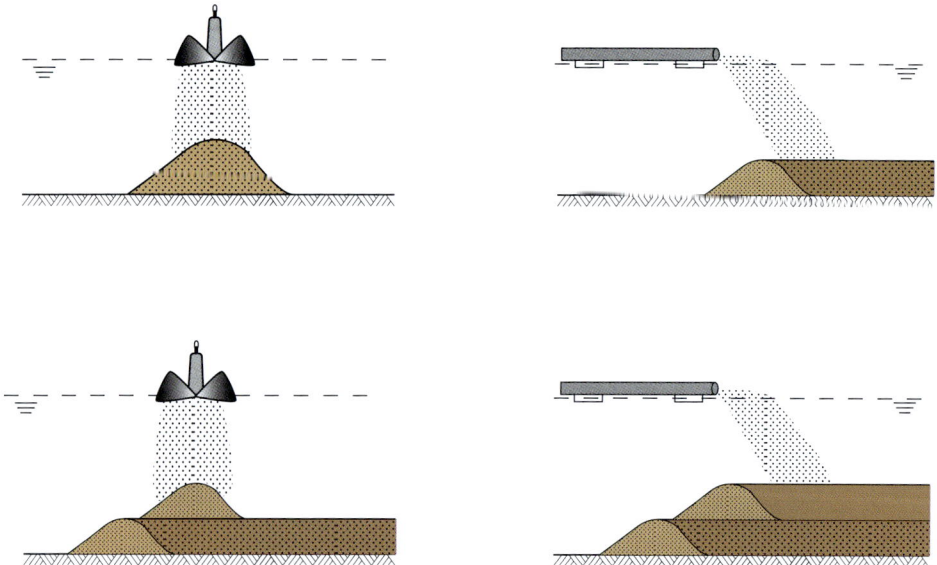

Figure 6.10    *Staged bund construction under water using a grab dredger.*

Figure 6.11   *Construction of bunds above water level.*

To protect the bund from erosion during hydraulic filling, plastic sheeting is occasionally placed on the inner slope of the bund.

## 6.5   Placement of fill material

### 6.5.1   *Underwater placement in bulk of fill material*

**General**
Filling by underwater placement is usually undertaken by TSHDs and barges. After sailing to the reclamation site, the vessels open their bottom doors or split their hulls allowing the mixture of fill and water to fall through the water column down to the seabed. After release from the hopper or pipeline, entrainment of surrounding water will dilute the mixture during its journey through the water column, hence reducing its concentration and increasing the seabed area that will be covered by the dumped material. Upon hitting the seabed, the fill-water mixture will create a crater with a turbulent mixture flow inside.

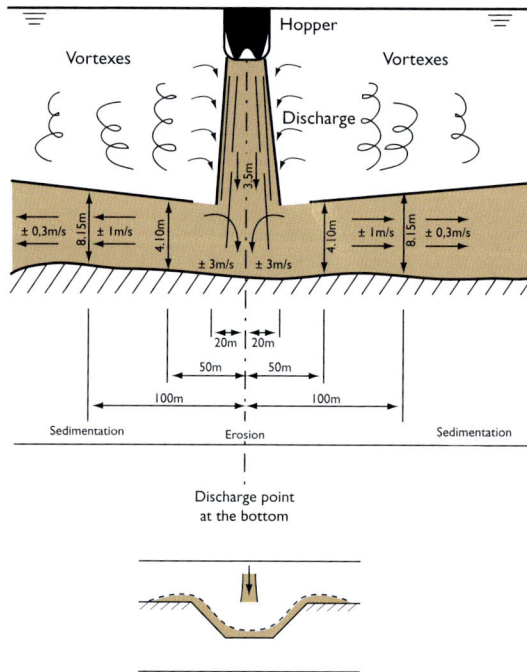

Figure 6.12  *Dilution of sand water mixture flow in the water column while discharging and crater formation.*

Figure 6.12 presents a sketch of both phenomena. For further details reference is made to the Report 152: Artificial sand fills in water (CUR, 1992). Due to these complex processes in combination with the possible presence of currents, it is difficult to predict the geometry of the fill mass after deposition.

Once unloaded the THSDs or barges require less draft, which may favour the operational conditions in the sand placement area.

Fill may also be placed under water through the discharge pipeline of a CSD which may be provided with a diffuser to control the soil mixture flow, see section 6.5.2.

**Filling control**

For optimal planning and management of the placement operations, the reclamation area is usually divided into grids. The grid size is generally slightly larger than the size of the hopper of the dumping vessels. For subsequent layers the designated dump locations may be staggered. Modern global positioning systems and computer-aided administration methods on TSHDs and barges are used to control filling operations by dumping. Regular interim bathymetric surveys are vital to guarantee a good result.

**Stability subsoil**

Should the existing subsoil consist of soft fine grained materials like silts and clays, the impact and load of the dumped fill may exceed the bearing capacity of the subsoil. The dumped fill will cause the soft materials to be squeezed aside. Subsequent mixing of the fill and the soft subsoil sediments may have an adverse effect on the quality and performance of the fill mass. In such a case dumping may not be a suitable filling method (depending on the thickness of the soft strata and the fill, the future loading conditions and required performance criteria of the reclamation area).

**Further considerations**

The TSHDs/barges must have access to the reclamation area. The site should not be fully enclosed by bunds. The water depth should also be sufficient to prevent grounding of the vessels after dumping.

In order to create a more continuous process of sand delivery into the reclamation, a CSD is sometimes used to re-handle fill that has been dumped by trailing suction hopper dredgers or barges in specially created underwater pits near the fill area, see also section 4.5.3.

### 6.5.2   *Placement of fill material using a discharge pipeline*

**General**

Filling by pipeline is the most common working method in reclamation projects. At its intake, the pipeline may be connected to a CSD, a TSHD, a plain suction dredger or a barge unloading dredger.

At the discharge end, the pipeline is occasionally fitted with a diffuser to better distribute the fill in the immediate vicinity of the outlet. The discharge end may be placed on land, positioned just above water or may be submerged.

The pipeline may also be connected to a spreader pontoon or a nozzle in order to spray or 'rainbow' the fill. These two techniques are discussed separately in the sections 6.5.3 and 6.5.4.

**Over water**

When the discharge end is positioned over water the pipeline may be mounted on pontoons or may be provided with floating jackets. Occasionally a diffuser is fixed to the end of the pipeline to more evenly distribute the fill in the water column and to reduce the kinetic energy of the mixture, see Figure 6.13.

Figure 6.13    *Diffuser.*

Without diffuser, erosion may cause craters in the existing seabed. When thickness control is required, it is possible to place the fill in lifts by moving the pontoon using winches and anchors. In this case the floating pipeline must have sufficient flexibility and slack to allow for the movements of the pontoon, see also Appendix A-8.5.

The use of a pipeline over water is restricted by the draught of the pontoon or pipeline and the auxiliary equipment for handling the pipeline. This technique is therefore not suitable to raise the fill above water.

**On land**
Onshore pipelines are usually laid on the ground. At the end the pipe may be raised 1 or 2 m above ground level to ensure a more or less free outflow of the mixture. The impact of the fill water mixture on the ground and subsequent erosion will result in craters, just like in the underwater situation. The use of a diffuser at the end of the pipeline may reduce this effect to a certain extent.

Bulldozers will continuously level the area in front of the pipeline during the pumping operations, until the whole area has been raised and a new pipe section can be coupled to the pipeline. By levelling the area the bulldozers also provide an initial compaction of the fill.

Fill can be raised to its final level in one or more lifts. The thickness of these lifts is determined by stability and compaction requirements and by economics, because the pipeline has to be removed each time before a next lift can be placed.

The large discharge capacities of the latest generation cutter suction dredgers and TSHDs require thick lifts (typically more than 2–3 m). Should this thickness be less, it may prove to be impossible for dry equipment to keep up with levelling the fill and coupling the pipe sections. In such a case branching of the pipeline to create alternative deposition areas may be a solution to avoid unwanted interruptions of the discharge operations of the dredgers.

**Pipeline arrangements and routings**
In general the most preferable route of the pipeline is a straight line between the dredger and the discharge point since bends, branching, control valves, snifters, etc. will all increase the hydraulic resistance. An increased hydraulic resistance will require a higher pump pressure and power to achieve a certain production rate.

However, the pipeline route and arrangement may also depend on many other factors such as:
 – the pumping capacity of the dredgers at the intake side;
 – the maximum pumping distance and the need for a booster pumping station;
 – the total length of pipeline available;
 – the type of fill supply (continuous/discontinuous);
 – the nature of the fill (particle size distribution, granular/cohesive, particle density, shape of particles, etc. and the homogeneity of these properties during filling);
 – the geometry of the reclamation area;
 – the thickness of the lifts;
 – the existing infrastructure in the vicinity of the reclamation;
 – limitations imposed by permits, permissions, etc.;
 – the location of the weir box and the requirements with respect to the discharge water;
 – the operations on the reclamation area.

**Examples**
If the fill supply is discontinuous (i.e. transport by a TSHD or barge) there may be sufficient time between the loads to rearrange the pipeline system without affecting the production. However, should the fill be supplied continuously by a large CSD, it may be more cost effective to include control valves and branches in the pipeline in order to avoid interruptions due to the shifting of the discharge point. Another option to reduce downtime may be the use of pipes with a quick coupling system.

Short pumping distances are to be preferred in order to reduce reclamation costs. Longer pumping distances require more pumping power and thus more fuel, cause extra wear of the pipelines, may require extra (expensive) crossings with existing infra-structure, may need time-consuming permits and/or permissions or interruptions of other activities on the reclamation area and should therefore be avoided.

The large discharge capacities of the dredgers may require more than one deposition area.

A typical work method for a beach nourishment project is shown in Figure 6.14.

The photographs show the following sequence of activities:

|       |                                                                                          |
|-------|------------------------------------------------------------------------------------------|
| 1     | preparation of sand field, bulldozer pushing front bund for pipeline installation (and to prevent backflow) |
| 2, 3  | preparation of sand field, bulldozer pushing guidance bund near waterline (to prevent uncontrolled flow of material out of fill area) |
| 3     | attaching new pipe segment (bolted)                                                      |
| 4     | sand field is ready for discharging, quick fit pipe is standby                          |
| 5     | start discharging                                                                        |
| 6     | bulldozer levelling discharged material in front of the pipeline                        |
| 7     | formation of crater and erosion channels near the outlet of the pipe                    |
| 8, 9  | bulldozer and excavator installing quick fit pipe in front of the discharge line        |
| 10, 11| quick fit pipe installed, levelling of reclaimed area by bulldozer                       |
| 12, 13| discharging completed, removal of quick fit pipe                                         |
| 14    | quick fit pipe is being replaced by bolted shore pipes                                   |
| 15, 16| re-shaping bund for next sand load                                                       |
| 17    | discharging next sand load, adding another quick fit                                     |
| 18    | etc.                                                                                     |

### 6.5.3  *Rainbowing*

The method is often employed in case of limited water depth for access by THSDs, which may be the result of initial dumping or because of its natural presence. Rainbowing is often used for beach replenishment where the vessel comes close to the shore but can also be used to create an island in open water. Once this island is in place, filling is continued from this raised island using pipelines. Other applications include the crossing of existing pipelines, dikes, etc.

The distance that can be bridged by rainbowing depends on the capacity of the pumps of the dredger, the flow rate and the concentration of the sand water mixture. This distance may be as large as 150 m.

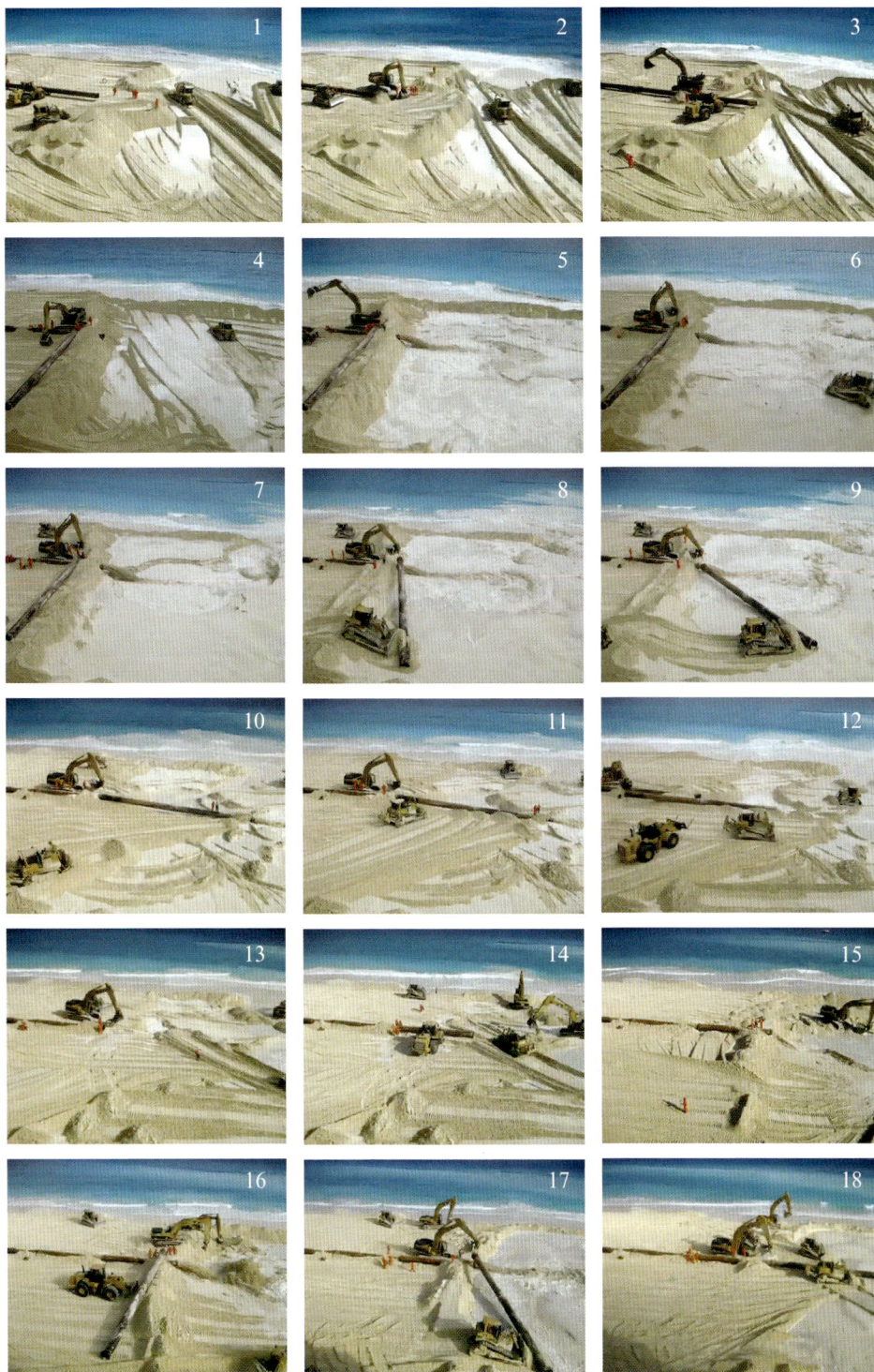

Figure 6.14 *Typical work method beach nourishment (Canon beach reclamation).*

Figure 6.15   *Rainbowing.*

Local authorities may not allow the use of this method as it often produces a spray of (salt) water mixed with fines that could cover a wide area, depending on the wind conditions.

### 6.5.4   *Spraying*

When the existing subsoil of the reclamation area consists of soft deposits with low shear strengths, the progressing face of the fill may cause mud waves, instabilities and/or inclusions of soft material within the fill see Figure 6.16.

Depending on the future use of the reclamation area and the thickness of the fill, mud waves may result in inadequate fill mass properties. Following measures can be taken to prevent these mud waves:

– Decreasing the induced shear stresses by reducing the lift thickness. Small layer thicknesses down to 0.25 m can be realised by means of a spreader pontoon. Note that spraying will require water depths that allow movement of the spreader pontoon. In particular with the latest generation of computer-controlled spreader pontoons it is possible to place a lift of a uniform, limited thickness. The speed of the spreader pontoon is continuously adjusted according to the flow rate and mixture concentration, in order to guarantee a uniform pre-set layer thickness. This is done by on on-board computer system which receives information from a flow meter and a concentration meter mounted in the last section of the pipeline. Subsequently, the same computer controls the winches on the spreader pontoon in order to obtain the required speed.
– Reducing the slope angle of the fill.
– Mechanical removal of the inclusions of soft compressible materials by hydraulic excavator.

Figure 6.16    *Mud wave on fill area.*

– Surcharging the areas containing the inclusions;
– Improvement of the quality of the soft materials by soil improvement techniques (see Chapter 7).

Alternatively, (part of) the poor quality materials may have to be removed prior
to filling.

## 6.6   Fill mass properties related to method of placement

Fill mass characteristics like density and grain size distribution are, to a certain
extent, determined by the adopted placement method. This will also result in some
variation in related geotechnical properties such as shear strength and compressibility (i.e. resistance against deformation). In most cases the density of fill placed
under water will be less than when placed above water level, see Figure 6.17. The
higher values above the water table are the result of a combination of hydraulic
deposition and bulldozer compaction.

Spraying by a moving spreader pontoon in order to place lifts of limited thickness
results generally in low densities. As a result of the difference in settling velocity
of the various grain sizes, segregation will occur leading upward to finer particle
size distribution in each lift. This may result in a relatively uniform, stratified fill
mass with low shear strength and a compressibility that may be higher than that
of clean sand.

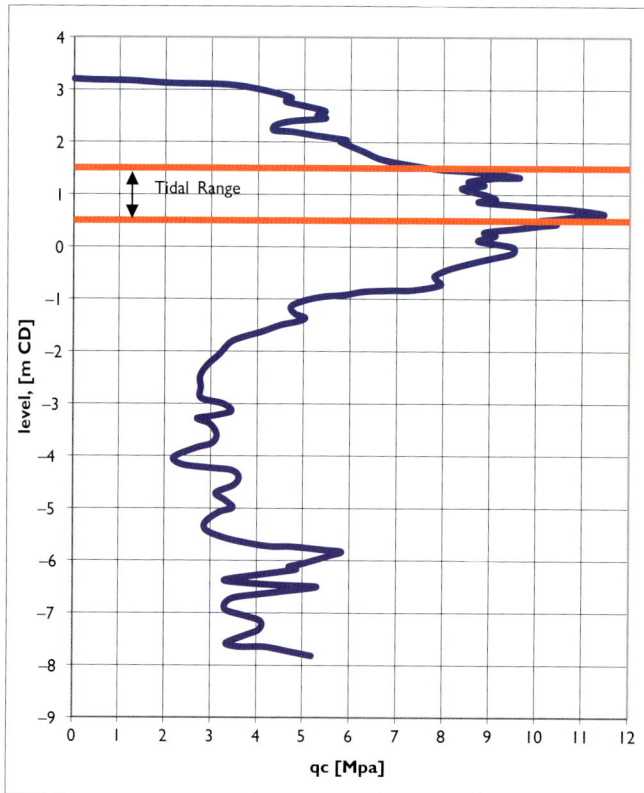

Figure 6.17    *Cone penetration test result showing a reduced cone resistance below the water level.*

Single point discharging from a pipeline, in particular above water, results generally in higher densities. The flow of the discharge water will entrain the individual particles over the surface of the fill until the coarse grains get trapped in a stable position within the soil matrix. The suspended fines will partly be caught in the pores between the coarse grains, but most will be transported until they settle after having reached calm or stagnant water. Without additional measures this will result in a fill mass with variable properties depending on the local depositional environment during placement. Typically this will result in:

- coarse grained deposits with relatively high densities (and related high shear strengths and low compressibility) near the point(s) of discharge;
- finer grained sediments with much lower densities (and corresponding low shear strengths and high compressibility) further away from this location.

In addition, the formation of craters near the outlet of the pipeline and erosion channels as a result of the flow of the sand – water mixture and the subsequent burial of these features with on-going filling may further increase the variability of the deposits.

Table 6.2   *Relative density Re as function of placement method.*

|  | Relative density (Re) | |
| --- | --- | --- |
| Placement method | Above water table | Below water table |
| Spraying | Not applicable | 20% < Re < 40% |
| Dumping | Not applicable | 30% < Re < 50% |
| Pipeline discharge (from above the waterline) | 60% < Re < 70% | 20% < Re < 45% |
| Rainbowing | no figures available | 40% < Re < 60% |

Table 6.2 presents the resulting mass properties corresponding to the various place-ment methods in terms of relative density ranges (see also section 8.3.6, Table 8.5). Because these figures depend on more variables than the placement method alone (discharge height, density initial sand-water mixture, mixture flow rate, etc.) the presented relative densities are very general and should be used with care as in practice large variations may be observed.

## 6.7   Management of poor quality materials

### 6.7.1   *Use of cohesive or fine grained materials*

Landfills are preferably constructed with well graded quartz sands. However, soil with inferior quality is often encountered during dredging. Especially silts and clays are considered as poor quality fill material.

In section 9.1 various solutions are discussed to make use of these materials and to improve their quality (ripening, use of additives, surcharge, vacuum consolida-tion, vertical drains, and sandwich structures). In view of the increasing environ-mental awareness and sand shortage in some regions, the (re-)use of silt and clays as fill material is becoming ever more important.

In case none of these methods apply, the subject material will be classified as unsuitable and will have to be disposed. The disposal can be done at offshore dumping grounds (often at large water depths) or at dedicated onshore disposal grounds. The settling ponds discussed in section 6.7.2 can be considered as such an onshore disposal ground.

### 6.7.2   *Settling ponds*

In case the fill contains too many fines and it is not allowed to return these fines with the discharge water into the surrounding environment, measures have to be taken. A common solution for hydraulic fills is to construct a settling pond adjacent to the fill area in which the fines can be collected. This is a bunded area large enough to sufficiently reduce the flow velocity of the discharge water to allow for settling of the fines. The objective is to reduce the concentration of the total suspended sediment in the return water to allowable values.

Fine particles are extremely difficult to trap. The following measures can be taken to optimise the settling within the pond:

– Raise the water level to increase the retention time. The water depth together with the width of the basin, determine the horizontal flow velocity in the pond.
– Construction of guidance bunds in the pond in order to increase the retention time of the discharge water in the pond (prevent short cuts for the flow). Such bunds will, however, also increase the flow velocity. It is therefore important to optimize such arrangements.

To reduce the settling of fines in the reclamation area the water level in this area should be kept as low as possible. This requires the installation of pumps between the reclamation area and the settling pond.

The water level in the reclamation area and settling pond is controlled with weir boxes, see Figure 6.3.

Excess water leaving the pond will freely flow (or is pumped through a pipeline) to the sea or any other receiving water body.

The area and volume required for an efficient reduction of the suspended solids can only roughly be calculated and depend on a number of parameters:

– the estimated fines content of the incoming discharge water;
– the flow rate of the incoming discharge water;
– the maximum suspended solids allowed at the outlet of the settling pond;
– the required residence time of the discharge water in the settling basin that should be sufficient for settlement of the fines;
– the critical velocity below which settling starts;
– the estimated volume and density of the fines after settling;
– the particle size and settling velocity of the suspended solids.

In defining the required volume of the settling pond the short term bulked volume is to be taken into account. This theoretical required storage volume needs to be increased with an additional volume to allow for sufficient retention time even at the end of the filling operation. Because of the complex nature of the 3D sediment transport, settling flow and self-weight consolidation, the theoretical approach is generally based on a simplified model. In addition deposition simulation tests are useful to gather insight in the settling properties of a mixture, see Appendix B.3.9.

Because of the long period required for ripening the fine material, it is not realistic to rely on the possibility of excavating settling ponds for the purpose of accommodating fines unless this is foreseen in the project planning (time, space).

In practice the volume change and catchment capability of fines of settling ponds are closely monitored and if necessary these ponds modified to satisfy the requirements. An example of a settling pond lay-out is given in Figure 6.18.

The allowable percentage of fines in return water depends on the environmental circumstances and therefore no fixed indication can be provided. As mentioned above silt ponds are meant to catch the fine fractions which do not settle initially. Additional measures may be a labyrinth lay out of the flow of tailing water to lengthen its discharge distance and enhance controlled sedimentation or add flocculants, which may be very costly.

Figure 6.18   *Example of settling pond lay-out.*

# CHAPTER 7

# GROUND IMPROVEMENT

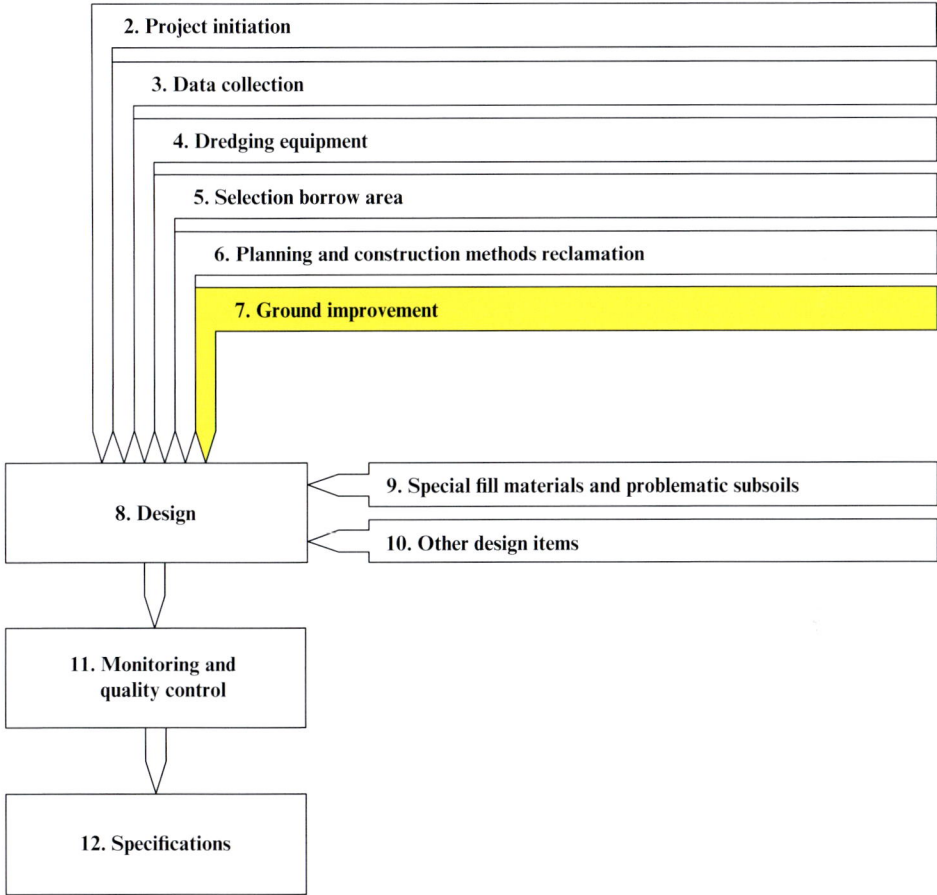

| 2. Project initiation |
| 3. Data collection |
| 4. Dredging equipment |
| 5. Selection borrow area |
| 6. Planning and construction methods reclamation |
| 7. Ground improvement |

8. Design

9. Special fill materials and problematic subsoils

10. Other design items

11. Monitoring and quality control

12. Specifications

**7 Ground Improvement**

**7.1 Introduction**

**7.2 Benefits of ground improvement**

**7.3 Overview of ground improvement techniques**

**7.4 Pre-loading with or without vertical drains**

| 7.4.1 Purpose and principle of pre-loading | 7.4.2 Vertical drains |

**7.5 Compaction**

| 7.5.1 Introduction | 7.5.2 Vibratory surface compaction | 7.5.3 Deep vibratory compaction |

| 7.5.4 Dynamic compaction techniques | 7.5.5 Explosive compaction |

**7.6 Soil replacement**

| 7.6.1 Introduction | 7.6.2 Soil removal and replacement | 7.6.3 Stone columns |

| 7.6.4 Sand compaction piles (closed end casing) | 7.6.5 Geotextile encased columns | 7.6.6 Dynamic replacement |

**7.7 Admixtures and in-situ soil mixing**

## 7.1 Introduction

When the behaviour of the fill mass (after deposition) and/or the underlying soil does not meet the required design criteria, ground improvement techniques can be implemented to improve the properties of the fill and/or subsoil.

Mitchell and Jardine (2002) gave the following definition for ground improvement (or ground treatment): "Ground treatment is the controlled alteration of the state, nature or mass behaviour of ground materials in order to achieve an intended satisfactory response to existing or projected environmental and engineering actions".

The selection of an adequate soil improvement technique requires understanding of the fundamental behaviour of soils, knowledge of various ground improvement techniques, comprehension of soil-structure interaction and acquaintance with the performance and limitations of available equipment (Chu et al., 2009). This specific combination of knowledge, expertise and experience may require consultancy of a Contractor specialised in ground improvement techniques.

Ground improvement of hydraulic fills may include the fill mass itself, the subsoil underlying the fill or both. It is typically carried out to:

- prevent excessive (differential) settlements of the surface of the reclamation when it is loaded by superstructures like buildings, roads, runways, bridges and other foundations;
- improve the shear strength of the fill and/or subsoil to ensure sufficient bearing capacity of the foundations or sufficient stability of the slopes;
- increase the density of the fill mass and/or subsoil to prevent liquefaction;
- improve the soil permeability in order to increase drainage capacity.

If design studies show that ground improvement does finally not result in the required fill mass behaviour, then the (technical) feasibility of the project may depend on the possibility to re-design the foundations of the superstructures. The input for such a design should then be based on the existing and/or improved soil conditions.

This chapter presents the purpose and principles of the most common ground improvement methods used in reclamation projects. For more techniques reference is made to Table 7.1 and the paper prepared by Chu et al. (2009).

## 7.2 Benefits of ground improvement

The feasibility of a land reclamation project often depends on the availability of sufficient, good quality fill material within a reasonable haulage distance of the site. Should this condition not be met, then ground improvement could offer a cost-effective solution to improve the properties of a fill mass constructed of otherwise non-compliant materials.

If quality control after the construction of a reclamation reveals that certain parts do not have the specified fill mass properties then it is occasionally required to remove the poor quality fill material. However, ground improvement could be a viable alternative (in respect of quality, time and cost) to repair these deficiencies without the need to replace the inferior fill by higher quality material.

Generally, ground improvement can be carried out at any location and at any time after construction of the reclamation. This makes it feasible to limit the improvement to those locations where it is actually required, for instance, along the alignment of a future road or runway or at the footprint of a structure. Should other locations also need better fill mass properties, then ground improvement can be undertaken where and whenever required.

Note that some of the ground improvement methods may not always be attractive, for instance:

- Techniques that are time-consuming: e.g., the use of vertical drains in combination with a temporary surcharge will accelerate the consolidation process considerably. This method can, however, still take considerable time which might not be available should the lay-out of a site suddenly be changed.
- Techniques that affect adjacent structures: e.g., dynamic compaction with large drop weights in the immediate vicinity of existing structures (e.g. quay walls) and infrastructure (e.g. pipelines) may induce esthetical or structural damage.

## 7.3 Overview of ground improvement techniques

Various techniques are available for improving the behaviour of the fill mass and/or the subsoil. Each method has its own advantages and disadvantages with respect to performance, time and cost. Selection of the most appropriate technique must be based on specific site conditions and the needs of the project. Specialist Contractors may have to be consulted to evaluate various options. A well-managed ground improvement operation, appropriate to the site, will enhance the prospects of in-time and safe project delivery (Das, 2009).

Ground improvement techniques may be classified in different ways. A distinction can be made between methods that can be applied from ground surface and techniques that have to be carried out from a certain depth below ground surface. Alternatively the classification is based on the nature of the material that can be treated by ground improvement methods. Often a distinction is made between cohesive (clay, silt) and granular (sand, gravel) materials. The Technical Committee 17 of the ISSMGE (Chu *et al.,* 2009) has presented another classification that uses two criteria: the behaviour of the ground to be improved and the use of admixtures. Table 7.1 presents an overview of the methods classified according to this system.

Table 7.1  *Classification of ground improvement methods adopted by TC17 of the ISSMGE (Chu et al., 2009).*

| Category | Method | Principle |
|---|---|---|
| A. Ground improvement without admixtures in non-cohesive soils or fill materials | A1. Dynamic compaction | Densification of granular soil by dropping a heavy weight from air onto ground. |
| | A2. Vibrocompaction | Densification of granular soil using a vibratory probe inserted into ground. |
| | A3. Explosive compaction | Shock waves and vibrations are generated by blasting to cause granular soil ground to settle through liquefaction or compaction. |
| | A4. Electric pulse compaction | Densification of granular soil using the shock waves and energy generated by electric pulse under ultra-high voltage. |
| | A5. Surface compaction (including rapid impact compaction). | Compaction of fill or ground at the surface or shallow depth using a variety of compaction machines. |
| B. Ground improvement without admixtures in cohesive soils (also see Table 4) | B1. Replacement/ displacement (including load reduction using light weight materials) | Remove bad soil by excavation or displacement and replace it by good soil or rocks. Some light weight materials may be used as backfill to reduce the load or earth pressure. |
| | B2. Preloading using fill (including the use of vertical drains) | Fill is applied and removed to pre-consolidate compressible soil so that its compressibility will be much reduced when future loads are applied. |
| | B3. Preloading using vacuum (including combined fill and vacuum) | Vacuum pressure of up to 90 kPa is used to pre-consolidate compressible soil so that its compressibility will be much reduced when future loads are applied. |
| | B4. Dynamic consolidation with enhanced drainage (including the use of vacuum) | Similar to dynamic compaction except vertical or horizontal drains (or together with vacuum) are used to dissipate pore pressures generated in soil during compaction. |
| | B5. Electro-osmosis or electro-kinetic consolidation | DC current causes water in soil or solutions to flow from anodes to cathodes which are installed in soil. |
| | B6. Thermal stabilisation using heating or freezing | Change the physical or mechanical properties of soil permanently or temporarily by heating or freezing the soil. |
| | B7. Hydro-blasting compaction | Collapsible soil (loess) is compacted by a combined wetting and deep explosion action along a borehole. |

(*continued*)

Table 7.1 (*Continued*).

| Category | Method | Principle |
|---|---|---|
| C. Ground improvement with admixtures or inclusions | C1. Vibro replacement or stone columns | Hole jetted into soft, fine-grained soil and back filled with densely compacted gravel or sand to form columns. |
| | C2. Dynamic replacement | Aggregates are driven into soil by high energy dynamic impact to form columns. The backfill can be either sand, gravel, stones or demolition debris. |
| | C3. Sand compaction piles | Sand is fed into ground through a casing pipe and compacted by either vibration. dynamic impact, or static excitation to form columns. |
| | C4. Geotextile confined columns | Sand is fed into a closed bottom geotextile lined cylindrical hole to form a column. |
| | C5. Rigid inclusions (or composite foundation, also see Table 5) | Use of piles, rigid or semi-rigid bodies or columns which are either premade or formed in-situ to strengthen soft ground. |
| | C6. Geosynthetic reinforced column or pile supported embankment | Use of piles, rigid or semi-rigid columns/inclusions and geosynthetic girds to enhance the stability and reduce the settlement of embankments. |
| | C7. Microbial methods | Use of microbial materials to modify soil to increase its strength or reduce its permeability. |
| | C8. Other methods | Unconventional methods, such as formation of sand piles using blasting and the use of bamboo, timber and other natural products. |
| D. Ground improvement with grouting type admixtures | D1. Particulate grouting | Grout granular soil or cavities or fissures in soil or rock by injecting cement or other participate grouts to either increase the strength or reduce the permeability of soil or around. |
| | D2. Chemical grouting | Solutions of two or more chemicals react in soil pores to form a gel or a solid precipitate to either increase the strength or reduce the permeability of soil or around. |
| | D3. Mixing methods (including premixing or deep mixing) | Treat the weak soil by mixing it with cement, lime, or other binders in-situ using a mixing machine or before placement |
| | D4. Jet grouting | High speed jets at depth erode the soil and inject grout to form columns or panels |
| | D5. Compaction grouting | Very stiff, mortar-like grout is injected into discrete soil zones and remains in a homogenous mass so as to density loose soil or lift settled ground. |
| | D6. Compensation grouting | Medium to high viscosity particulate suspensions is injected into the ground between a subsurface excavation and a structure in order to negate or reduce settlement of the structure due to ongoing excavation. |

(*continued*)

Table 7.1   (*Continued*).

| Category | Method | Principle |
|---|---|---|
| E. Earth reinforcement | E1. Geosynthetics or mechanically stabilised earth (MSE) | Use of the tensile strength of various steel or geosynthetic materials to enhance the shear strength of soil and stability of roads, foundations, embankments, slopes, or retaining walls. |
| | E2. Ground anchors or soil nails | Use of the tensile strength of embedded nails or anchors to enhance the stability of slopes or retaining walls. |
| | E3. Biological methods using vegetation | Use of the roots of vegetation for stability of slopes. |

Table 7.2 shows a summary of the techniques that are most relevant for reclamation projects. Some of these techniques will be discussed in more detail in this chapter.

## 7.4   Pre-loading with or without vertical drains

### 7.4.1   *Purpose and principle of pre-loading*

Pre-loading (or surcharging) of a fill mass or subsoil is generally applied to minimise post-construction settlements when (highly) compressible, soft soil layers are encountered at the project location and significant settlements are expected as a result of the weight of the fill and future structural loads imposed on the reclamation. Pre-loading can be achieved by the application of a temporary surcharge with fill material to accelerate the settlement process which is the most effective method for clayey and organic soils. During the execution of the works, the thickness of the preload can be adjusted if required (based on monitoring results) which makes surcharging a flexible solution. The principle of a surcharge and the subsequent increase of the settlement rate are presented in Figure 7.1. Further reference is made to section 8.5.3.4.

**Surcharge with sand or other material**
A surcharge usually consists of a temporary load of sand or other material, which is placed on top of the reclaimed area that needs to be consolidated. Once sufficient consolidation (usually expressed in terms of settlement or dissipation of excess pore pressure) has taken place, the surcharge can be removed and construction activities may start. The thickness of a surcharge may vary considerably, but will generally be in the range of 2–10 m (i.e., a surcharge of approximately 35 kPa – 180 kPa when placed above the water table).

A surcharge consisting of sand is generally easy to place and the most cost-effective method to accelerate the settlement process. This is particularly true if the material

Table 7.2  Overview of ground improvement techniques most relevant for reclamation projects.

| Method | Techniques | Soil types | Application depth | Treatment depth | Suitable for | | Improvement | | | |
|---|---|---|---|---|---|---|---|---|---|---|
| | | | | | Subsoil | Fill | Settlement behaviour | Strength/ stability | Liquifaction | Drainage capacity |
| Consolidation (section 7.4) | Pre-loading with or without vertical drains | clay, peat, silt, but also compressible materials such as carbonate sands | Drains at depth, surcharge at the surface (sand) or at depth (atmospheric pressure) | Up to 30–60 m | • | • | • | • | | •(PVD) |
| Compaction (section 7.5) | Vibratory compaction: vibratory roller | granular material | At the surface | Up to 0.5–1.0 m | | • | • | • | | |
| | polygonal drum compactor | granular and cohesive materials | At the surface | Up to 1.5–3.0 m | | • | • | • | • | |
| | vibroflotation | granular material (<15% fines) | At depth | >30 m | • | • | • | • | • | • |
| | vibratory probes | granular material | At depth | 10–15 m | • | • | • | • | • | |
| | Dynamic compaction techniques: dynamic compaction | granular material | From the surface | Up to 8–12 m | • | • | • | • | • | |
| | rapid impact compaction | granular material | From the surface | Up to 6–7 m | • | • | • | • | • | |
| | high energy impact compaction | granular material | From the surface | up to 2–4 m | • | • | • | • | • | |

| | | | | | | | | | | | | | | |
|---|---|---|---|---|---|---|---|---|---|---|---|---|---|---|
| Soil replacement (section 7.6) | Soil removal and replacement | (very) soft cohesive soil | from seabed/surface | 0–30 m | • | • | • | • | • | • | • | • | • | • |
| | Stone columns | gravel, sand, silt and clay | at depth | 20–30 m | • | • | • | • | • | • | • | • | • | • |
| | Sand compaction piles | gravel, sand, silt and clay | at depth | 20–30 m | • | • | • | • | • | • | • | • | • | • |
| | Geotextile encased sand columns | clay, peat | at depth | typically 10–15 m | • | • | • | • | • | • | • | • | • | • |
| | Dynamic replacement | gravel, sand, silt and clay | at depth | up to 6–7 m | • | • | • | • | • | • | • | • | • | • |
| | Soil removal and replacement | all, mainly very soft soils | at the surface | n/a | • | • | • | • | • | • | • | • | • | • |
| Soil mixing (section 7.7) | Admixtures (e.g. cement and lime stabilization), in-situ soil mixing | | | | | | | | | | | | | |
| | shallow soil mixing | sand, soft clay, silt and organic | both at depth and at the surface | ≤12 m (SSM) | • | • | • | • | • | • | | • | • | • |
| | deep soil mixing | sand, soft clay, silt and organic | both at depth and at the surface | 3–50 m (DSM) | • | • | • | • | • | • | | • | • | • |

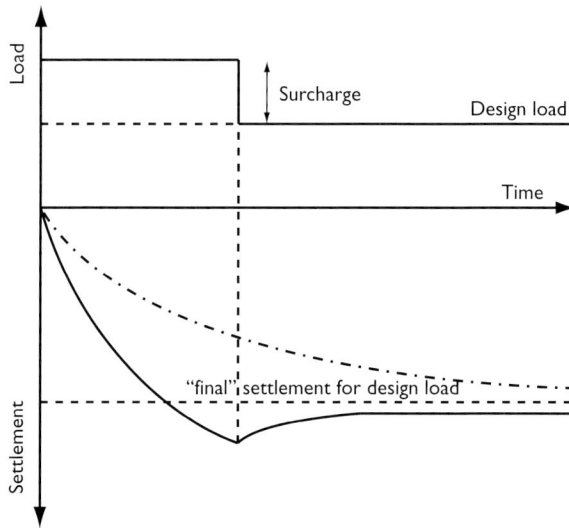

Figure 7.1    *Principle of accelerating the settlement process by pre-loading.*

can be re-used in the next project phase or when it can be sold after completion of the consolidation process. However, because of the increased load, instability of the underlying soft subsoil may be induced during placement of the surcharge and – as a consequence of its presence – it may also prevent other construction activities on the reclamation site.

During the design phase analyses have to be made to predict the settlements (see section 8.5.3.2) as a function of the time and the surcharge load and, if applicable, the spacing of the vertical drains (see section 7.4.2). Input for these calculations has to be derived from the results of site investigations and laboratory testing (see section 3.4.2.2).

During execution the actual thickness of the surcharge and the settlement has to be monitored (see section 11.3.3.2). The monitoring results need to be compared with the predictions made during the design phase. If required, parameters have to be adapted in order to fit the predicted time – settlement behaviour with the measured data. Timely removal of the surcharge is imperative to prevent more settlement (hence reduction in fill volume) than strictly required to achieve the fill profile as the surcharge load is likely to exceed the design load. On the other hand, removing the surcharge too early will result in excessive residual settlements. The decision to remove a temporary surcharge is based on the achievement of a preferred degree of consolidation of the cohesive subsoil or the realization of a certain percentage of the expected total settlement. It is therefore important to frequently monitor the settlements during placement of as well the fill as the temporary surcharge. The effect of a surcharge should be such that the residual settlement after construction is limited to an allowable value that suits the operation during the design life of the land reclamation. Future permanent or live load conditions may also be considered in the design of a temporary surcharge.

**Vacuum consolidation**

An interesting alternative surcharge could be the application of atmospheric pressure. A technique known as vacuum consolidation mobilises the atmospheric pressure by actively removing the pore water from the soil mass by pumping. As long as the treated soil mass is sealed from atmospheric conditions, pumping will reduce the pore pressures while increasing the effective stress in the soil mass. Although this technique is generally more expensive, less robust and more vulnerable to operational problems (breakdown pumps, leaking membranes, presence of aquifers in the soil mass to be treated, etc.), it may have some advantages above a traditional surcharge:

– the atmospheric pressure will exert an isotropic load without introducing shear stresses in the subsoil and will, therefore, not affect the stability of the soft strata;
– a temporary surcharge is not necessary: just switching on the pumps will introduce the load and once the pumps are switched off the surcharge will disappear;
– some vacuum consolidation methods allow for immediate access to the site after installation of the system, which enables a combination with a temporary surcharge of sand or the start of construction activities.

The maximum theoretical surcharge that can be achieved by any vacuum system equals 100 kPa, but in practice this value will be significantly less as a result of losses in the system. Most methods operate in combination with vertical drains and may differ from each other in the way the soil mass is isolated from atmospheric conditions. Reference is made to Chu *et al.* (2009).

### 7.4.2   *Vertical drains*

Pre-loading alone has the disadvantage that the surcharge must remain in place for months or even years, often delaying further construction operations. A long consolidation period is generally the result of the thickness and the low permeability of compressible strata hampering the dissipation of excess pore pressures (see Section 8.5.3.4).

By installing vertical drains the length of the drainage path for the pore water can be reduced considerably which will accelerate the consolidation process and, hence, the settlement rate (see Figure 7.2).

Appendix D, section D.3.2 presents the calculation of the consolidation period as a function of the coefficient of consolidation of the compressible strata, the drain spacing and the drain diameter. To a certain extent the consolidation rate can be controlled by adjusting the drain spacing.

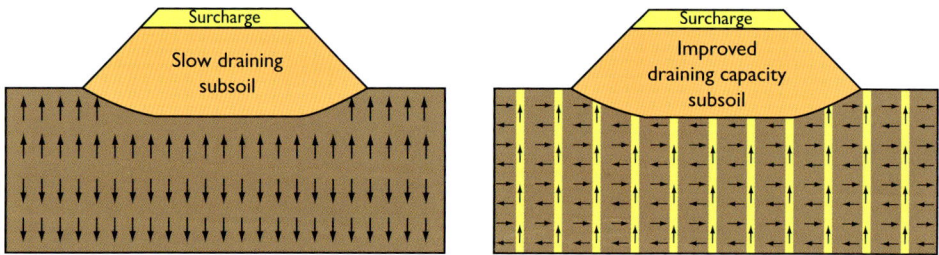

Figure 7.2 *Pre-loading of subsoil and effect of vertical drains (after Stapelfeldt, 2006).*

Vertical drains may comprise prefabricated band drains or wick drains (maximum installation depth of approximately 60 m) or bored columns filled with sand (maximum installation depth approximately 30 m). The drains are normally installed in a triangular grid with a grid spacing between 1 to 3 m.

The drains must be able to discharge the extracted pore water into a free-draining medium (sand or gravel layer, horizontal drains, etc.), usually—but not necessarily—located at the top of the vertical drain. It is essential to ensure that such a drainage blanket has sufficient permeability to handle the extracted water during all stages of the construction works. The hydraulic head within this horizontal layer or drain must be less than the hydraulic head in the vertical drain in order to allow the pore water to be removed from the compressible soil mass.

Buckling of prefabricated drains or shearing of sand columns as a result of (large) settlements may reduce the discharge capacity of the drains. Most product specifications of prefabricated drains include minimum discharge capacity at buckling.

Installation of prefabricated vertical drains is not only carried out on land (see Figure 7.3), but also over water using floating pontoons or barges to accommodate the drain stitchers (see Figure 7.4). The installation process over water is, however, slower than drains installed on land.

Vertical drains are normally installed when very soft to soft subsoil conditions are encountered at the reclamation site. If such a material is present at ground surface (above the water table) then it is recommended to first install a fill layer of minimum 1 to 2 m thickness in order to allow access for the drain installation equipment. The stability of the equipment used for the installation of the drains shall be checked before the start of the works in order to avoid accidents.

The European Standard "EN 15237:2007, Execution of Special Geotechnical Works, Vertical drainage" describes the installation and testing procedures of prefabricated vertical drains.

Figure 7.3 *Drain installation on land (left) and extracted pore water escaping at top of drain (right).*

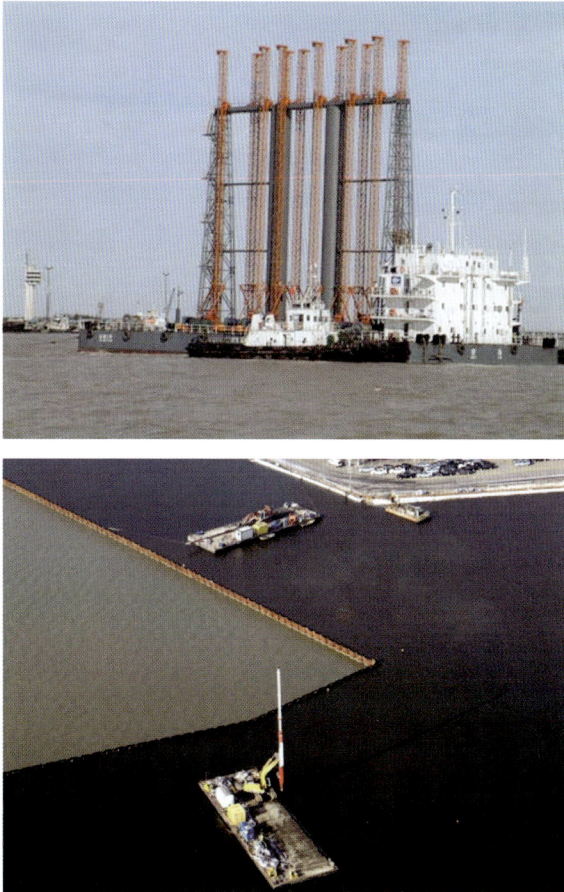

Figure 7.4 *Offshore vertical drain installation barge (upper picture, after Yan et al., 2009), and vertical drain installation over water (lower picture) at Osthafen, Bremerhaven, using floating pontoons.*

## 7.5 Compaction

### 7.5.1 Introduction

The engineering properties of a granular soil, such as compressibility, shear resistance and resistance against liquefaction, mainly depend on the state of compaction, typically expressed in terms of (relative) density and effective stress. Increasing the density by compaction will generally improve these mass properties. Various methods exist to compact the soil which can be divided into:

– vibratory techniques including surface compaction and deep compaction;
– dynamic compaction techniques based on (heavy) impact at the surface;
– explosive compaction.

### 7.5.2 Vibratory surface compaction

**Purpose and principle**
Vibratory surface compaction is a compaction method undertaken at the surface by vibratory rollers, plates and/or tampers to increase the density (and thus the shear strength) and, in particular, the stiffness of the fill. It is often used in road and airfield construction to compact the sub base and base courses consisting of granular material such as rock fill and sand or in foundation construction to compact the soil underneath shallow foundations. While moving forward, the vibratory rollers combine static pressure and dynamic loading (vibrations with low amplitude and high frequency) that re-arranges the grains into a denser state. Non-vibrating pad foot rollers and tyre rollers are more effective in cohesive materials as a result of their kneading action.

**Compaction method**
Various types of equipment are available: single-drum vibratory rollers, towed single-drum vibratory rollers, vibratory tandem rollers, hand-held tampers and vibratory plate compactors. Most of the vibratory rollers have a smooth drum and are suitable for granular material only. Vibratory rollers have weights that typically range between 4 and 25 tonnes and generally move at speeds not exceeding 6 km/hr.

A special development is the polygonal drum roller consisting of three octagonal drum elements placed next to each other (see Figure 7.5). According to the manufacturer the combination of this special drum shape (weight 14 to 26 tonnes) with the high dynamic forces enables an optimum introduction of the compression and shear waves into the soil.

Figure 7.5    *A polygonal drum compactor (after Mengé, 2007): Observe the impacted surface of the ground.*

**Effects and limitations**

The influence depth of vibratory rollers is limited and varies between 0.30 m and 0.80 m, depending on the type of soil to be compacted and the equipment used. In practice vibratory rollers are only used for fill layers of limited thickness. However, discharge capacities of modern dredgers result in layer thicknesses not less than 2–3 m. This implies that the use of vibratory rollers is limited to the uppermost layer only.

Depth compaction trials with the polygonal drum roller (Weingart *et al.,* 2003) indicated an increase of density to a degree of compaction of 95% MDD at a depth of more than 1.30 to 1.50 m. Compaction performance was reported as 500–1500 m³/hr. As a result of the special drum shape, the roller is suitable for granular, mixed and cohesive soils.

### 7.5.3    *Deep vibratory compaction*

#### 7.5.3.1    General

The principle of all deep vibratory compaction techniques is based on the re-arrangement of grains into a more dense state as a result of vibrations transmitted by vibrating probes that are inserted into the ground. Deep vibratory compaction has been developed to compact granular soils to depths of more than 30 m. The deep vibratory methods are limited to granular soils, usually with a maximum fines content of 10–15%. Figure 7.6 indicates the compactability of soils based on results of Cone Penetration Tests.

Compaction points (e.g., the locations where the probes are inserted into the ground) are most commonly arranged on a triangular grid pattern over the area to be compacted (see Figure 7.7).

Operational parameters such as the grid spacing of the compaction points, vibration time and the rate of withdrawal of the probe are in most cases determined during compaction trials undertaken at the project site itself.

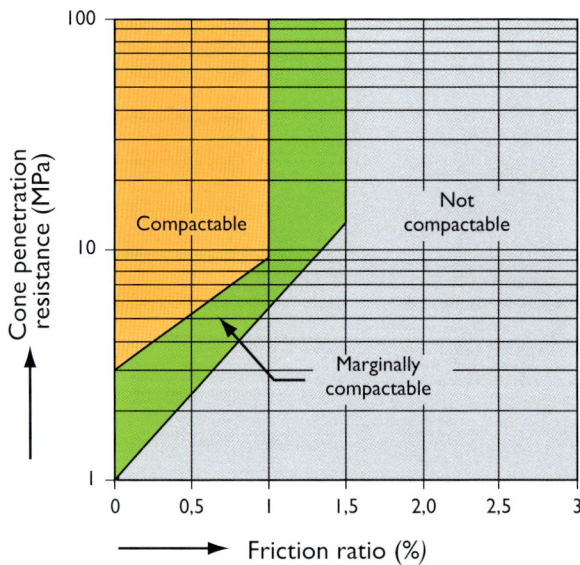

Figure 7.6    *Compactability chart for deep compaction based on CPT data (Massarsch et al., 2005).*

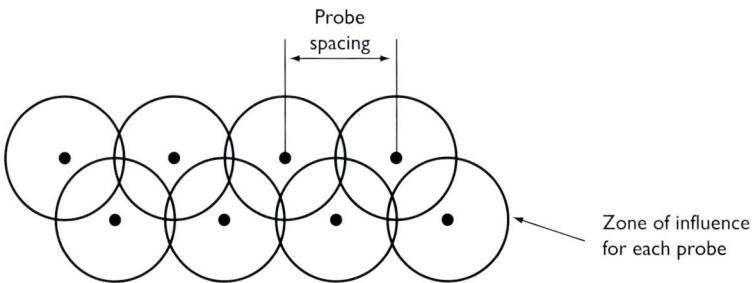

Figure 7.7    *Zone of influence and probe spacing for vibroflotation.*

The main difference between the various techniques of deep vibratory compaction is the design of the probe. Based on this design deep vibratory compaction techniques can be divided into:

–  vibratory probes without water or air jets vibrating in vertical direction;
–  vibroflotation with water and/or air jets vibrating in horizontal direction.

For a first rough estimate of the grid spacing sometimes simple charts showing the achievable sand densification (frequently given in Relative Density) as a function of the vibrator grid spacing are used. An example of such a chart is given in Figure 7.8.

Some of these charts may work well for a specific machine type in specific soil conditions. However, the variations in soil conditions and compaction equipment are so large that the use of these charts for the purpose of designing and specifying a project is discouraged. Special precautions need to be taken with compressible

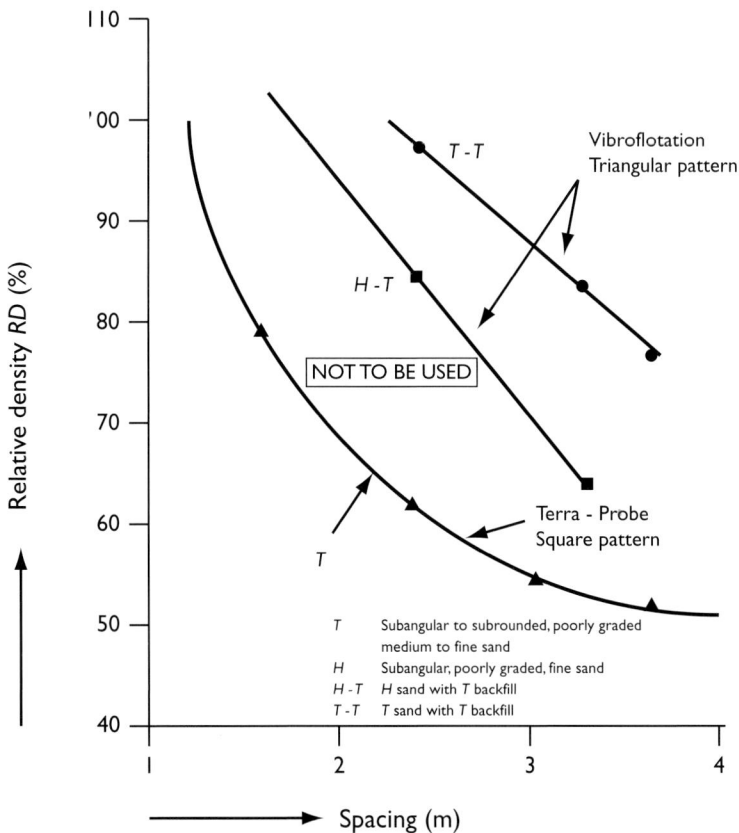

Figure 7.8  *Design graph for a preliminary assessment of the probe spacing (Brown, 1976).*

159

Pattern No. A          B          C          D          E

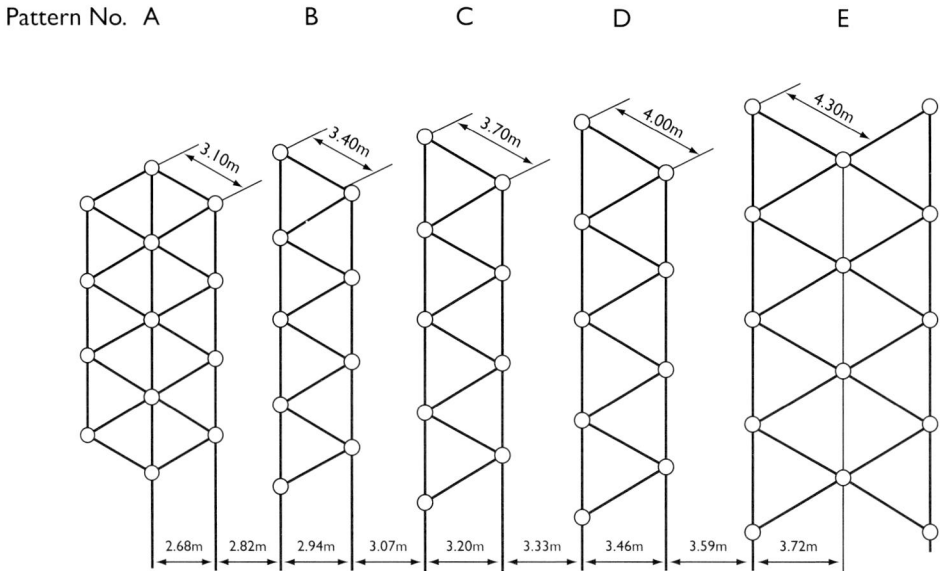

Figure 7.9    *Example trial compaction.*

or crushable soils such as carbonate sands which may even require a more closely spaced grid (see section 9.2). It is therefore strongly recommended to perform a trial compaction rather than guessing the grid from charts. An example for such a trial compaction is presented in Figure 7.9.

**Quality control**
As a consequence of the use of probes, the compaction method will create vertical columns of compacted material. The degree of compaction will decrease with increasing horizontal distance to the insertion point of the probe resulting in non-homogeneous compaction over the treated area. In situ tests, such as the Cone Penetration Test may be carried out to verify the required quality of the soil improvement. Other methods are described in the European Standard EN 14731:2005 Execution of special geotechnical works—ground treatment by deep vibration.

Quality control may be based on a weighted profile of the cone resistances measured in CPTs executed at, for instance, a number of two specified positions within the compaction grid. Figure 7.10 shows examples of such locations. The use of a moving average value (e.g. over a depth of 0.5 to 1.0 m) of the mean cone resistance ensures a rational evaluation of the compaction results as it averages the effect of incidental thin horizons of less compacted fill in the quality assessment. The actual positions of the CPTs for verification of the achieved improvement should preferably be specified in the Contract documents in order

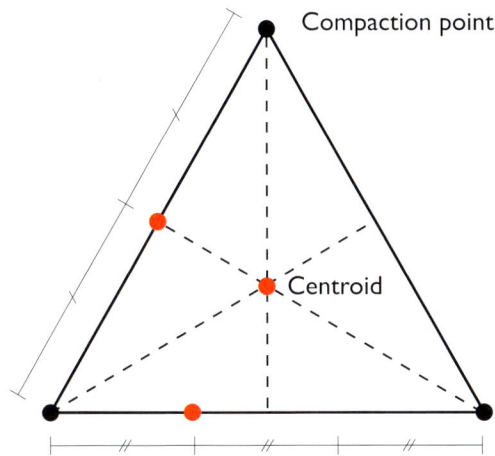

Figure 7.10  *Example of possible post-vibroflotation
test locations (coloured dots) related to a
triangular compaction grid (black dots).*

to avoid discussions during the execution of the works. The evaluation report regarding the compaction efforts shall also include the unprocessed data of the CPTs for reference. It is further important to accurately localize the actual coordinates of the pre- and post-compaction test locations.

## 7.5.3.2   Vibratory probes without jets

**Purpose and principle**

The principle of vibratory probes is based on compaction of granular material as a result of vertically polarised waves transmitted into the soil by a long vibrating probe. The vibrations are generated by a heavy vibrator clamped at the upper end of the probe exciting the probe over its full length in a vertical direction. The probe is usually suspended from a crane or guided mast. Modern vibrators are hydraulically driven and often the vibration frequency can be varied during operation. Vibratory probes are developed to compact granular material to depths ranging from 10 to 15 m.

Different types of compaction probes have been developed in Japan, North America and Europe (Massarsch 1991). The geometric shapes of simple probes such as steel tubes or H-beams are not very efficient for soil compaction. Therefore, special probe shapes such as the Tristar probe and double Y-probe (see Figure 7.11) have been constructed to improve the compaction effect of the vibratory probe.

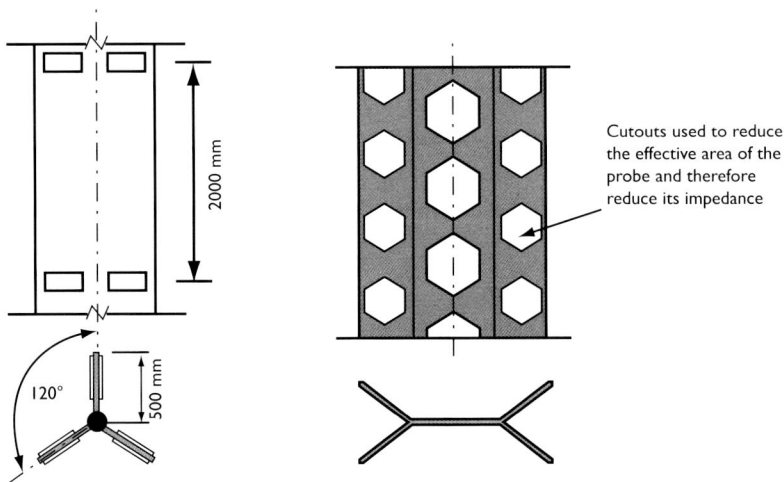

Figure 7.11    *Left: Star profile (Van Impe, 1989); Right: Double Y-probe (Van Impe et al., 1994).*

Research by Massarsch has led to an improved energy transmission using resonance vibro-compaction. The key features of the resonant compaction technique are the use of a specially designed compaction probe (double Y with adapted impedance) and a heavy vibrator with variable operating frequency on top of the probe. After probe insertion, the frequency of the vibrator is adjusted to the resonance frequency of the soil layer, thereby amplifying the ground response (Massarsch, 2002).

**Compaction method**
The probe is inserted into the soil and vibrated down to the maximum depth. Withdrawal of the probe is realised in a vertically oscillating manner. The optimum duration of the vibration, rate of withdrawal and the spacing between the compaction points are generally determined during compaction trials undertaken prior to the execution of the works. The maximum depth of compaction depends on the capacity of the vibrator and size of the crane or piling rig. The spacing between insertion points of the probe ranges typically between 1.5 m and 4.0 m and is generally smaller than for vibroflotation.

**Effects and limitations**

The results of the compaction are generally measured by Cone Penetration Testing (or Standard Penetration Testing) carried out before and after the compaction activities.

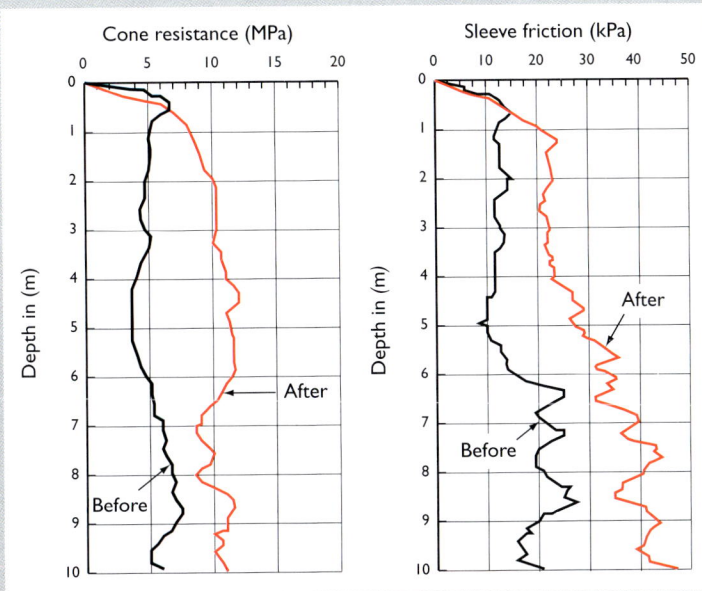

Figure 7.12   *Example of filtered average values of the cone resistance and sleeve friction before and after vibratory compaction of a hydraulic fill (Massarsch et al., 2002).*

### 7.5.3.3   Vibroflotation

**Purpose and principle**

Vibroflotation relies on the fact that particles of non-cohesive soil can be re-arranged into a denser state under the influence of vibrations from specially designed vibratory probes. The action of the vibratory probe, usually accompanied by water (and air) jetting to temporarily reduce the inter-granular forces between the soil particles, allows the grains to move into a more compact configuration. Vibroflotation is used for in-situ compaction of thick layers of loose granular soil deposits to depths that may exceed 30 m.

In loose to medium dense saturated sands, the strong ground vibrations in the imme-diate vicinity of the probe will result in an increase of pore-water pressure in the soil column surrounding the vibratory probe, which is considered to lead to a state of cyclic mobility of the soil mass. Compaction with vibratory probes is limited to granular materials only. An increase of the fines content will reduce the permeability and, hence, the efficiency of the compaction method. The efficiency of vibroflotation is generally higher than the efficiency of vibratory probes (without water or air jets).

Figure 7.13   *Crane mounted with tandem vibratory probes.*

## Compaction method

The process of vibroflotation involves the use of a vibro-flot (see Figure 7.14), which is the vibratory unit at the end of a probe that generates horizontal vibrations. The probe weighs 15–40 kN, has a length of 2 m to 5 m and is usually suspended from a crane. At the side and the tip of the vibrating unit openings are located for water or air jets. The compaction process can be divided into the following stages:

– The water jet at the bottom of the vibro-flot is turned on and creates a "quick sand" condition in the soil allowing the vibrating probe to sink to its anticipated depth under its own weight;
– The probe is raised stepwise. At prescribed vertical intervals the probe is halted for a short time period (approximately 30 to 60 seconds) to allow for vibrating. This process compacts the soil into a denser state, in order to achieve a required density.

The zone of compaction around a single probe, and thus the required probe spacing, will vary with a radius ranging from 3 to 5 m according to the nature of the subsoil and the type of vibroflot used. As for most compaction techniques, a trial must be executed at the actual project site to establish the optimum grid spacing, frequency and amplitude, and the vertical vibration interval to achieve the required degree of compaction.

Figure 7.14    *Crane mounted with vibratory probe, the vibroflot (info Keller).*

The European Standard "EN 14731:2005, Execution of Special Geotechnical Works, Ground treatment by deep vibration" provides information on the planning, execution, testing and monitoring of Vibroflotation.

**Effects and limitations**

The effect of soil compaction with vibroflotation can be checked with a Standard Penetration Test (SPT) or a Cone Penetration Test (CPT) (see Figure 7.15).

Depending on the initial density of the material and the vibroflot used, typical values for the expected settlement of the treated layer may range between 5% and 10% of its untreated thickness. However, the success of in-situ compaction by vibroflotation depends on several factors. The most important of these is the grain-size distribution of the soil. Only soils with a fines content less than approximately 15% can generally be compacted by vibro-compaction (see also Figure 7.22). If the fines content exceeds this percentage it may become more difficult to achieve the required densities. Compaction of soil with only thin intermediate layers of fines is generally not problematic because of the redistribution of particles during the compaction process.

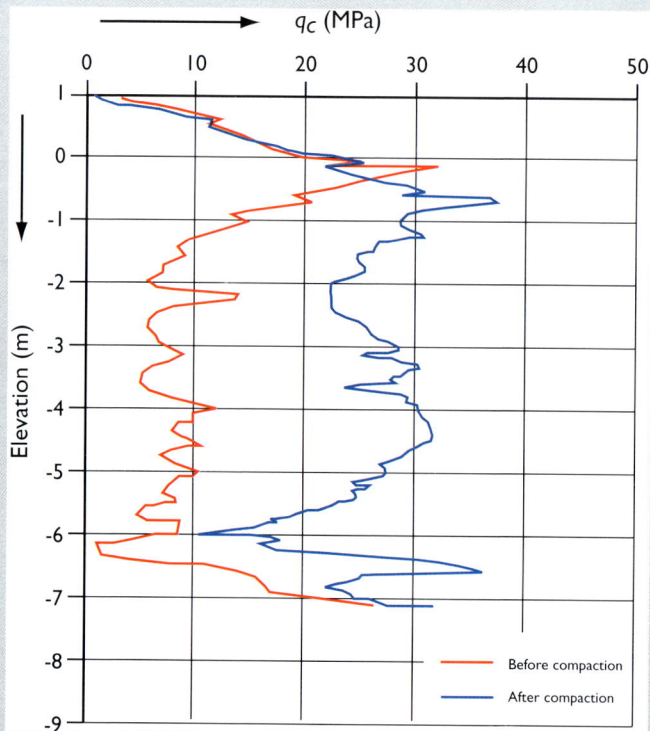

Figure 7.15  *Example of the effect of vibroflotation on the cone resistance of a hydraulic fill (Mecsi et al., 2005).*

Figure 7.16  *Compaction of the slope of a reclamation in Singapore (info Keller).*

Compacting the material in a slope along the boundary of a reclamation by means of vibroflotation can prove to be very difficult. Special measures may have to be taken such as the placement of a temporary fill layer overlying the slope which has to be removed after the compaction activities.

The vibroflotation technique enables compaction of deeper layers. The top layer (upper 1 to 2 m), however, will not be compacted sufficiently. Additional surface compaction is normally required after vibroflotation operations.

### 7.5.4  Dynamic compaction techniques

**Purpose and principle**
Dynamic compaction is a technique used for in-situ densification of granular soil deposits by heavy impact. This technique primarily involves dropping a heavy weight repeatedly on the ground surface at a regular grid of impact points. Compaction of the fill or subsoil is achieved by a re-arrangement of the grains as a result of the shear strains induced by the impact of the heavy drop weight. Physical displacement of particles and, to a lesser extent, low-frequency excitation will reduce the void ratio and increase the relative density. The stress waves generated by the weight contribute to the densification process. The effectiveness of the method decreases with increasing compressibility (or crushability) of the materials to be compacted.

**Compaction method**
Common dynamic compaction techniques include the Dynamic Compaction (or Heavy Tamping) method, the Rapid Impact Compaction (RIC) and the High Energy Impact Compaction (HEIC) techniques.

Dynamic Compaction (DC) or Heavy Tamping is mainly used for deep compaction and literally involves dropping a weight (or pounder) that is suspended from a crane onto the ground. The depth of treatment depends on the soil type, the weight of the pounder in tonnes and the drop height in metres. The drop weight varies from 8 to 35 tonnes and the drop height varies between 7 m and 30 m. The degree of compaction achieved also depends on the grid spacing between the impact points and the number of impacts per location (Das, 2009). The maximum depth of treatment may be approximately 12 m (see Figure 7.17) but will depend on the soil characteristics and operational variables such as drop height, drop weight, number of impacts, etc. Generally, trials must be undertaken to determine the optimum operational variables.

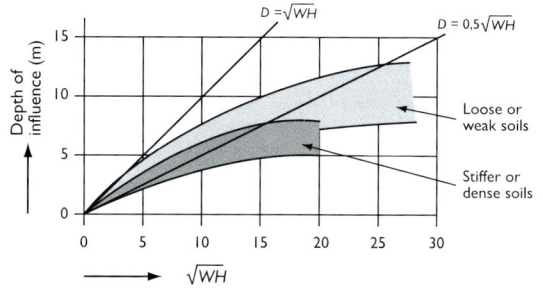

Figure 7.17 *Depth of influence D (m) of dynamic compaction as a function of the drop weight W (tonnes) and fall height H (m) (Moseley, 2004).*

Figure 7.18 *A rapid impact compaction machine with 16-tons hammer shifting between impact locations.*

Dynamic Compaction tends to disturb and loosen the top layers even though it compacts the deeper layers. Often different passes are required to compact the deeper layers with heavy impact on a wide grid, followed by lower impact in a narrower grid to improve layers closer to the surface.

Rapid Impact Compaction (RIC) is a surface compaction technique involving a hydraulic hammer that is dropped on a steel footing placed on the ground surface (see Figure 7.18). The hydraulic hammer is attached to a crane that provides the hydraulics for the hammer. The weight of the hammer ranges between 9 and 18 tonnes, while the diameter of the footing may vary between 1.5 m and 2.5 m. The stroke of the hammer may be adjustable, but the maximum value of the latest generation is limited to approximately 1.5 m. Impact frequency varies from 40 to 80 impacts per minute. The influence depth may be 6 to 7 m but will strongly depend on the type of material to be compacted, the soil stratification, the size of the footing, the weight and stroke of the hammer, grid spacing, number of impacts per point, etc. Should a second pass be required, this will generally be undertaken on a staggered grid. Productions depend on the requirements and site conditions, but generally range between 1,500–5,000 m²/12 hrs. Like for other compaction techniques, trials need to be carried out to establish the optimum configuration and dimensions.

High Energy Impact Compaction (HEIC) is used for surface compaction and consists of providing repeated high energy impacts at the ground surface by towed, rotating, cam-shaped, rectangular or pentagonal steel drums of 8 to 16 tonnes in weight from drop heights ranging from 0.15 to 0.25 m (see Figure 7.19). Although best suited for granular soils, the HEIC method has been effective in clayey subgrades as well (Chu *et al.*, 2009). The optimum towing speed (usually 10–12 km/hr), shape and weight of the steel drum and the number of passes must be established by trials on site. The maximum achievable depth of treatment ranges generally from 1.5 to 2.5 m. Production rates may be as high as 6000–9000 m²/12 hours. Because the magnitude of the impact depends, to a certain extent, on the towing speed, this technique is not useful for irregularly shaped or small reclamation areas.

Figure 7.19   *HEIC with rotating cam shaped drums.*

**Effects and limitations**

The response of granular and cohesive soils when subjected to high-energy impacts is fundamentally different. Dynamic compaction literally squeezes water out of the soil to effectively pre-load the ground (Moseley *et al.*, 2004). Granular and dry cohesive fills respond well to dynamic compaction. The response of clays, however, is more complex than that of granular soils. Therefore care must be exercised in the treatment of weak natural clayey soils or clay fills below the water table. Their low permeability will prevent pore water to escape from the soil mass. Nevertheless, claims have been made that, in particular, HEIC may be effective in clayey soils as well (Chu *et al.*, 2009).

### 7.5.5 *Explosive compaction*

Explosive Compaction (EC) involves placing a charge at depth in a borehole in loose soil (generally sands to silty sands or sands and gravels), and then detonating the charge (Gohl *et al.*, 2000). Several charges are fired at one time, with delays between each charge to enhance cyclic loading while minimizing peak acceleration. Often several charges will be stacked in one borehole with gravel stemming between each charge to prevent sympathetic detonation. Although not that often used in land reclamation projects, EC can be attractive, as explosives are an inexpensive source of readily transported energy and allow densification with substantial savings over alternative methods. Only small-scale equipment is needed (e.g. geotechnical drill or wash boring rigs), minimizing mobilization costs and allowing work in confined conditions. Compaction can be carried out at depths beyond the reach of conventional ground treatment equipment. Most EC has been driven by concerns over liquefaction, and has been on loose soils below the water table to depths of nearly 50 m.

Like many other geotechnical processes, explosive compaction has been designed largely on experience rather than theory. It is common to carry out a trial before starting full-scale treatment. This empirical design basis appears to be an obstacle to the widespread use of an otherwise inexpensive and effective compaction method: owner's review boards are often reticent in approving proposals, Contractors are unsure of risk factors when bidding work, and consulting geotechnical engineers lack familiarity with the method.

Following detonation of the charges, some immediate ground heave or small settlements occurs around individual blast holes. Nothing is then apparent at ground surface for at least several minutes or, in the case of fine sand, tens of minutes; then the ground starts to settle, and continues to settle for upwards of an hour. Compaction induced by explosives is not over the few seconds of explosive detonation but is rather an induced consolidation over several hours, even with sandy gravels. Explosives generate residual excess pore water pressure that dissipates and results in consolidation related compaction.

Penetration resistance of the compacted sands shows pronounced time dependence (Gohl *et al.,* 2000). Some case histories show no increase or even a decrease in penetration resistance immediately after blasting, while other case histories show only a modest increase in resistance. However, the penetration resistance two weeks after blasting is often double the pre-compaction value. Delayed strength gain must be allowed for in developing an explosive compaction. An example of delayed strength gain and achievable penetration resistance is shown in Figure 7.20.

**Practical considerations**
Induced vibrations on nearby structures need to be controlled where blast densification is carried out in developed areas. Blasting within 30–40 m of existing structures requires a reduction in the charge weights per deck (involving a reduction in

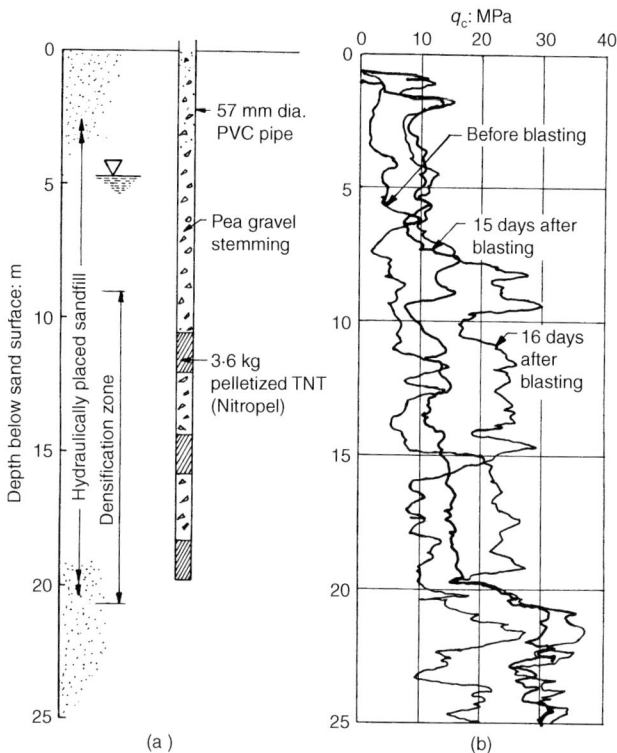

Figure 7.20   *Example of (a) explosive compaction loading (after Stewart & Hodge, 1988) and (b) achieved change in CPT resistance (after Rogers et al., 1990).*

blast hole spacings) and in the number of holes detonated at any one time. Also, when blasting is carried out on or adjacent to slopes, blast patterns are adjusted to restrict the zone of residual pore water pressure build-up and minimize the risk of slope instability.

Design of the appropriate charge delays between adjacent decks in each borehole and between adjacent boreholes is carried out using the following process:

a. Ground vibration patterns (peak particle velocities and frequency content) are determined at a particular location of concern remote from the blast point due to a single charge. This is best done using field measurements.
b. The frequency range of potentially damaging vibrations is selected based on structural vibration theory or other considerations.
c. The effects of sequential charge detonation from a decked array of boreholes are assessed by a simple linear combination of the single charge wave trains in which time delays between decks and between adjacent boreholes are varied. Optimum blast delays are then determined to minimize the peak particle velocity or, alternatively, the vibrational energy content in the frequency range of interest.

### 7.6 Soil replacement

#### 7.6.1 *Introduction*

Soil replacement techniques are used to improve the strength of the soil and to reduce settlements resulting from future loading conditions. Soil replacement methods derive their improvement effects from a total removal and replacement of the existing soil or a combination of soil displacement and reinforcement through the insertion of granular material (see Moseley *et al.*, 2004). The methods described below classify as soil replacement techniques and will be discussed in the following sections:

− soil removal and replacement;
− stone columns;
− sand compaction piles;
− geotextile encased sand columns;
− dynamic replacement.

#### 7.6.2 *Soil removal and replacement*

One of the oldest and simplest methods to improve the subsoil is the removal of problematic soils and their replacement by fill material with better properties (i.e., sand). Soils that are often replaced include contaminated or organic soils. The removal and replacement of soil is performed with conventional earth-moving equipment or dredging equipment.

An important aspect of soil replacement, and often a limiting factor, is the need for transport and disposal of the removed material. If the soil cannot be recycled as a construction material in the fill itself or disposed of at a nearby location, the costs of removal can be significant, especially if the soil is contaminated. In these cases other soil improvement techniques, which immobilise or stabilise contaminants in the soil, may be more economical.

### 7.6.3   Stone columns

### 7.6.3.1   Purpose and principle

The stone column technique comprises the insertion of compacted columns of granular material into the soils to be treated. These columns displace the existing deposits and are tightly interlocked with the surrounding soil.

Stone colums can be used in a very wide range of soil conditions and are aimed to improve drainage, shear strength and stiffness. The design methods are linked to the type of soil to be treated (cohesive or granular) and the effect to be achieved (e.g. bearing capacity).

Typical execution methods are based on the vibro-replacement technique or on the insertion of a closed-ended casing by vibration or driving. During extraction of the probe or casing the gravel is compacted and a further enlargement of the column diameter is realised. Only the more frequently used vibro-replacement technique will be further discussed in this section.

### 7.6.3.2   Execution of stone columns by the vibro-replacement technique

To construct stone columns, a vibro-flot (see Section 7.5.3.3) is allowed to penetrate to the design depth and the resulting cavity is during retrieval of the probe filled with gravel or crushed rock, free of clay and silt. Several techniques have been developed to install stone columns in soils with different groundwater levels, ranging from completely saturated (wet, top-feed process) to dry (bottom-feed process).

The wet vibro-replacement (see Figure 7.21) based on both replacement and displacement comprises the following steps (see www.keller.co.uk):

– Assisted by jetting water, the oscillating vibrator (vibroflot) penetrates under its own weight to the design depth.

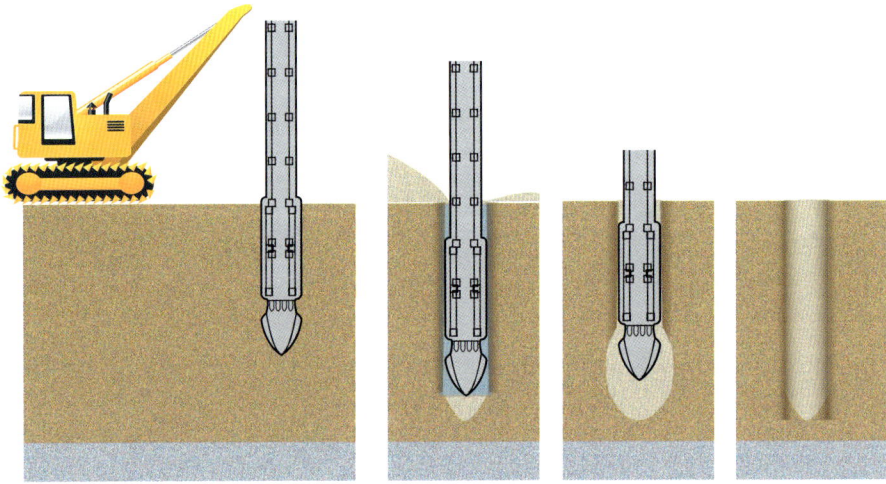

Figure 7.21    *Vibro-replacement: Wet process.*

- Then the water jets are adjusted in such a way that a space remains open around the vibrator and its extension tube. Once at depth a charge of coarse grained backfill material is placed from the surface into the hole down to the tip of the vibrator.
- By moving the vibrator in small steps up and down and due to the horizontal forces introduced by the vibrator, the backfill material is compacted and forced into the surrounding soil mass. By adding successive charges of gravel or rock and by compaction of each of them, a column of very compact stone is built up to ground level.

The dry, bottom-feed method uses a vibro-probe with a separate tremie pipe along the probe feeding the gravel to the tip of the vibrator. The probe penetrates to the required depth by vibration and air jets. Once at depth, gravel is fed into the ground through the tremie pipe and compacted by the probe. This process is repeated while the probe is withdrawn.

Various methods exist to construct stone columns over water. Traditionally a blanket of gravel with a thickness typically varying between 2 to 3 m is placed on the seabed. The vibro-probe penetrates the layer to its final depth. As with the top feed method, the hole is supported by water jets and gravel drops in the annular space between the existing ground and the probe to the bottom of the hole. The gravel is compacted by the probe.

Nowadays systems exist that resemble the dry-feed method, feeding the gravel from a (single-batch) hopper or by a pump or compressor through a tremie pipe to the bottom of the hole.

For design aspects related to stone columns reference is made to Priebe, 1995.

## Effects and limitations

Stone columns increase the shear strength and stiffness of soft, cohesive soil strata and can also be used to improve the resistance of granular soils against liquefaction.

The grain size distribution envelope in Figure 7.22 indicates when vibro-replacement should be used instead of vibro-compaction.

Stone columns are not suitable in soft soils having a very low undrained shear strength since the lateral support for the columns may be too small. This can result in lateral bulging of the columns.

The wet top-feed method will generate slurry and eroded soil that is washed out of the hole. This may be a limitation when operating in small confined areas or in environmentally sensitive areas.

Depending on the soil conditions and water depths, stone columns can be installed on land and over water to depths exceeding 40 m.

A high area replacement ratio (i.e., ratio of the area occupied by stone columns to the overall area of the improved ground) may cause heave of the surrounding ground.

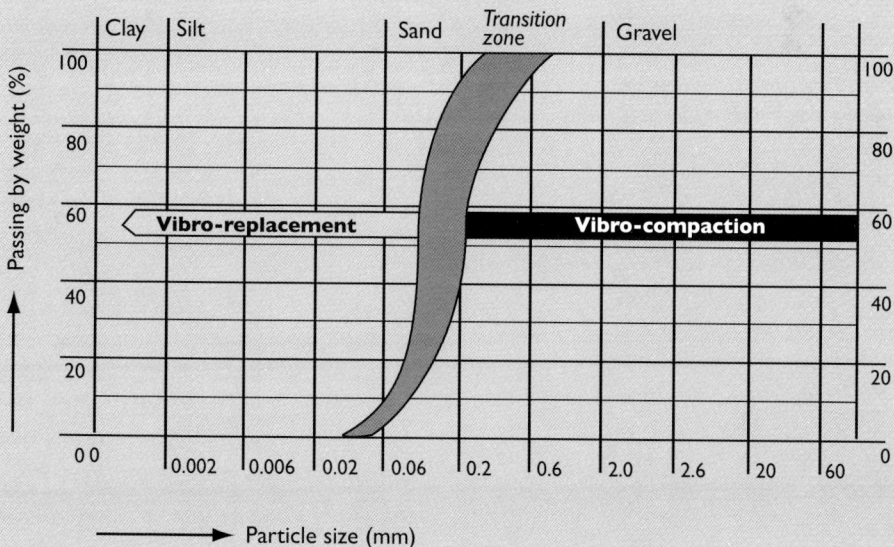

Figure 7.22   *Fields of application for vibro methods (Moseley et al., 2004).*

### 7.6.4   Sand compaction piles (closed end casing)

The sand compaction pile method is used to form compacted, large diameter (1.5–2.5 m) sand piles by vibration in soft ground. Although originally developed to improve the shear strength, reduce the compressibility and prevent liquefaction of loose sands it is also applied in soft cohesive strata to increase the shear strength and reduce the settlements. Lateral support is provided by the surrounding soil. For design aspects related to sand compaction piles reference is made to Priebe, 1995.

Sand compaction piles are typically installed by inserting a closed-ended casing filled with sand into the soil mass that needs to be treated. The casing is driven to its required depth by means of a vibrator mounted on top. During partial withdrawal of the casing, its closed-end opens up, releasing the sand. The casing is re-driven and the sand is compacted by the vibrations of the casing. This process is repeated until completion of the pile. During the whole operation, additional sand is continuously being fed to the top of the casing. Compressed air may be used to keep the hole open when the casing is partially withdrawn.

**Effects and limitations**

The maximum improvement depth of the sand compaction pile technique is reported to be 70 m (Kitazume, 2005).

Sand compaction piles can be constructed both on land and over water to improve the properties of soft subsoils. In reclamation projects, sand compaction piles are often used along the boundaries of fill areas to ensure stability of the slopes.

High area replacement ratios generally cause heave of the ground in the immediate vicinity of the sand compaction piles.

### 7.6.5   Geotextile encased columns

Another soil replacement method uses Geotextile Encased sand Columns (GEC). These columns can be installed in very soft soils. They can accelerate the consolidation process, increase the shear strength and reduce the settlements of soft subsoil.

At first sight geotextile encased columns do not differ much from common sand drains. However, in contrast to sand drains, the geotextile encasing

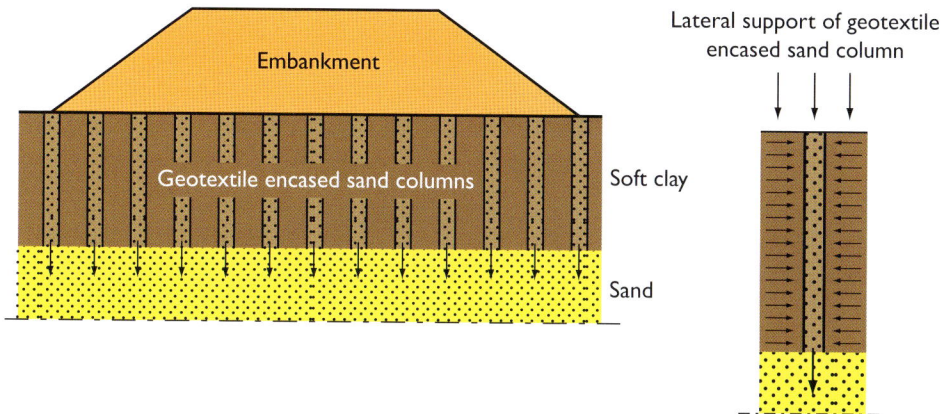

Figure 7.23    *Principle of Geotextile Encased Columns (GEC).*

restricts horizontal displacement of the sand column. The lateral support from the surrounding soil in combination with the radial supporting effect of the geotextile results in a high stiffness of the GEC. This enables the columns to function as bearing support by transferring loads to deeper more competent soil layers.

Two installation methods can be distinguished:

–   The excavation method: an open steel pipe is driven into a more competent foundation layer, its content is then removed by an auger and, after lowering a geotextile casing in the pipe, the casing is filled and the steel pipe pulled;
–   The vibro-displacement method: this method, which is more economical, is also more commonly used. A steel pipe with two base flaps (which close upon contact with the soil) is vibrated down into the bearing layer, displacing the soft soil. The geotextile casing is installed and filled with sand. At this stage, the sand in the column is in a loose condition. During retrieval of the casing, the flaps at the bottom open and by vibrating the pipe the sand is compacted. This will result in a geotextile-encased column of medium density being left behind in the soft soil (Raithel *et al.,* 2002).

The resulting sand piles are usually 0.4–0.8 m in diameter and installed at typical grid spacings ranging between 1.5 and 3 m.

With a geotextile encased column, the horizontal support of the soft soil can be much lower than for non-encased columns due to the radial supporting effect of the geotextile casing. The columns act simultaneously as vertical drains, but the main effect is the transport of the load to a deeper bearing layer. To carry the high ring tension forces, the geotextile casings are manufactured seamlessly (Raithel *et al.*, 2002).

### Effects and limitations

Geotextile encased sand columns can be used as a ground improvement and bearing system in very soft soils, for example, peat or sludge with undrained shear strengths less than 15 kN/m². The effect of the soil improvement is usually expressed by an improvement factor $\beta$, which is the ratio between the settlement of unimproved soil and the settlement of improved soil. The factor $\beta$ of ground improvement in soft soil amounts to about $\beta = 2.5$ to 4. The soil improvement factor of GEC's shows a significant increase with increasing geotextile stiffness (Kemfert, 2003).

### 7.6.6   *Dynamic replacement*

Dynamic replacement can be used to increase the strength of (saturated) soft cohesive and organic soils and to reduce the settlements during future loading conditions.

Large diameter (2.5–4.0 m) columns are formed by placing a blanket of aggregate over the site and driving the aggregate into the soft soil by dropping a 15 to 30 tonne weight from heights ranging from 10 to 40 m. Alternate filling of the impact crater and subsequent tamping until completion of the column will result in pillars with a typical mushroom shape.

### Effects and limitations

Depending on the consistency of the subsoil replacement ratios of 20–25% can be achieved, increasing both the strength and stiffness of the treated strata considerably. As a result of the large diameter, significant loads can be supported by individual columns.

During compaction, heave of the surface in the immediate vicinity of the columns can occur and may have to be cut down. Figure 7.24 shows the exposed top of a number of dynamic replacement columns after installation and subsequent excavation.

The method is considered to be cost effective and fast, but can only be undertaken in soft soils and to limited depths.

Figure 7.24 *Top of dynamic replacement columns after installation and excavation (info Menard).*

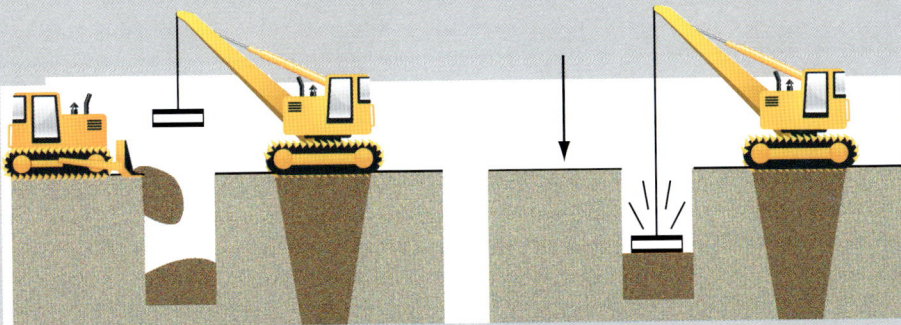

Figure 7.25 *Principles of dynamic replacement with pre-excavation (info Menard).*

Should long columns be required, then the location of a column can be pre-excavated and (partly) filled with aggregate before tamping starts. This will also reduce the heave of the surrounding soil.

Although generally used on land, underwater applications have been reported by Hamidi *et al.* (2010) installing dynamic replacement columns at water depths of up to 30 m.

## 7.7 Admixtures and in-situ soil mixing

Admixtures are used to stabilise soils in the field, particularly fine-grained soils. The most common admixtures are lime, cement and fly ash. The main purpose of soil stabilisation is to:

- improve strength and stiffness;
- increase workability of clayey soils;
- reduce shrinking and swelling of the soil;
- immobilize contaminants in case of contaminated soils.

The improvement is based on in-situ mixing of soil with admixtures, which chemically react with the soil and/or the groundwater. Generally the resulting stabilised soil mass has a higher strength, lower permeability and lower compressibility than the native soil. For environmental treatment, admixtures can be enriched with chemical oxidation agents or other reactive materials to immobilise or neutralise contaminants (Moseley et al., 2004).

The most important techniques to create stabilised columns or walls in the soil are in-situ Deep Soil mixing Methods (DSM) and shallow mixing methods (SSM). Both methods are based on the construction of stiff columns in soft ground. These columns consist of a mixture of the natural soil with stabilizing additives which are added as dry or wet components and mechanically mixed with the soil (Moseley et al., 2004).

The admixtures are injected into the soil in dry or slurry form through hollow rotating mixing shafts tipped with various cutting and mixing tools (see Figure 7.26). In some methods, the mechanical mixing is enhanced by

Figure 7.26 *Deep soil mixing equipment (Moseley et al., 2004).*

simultaneously injecting fluid grout at high velocity through nozzles in the mixing or cutting tools.

## Effects and limitations

Strength improvement can be determined from pre- and post-treatment CPTs or laboratory tests on undisturbed samples. Proper care should be given to the selection of the stabilising material:

- lime/cement stabilization is most suitable for granular soils and clayey soils with low plasticity;
- calcium clays are more easily stabilized by the addition of cement;
- sodium and hydrogen clays, which are expansive in nature, respond better to lime stabilization (Das, 2009).

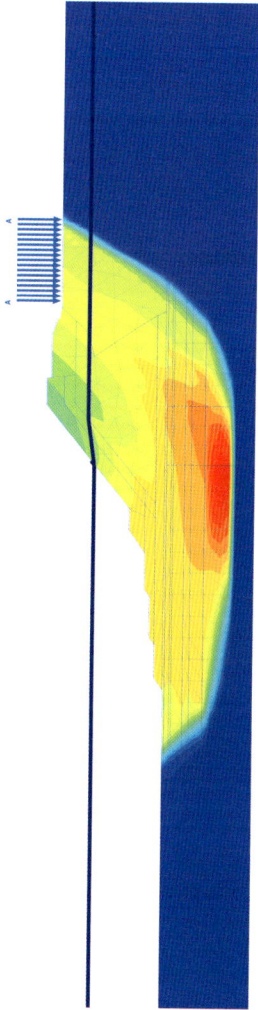

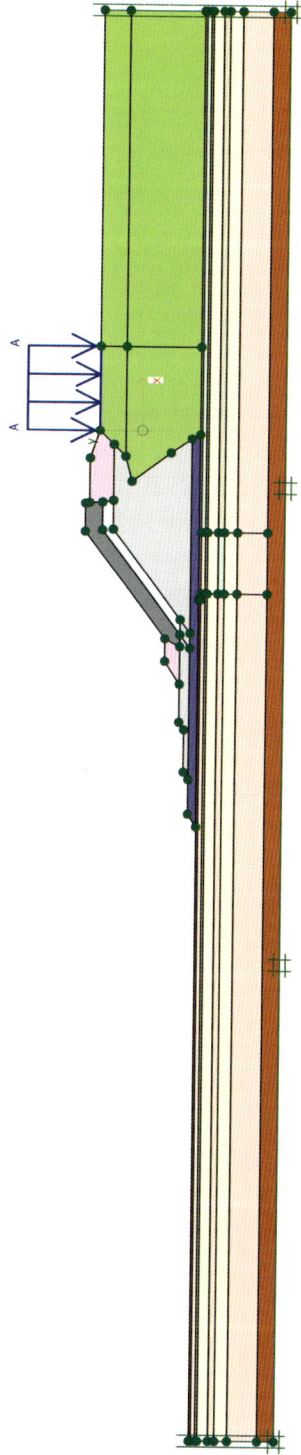

# DESIGN OF RECLAMATION AREA

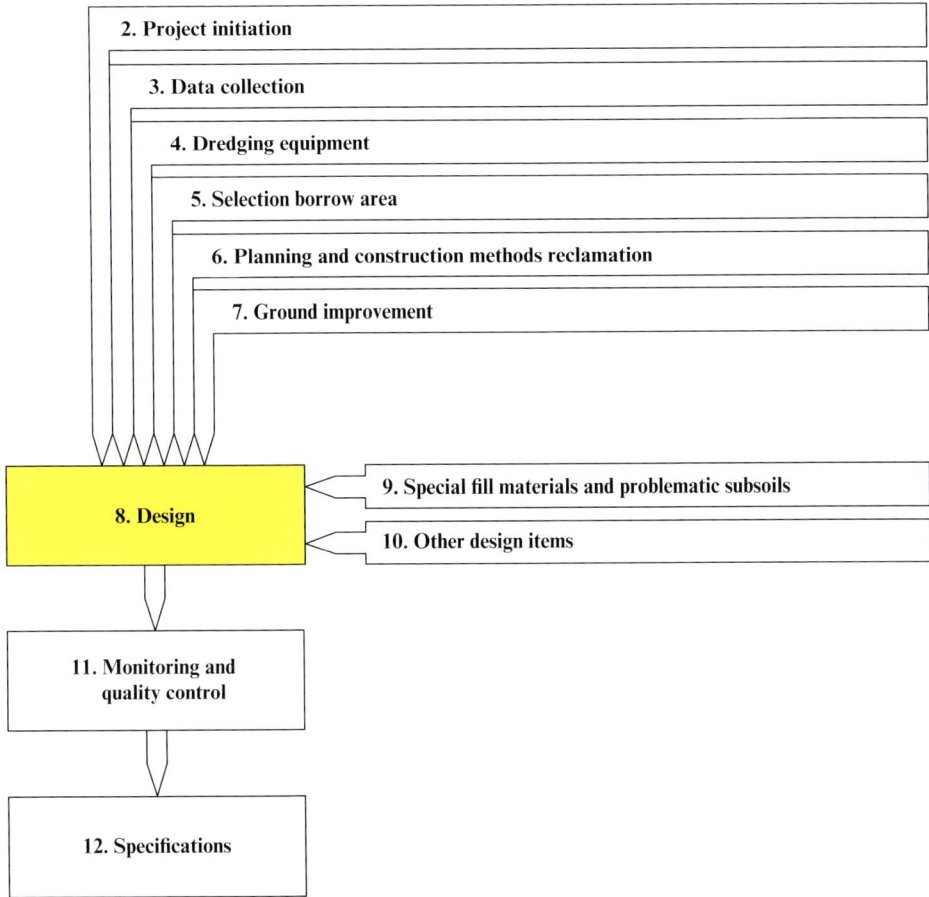

```
┌─────────────────────────────────────────────────────────────┐
│ 2. Project initiation                                        │
│   ┌─────────────────────────────────────────────────────────┐
│   │ 3. Data collection                                       │
│   │   ┌─────────────────────────────────────────────────────┐
│   │   │ 4. Dredging equipment                                │
│   │   │   ┌─────────────────────────────────────────────────┐
│   │   │   │ 5. Selection borrow area                         │
│   │   │   │   ┌─────────────────────────────────────────────┐
│   │   │   │   │ 6. Planning and construction methods reclamation │
│   │   │   │   │   ┌─────────────────────────────────────────┐
│   │   │   │   │   │ 7. Ground improvement                    │
```

┌─────────────────────┐        ┌──────────────────────────────────────────────┐
│                     │ ◄──────│ 9. Special fill materials and problematic subsoils │
│   **8. Design**     │        └──────────────────────────────────────────────┘
│                     │ ◄──────┌──────────────────────────────────────────────┐
└─────────────────────┘        │ 10. Other design items                       │
                               └──────────────────────────────────────────────┘

┌─────────────────────┐
│ **11. Monitoring and│
│    quality control** │
└─────────────────────┘

┌─────────────────────┐
│ **12. Specifications** │
└─────────────────────┘

**Chapter 8 Design of reclamation areas**

**8.1 Design philosophy**

**8.2 Basic design mass properties**

| 8.2.1 Strength of fill mass | 8.2.2 Stiffness of fill mass | 8.2.3 Density of fill mass and subsoil | 8.2.4 Permeability of fill mass |

8.2.5 Platform level

**8.3 Density**

| 8.3.1 Definition of key parameters | 8.3.2 Density ratios | 8.3.3 The use of densities or density ratios in specifications |

| 8.3.4 Effect of grain size distribution on the density of a soil sample | 8.3.5 Density measurement | 8.3.6 Typical relative density values of hydraulic fill before compaction |

**8.4 Strength of the fill mass and subsoil (bearing capacity and slope stability)**

| 8.4.1 Introduction | 8.4.2 Shear strength | 8.4.3 Relative failure modes |

**8.5 Stiffness and deformation**

| 8.5.1 Introduction | 8.5.2 Stiffness | 8.5.3 Deformations | 8.5.4 Techniques for limiting settlement |

**8.6 Liquefaction and earthquakes**

| 8.6.1 Overview | 8.6.2 History of understanding | 8.6.3 Flow slides versus cyclic softening | 8.6.4 Assessing liquefaction susceptibility |

| 8.6.5 Movement caused by liquefaction | 8.6.6 Fill characterization for liquefaction assessment | 8.6.7 Note on soil type (calcareous and other non-standard sands) |

## 8.1  Design philosophy

The design philosophy will be based on the rational process of matching the anticipated loading response of the fill mass to the requirements imposed by the future use of the reclaimed land, all within the technical boundary conditions of the project. Obviously the function of a beach nourishment differs from that of a reclamation area intended to accommodate an airport or a container terminal. It is therefore reasonable to assume that the corresponding technical specifications resulting from the design process will also differ. This philosophy must ensure that reclamation is not over-designed and that geotechnical specifications are reasonable, measurable and feasible.

An essential element of the design philosophy is an iterative design procedure: this means that during the development process the structural design of the super structures may have to be re-considered.

### Boundary conditions

Once the functional requirements are defined, an adequate inventory of all boundary conditions has to be made. Boundary conditions may be external, often uncontrollable conditions that could affect the performance of the hydraulic fill like the quality and quantity of the fill material available, the nature of soils underlying the future fill area, earthquakes and other natural hazards, tides and waves.

### Construction methods

Usually the Contractor selects or proposes the most cost-effective and suitable construction method to realise the fill mass properties according to the performance requirements and within the boundary conditions. It is essential, however, that the designer of a hydraulic fill also has a thorough understanding of the possibilities and limitations of the various construction methods including dredging and reclamation operations and soil improvement techniques.

### Design cycle

A design of the fill mass is part of the design cycle of a total project:

> Should it not be possible to identify a construction method (and—if required—a soil improvement technique) that results in a fill mass that meets the specified performance requirements without excessive costs, the engineer may have to reconsider the starting points of the fill design. This will generally imply adjustments of the design of the superstructures resulting in adapted performance requirements. Design changes of the structures may include modifications to foundations, slope angles, drainage facilities and/or revetments, alterations of final grade levels, the introduction of soil retaining structures, etc.

Whether to apply a soil improvement technique or to adjust the design of the superstructures may often be decided on the basis of costs, but technical feasibility and low failure probability may be important criteria as well.

Soil improvement methods include techniques such as surcharging (in combination with vertical drains), compaction, deep soil mixing and vibro replacement (see Chapter 7).

Economics and technical feasibility of the reclamation and soil improvement techniques may affect the geometry of a reclamation area as well. Long pumping distances, for instance, may result in the need for a booster pump station. Adjustment of the geometry of the proposed reclamation area could reduce the pumping distance and, hence, lower the costs.

In this Manual the design of a hydraulic fill will mainly relate to the geometry of the reclamation area and the geomechanical properties of the fill mass, because the results of the environmental impact assessment (EIA) are considered to be part of the boundary conditions. Note that in the "Building with Nature" philosophy, the findings of an EIA may be additional starting points of the design rather than purely boundary conditions.

Figure 8.1 presents a flow diagram of the simplified design process of a hydraulic fill.


**Failure probability and testing**

To account for the uncertainties involved, a design generally includes safety margins. Large safety margins will generally result in high construction costs, but a low probability of failure, while low safety margins will reduce the construction costs but raise the failure probability.

The Client is responsible for defining the failure probability that is acceptable.

As reduction of the failure probability at design level may prove to be very costly, the reduction of the risk of failure may be made possible by improving the nature of the testing scheme of the fill mass (i.e., quality control). Such an improvement could relate to the quality of the testing (i.e., technical specifications must specify tests that better simulate the future loading conditions; see Chapter 12), but may concern the frequency of the testing regime as well. This approach will result in a reduction of the uncertainties and, hence, will allow for a reduced safety margin without affecting the failure probability. Should this intensified testing programme identify localised areas that do not meet the specifications, soil improvement may be used to rectify these non-conformities.

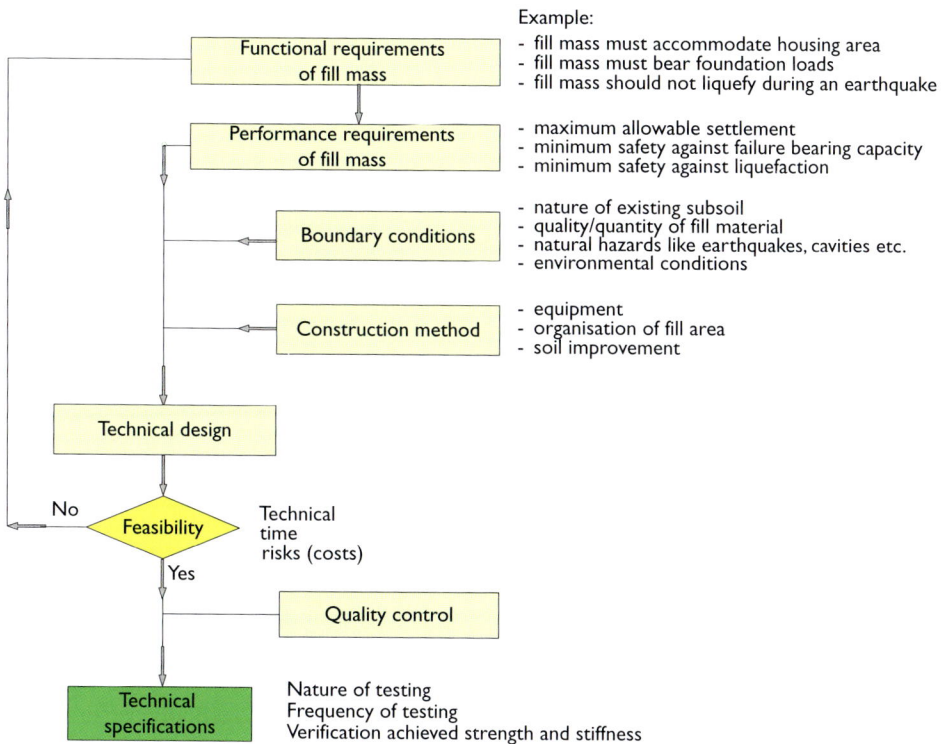

Example:
- fill mass must accommodate housing area
- fill mass must bear foundation loads
- fill mass should not liquefy during an earthquake

- maximum allowable settlement
- minimum safety against failure bearing capacity
- minimum safety against liquefaction

- nature of existing subsoil
- quality/quantity of fill material
- natural hazards like earthquakes, cavities etc.
- environmental conditions

- equipment
- organisation of fill area
- soil improvement

Technical
time
risks (costs)

Nature of testing
Frequency of testing
Verification achieved strength and stiffness

Figure 8.1   *Flow chart of design process of a hydraulic fill.*

## 8.2   Basic mass properties

Performance requirements follow from the definition of land use of the hydraulic fill set out in the functional requirements and determine the required basic properties of the fill mass. These properties include:

− strength of fill mass and subsoil: bearing capacity and slope stability;
− stiffness of fill mass and subsoil: settlement, horizontal deformation;
− density of the fill mass and subsoil: resistance against liquefaction;
− permeability of the fill mass: drainage capacity;
− elevation of the fill mass: safety against flooding.

### 8.2.1   *Strength of fill mass: Bearing capacity and slope stability*

The ability of a fill mass to bear foundations or to sustain slopes is determined by the shear strength of the fill mass and the soil below the fill mass (see Section 8.3).

A fill often serves as the foundation for buildings, infrastructure, container storage and other structures and must safely transfer these loads to the subsoil. Foundations may consist of concrete footings and slabs below buildings, but may also be composed of load distributing layers of selected fill and gravel below roads, railways, etc. The shear strength required to bear the foundation loads follow from the weight of the structures, equipment and/or stacked containers and the design of the foundations.

The most relevant slopes of a reclamation area are usually the slopes at the edge of the fill area that often need to be protected against waves and currents. The angle of these slopes and, hence, the required shear strength of the fill, is often determined by the design of the shore protection, but may also be related to boundary conditions such as the nature of the existing subsoil, the quantity of fill available, existing water depths or the presence of adjacent facilities.

Although the shear strength of a granular hydraulic fill mass is generally sufficient to bear the foundations of the superstructures or to construct slopes of the required gradient, various soil improvement methods such as compaction, soil replacement and soil reinforcement are available to increase the shear strength. Should soft, cohesive strata underlying the fill mass affect the stability of the fill mass, it may be possible to accelerate the consolidation process by surcharge and vertical drains. Reference is made to Chapter 7.

If these measures fail to produce the desired results at acceptable costs the design of the superstructures may have to be changed. Such changes could include, for instance, replacement of shallow foundations by piled foundations or the design of soil retaining structures to achieve the required slopes.

### 8.2.2 *Stiffness of fill mass: Settlements, horizontal deformations and tolerances*

The stiffness of the fill mass is defined as its resistance to stress-induced deformation and is usually expressed in terms of a modulus of elasticity, shear modulus or compression index (see Section 8.5).

The deformations of a fill mass are often governed by the stiffness of the subsoil. In particular the presence of a soft, cohesive subsoil below the fill may cause not only long term consolidation settlements of the fill surface, but also horizontal deformations near the edges of the fill mass. These deformations may adversely affect the integrity of structures built on or in the immediate vicinity of the fill mass.

Deformations occurring in a granular subsoil and/or fill mass are generally much smaller, but, nevertheless, may not be acceptable for structures that are very sensitive to deformations.

A design of a reclamation area should include predictions of the expected deformations of the subsoil and fill mass in order to:

- Determine the acceptability of these deformations or the need for soil improvement by comparing the predicted deformations with the relevant performance requirements. Soil improvement could include a surcharge and vertical drains to accelerate the consolidation process, compaction of the granular fill or soil replacement (see Chapter 7).
- Determine the required thickness of the fill taking into account the specified final surface level of the area, the settlements occurring during construction (construction settlements) and the settlements that the area may experience after constructions (residual settlements).

Tolerances of the required final platform level of the fill area must be defined considering not only the requirements that originate from the design of the superstructures, but also the accuracy with which the settlements can be predicted.

If design calculations indicate that despite soil improvement the expected deformations are still too much to safely accommodate the superstructures, the foundation design may have to be adapted, the lay-out of the site changed or, ultimately, the proposed site relocated.

### 8.2.3 *Density of the fill mass and subsoil: Resistance against liquefaction*

The fill mass must have sufficient resistance against liquefaction, which is defined as the sudden, substantial reduction in shear strength of a granular soil mass caused by the development of excess pore pressures. Monotonically or cyclically induced shear strains may generate these excess pore pressures, when the soil assembly has the tendency to contract under undrained conditions. Contractive behaviour of the soil skeleton is usually associated with low (relative) density, possibly combined with a high mean effective stress. Reference is made to Section 8.6. The loss of shear strength during liquefaction will adversely affect the bearing capacity of foundations and the stability of slopes.

Performance criteria concerning liquefaction should be based on the acceptable low probability of soil failure. The corresponding requirements are usually expressed in terms of (relative) density (as function of the effective stress) and the particle size distribution. Possible mitigating measures include compaction of the subsoil and the installation of draining elements.

### 8.2.4  *Permeability of fill mass: Drainage capacity*

Water from natural precipitation, overtopping and other sources may hamper the intended use of the reclamation area if it does not readily drain away. Natural drainage of a reclaimed area includes the infiltration of the water into the fill in combination with the natural surface runoff. An important property that influences the infiltration capacity is the permeability of the fill mass.

If the natural drainage capacity is not sufficient to cope with the water supply by precipitation, overtopping, etc., an artificial drainage system may have to be designed.

Further reference is made to Chapter 10.2.

### 8.2.5  *Platform level: Safety against flooding and erosion*

The platform level of fills placed in coastal areas or along rivers must be sufficiently high to prevent them from flooding and limit surface erosion.

The design of the fill height takes into account the tidal variation, wind set up, wave run up and wave overtopping. Common practice is to express the design platform level in terms of the probability of occurrence of a water level exceeding this design platform level. The same concept of probability of occurrence is used to specify the maximum wave run up usually defined as a maximum volume of overtopping water per unit of time (litres/s).

The allowable occurrence is expressed as an acceptable number of times per year, usually depending on the consequences (loss of human life, the value of the infrastructure on the reclamation area).

Except for beach nourishments, fill slopes exposed to wave and current action must be provided with a shore protection (see Section 10.4).

## 8.3  Density

The density of a soil is very often used as an indirect quality description of the soil. In most technical specifications dealing with reclaimed material, there is a clause dealing with the density of the soil. However, it should be kept in mind that the density of a soil is not a mechanical property. Since the density of the soil is a parameter which is generally easy to measure, it is often used as a correlating parameter to determine mechanical properties of a soil.

E.g., The relative density of a sand mass can be used to derive the angle of internal friction through correlations.

The notion "density of a soil" appears to be a simple one at first sight. However, since soil is generally a bulk material containing voids, which are partly filled with water and air, things are slightly more complicated.

Confusion may be created when the used parameters are not accurately defined, especially when dealing with terms such as relative densities and degree of compaction.

### 8.3.1 *Definition of key parameters*

Following density-related parameters are used in this manual:

$\rho$    density: mass per unit volume.            (t/m³) or (kg/m³)

$\gamma$    volumetric weight $(= \rho \times g)$           (kN/m³)

$\rho_s$:  density of solids (volumetric mass of a single solid soil particle without pores) (t/m³)

$\rho_g$:  particle density (volumetric mass of a single solid particle/grain containing enclosed pores) [t/m³]

In British and American literature, the definition specific gravity is often used. The specific gravity is the ratio of density of solids over the density of water.

$G_s$:  specific gravity $(= \rho_s/\rho_w)$           (dimensionless)

$\rho_d$:  dry density (volumetric mass of a dry soil sample)        (kg/m³)

$\rho_b$:  bulk density (volumetric mass of a bulk soil sample)        (kg/m³)

$\rho_{sat}$:  saturated density (volumetric mass of a saturated soil sample)     (kg/m³)

$S$:  degree of saturation (ratio of volume of water to the volume of voids in a soil sample)

$w$:  water content (ratio of the mass of the water fraction to the mass of the solid fraction in a soil sample)           (–)

$e$:  void ratio (ratio of the volume of void space to the volume of solids in a soil sample)                                                                                (−)

$n$:  porosity (ratio of the volume of void space to the total volume of soil)    (−)

$\rho_d = W_s / V$

$\rho_b = W / V$

$S = V_w / V_v$

$e = V_v / V_s$

$n = V_v / V$

$V$  = total volume of soil (m³)
$V_s$  = volume of solids (m³)
$V_v$  = volume of voids (m³)
$V_w$  = volume of water (m³)
$V_a$  = volume of air (m³)
$W$  = total mass of soil (t)
$W_s$  = mass of solids (t)
$W_w$ = mass of water (t)

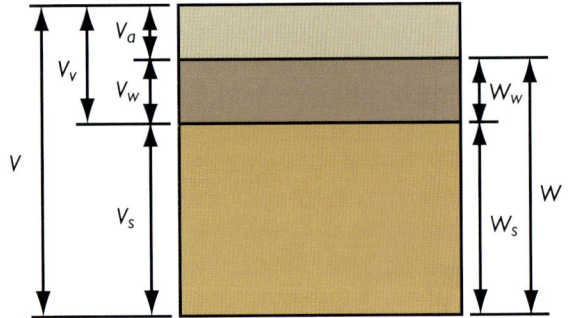

### 8.3.2  Density ratios

For non-cohesive coarse-grained soils, laboratory tests have been developed to determine the loosest and densest state of the subject soil. Reference is made to section 8.3.5 for a brief description of these tests.

When compacting a soil sample to its densest state (*), the following parameters can be derived:

$\rho_{d\,max}$:  maximum dry density (kg/m³)

$e_{min}$:  minimum void ratio (kg/m³)

$n_{min}$:  minimum porosity (−)

When fabricating a soil sample in its loosest (dry) state (*), the following parameters can be derived:

$\rho_{d\,min}$:  minimum dry density (kg/m³)

$e_{max}$:  maximum void ratio (−)

$n_{max}$:  maximum porosity (−)

192

(*) Please note that the minimum and maximum dry density is not a unique soil property, but depends on the test method adopted. For example, for the determination of the maximum dry density, a standard compaction energy is used. If more compaction energy were to be used in another method, a denser state can possibly be obtained.

In geotechnical literature and standards, the terms relative density and density index are often used. In view of the contradicting definitions in various standards, these terms are replaced in this manual by relative void ratio $R_e$ and relative porosity $R_n$. The term relative density is only used in a general manner and does not refer to a specific definition.

When comparing an in situ soil sample to its loosest and densest state, the following ratios can be defined:

$R_e$: relative void ratio

$$R_e = \frac{e_{max} - e}{e_{max} - e_{min}} 100\% = \frac{\rho_d - \rho_{d\,min}}{\rho_{d\,max} - \rho_{d\,min}} \cdot \frac{\rho_{d\,max}}{\rho_d} \cdot 100\%$$

$R_n$: relative porosity

$$R_n = \frac{n_{max} - n}{n_{max} - n_{min}} 100\% = \frac{\rho_d - \rho_{d\,min}}{\rho_{d\,max} - \rho_{d\,min}} \cdot 100\%$$

$R_c$: degree of compaction

$$R_c = \frac{\rho_d}{\rho_{d\,max}} 100\%$$

International confusion about these parameters often arises because of different naming in various standards. Especially $R_e$ and $R_n$ are often confused, while it is clear that $R_e$ will always lead to higher values compared with $R_n$. Table 8.1 gives an overview of the various definitions in the major international standards.

Above parameters can be correlated to each other (**)

$$R_e = \frac{R_c \times \rho_{d\,max} - \rho_{d\,min}}{\rho_{d\,max} - \rho_{d\,min}} \times \frac{1}{R_c} \times 100\%$$

$$R_n = \frac{R_c \times \rho_{d\,max} - \rho_{d\,min}}{\rho_{d\,max} - \rho_{d\,min}} \times 100\%$$

$$R_e = \frac{R_n}{R_c}$$

Table 8.1    *Definition of density ratios.*

| Ratio | Formula | Eurocode | BS (British Standard) | ASTM | EAU (DIN) |
|---|---|---|---|---|---|
| $R_e$ relative void ratio | $R_e = \dfrac{e_{max} - e}{e_{max} - e_{min}} \cdot 100\%$ | $I_D$ Density Index | $I_D$ Density Index | $D_d$ Relative density | $I_D$ Specific degree of density |
| $R_n$ relative porosity | $R_n = \dfrac{n_{max} - n}{n_{max} - n_{min}} \cdot 100\%$ | / | / | $I_d$ Density index | $D$ Degree of density |
| $R_c$ Relative compaction | $R_c = \dfrac{\rho_d}{\rho_{d\,max}} \cdot 100\%$ | % Proctor density | Relative compaction | $R_c$ Relative compaction | $D_{pr}$ Degree of compaction |

(**) Correlations are only valid if the maximum and minimum reference densities are measured with the same method for each parameter.

A typical graphic example showing the relationship between above-mentioned parameters is presented in Figure 8.2.

Other relationships between density-related parameters can be found in Appendix C and D.

### 8.3.3    *The use of densities or density ratios in specifications*

From an engineering perspective, the specification of the density of a material should be based on the future use of the fill. The relative void ratio or relative porosity (also called relative density) is often used to describe sand deposits. Correlations have been developed relating the relative density to the angle of internal friction, liquefaction potential, etc.

Some manuals present guidelines for the relative void ratio or relative porosity. These values can be used as a guideline in the preliminary design of a hydraulic fill. It should be kept in mind that the relative density is an indirect way of defining the strength (see Section 8.3). The required relative density of non-cohesive soils in hydraulic fills depends on the type of utilisation, which is laid down in the functional requirements.

| | $n$ | $e$ | $\rho_w$ | $\rho_s$ | $\rho_{sat}$ | $\rho_d$ | $R_e$ | $R_n$ | $R_c$ | Vol .[m³] |
|---|---|---|---|---|---|---|---|---|---|---|
| maximum density | 35% | 0.54 | 1.025 | 2.65 | 2.08 | 1.723 | 100% | 100% | 100% | 0.85 |
| minimum density | 45% | 0.82 | 1.025 | 2.65 | 1.92 | 1.458 | 0% | 0% | 85% | 1.00 |
| in situ density | 40% | 0.67 | 1.025 | 2.65 | 2.00 | 1.590 | 54% | 50% | 92.3% | 0.92 |

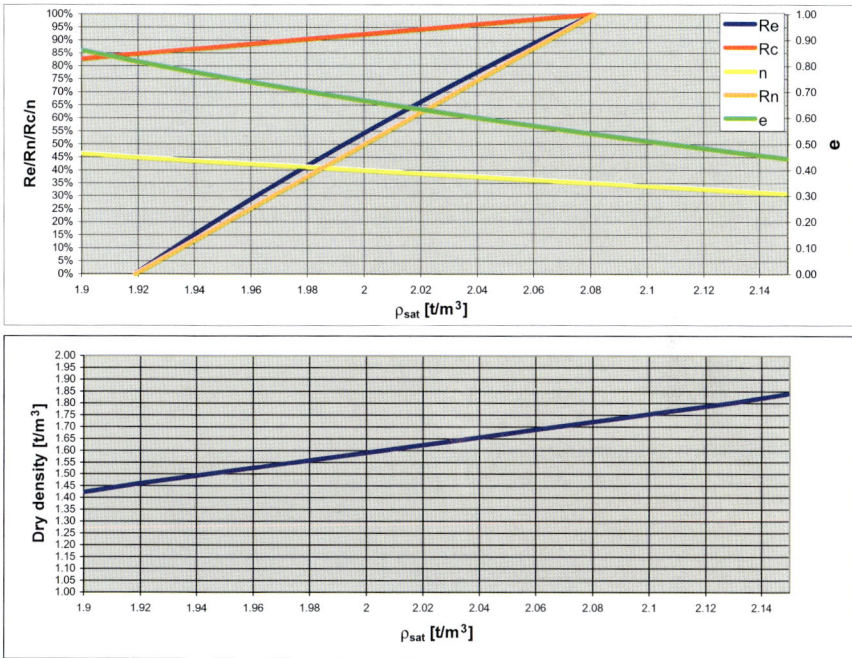

Figure 8.2  *Typical relationship density related soil properties.*

| | Relative porosity $R_n$ (ref. EAU 2004) | |
|---|---|---|
| | Fine sand | Medium sand |
| Type of utilization | $d_{50}$  0.15 mm | $d_{50} = 0.25$ to 0.50 mm |
| Storage areas | 0.35–0.45 | 0.20–0.35 |
| Traffic areas | 0.45–0.55 | 0.25–0.45 |
| Structure areas | 0.55–0.75 | 0.45–0.65 |

Eurocode 7 geotechnical design mentions the following note concerning the relative density:

Structural fills on which foundations are built shall be made with suitable materials for which an appropriate density, a 100% Proctor density (2.5 kg rammer, height of fall 0.3 m) as an average and a 97% Proctor density as a lower limit shall be assured and the risk of collapse and excessive differential settlements shall be prevented.

The use of more direct measurements of strength properties for the detailed design (such as cone penetration values, angle of internal friction) or the state parameter $\psi$ as a measure of in situ state (Jefferies and Been, 2006) is, however, strongly advised, see section 8.6.2.

### 8.3.4 *Effect of grain size distribution on the density of a soil sample*

Several factors are known to affect the density of a soil. Since the density of a soil sample is related to its degree of compaction, comparing it with a reference density state is easier. Usually the maximum dry densities of soil samples are compared.

Following factors are known to influence the density of a soil:

– Clay content or fraction <2 $\mu$m
  Particles smaller than 2 $\mu$m can fill up the voids in between silt, sand and gravel particles. The fraction smaller than 2 $\mu$m can therefore drastically influence the soil density.
– Median of the grading curve $d_{50}$
  The grain size should theoretically not have an influence on the soils density, since a soil with small particles is only a scale model from a soil with large particles. However, natural non-cohesive soil deposits tend to have larger bulk densities with increasing average grain size. This is the result of better grading of the coarser soils.
– Grain size distribution
  The grain size distribution has a significant influence on the density of a soil. The "better" the grading, the higher the density. The grading of a soil is generally described by the uniformity coefficient $C_u = D_{60}/D_{10}$, and the coefficient of curvature $C_c = (D_{30})^2/(D_{10} \cdot D_{60})$. $D_x$ represent the sieve size for which x% of the sample is passing.
– In the Unified Soil Classification system, a sand is defined as 'well graded' if it meets the following criteria:
  ○ less than 5% of fines
  ○ $C_u > 6$ (for well graded gravel, $C_u > 4$ can be applied)
  ○ $1 < C_c < 3$

In the European standards, the grading of a soil is defined differently (Table 8.2).

– Various researchers have investigated the relationship between the uniformity coefficient and the maximum Proctor density. An overview of these correlations is plotted in appendix C, Correlations and correction methods.
– Fines fraction (<63 or 74 micron)

  A uniform relationship between the fine fraction of a sand and the maximum Proctor density cannot be determined. As described, the

Table 8.2   *Source EN-ISO14688–2.*

| Shape of grading curve | $C_U$ | $C_C$ |
|---|---|---|
| Multi-graded | >15 | $1 < C_C < 3$ |
| Medium-graded | 6 to 15 | <1 |
| Even-graded | <6 | <1 |
| Gap-graded | Usually high | Any (usually <0.5) |

coefficient of uniformity is also influencing the maximum Proctor density. However it is generally acknowledged that adding fines to sand will lead to larger maximum densities since the fine particles will fill up the voids of the sand matrix. When adding too much fines, the sand matrix will not be dominant anymore and the sand/silt mixture will become difficult to compact. This will lead to smaller maximum densities. This effect is clearly demonstrated in Figure 8.3. In this figure, the highest maximum density is achieved with a silt content of 22%.

According to Floss *et al.* [1968] the optimum density for gravels and sand gravel mixtures can be reached at a percentage of fines (<63 micron) of 15% to 30% as in Figure 8.4. As can be observed on the graph, the largest maximum density is reached for samples 7 and 8, having a silt fraction of 25% and 32% respectively.

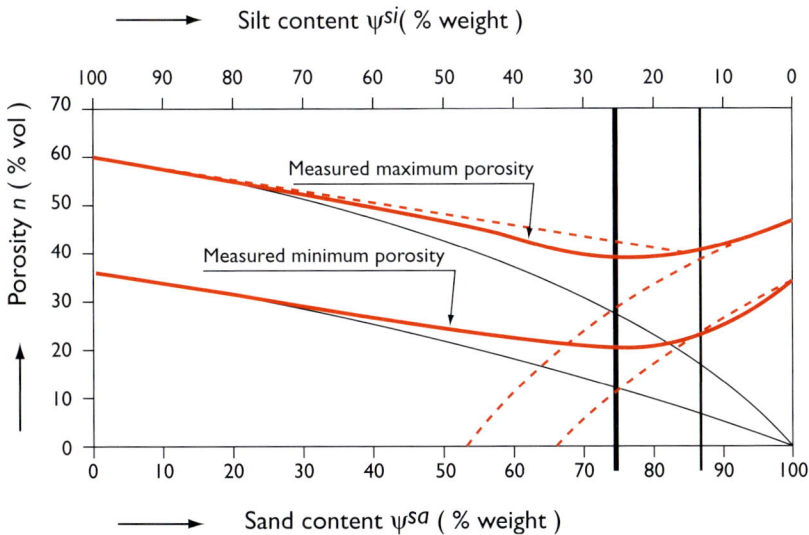

Figure 8.3   *Minimum and maximum porosity for sand/silt mixtures (Winterwerp 1996).*

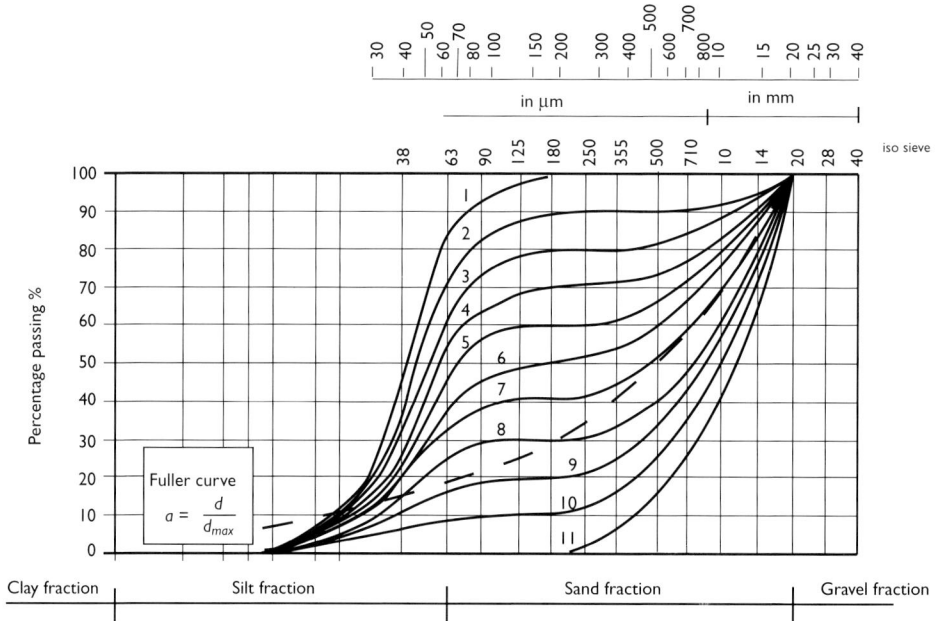

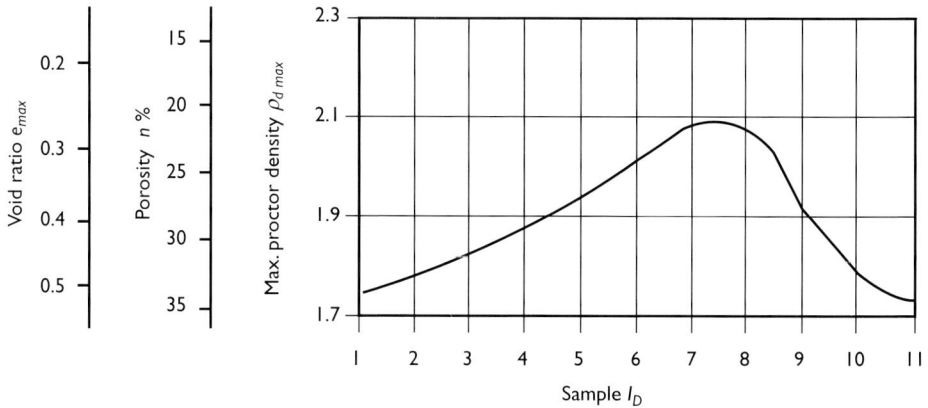

Figure 8.4 *Optimum density as function of fine content (Floss et al., 1968).*

### 8.3.5 *Density measurement*

#### 8.3.5.1 Measurement of reference densities (minimum and maximum density)

The minimum dry density of a soil is generally determined by pouring the soil through a funnel into a calibrated volume. The maximum dry density can be determined

198

by the Proctor test or by means of a vibrating hammer or vibrating table. These densities are often used as a reference level when evaluating the in situ dry density. Reference is made to Appendix B for a description of these laboratory tests.

### 8.3.5.2   Direct measurement of in situ density

Several methods exist to determine the in situ density of soil:

- the sand replacement or sand cone method (BS1377, ASTM D1556 and AASHTO T191);
- the core cutter method or drive cylinder method (BS 1377, ASTM D2937 and AASHTO T204);
- the weight in water and water displacement method (BS 1377 and AASHTO T233);
- the rubber balloon method (ASTM D2167 and AASHTO T205);
- nuclear density tests (BS 1377, ASTM D6938).

An overview of these methods is given in appendix B.

### 8.3.5.3   Indirect measurement of relative density by cone penetration testing

The relative density of the hydraulic fill can also be determined by indirect testing. Especially for sand layers deeper than 2 m below the surface or below the phreatic surface, it may be virtually impossible to determine the in situ density using the direct measurement methods. In granular material, the only option for taking undisturbed samples is by freezing the soil prior to the sampling. Often the Cone Penetration Test (CPT) is used to determine the relative density of those submerged or deep sand layers. Over the years several correlations have been established between cone resistance ($q_c$) and relative density.

Those correlations were based on calibration chamber tests on sands. Some commonly used reference 'research' sands along with their main properties are listed in Table 8.3.

It is noted that parameters can differ considerably depending on the type and mineralogy of the subject sand.

The resulting correlations are only valid for the type of sand that has been tested in the calibration chamber. Additionally, special attention should be paid to:

- The adopted test method for the determination of minimum and maximum densities. Proctor tests or vibrating table tests might give very different results.
- The definition of the term relative density.

Table 8.3   *Characteristics reference sands.*

| Sand | Ticino (TS) | Hokksund (HS) | Mol (MolS) | Quiou (QS) | Dogs Bay (DBS) |
|---|---|---|---|---|---|
| $D_{50}$ | 0.5 | 0.44 | 0.195 | 0.61 | 0.6 |
| $D_{10}$ | 0.41 | 0.27 | 0.13 | 0.1 | |
| $C_u$ | 1.58 | 2.2 | | 1.3 | 1.92 |
| %fines | 0 | 0 | | | |
| $\rho_s$ | 2.67 | 2.7 | 2.65 | 2.697 | 2.72 |
| $e_{max}$ | 0.915 | 0.906 | 0.918 | 1.303 | 1.72 |
| $e_{min}$ | 0.568 | 0.539 | 0.585 | 0.786 | 1.08 |
| Angularity | Sub-angular to angular | Angular | Sub angular | Sub-angular to sub rounded | |
| Mineralogy | 30% quartz 5% mica 65% feldspar | 35% quartz 10% mica 45% feldspar | >95% quartz | 12% quartz 73.5% shell fragments 14.5% $CaCO_3$ aggregates | 93% $CaCO_3$ |

To understand the basis of the correlations between $q_c$ and $R_e$ or $R_n$ the parameters which influence the measured cone resistance should be distinguished:

– in situ density of tested materials;
– in situ vertical and horizontal stresses;
– compressibility of the tested material;
– crushability of the tested material.

For a particular sand with a given compressibility and minimum and maximum void ratio, the relationship between cone resistance $q_c$, effective stress $\sigma'$ and relative density $R_e$ will be as follows:

$$R_e = A \cdot \ln\left(\frac{q_c}{B \cdot \sigma'^C}\right)$$

Where $\sigma'$ is either vertical effective stress or mean stress. A, B and C are soil constants, which are mostly determined by the compressibility of the sand.

A typical example of this formula is the Baldi correlation. See also appendix C, Correlations and correction methods.

The Baldi correlation (or other appropriate correlations) should be corrected in case of:

– compressible sands;
– calcareous or carbonate sands;
– saturated sand deposits;

- fine content >5%;
- CPT readings close to the surface (first 1 to 2 m).

More details about these correction methods can be found in appendix D, geo-technical principles.

8.3.5.4   Indirect measurement of relative density by SPT testing

Several correlations between the Standard Penetration Test (SPT) value and the relative density exist. One of the most commonly used correlations is the Gibbs and Holz correlation.

It should be noted that the British Standard classification of the relative density of a sand or gravel is based on the SPT value, see Table 8.4. This means that some correlation has been adopted already.

Table 8.4   *Sand and gravel: classification according to relative density.*

| | BS 5930 | | ISO 14688 | |
| | SPT | | Re | |
| Description | From | To | From | To |
| --- | --- | --- | --- | --- |
| very loose | 0 | 4 | 0 | 15 |
| loose | 4 | 10 | 15 | 35 |
| medium dense | 10 | 30 | 35 | 65 |
| dense | 30 | 50 | 65 | 85 |
| very dense | 50 | ... | 85 | 199 |

8.3.5.5   Measurement of in situ state parameter $\psi$ by cone penetration testing

In critical state soil mechanics, the state parameter $\psi$ is used to express the difference between void ratio and void ratio at critical state, for the same effective stress. This concept is used, among others, in liquefaction analysis and reference can be made to Section 8.6.2 for a basic explanation of this parameter.

State parameter $\psi$, the preferred measure of in situ state within a critical state framework, can be measured (interpreted) from cone penetration tests with reasonable precision. Without measuring the critical state properties of the sand (e.g. $M$, $\lambda$), shear modulus ($G_0$) or plastic hardening, interpretation relies on the friction ratio and $B_q$ parameters as indicators of material type and a precision in $\psi$ of $\pm0.08$ is possible. With supporting laboratory testing to determine the critical state parameters and measurement of $G_0$ with a seismic CPT, a precision in $\psi$ of $\pm0.02$ can be achieved. Plewes *et al.* (1992), Been and Jefferies (1992) and

Robertson (1998) describe slight variations on the approaches using the CPT alone, while Been and Jefferies (2006) and Shuttle and Jefferies (1998) describe the more precise methods.

### 8.3.6 *Typical relative density values of hydraulic fill before compaction*

The resulting relative density of a hydraulic fill prior to compaction is strongly influenced by the disposal method and disposal circumstances. An indication of the range of relative density which can be achieved can be found in Table 8.5.

In the EAU (2004 ) following figures can be found:

– Hydraulic filling below the waterline:
– $R_n = 0.35\text{–}0.55$ for fine sand with a mean grain size $d_{50} < 0.15$ mm
– $R_n = 0.15\text{–}0.35$ for fine sand with a mean grain size $d_{50} = 0.25$ to 0.5 mm

Table 8.5  *Typical relative densities—interpreted from Handboek Zandboek (2004).*

| Placement Method | Relative Density, $R_e$ |
|---|---|
| Saturated soil | |
| – discharged under water (spraying) | 20–40% |
| – discharged under water (dumped) | 30–50% |
| – discharged under water (overflow) | 20–40% |
| – discharged under water (rainbowing) | 40–60% |
| – discharged above water (free flow through pipe) | 60–70% |
| – discharged above water (rainbowing) | 60–80% |
| – inside a big hopper | 30–50% |
| – inside a small hopper | 10–30% |
| Dry soil | |
| – on a dump truck (filled from a funnel) | 10–20% |
| – on the discharge area (compacted by bulldozers) | 50–60% |
| – on the discharge area (compacted with compacting equipment) | up to 100% |

## 8.4  Strength of the fill mass and subsoil (bearing capacity and slope stability)

### 8.4.1 *Introduction*

Shear strength is an important parameter in view of overall safety with respect to relevant geotechnical failure modes of the structure. When speaking of the "strength" of a reclamation area, both the shear strength of the hydraulically placed fill material as well as the shear strength of the existing subsoil are meant.

In this section, the shear strength which can be expected for a hydraulic fill mass made of "suitable" and "unsuitable" material will be discussed. In further sections, attention is given to the strength required to fulfil the design/performance requirements of the fill mass.

To be able to specify these requirements, an understanding of the state of the fill material after hydraulic placement is necessary. In case the fill requires further improvement, knowledge of the potential improvement techniques and their results is required. It is finally essential to understand the potential failure mechanisms so that the designer can define a minimum shear strength required to guarantee the desired safety level.

Discussing the "strength" of a fill mass primarily refers to the Ultimate Limit State (ULS), while "stiffness" (section 8.5) refers to Serviceability Limit State (SLS). Both "Limit States" should be checked during the geotechnical design of the reclamation.

In the following paragraphs considerations will be given to "suitable" and "unsuitable" fill material. The definition of such materials is normally made by the designer and is defined in the technical specifications of a project. Without going into detail on this subject, suitable versus non-suitable fill is considered to be the difference between granular material such as sand and finer materials such as silt and clay. Detailed requirements on the allowable percentage of fines depend on many factors, such as the risk for segregation and the need for compaction, which are both discussed elsewhere. The effect of the fines content on the shear strength (and also on the stiffness) of granular material should be investigated before defining the maximum allowable percentage of fines in the specifications.

An important issue in dredging and hydraulic filling is the change of the material originally encountered in the borrow area as compared to the material which arrives at the reclamation site as a result of the dredging process. The original structure of the in-situ material is completely destroyed by dredging and a new structure is formed in the reclamation area. Apart from these structural changes, the particle size distribution may also change caused by loss of fine material during the dredging operation (at the cutter head or in the overflow of the hopper) or from the reclamation (process water runoff).

Segregation in the reclamation area cannot easily be avoided. This may change the particle size distribution of the material in the reclamation. Under some circumstances the fines content of the material placed in the reclamation area may become higher than the fines content of the material originally found in the borrow area. This can be caused by degradation of the material during dredging and hydraulic transport. Degradation of dredged material is discussed in section 9.2.

### 8.4.2 *Shear strength*

The shear strength of a reclamation area is primarily related to the type of fill material used. Distinction is made between "high quality fill material" and "poor quality fill material".

It should be noted that the terms "high quality" and "poor quality" do not refer to the quality of the resulting reclamation area. Even with poor quality fill material, a high quality reclamation area can be achieved, given a proper design.

#### 8.4.2.1 High quality fill material

In general, a "suitable" fill material is a granular material with a limited percentage of fines and stone-sized fragments. Additional requirements on chemical characteristics such as organic content, sulphates might be set, but such specifications are project dependent. The definition of "suitable material" will always be a project specification and for this reason the more general description of "high quality fill material" is used in this Manual.

The shear strength of a soil is defined by the Mohr-Coulomb equation:

$$\tau = c' + \sigma'_n \cdot \tan(\varphi')$$

Where:

| | |
|---|---|
| $\tau =$ shear strength (kPa) | $\sigma'_n =$ effective normal stress (kPa) |
| $c' =$ effective cohesion (kPa) | $\varphi' =$ effective friction angle (°) |

When shear stress is applied to a sand mass, the material will change in volume. Depending on its state of compaction and stress level, the sand will dilate (dense sand) or contract (loose sand), both converging to the "critical state" of shear without volume change. This is a very important characteristic of a granular medium that influences both the dredging process itself and the behaviour of the material in its final application. Figure 8.5 presents the principles of dilatation and contraction.

During shearing, dense sands with dilatant behaviour will demonstrate a peak shear strength followed by a lower residual (constant volume) shear strength. Loose sands will not demonstrate such behaviour and during shearing their shear strength will gradually increase to the residual (constant volume) shear strength at large deformations (see Figure 8.6).

For a granular material, factors affecting the shear strength are particle size distribution, particle shape, mineralogy, angularity, degree of compaction and stress

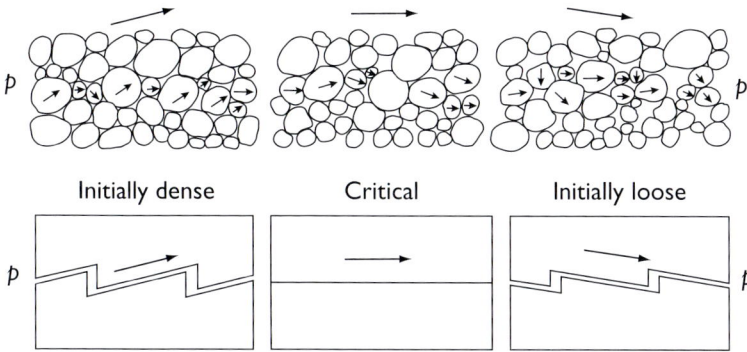

Figure 8.5   *Principle of dilatation (left) and contraction (right) demonstrated in granular material model and saw-blade model.*

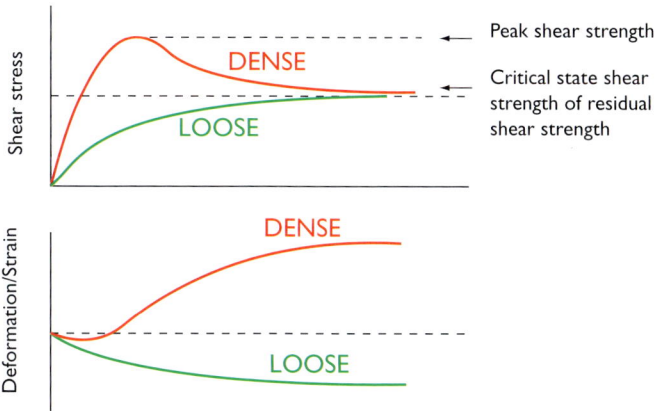

Figure 8.6   *Effect of volume change behaviour on shear strength measurements.*

level. For high quality fill materials, the shear strength is mainly dominated by the effective friction angle $\varphi'$. In the next section, three approaches are discussed for the prediction of the effective friction angle.

It should be stressed that these 3 approaches can be used for the preliminary design only. The adopted values should be validated by laboratory tests.

| a. | $\varphi'$ as function of effective normal stress & porosity | Figure 8.7, Terzaghi 1996 |
|----|----|----|
| b. | $\varphi'$ as function of mean effective stress & relative density | Figure 8.8, Bolton 1986 |
| c. | $\varphi'$ as function of SPT-value + angularity + grading | Ref. BS8002 |

## a. $\varphi'$ as function of effective normal stress & porosity

In Figure 8.7, the secant friction angle $\varphi'_s$ is plotted as a function of the effective normal stress $\sigma'_n$ and porosity n. This figure shows that in granular material the friction angle increases with increasing density and decreases with increasing effective normal stress.

The secant friction angle $\varphi'_s$ is the friction angle obtained for a failure point assuming no effective cohesion.

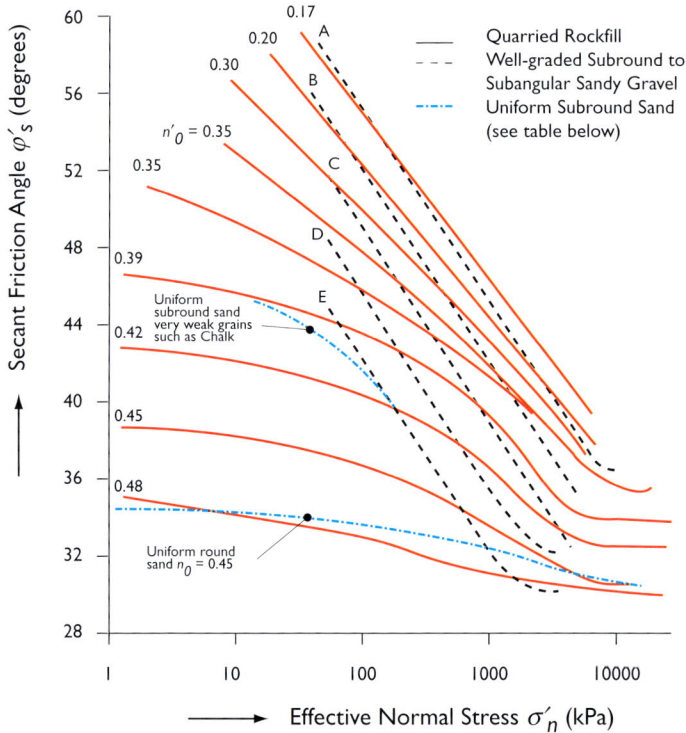

Figure 8.7  *Values of secant friction angle for granular soils (After Terzaghi et al., 1996).*

| Rockfill grade | Particle $q_u$ (MPa) |
|:---:|:---:|
| A | ≥220 |
| B | 165–220 |
| C | 125–165 |
| D | 85–125 |
| E | ≤85 |

Similar relations were defined for Mol sand and some other types of sand by De Beer (1964).

### b. $\varphi'$ as function of mean effective stress & relative density

Bolton has demonstrated that the effective peak friction angle of sand $\varphi'_{max}$ is dependent on the critical state friction angle $\varphi'_{crit}$ which is defined by the mineralogy (33° for quartz for example) and angularity, to be increased with a term dependent on relative density, stress level and plane strain or triaxial conditions. The term to be added is called $\varphi'_{max}-\varphi'_{crit}$. An empirical relation for this term is given in the literature (Bolton, 1986).

In Figure 8.8 an example of the values found for sand in triaxial condition is given. In this figure $p'$ is the mean effective stress and $R_e$ is the Relative void ratio (relative density). $\varphi'_{max}-\varphi'_{crit}$ is the difference between the effective friction angle for the material at its degree of compaction and stress state and the critical friction angle which is a material parameter. Again this figure illustrates that the effective friction angle increases with increasing compaction and decreasing effective stress.

### Remark

$\varphi'_{crit}$ is also called the constant volume friction angle $\varphi'_{cv}$.

When a hydraulic fill is realised with sand, below or above the water table, the sand particles settle quickly, forming a grain skeleton structure. The resulting relative density depends on the method of placement (see Section 8.3.6). As such, the

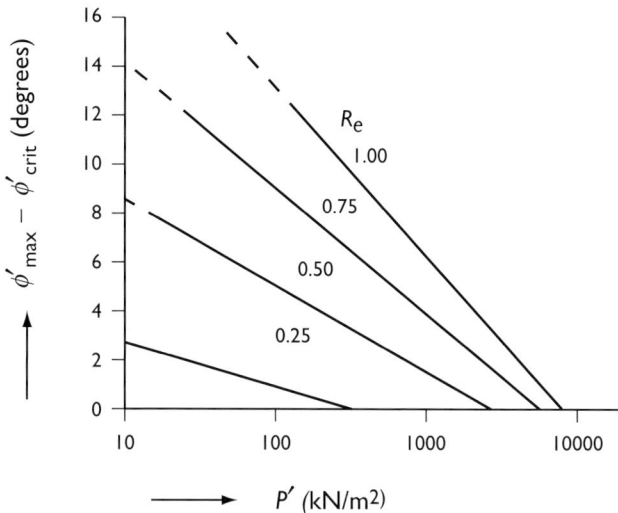

Figure 8.8 $\varphi'_{max} - \varphi'_{crit}$ for sands in triaxial testing (Bolton, 1986).

sand mass will almost immediately obtain a minimal shear strength equivalent to the "critical friction angle" which is defined by mineralogy, grain shape and angularity. Figure B-17 in Appendix B presents the classification of grain angularity according to the Powers (1953) scale.

For a clean silica sand, $\varphi'_{crit}$ can easily be chosen as minimal 32° (33° with a precision of 1° as given by Bolton). Consequently, the effective friction angle of recently deposited quartz sand, is never lower than 32°. For a different mineralogy or when the sand contains fines, other values apply.

Figure 8.7 and Figure 8.8 demonstrate clearly that compaction will increase the effective friction angle and thus the shear strength. This also explains why sometimes compaction is required to achieve certain shear strength, even when deformations are not a critical issue. This may, for example, be the case for sand fill in the zone of influence of a quay wall or to avoid liquefaction when fill is placed in earthquake areas.

### c. $\varphi'$ as function of SPT-value, angularity and grading
BS 8002:1994 (2001) is presenting a method to estimate the peak friction angle based on the findings of Bolton.

Please note that BS 8002:1994 has the status of 'superseded, withdrawn' and has been replaced by Eurocode 7: Geotechnical design - BS EN 1997-1.

This method requires the input of the SPT-value, angularity and grading of the granular material. Indirectly, the effective stress and state of compaction are also taken into account, since they will affect the measured SPT-values.

The estimated peak effective friction angle is given by:

$$\varphi'_{max} = 30 + A + B + C$$

The estimated critical state friction angle is given by:

$$\varphi'_{crit} = 30 + A + B$$

The factors A, B and C depend on different parameters:

- A: angularity of the particles;
- B: grading of the sand/gravel;
- C: results of the Standard Penetration Test,

and are given in Table 8.6.

Table 8.6  *Shear strength for siliceous sand and gravel ( BS 8002 ).*

| Constant | Values | Remarks |
|---|---|---|
| A: Angularity | A (in degrees) | Angularity is estimated from visual description of soil |
| Rounded | 0 | |
| Sub-angular | 2 | |
| Angular | 4 | |
| B: Grading of soil | B (degrees) | Grading can be determined from grading curve by use of the uniformity coefficient $C_U = D_{60}/D_{10}$ as specified below* |
| Uniform | 0 | $C_U < 2$ |
| Moderate grading | 2 | $C_U$ 2 to 6 |
| Well graded | 4 | $C_U > 6$ |
| C: $N'$ (blows/300 mm) | C (degrees) | $N'$ from results of Standard Penetration Test modified where necessary by Figure 8.9. |
| <10 | 0 | |
| 20 | 2 | |
| 40 | 6 | |
| 60 | 9 | |

* Intermediate values of A, B and C by interpolation.
* * A step-graded soil should be treated as uniform or moderately graded soil according to the grading of the finer fraction.

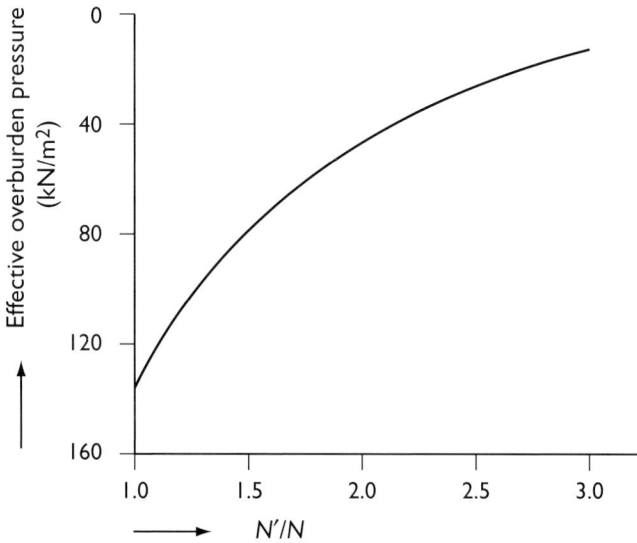

Figure 8.9  *Derivation of N' from SPT value N ( BS 8002 ).*

All friction angles described in the above figures and theories are "secant friction angles". In general effective cohesion $c'$ is considered to be zero ($c' = 0$ kPa) for recently deposited sand, although sometimes a certain "cohesion" is found for granular material in shear box and triaxial tests. This can be caused by cementation or interlocking of the grains. Calcareous sands can exhibit some cementation even very soon after placement of the hydraulic fill.

Cohesion can also be a result of partial saturation of the sand caused by suction pressure and often described as "apparent cohesion" (Fredlund and Rahardjo, 1993). This issue will not be discussed here but it can be of importance in fresh hydraulic fill above the water table which is partially saturated. On the other hand, since the degree of saturation might change in the course of weeks or months after placement of the fill, it might be unreliable to take this into account.

### 8.4.2.2 Poor quality fill material

In some cases, poor quality fill material will have to be used. Reference is also made to section 9.1 Cohesive or fine-grained fill materials for a more detailed description on the use of poor quality fill material. A shortage of sand sources in the close vicinity of a project or environmental constraints may have as a consequence that one is forced to work with cohesive materials for the land reclamation. In these cases the shear strength of the fill, and its increase in time, is much more difficult to predict compared to the use of high quality fill materials. A good knowledge of dredging and deposition methods is necessary.

In general, poor quality fill material can be everything which is not "suitable"; i.e., sand with too high fine content, silt or clay. The shear strength of these materials is influenced by the consolidation behaviour and can be defined by the effective strength characteristics $\varphi'$ and $c'$ (in combination with a time dependent excess pore pressure) or by the undrained shear strength $c_u$. The choice between undrained or effective shear strength parameters depends on the type of material and the rate of loading or unloading.

Sometimes, for the undrained strength, the parameter $\varphi_u$ ($\neq 0$) is used. This approach is, however, not well defined (how to derive this parameter?). This can lead to misunderstanding and sometimes even to mixing up with effective strength parameters. This approach is not recommended and therefore the undrained shear strength $c_u$ will be used in this Manual, in combination with $\varphi_u = 0$.

**Implications of using poor quality fill material**

The use of poor quality fill materials can have serious implications such as:

When, for example, very silty fine sand is dredged and hydraulically placed, the material will dewater rather slowly. During the filling operation itself, driving with vehicles over this type of material is not possible. However, when the fill becomes partially saturated or fully dry, it will become a very stable layer. Working with very fine non-cohesive saturated material can be dangerous since this material is sensitive to vibrations and may liquefy locally as a result of the vibrations of bulldozers and excavators working on site.

When silt or very silty sand is pumped as backfill material behind a quay wall or other retaining structure, it is imperative to realise that, in the short term, this material could behave like a "heavy fluid" having a negligible strength and fluid density equal to the density of the pumped mixture. This should be taken into account in the analysis of the temporary stability of the retaining structure.

When clay is pumped into the reclamation area after hydraulic dredging, its shear strength is very difficult to predict. The development of the strength increase can take a long time unless special measures are taken for dewatering and/or soil improvement (reference is made to Chapter 7 and section 10.2). As a rule of thumb the water content of the sedimented clay material can be expected to be 1.5 to 2 times the liquid limit of the clay after some weeks (Skempton, 1970; Buchan and Smith, 1990; Tsuchida and Gomyo, 1995).

In some cases, mechanical dredging of the soft material can be a more advantageous solution since the shear strength of the relocated clay material will start from its remoulded/residual shear strength, which is higher than the shear strength of hydraulically dredged and placed slurry.

The sensitivity of clay is defined as the ratio of the remoulded undrained shear strength to the peak undrained shear strength.

The behaviour of a hydraulically dredged and placed slurry depends very much on its grain size and mineralogy. Clay that has been dredged and transported hydraulically will generally have a very low density and clay particles may be completely in suspension. When pumping such a slurry into a closed reclamation area, the first phenomenon to appear is sedimentation. Sedimentation can occur rather quickly (in a period of days) and the free water can be drained off easily. Once the clay particles form a soil skeleton, self-weight consolidation and–at the surface–desiccation starts (Stark, 1984). This can be a very lengthy process (periods of months to years). In this stage, the clay has a very low shear strength, and it will be impossible to work on it.

After full consolidation the effective strength characteristics of the clay can be linked to the plasticity index of the clay (Figure 8.10).

Similarly, the undrained shear strength of a normally consolidated clay can be linked to its consistency, expressed as the liquidity index as shown in Figure 8.11. (see also box soil consistency in section 9.1.3.2).

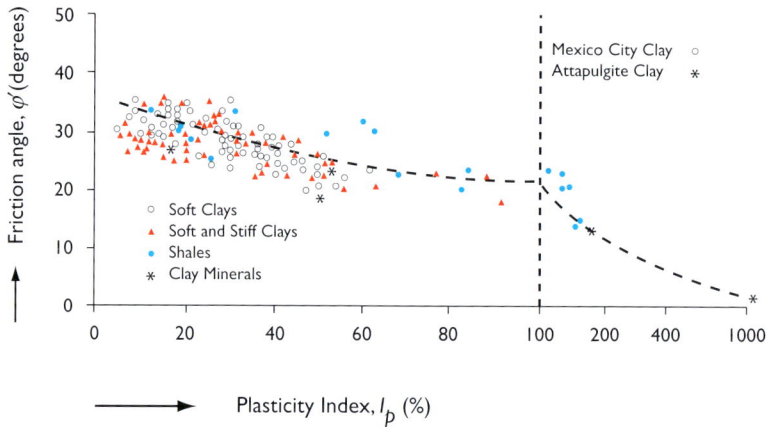

Figure 8.10    *Effective friction angle for normally consolidated clays (Terzaghi et al., 1996).*

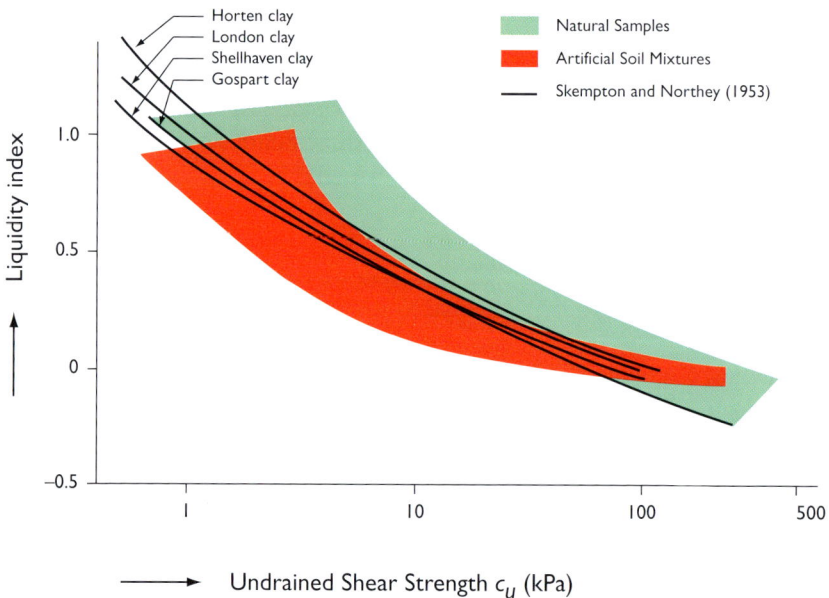

Figure 8.11    *Shear strength of soft normally consolidated clay as a function of liquidity index (Wasti and Bezirci, 1986 after Skempton and Northey, 1953).*

The undrained shear strength of a soft clay is often expressed as a function of the effective consolidation stress. For normally consolidated cohesive soil the following formula can be used:

$$c_u = \alpha \cdot \sigma'_{v0}$$

Where

| | | |
|---|---|---|
| $c_u$ | = Undrained shear strength | (kPa) |
| $\sigma'_{v0}$ | = Effective vertical stress | (kPa) |
| $\alpha$ | = Coefficient, often $\alpha = 0.22$ is taken when no information is available | (–) |

The factor $\alpha$ is sometimes expressed in terms of plasticity index as given in formula 8-1 (Skempton, 1957).

$$\alpha = (0.11 + 0.0037 * PI) \qquad (8\text{-}1)$$

The undrained shear strength of a clay not only depends on the consolidation stress of the sample, but may also depend on the testing method and the stress history of the material as illustrated in Figure 8.12. The factor $\alpha$ found from $c_u / \sigma'_v$ is not a unique value. Theoretically an $\alpha$-value must be defined/selected depending on the loading condition as illustrated in the figure. This is not, however, always possible and often an average $\alpha$-value is used.

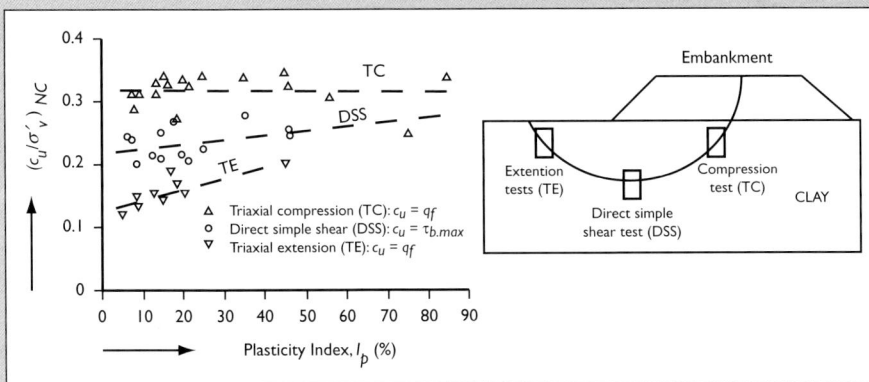

Figure 8.12 *Undrained shear strength related to effective stress, plasticity index and testing method (Ladd, 2003).*

Based on this model, the strength increase under new loading conditions can be predicted. As consolidation progresses, the effective vertical stress increases, resulting in an increased undrained shear strength. This approach is often used in the design of a staged reclamation fill on top of soft clay.

The following general formula can be used:

$$c_u(t) = c_{u,0} + \alpha \cdot U(t) \cdot \Delta\sigma'_v$$

Where:

| | | |
|---|---|---|
| $c_u(t)$ | = undrained shear strength at time $t$ after loading | (kPa) |
| $c_{u,0}$ | = undrained shear strength of the soil before loading | (kPa) |
| $U(t)$ | = degree of consolidation at time $t$ after loading | (–) |
| $\Delta\sigma'_v$ | = increase of effective vertical stress due to loading after full consolidation. | (kPa) |

This model is valid for clays only. Using the above presented formula, adopting a factor $\alpha$ for silt, results in a very low undrained shear strength for this material. This may not always be correct. For silt, the concept of undrained shear strength must be used with due consideration. A judgement for each individual situation should be made whether this would be a correct approach or not.

**The shear strength of a deposit of poor quality fill materials can be increased in various ways.**

When the top level of a fresh (very) soft clay deposit is below the water table, a sand layer can very gently be installed upon it by means of a spraying pontoon. This will increase the stress and improve the self-weight consolidation process. Such a sand layer will allow for access of light equipment or for the deposition of subsequent sand layers by means of land pipelines, using the advantage of gentle slopes realised by hydraulic fill (see Section 9.1.3.5 and 9.1.3.6).

The most evident technique of soil improvement of soft material is the use of a surcharge in combination with prefabricated vertical drains (PVDs) in order to improve the rate of consolidation.

When large deposits with soft material have to be realised, a sandwich construction technique can be implemented with alternating clay and sand layers placed upon each other (in this case the clay is mechanically dredged). An example of this technique can be found in Muthusami et al. (2004).

Alternatively, the use of binders can be considered in order to obtain a stabilised mass, although after hardening of the binder further consolidation should not be expected. Description of the cement mixing method, which can even be realised during the pumping process, is given in the Premixing Method, Principle, Design and Construction (2003).

A technique being researched by the dredging industry is the use of flocculants to accelerate sedimentation and consolidation processes which can further be improved by PVDs and surcharge.

**Stiff to hard clay**

Stiff to hard clay dredged and pumped into a reclamation site will result in the creation of "clay balls", which are lumps of original clay that have not disintegrated. The part of the clay that disintegrates forms a (dense) clay slurry as it mixes with the process water. This leads to a heterogeneous fill with clay balls which settle close to the end of the pipeline and a slurry that partly remains within the macro pores in between the clay balls.

Various researchers have concluded that most of the inter-lump voids in a lumpy clay fill will close up after surcharging. However, the resulting fill will still show a considerable heterogeneity when it comes to strength parameters. It is believed that the highest measured shear strength values correspond to the original intact clay balls or lumps, while the lower shear strengths values correspond to the interface between those clay balls. In situ testing is the only method to correctly assess the shear strength in this case.

Some examples of working with "poor quality fill material" are discussed in section 9.1.

### 8.4.2.3   Assessment of shear strength

The assessment of the shear strength of a fill is an evolving process during the design and execution of a hydraulic fill project. A different approach is followed for high and poor quality fill materials. Generally, following phases can be identified:

**A. Preliminary Design**
Based on the particle size distribution and the mineralogy of a high quality fill material, its shear strength can be predicted following methods such as described in section 8.4.2.1. Bear in mind that most of the literature data applies for silica sand only.

For low quality fill materials, the resulting shear strength is more difficult to predict, and depends on the placement methods. For hydraulic fill it is a safe assumption to assume an initial shear strength of 0 kPa. During the consolidation process, the shear strength of the fill can be assessed as discussed in section 8.4.2.2.

For mechanical dredging, the remoulded shear strength can be used as initial shear strength.

**B. Detailed design**
The shear strength of the envisaged fill material should be known during the design phase of the reclamation. Laboratory testing on the material encountered within

Table 8.7 *Shear strength analysis - an evolving process.*

| | Project phase | High quality fill Material (sand, gravel) | Poor quality fill Material (silt, clay) |
|---|---|---|---|
| A | Preliminary design | Estimation of shear strength (or $\varphi'$) by means of Figure 8.7, Figure 8.8, BS8002 approach or similar. | – Estimate shear strength based on remoulded shear strength for mechanical dredging.<br>– assume initial shear strength = 0 kPa for hydraulically dredged slurry<br>– after consolidation, estimate shear strength based on Figure 8.10 or Figure 8.11 (or similar) |
| B | (Detailed) design | Lab testing: determine correlation $\varphi'$, relative density at various stress conditions | Lab testing: simulate sedimentation, consolidation and desiccation process in laboratory, and determine appropriate shear strength. |
| C | Setup of specifications | – specify minimal $\varphi'$ for sand, or<br>– set minimal relative density level based on correlation $\varphi'$, relative density | – specify minimum required shear strength for subsequent building phases and/or final completion. |
| D | Execution/ Monitoring | – measurement of relative density (ref 7.3 density)<br>– used direct correlation between $\varphi'$ and CPTU or SPT measurement<br>– other | – assessment of undrained shear strength by field vane testing and/ or correlation to COPTU testing<br>– other |

the borrow area is a first requirement. The change of composition of the material caused by the dredging process should also be taken into account.

**High quality fill materials (sand)**

For example, dredging sand with a trailer hopper suction dredge will, owing to the process of overflow, result in a certain loss of the finer particles. The material selected for laboratory testing will have to be adapted by reducing the fines content in order to account for this effect. The fines can be washed out by wet sieving on a 63 $\mu$m sieve. If required a limited amount of fines can be re-added to the sample.

During the design of a project, laboratory shear strength tests can be performed on reconstituted (adapted) sand samples at different relative densities and at different stress conditions. The results will allow for a much more reliable prediction of the shear strength of the fill material. Such an investigation is also important to evaluate if compaction of the fill material after placement will be necessary. When performing laboratory tests, a sensitivity analysis can be made to investigate the influence of the fines content on the shear strength of the fill material.

**Poor quality fill materials (clay, silt)**
The destruction of the soil matrix during dredging can be simulated by mixing a soil sample with water. The amount of water to be added depends on the dredging method. When dredging with a cutter suction dredger, a mixture of 35% of soil to 65% of water is appropriate. After artificial consolidation, the resulting undrained shear strength can be determined. The desiccation process can also be simulated (reference is made to section 9.1).

In case segregation of the silt & clay fraction is expected on the reclamation area, the properties of this fraction should also be determined. This can be achieved by sieving of the fraction smaller than 63 μm. With this fraction, a slurry can be prepared with a void ratio similar to what is expected on reclamation site. Subsequent tests on this slurry are: deposition simulation testing, consolidation testing and testing of the undrained shear strength.

## C. Specifications
**High quality fill materials (sand)**
Having defined the correlation between relative density, stress level and shear strength for a certain sand material, the specification for the hydraulic fill comes back to a relative density requirement. Alternative, a direct correlation between the shear strength (or angle of internal friction) and the adopted monitoring test can be used. (reference is made to Appendix C, Correlations and correction methods).

**Remark**

When defined or specifying the required strength of a fill mass, a fundamental choice has to be made. A choice can be made for a high strength. This may result in a lot of compaction effort, but will possibly lead to a smaller and/or cheaper foundation design.
On the other hand, when choosing for a lower strength, compaction may be limited or even not required. This will mean an immediate cost saving, but may result in more expensive foundation structures.
The specified strength ($\varphi$) should however be feasible, hence the importance of testing.

**Poor quality fill materials (clay, silt)**
The required shear strength can be measured directly via a field vane test or can be deduced from a CPT(U) test (Lunne *et al.*, 1997). In such case, it is strongly advised to setup a site specific correlation between the CPT(U)-test and the shear strength. This can be realised when at the same location of some CPT(U) tests also in situ van shear tests are performed.

### D. Execution/Monitoring

The relative density of the fill can be tested by several methods after realisation of the fill. The most common method is the use of the Cone Penetration Test (CPT). The Standard Penetration Test (SPT) is an alternative method although less frequently used. The CPT test has the advantage to provide a continuous profile. Some correlations between CPT or SPT results and relative density can be found in appendix C, Correlations and correction methods.

Direct correlations to define the effective friction angle for granular soil based on CPT or SPT results also exist. Some examples are presented in appendix correlations. Alternatively, correlations applicable for CPTs can be used after estimation of the cone resistance based on the SPT-$N_{60}$ value by means of the $q_c$-SPT correlations. A disadvantage of using SPT-N values is that it is very often unclear whether the reported value is the measured or corrected $N_{60}$-value. Information on how to define the appropriate correction factor is also often missing. In Table 8.8 a correlation between SPT-$N_{70}$, relative density Re and shear strength is given. In this table $N_{70}$ is used instead of the more common $N_{60}$.

The undrained shear strength $c_u$ of cohesive material can directly be measured on site by means of the in-situ vane shear test. This test is also used to calibrate the correlation between the undrained shear strength and the cone resistance of a CPT or the N-value of an SPT.

A frequently used correlation to derive $c_u$ from a CPT is $c_u = \dfrac{q_c - \sigma_{v0}}{N_k}$, where $\sigma_{v0}$ is the total stress and $q_c$ the cone resistance. $N_k$ is an empirical factor that can vary between 11 and 19 for marine clays, with an average of 15 (Lunne and Kleven, 1981). For more detailed information on the factor $N_k$ and its relation with other soil parameters such as for instance plasticity, reference is made to Lunne *et al.* (1997).

Correlations for the undrained shear strength $c_u$ with the SPT-N value exist as well, although for very soft to soft clays this test is often not accurate.

Table 8.8   *Empirical values for φ, relative density and unit weight of granular soil based on the SPT $N_{70}$ at about 6 m depth and normally consolidated (table 3–4 Bowles 1997).*

| Description | | Very Loose | Loose | Medium | Dense | Very Dense |
|---|---|---|---|---|---|---|
| Relative density Re | | 0% | 15% | 35% | 65% | 85% |
| SPT $N_{70}$ | Fine | 1–2 | 3–6 | 7–15 | 16–30 | ? |
| | Medium | 2–3 | 4–7 | 8–20 | 21–40 | >40 |
| | Coarse | 3–6 | 5–9 | 10–25 | 26–45 | >45 |
| $\varphi'$ [°] | Fine | 26–28 | 28–30 | 30–34 | 33–38 | <50 |
| | Medium | 27–28 | 30–32 | 32–36 | 36–42 | |
| | Coarse | 28–30 | 30–34 | 33–40 | 40–50 | |
| $\gamma$ [kN/m³] | | 11–16 | 14–18 | 17–20 | 17–22 | 20–23 |

Table 8.9    *Correlation SPT-N and $c_u$ (Terzaghi et al., 1996).*

| N-value | Approximate $c_u$ [kPa] |
|---------|------------------------|
| 0 to 2 | 0 to 12.5 |
| 2 to 4 | 12.5 to 25 |
| 4 to 8 | 25 to 50 |
| 8 to 15 | 50 to 100 |
| 15 to 30 | 100 to 200 |
| >30 | >200 |

In Table 8.9 (Terzaghi 1996), an almost constant relationship of $c_u/p_a = 0.06$ . N (with $p_a$ = the atmospheric pressure) is used. However, this is not a unique relationship, but depends of the plasticity of the clay, as illustrated in Appendix C, Correlations and Correction Methods.

The effective shear strength can be tested in the laboratory by means of different test procedures such as a the direct shear test, simple shear test and the triaxial test. Normally testing is performed on saturated material. For every test setup, a clear distinction must be made between testing the material under drained or undrained conditions. From this point of view triaxial tests, type CU (Consolidated Undrained) or CD (Consolidated Drained) allow for a better control of the drainage conditions and porewater pressures than for instance shear box tests. For further elaboration on these types of tests, reference is made to textbooks on Soil Mechanics and laboratory testing (Head, 2006).

An issue very often neglected is the difference in shear strength parameters for soils in triaxial and plane strain conditions. Laboratory testing in plane strain conditions is very exceptional and limited to specialised research laboratories. According to the literature (e.g. EAU, 2004), for granular soils the effective friction angle is in plane strain conditions 10% higher compared to triaxial conditions.

Undrained shear strength testing of cohesive material (e.g. clay) can be realised in the laboratory by means of the vane shear test or the triaxial test, type UU (Unconsolidated Undrained) or UCS (Unconfined Compression Strength) on undisturbed samples. In such triaxial tests, sample preparation and sample size may influence the results.

**Correction of field vane results**

A specific point of attention is that the results of vane shear tests need to be corrected for plasticity, anisotropy and rate effects according to the formula $c_u = \mu \cdot c_{vane}$. Such a correction factor is based on experience. Bjerrum (1972) has published a figure with the correction factor $\mu$ as a function of the plasticity index $I_p$ while other authors have suggested similar correction factors as a function of the liquid limit $w_1$. Figure 8.13 presents an example of such a correction factor (Eurocode 7, Part 2, 2006—Annex I). This correction is valid for both field and laboratory vane tests.

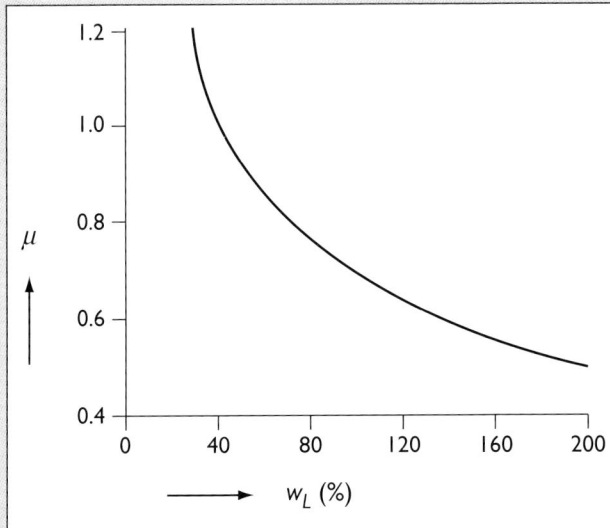

Figure 8.13   *Correction of vane shear test results based on the liquid limit ($w_L$) of the clay (EC7, Part 2, 2006).*

Effective strength testing of soft clay by means of triaxial testing sometimes leads to unexpected results with effective friction angles much higher than what would be expected based on, for example, Figure 8.10. This is caused by the large deformations which are allowed in triaxial tests (up to 15% axial strain), while no peak strength may be found in this range. For the design of structural elements the peak (or maximum) shear strength is not always used. Frequently only the shear strength is mobilised that belongs at a certain level of deformation (e.g. 2% or 5% axial deformation in the triaxial test). This approach seems logical for the design of a structure which allows for limited deformations only. It is, on the other hand, not in line with the approach of the Ultimate Limit State (ULS) where the stability is analysed in the failure mode, irrespective of the magnitude of the deformations. As such, two approaches are mixed and the safety approach in the ULS method cannot be used when working with such (reduced) shear strength values.

### 8.4.3  *Relevant failure modes*

#### 8.4.3.1  Introduction

The shear strength of a hydraulic fill material is crucial in the design and execution of reclamation projects. During the execution stage, it is essential to be able to access the fresh fill with bulldozers and backhoes, in order to advance with land pipelines and to construct containment bunds. The initial stability of the working platform is very important with respect to the feasibility of working methods (lift thickness, compaction in between subsequent lifts, surcharge procedures, timing and execution of soil improvement techniques such as the installation of PVDs, etc.).

After completion of the filling works, the stability of the structures to be built on the reclamation area has to be guaranteed. Sufficient safety for temporary situations during construction should, however, also be satisfied. Therefore sufficient attention must be paid to the work method in relation to stability issues during intermediate situations (slopes of hydraulic fill or surcharge, etc.). These aspects are normally not verified by the hydraulic fill designers, but these are important issues for the hydraulic fill contractor. Sometimes containment bunds are designed and constructed by a main civil contractor while the dredging sub-contractor only has to pump in hydraulic fill material. If stability becomes a problem, it could delay the project schedule to a great extent.

When fill is to be placed behind a vertical quay wall, the structure will be designed based on a certain assumption for the shear strength of the fill material. The shear strength has to be guaranteed during and after the execution of the reclamation. In case this shear strength might not be practically achievable, an adjustment in the design of the structure is inevitable. On the other hand, when a higher shear strength can be guaranteed for the fill material, then the design of the structure can be optimised. In order to obtain an economical design and to avoid problems during the execution of the project, it is important to have sufficient information on the quality of the fill material (already during the design phase) and to stimulate a close cooperation between the structural designer and the dredging engineer.

Bearing capacity is often a requirement in specifications for hydraulic fills. This is normally related to applied load or bearing pressure from equipment to work on the fill, shallow foundations or the future loads to be carried by the reclamation. However, when a fill layer of limited thickness is to be placed over soft subsoil, a verification of other possible failure mechanisms, such as punch through and squeezing should be carried out as well. Furthermore, slope stability is one of the main issues in land reclamation projects. Sufficient safety against loss of stability should be guaranteed during the execution (temporary boundary slopes and bunds) and after the completion of the works. These subjects will be discussed in the following sections.

### 8.4.3.2 Safety approach

A diversity of failure mechanisms need to be analysed during the design phase of a reclamation project in order to guarantee sufficient safety. One of the main failure modes to be checked is the Ultimate Limit State (ULS). The purpose of ULS calculations is to determine the minimum safety factor in case the available strength is fully mobilised.

The different failure modes discussed in this chapter all relate to the Ultimate Limit State. When checking the safety, the maximum available strength of the soil to be mobilised is defined. The corresponding safety factor is then calculated for the actual condition. Generally, the required value of the safety factor reflects both the uncertainty with respect to the soil parameters and loads as well as the extent of the consequences should a failure occur.

Two safety approaches exist which are briefly discussed in this section. In principle, one of the following safety approaches can be followed:

– an analysis based on partial safety factors;
– an analysis based on an overall safety factor.

**Partial safety approach**
In the partial safety approach the uncertainties of all parameters are taken into account: this principle is not only used for soil mechanics but also for other disciplines. The first step is to define characteristic values for the soil parameters. This already is a difficult exercise and requires extensive testing and analysis of the soil in geotechnically homogeneous layers.

The characteristic value is a safe estimation of the average value over a large homogeneous soil zone or the 5% lower bound fractile of all values measured. The first assumption is valid when the failure mode studied mobilises a large soil volume, while the second assumption applies for small foundation elements only. Eurocode 7, Part 1 (2004) states in this respect: "*A cautious estimate of the mean value is a selection of the mean value of the limited set of geotechnical parameter values, with a confidence level of 95%; where local failure is concerned, a cautious estimate of the low value is a 5% fractile.*"

Partial material factors are applied on the characteristic strength values ($\tan(\varphi'_k)$, $c'_k$ and $c_{u,k}$) in order to arrive at the "design values". The characteristic values have to be divided by 1.25 for the effective shear strength parameters and by 1.40 for the undrained shear strength.

Partial safety factors are also applied on loads (with distinction of permanent loads and mobile loads) in order to cover the uncertainties of these loads (actions). Permanent loads also include the soil weight itself. In EC7, sometimes a model factor is introduced as well. This needs to be explicitly specified in the National Annexes to EC7 applicable for that country. Also dimensional elements (e.g. depth of excavation) and water levels may have to be increased or decreased to cover possible uncertainties.

Taking into account all the above discussed partial factors, the ratio of resisting over driving forces obtained from the calculation model needs to be $\geq 1.0$.

For detailed information reference is made to Eurocode 7 and related international literature. In EC7, different "Design Approaches" are possible. Within a Design Approach different sets of partial safety factors exist in which the safety is either more focussed on the soil at one side or on the loads on the other side. Table 8.10 presents the partial safety factors according to EC7. For design approach one, two different combinations of partial safety factors must be verified. For more specific information on the Design Approach and the partial safety factors applicable in a certain country the National Annex for that country should be consulted.

**Overall safety approach**
The overall safety approach is a more classical approach in which the calculations are performed using characteristic values for soil parameters and expected loads. The calculation model results in a Factor of Safety (FS) which has to be higher than a certain specified value (larger than 1.0). This factor depends on the type of Limit State to be checked and may also depend on the accuracy of the soil parameters.

For bearing capacity calculation for shallow foundation a FS = 2 is often adopted.

When using the overall safety approach in slope stability analysis, the minimum required factor of safety varies in between 1.3 and 1.5 for permanent conditions and in between 1.1 and 1.3 for temporary conditions during construction. (see Table 8.11) Some standards or guidelines mention safety factors to be respected but it is not always clear to which calculation model they apply. The required factor of safety should be related to several aspects:

− Quantity and quality of soil investigation.
− Loading conditions.
− Consequences of failure (human life, material damage, etc.). For example, the failure of an underwater slope during the hydraulic fill operations may often not be critical.
− Uncertainty linked to the model used for the analysis.

Table 8.10  Partial Safety factors according to EC7.

| | | Design Approach 1 | | | | | | | | | | Design Approach 2 | | | | | Design Approach 3 | | |
|---|---|---|---|---|---|---|---|---|---|---|---|---|---|---|---|---|---|---|---|
| | | Combination 1 | | | Combination 2 and anchors | | | Combination 2 - piles | | | | Combination 1 | | | Slopes | | Combination 1 | | |
| | | A1 | M1 | R1 | A2 | M2 | R1 | A2 | M1 .. or | M2 | R1 | A1 | M1 | R2 | A1 | M=R2 | A1* or A2** | M2 | R3 |
| **Actions** | Permanent Unfav | 1.35 | | | | | | | | | | 1.35 | | | 1.35 | | 1.35 | | |
| | Fav | | | | | | | | | | | | | | | | | | |
| | Variable Unfav | 1.5 | | | 1.3 | | | 1.3 | | | | 1.5 | | | 1.5 | 1.3 | 1.5  1.3 | | |
| **Soil** | tan $\phi'$ | | | | | 1.25 | | | | 1.25 | | | | | | | | 1.25 | |
| | $c'$ | | | | | 1.25 | | | | 1.25 | | | | | | | | 1.25 | |
| | $cu$ | | | | | 1.4 | | | | 1.4 | | | | | | | | 1.4 | |
| | UCS | | | | | 1.4 | | | | 1.4 | | | | | | | | 1.4 | |
| | Unit weight | | | | | | | | | | | | | | | | | | |
| **Spread footing** | Bearing | | | | | | | | | | | | | 1.4 | | | | | |
| | Sliding | | | | | | | | | | | | | 1.1 | | | | | |
| **Driven piles** | Base | | | 1.25 | | | | | | | 1.3 | | | 1.1 | | | | | |
| | Shaft (compression) | | | 1.25 | | | | | | | 1.3 | | | 1.1 | | | | | |
| | Total/combined | | | | | | | | | | 1.3 | | | 1.1 | | | | | 1.1 |
| | Shaft in tension | | | | | | | | | | 1.6 | | | 1.15 | | | | | |
| **Bored piles** | Base | | | 1.25 | | | | | | | 1.6 | | | 1.1 | | | | | |
| | Shaft (compression) | | | 1 | | | | | | | 1.3 | | | 1.1 | | | | | |
| | Total/combined | | | 1.15 | | | | | | | 1.5 | | | 1.1 | | | | | 1.1 |
| | Shaft in tension | | | 1.25 | | | | | | | 1.6 | | | 1.15 | | | | | |

| | | R1 | R2 | R3 | |
|---|---|---|---|---|---|
| **CFA piles** | Base | 1.1 | 1.45 | 1.1 | |
| | Shaft (compression) | 1 | 1.3 | 1.1 | |
| | Total/combined | 1.1 | 1.4 | 1.1 | |
| | Shaft in tension | 1.25 | 1.6 | 1.15 | 1.1 |
| **Anchors** | Temporary | 1.1 | 1.1 | 1.1 | |
| | Permanent | 1.1 | 1.1 | 1.1 | |
| **Retaining wall** | Bearing capacity | | | 1.4 | |
| | Sliding resistance | | | 1.1 | |
| | Earth resistance | | | 1.4 | |
| **Slopes** | Earth resistance | | | | 1.1 |

* on structural actions
** on geotechnical actions

For slope stab, actions on the soil (e.g. structural actions, traffic load) are treated as geotechnical actions

A1, A2 : sets of partial safety factors on actions
M1, M2 : sets of partial safety factors on soil parameters
R1, R2, R3 : sets of partial safety factors on resistance

Table 8.11 *Suggested overall safety factors for slope stability.*

| Environmental and economic risk | Type of structure | Safety risk | | |
|---|---|---|---|---|
| | | Negligible | Low | High |
| Negligible | Permanent | 1.30 | 1.35 | 1.40 |
| | Temporary* | 1.10 | 1.15 | 1.20 |
| Low | Permanent | 1.35 | 1.40 | 1.45 |
| | Temporary* | 1.15 | 1.20 | 1.25 |
| High | Permanent | 1.40 | 1.45 | 1.50 |
| | Temporary* | 1.20 | 1.25 | 1.30 |

* Temporary: also valid during construction

Table 8.11 presents overall safety factors against loss of stability. The suggested values account for safety, environmental and economic risks for different type of structures.

> Within the Limit Equilibrium approach, state-of-the-art software also allows analytical methods to search for the most critical (non-)circular failure surface. The question can be raised whether a safety factor, which was defined decades ago when calculation and soil investigation methods were less accurate, should still be applicable.
>
> In fact the actual more powerful calculation means allow to find more correct results and thus have a lower 'model uncertainty' compared to the formerly used methods. This should reflect in a lower required (overall) safety factor.

**Additional considerations**

For dredging and reclamation projects, characteristic soil parameters can often not be defined based on a statistical analysis. A sufficient number of laboratory tests on samples from individual soil layers encountered at the project site is rarely available for such an analysis. Frequently, soil parameters have to be selected based on "Engineering judgement". This is difficult to justify in view of the Eurocode design principle for important structures, which promotes very strongly the execution of an extensive soil investigation in order to arrive at a well-founded selection of parameters.

When defining characteristic values for individual soil layers, an appropriate geological model must be established in which homogeneous geotechnical layers can be identified. If such a model is not carefully prepared, then the scatter in results may be too large to be able to apply a statistical analysis. This can result in a very large standard deviation and absurdly low characteristic design parameters.

Again this requires an adequate and good quality soil investigation programme supported by a geological soil model.

Eurocode 7 does not provide guidelines for the selection of partial factors for temporary conditions. Reference is made to EAU 2004 for an example of an approach in which also temporary conditions are covered.

8.4.3.3   Analytical calculation models versus Finite Element Method (FEM)

In the following paragraphs, several analytical calculation methods will be discussed for the verification of the failure mode and the Ultimate Limit State.

These analytical methods have proven to give reliable results for the limit states discussed. The factors of safety normally adopted in the analytical analyses cover, to some extent, also the uncertainty with respect to the theoretical calculation model used.

Another technique used for such calculations is based on the Finite Element Method (FEM). In this calculation method the soil is subdivided into small elements and each element gets a soil model attributed which expresses the stress-strain behaviour of the soil. In fact, the FEM is primarily a method to calculate deformations (Serviceability Limit State) but it is also frequently used to verify the Ultimate Limit State.

Checking for the ULS in a deformation-related model is normally performed by reducing the shear strength in small iterative steps until large deformations occur in the model. This may, however, lead to numerical problems with the FEM and requires numerical manipulation to allow for such calculations.

FEM is a very powerful calculation method that has many advantages but also some disadvantages:

**Advantages:**
– Using the FEM gives a clear insight in the deformations (vertical and horizontal);
– FEM allows for modelling of complex geometries, if required with structural elements;
– FEM allows for studying multiple loading cases with loading and unloading phases;
– FEM allows for studying consolidation processes as a function of time;
– The offset of failure mechanisms (such as squeezing and punching) can be shown when large deformations are allowed in the model;
– FEM Allows for visualization of governing failure mode;
– 3D modeling is possible.

**Disadvantages:**
- Many different constitutive models exist. The choice of parameters for these sophisticated models is not an easy task. Practical geotechnical engineers may have insufficient "feeling" for these parameters;
- Calculated deformations are not always reliable. A correct choice of the soil model and the related parameters is of utmost importance: a calibration might be needed.
- The modelling of structural elements needs special attention such as the use of reduced interface friction and how the element behaves when performing ULS simulations;
- The user needs to have a good understanding of the theory and technique of modelling used in the software since calculation results must be "interpreted" (e.g., reject irrelevant results);
- Expectations using FEM may be too high and do not necessarily give a correct answer to each problem. Skilful engineering judgment is still necessary and might get lost when using the FEM "blindly".

When using the Finite Element Method to check the Ultimate Limit State, the reduction factor on the shear strength parameters to introduce failure is often considered as the factor of safety. This factor does not necessarily have to be the same as defined with analytical methods since the uncertainty linked to the method/model used is different. Normally the partial safety factors as required in the Eurocode 7 approach are the required values.

In the following sections only analytical methods will be discussed and useful remarks concerning the use of the FEM will be given.


8.4.3.4   Bearing capacity

The term "bearing capacity" is very often used without a theoretically correct definition of what is meant. Bearing capacity of a foundation is the maximum load that can be applied on a foundation, leading to failure. Failure is defined as the development of large uncontrolled deformations under a small load increment.

"The" bearing capacity of a fill does not exist: The bearing capacity of a shallow foundation constructed on or in the fill can be defined. The dimensions of the foundation, the foundation depth as well as the thickness and shear strength of the fill layer are important parameters in such an analysis. In addition, the natural foundation soil, if present within the zone of influence of the foundation, and the groundwater table are important as well.

An example of a bearing capacity calculation method for shallow foundations is the theory developed by Meyerhof (1951, 1953, 1963): A short overview of this theory can be found in Appendix D Geotechnical Principles.

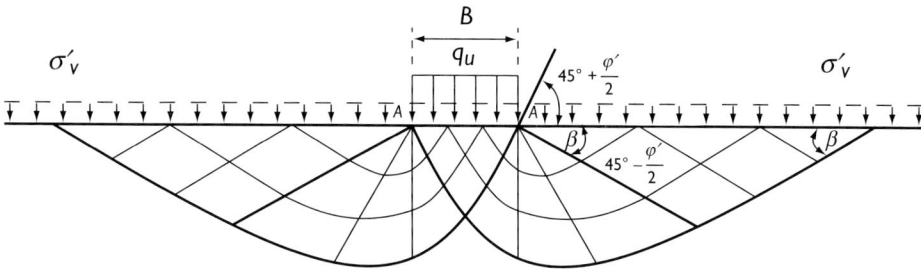

Figure 8.14    *Failure mode for bearing capacity analysis under a shallow foundation (Bowles, 1997).*

It is noted that the maximum depth of the shear planes underneath a footing is limited which means that the fill material below a certain depth does not contribute to the bearing capacity of the foundation. The following formula gives an indication of the maximum depth of the active wedge below the foundation level:

$$H_{max} = B \cdot \tan (45° + \varphi'/2)$$

where:
$H_{max}$ = maximum depth failure plane below foundation level (m)
$B$     = width of the footing (m)
$\varphi'$     = effective friction angle (degrees)

This indicates that in case of a high value for the friction angle $\varphi = 40°$ of the fill, the shear planes will not reach a depth of much more than 2.0–2.5 times the width of the foundation. Note that this rule of thumb may not apply if a relatively thin layer of fill is covering soft soil strata. Punching through might become the governing failure mechanism (see Section 8.4.3.5) in that case.

The use of Finite Element Methods (FEM) in order to predict the bearing capacity of a foundation is also commonly accepted and will not be discussed in detail in this Manual. However, it should be realised that bearing capacity is often a 3D problem while most commercial FE software is limited to a 2D analysis only. Although 3D FE analysis methods do exist, they are not commonly used for analysing the bearing capacity of shallow foundations.

Sometimes the term "bearing capacity" is used to define the uniform life load that can be carried by the reclamation area (e.g. 60 kN/m² for a container terminal). Such a life load not only affects the design of, for instance, a quay wall or other structures, but it also has an effect on the allowable residual long-term settlement of the fill mass after construction (which is frequently limited by the Specification as well). The so specified 'bearing capacity' (e.g. 60 kN/m²) is not a bearing capacity criterion but in fact information to calculate long-term overall settlements and horizontal earth pressures. It would be better not to use the term 'bearing capacity' in this case but 'uniform life load'. When specified as information for settlement assessment, it is advised also to specify which fraction of this life load will quasi permanently be present (e.g. 50%) and thus should be taken into account for the settlement calculation.

---

**Example of Correct specification of Bearing Capacity**

The unit bearing capacity for a shallow footing (dimensions A m by B m), applied at a depth of C m below the final ground surface, has to be minimum xx kPa (vertical loading), with an overall Safety Factor of yy.

A, B and C are dimensional values expressed in length units (m);

xx is a loading value (e.g. 150 kPa);

yy is an overall Factor of Safety (e.g. FS = 2)

---

Checking the ground bearing capacity for a shallow foundation or a stack of containers can easily be done adopting the above theories and it will seldom lead to bearing capacity problems, provided the fill layer thickness is large enough to prevent punch trough or other type of failure modes.

For temporary conditions during construction, the fresh fill material must be able to support construction plant and equipment. Wet cohesive soils particularly are prone to rapid softening so may not be able to support wheeled or tracked vehicles i.e. the site easily becomes a quagmire. Hence checking the trafficability of the fresh fill is essential.

The bearing capacity may become an issue when the subsoil consists of cohesive fine-grained material.

For fine-grained cohesive soil with an undrained shear strength $c_u$, the general bearing capacity formula can, in the case of a centrically applied vertical load, be simplified to:

$$q_u = \sigma'_{v,0} + s_c \cdot N_c \cdot c_u \qquad (8\text{-}2)$$

where:
$q_u$ = ultimate bearing capacity (kPa)
$\sigma'_{v,0}$ = effective vertical stress at the foundation level (kPa)
$N_c$ = $2 + \pi = 5.14$ (Meyerhof) (−)
$s_c$ = shape factor (−)
$c_u$ = undrained shear strength (kPa)

In principle the peak shear undrained strength $c_u$ can be used; although the geotechnical engineer needs to use his 'engineering judgement' to decide when using the remoulded undrained shear strength would be more appropriate. This could occur when the soil becomes heavily disturbed by the operations or the traffic.

## Zone load testing

A direct method to verify the bearing capacity in the field is the execution of a Zone Load Test (ZLT) (ICE, 1987) which, in fact, is a large Plate Load Test (PLT).

In such a test, a footing with more realistic dimensions (e.g. a 3 m by 3 m plate) is loaded up to it's design load (or more) while the settlement behaviour is monitored.

The actual safety factor with respect to the Ultimate Limit State can be verified by means of extended load procedures. The prediction of the long-term behaviour by extrapolating the settlement measured during a 24h or even 48h time interval is also possible (see paper by Briaud and Gibbens, 1996).

When Zone Load Tests are used, (see 11.3.3.1 and 12.4.13.5) the specifications often require that the settlement under the design load (e.g. 150 kPa) has to be limited to a certain long-term settlement (e.g. 25 mm). This is, in fact, an indirect specification for the bearing capacity since the limited settlement can only be guaranteed when the bearing capacity is fulfilled. In this case no factor of safety for the bearing capacity is specified.

The ZLT and PLT are discussed in more detail in 11.3.3.1.

## Safety Factors

When using the above mentioned analytical formulas, an overall safety factor of 2.0 against bearing capacity failure is generally adopted in an overall safety approach. When using the partial safety approach, reference is made to EC7 and the national annexes to the EC7. In the partial safety approach the overall safety factor seems to become smaller, depending on the contribution of permanent load and live loads, each having a different partial factor.

Since the Eurocode 7, does not give guidance for safety factors to be adopted during temporary conditions, some freedom exists to choose lower safety factors. In the overall safety approach often a safety factor of 1.5 is used for the bearing capacity during temporary conditions. In the German standards some guidance is given for temporary conditions depending on different loading cases. Reference is made to DIN 1045 (2005) and EAU (2004).

By selecting a safety factor, the allowed deformations are also indirectly defined. A smaller safety factor results, in principle, in larger deformations. An overall safety factor of 2 against loss of bearing capacity will generally result in a design where deformations remain limited within commonly accepted allowable values. When deformations are less important, even a smaller safety factor might be acceptable.

When using FEM models to check the bearing capacity, it is difficult to define the maximum load that can be applied on a foundation with given dimensions. Basically the FEM is a method in which the stress-strain behaviour of the soil is modelled as realistically as possible. However, failure occurs if deformations become uncontrollable and this cannot be modelled with standard FEM software. In general, the dimensions and foundation level are chosen and the design load is applied. The factor of safety is then calculated by means of a $\varphi'/c'$-reduction method (see 8.4.3.7.3) until the deformations become unrealistically large due to instability of the system.

Alternatively, the load on a foundation may be increased until failure occurs, while using the actual design shear strength characteristics for the soil layers in the model.

Such a failure analysis using Finite Element Methods requires some further attention since, according to the applicable Standards, different load factors apply for permanent and live loads respectively. Reference can be made to Potts and Zdravkovic (1999) for how to use partial safety factors with FEM.

8.4.3.5   Punch through

When a fill layer of limited thickness is placed on top of a soft soil (e.g. very soft clay), a slice of the fill can be pushed downward into this soft layer. This punch through type of failure is sketched in Figure 8.15. Punch through may occur when

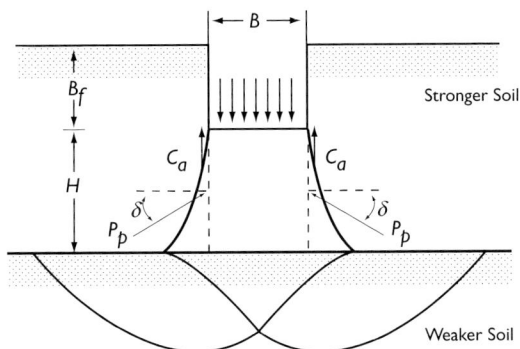

Figure 8.15   *Principle of punching failure (from Das, 2009).*

the influence zone of the failure mechanism is larger than the thickness of the fill layer.

Since shear stresses will also develop in the fill material, it might be too conservative not to consider some spreading of the introduced load. Load spreading can be taken into account by assuming a $1_H$ on $3_V$ to $1_H$ on $1_V$ spreading angle. Consequently, the load increase on the soft layer is assumed to be distributed over a larger surface. The bearing capacity at this level can be checked using the above presented bearing capacity formula 8–2.

More detailed and theoretically correct methods are described in Das (2009) where different bearing capacity formulas are presented which do not allow for load spreading but which account for the shear resistance along the vertical dashed lines shown in Figure 8.15.

This failure mode can be critical during intermediate phases such as after the installation of the first layer of fill material over a soft subsoil. Often construction equipment (e.g. bulldozers or excavators equipped with a mast to install PVDs) is moving over the site in an early stage. Punch through can occur suddenly and be extremely dangerous for labour handling the equipment.

Figure 8.16   *Example of punch through failure.*

It should be checked if such a classical failure mechanism can fully develop in the fill. This means that the ratio between the width of the foundation (or loaded area) and the thickness of the fill layer should be checked. As a rule of thumb the punch through failure mechanism needs to be checked when the ratio H/B is less than 2.0 to 2.5 (with H and B defined in Figure 8.15).

The same safety approach can be used as discussed with respect to the bearing capacity of foundations in general.

### 8.4.3.6  Squeezing

Squeezing can occur when a relatively thin layer of very soft to soft soil is enclosed by more competent (sand) layers. As a result of the applied load, a plastic zone occurs at the base of the most probable slip circle (see section 8.4.3.7). This plastic zone rapidly increases toward the edges of the loaded area caused by a further weakening (remoulding effect) of the soft layer. Finally, the weakened material is not capable any longer to deliver sufficient shear resistance and subsequently the layer squeezes out horizontally (Figure 8.17). As a result, the (soil) structure may sink as much as the original thickness of the squeezed layer.

The mechanism of squeezing is a rather complicated phenomenon to analyse, but relative simple approximation methods exist which suffice in most cases. Care has to be taken with more sophisticated methods such as Finite Element Methods giving the illusion of more precision. Current FEM's are not well suited to model large deformation problems like squeezing as excessive mesh distortions generally cause numerical instabilities. The subject FEM model should only be used to iden-tify the failure mechanism and not to model the resulting deformations.

New promising developments, however, may solve these problems in the future: the Material Point Method (MPM) originating from fluid mechanics simulates large deformations by moving material points through a fixed finite element grid rather than using the classical updated Langrangian FEM (Beuth *et al.*, 2008).

One of the analysis methods regarding squeezing has been defined by Matar-Salençon (CUR 162, 1996). A short description of this method can be found in Appendix D, Geotechnical Principles.

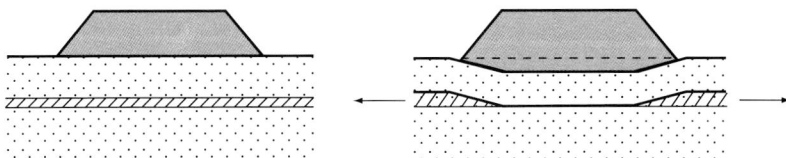

Figure 8.17  *Principle of squeezing.*

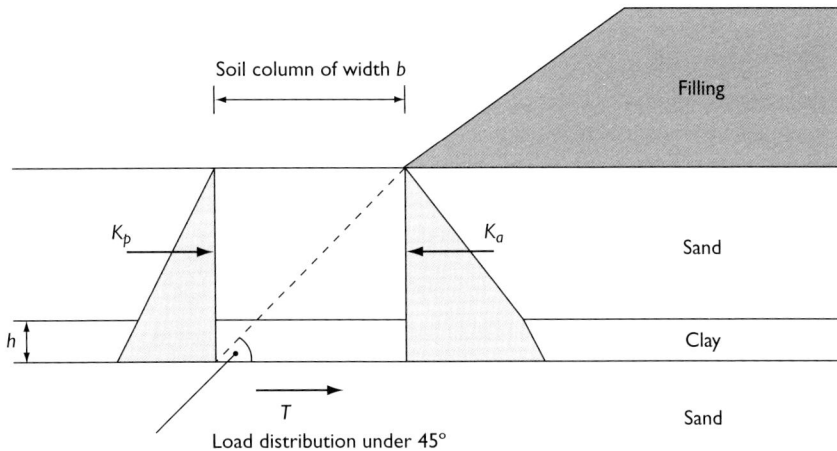

Figure 8.18    *Principle of manual calculation method for horizontal equilibrium.*

Another calculation method is the calculation of horizontal equilibrium. This method is based on the assumption that no squeezing or horizontal sliding will occur in case of a horizontal equilibrium of the soil column.

The condition of horizontal equilibrium is satisfied if the integrated active soil pressure over the soil column is less than the integrated passive soil pressure plus the shear force at the bottom of the layer. This equilibrium condition is displayed in Figure 8.18.

This calculation method is also relatively simple and only requires information on a few, often well known, input parameters. An additional safety margin is obtained by ignoring cohesion.

With respect to squeezing, no recommendations exist with respect to the minimum required overall safety factor to be adopted in the design. In principle similar safety factors are used as for a conventional slope stability analysis as discussed in the next section.

**Squeezing as a construction technique**

Although squeezing is described as a potential failure mechanism, it can also be considered to be a construction method when a bund is constructed on a soft subsoil and squeezing is used to intentionally displace the soft material. In that case a prediction must be made at which loading squeezing will actually occur. Moreover, the deformations have to be monitored and controlled during the construction.

When using squeezing as a construction technique, the following issues should be considered:

To calculate progressive squeezing, consider using the remoulded shear strength of the soil;

After partial squeezing, the soil fill is at a factor of safety of 1. Further construction activities might require a higher factor of safety. How can this situation be improved?

A thin layer of soft material remaining below the fill mass may lead to slope stability problems in later construction phases? This can be checked by soil investigation after squeezing operations.

### 8.4.3.7  Slope stability of fill and subsoil

#### 8.4.3.7.1  Design methods

Each reclamation project requires an analysis of the slope stability along the boundaries of the fill during the subsequent phases of the project. Failure can occur in the soil mass itself or/and in the underlying subsoil. Figure 8.19 shows some examples of slope failure.

All Ultimate Limit State (ULS) calculations are based on the assumption of a sliding mechanism and define the (overall) safety factor against failure as the ratio of the maximum available shear strength and the mobilised shear stress.

Failure surfaces can have different shapes such as circular, logarithmic spiral, planar, combined, etc. Since different shapes can occur in nature, selecting a surface that is representative for the failure mode expected in the field is of utmost importance. This can be influenced by many aspects such as homogeneity of the soil, geometry of the slope, and loading conditions.

In Figure 8.20, a specific situation of slope failure is given in order to demonstrate that a specific soil layering can result in a failure mode which is different from the standard "circular" sliding surface that is most often used in a slope stability analysis.

The stability of slopes can be verified by means of two different approaches:

− Limit Equilibrium Methods which are based on the solution of one or more equilibrium equations (moment, vertical, horizontal) for an assumed failure surface;
− Finite Element Methods in which the $\varphi/c$-reduction method is used.

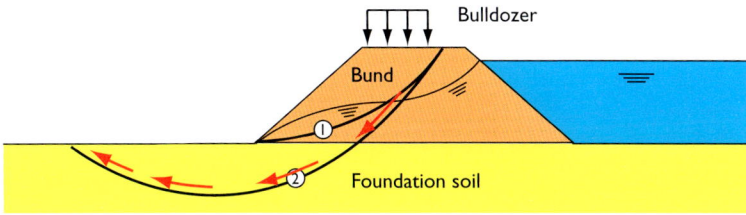

a. Bund stability

① Failure in bund mass only
② Failure in bund mass and foundation soil

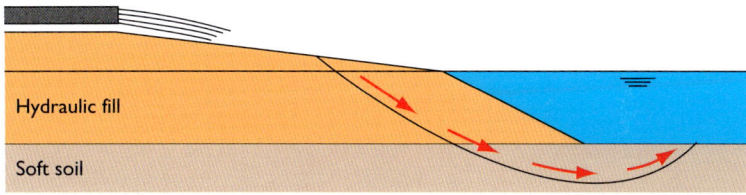

b. Hydraulic fill front

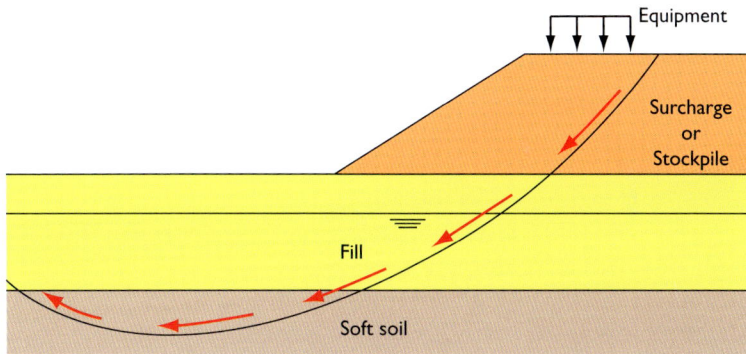

c. Surcharge on hydraulic fill

Figure 8.19  *Examples of slope failure modes.*

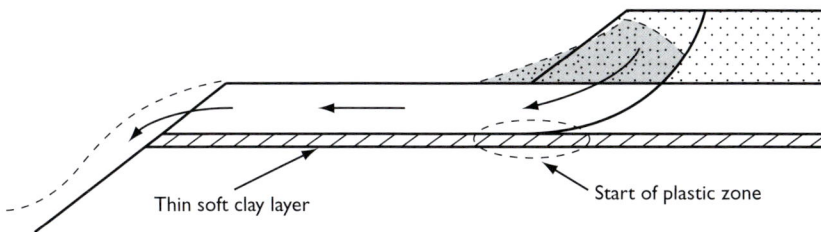

Figure 8.20  *Example of a slope failure influenced by a thin soft layer.*

237

Both methods will briefly be discussed. For detailed discussion on these methods reference is made to specific literature or to the manuals of commonly used software such as the Geo-Slope, SLOPE/W manual (2012).

### 8.4.3.7.2   Limit Equilibrium Methods

These methods consider instability caused by shearing along a slip surface. Failure by shearing occurs when the soil cannot mobilise the shear strength required for equilibrium along a specific sliding surface. Failure surfaces or interfaces between rigid bodies may have a variety of shapes including circular, planar and more complicated shapes. The mass of soil or rock bounded by the failure surface can be treated as a rigid body or as several rigid bodies moving simultaneously.

The advantage of Limit Equilibrium Methods is that they can be translated into analytical equations. The more basic methods can relatively simply be worked out by hand for one or just a few defined failure surfaces. However, by using modern software, thousands of potential failure surfaces can be verified in order to find the governing mechanism with the lowest factor of safety.

For the selection of a calculation method and/or type of slip surface, the following guidelines could be considered:

– Where subsoil or fill material is homogeneous and isotropic, normally circular failure surfaces can be assumed.
– For layered soils with considerable variations in shear strength, special attention should be paid to the layers with lower shear strength. This may require the analysis of noncircular sliding surfaces.
– In jointed materials, including hard rock and layered or fissured soils, the shape of the failure surface can partly or fully be governed by discontinuities. In such cases an analysis with three-dimensional wedges should normally be made.
– Existing failed slopes, which can potentially be reactivated should be analyzed, considering circular as well as non-circular sliding surfaces.
– If the failure mode cannot be assumed to be two-dimensional, the use of three-dimensional sliding surfaces should be considered.
– Groundwater flow and pore water pressures are very important in a slope stability analysis: these aspects have to be modeled as realistically as possible.
– Short-term and long-term stability can be different as a result of consolidation effects. Both have to be checked.

**Methods of slices**
Different Limit Equilibrium Methods exist. The most frequently used is the method of slices in which the soil within the failure surface is subdivided into "slices" with a certain width. The equilibrium of these slices is written in terms of moment, vertical and horizontal stability (Figure 8.21). A slope stability analysis should, as a

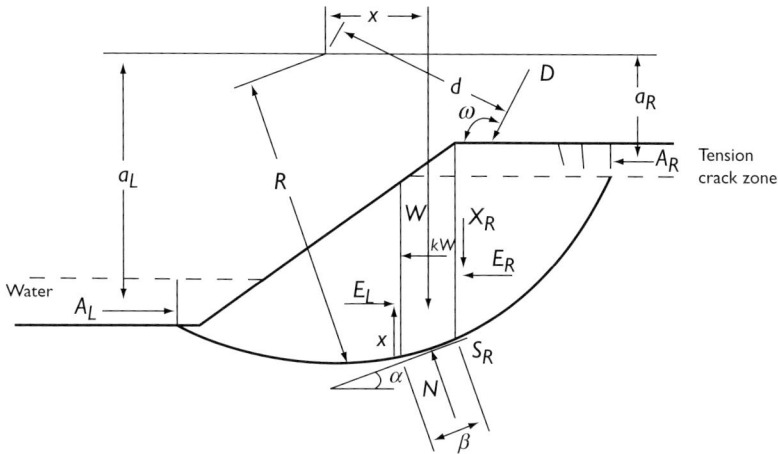

Figure 8.21    *Method of slices, forces acting on slices with circular slip surface (from GeoSlope, SLOPE/W manual, 2010).*

minimum, verify the overall moment and vertical stability of the sliding mass. If the horizontal equilibrium is not checked, inter-slice forces are neglected or assumed to be horizontal.

In the method of slices the factor of safety is usually expressed as the ratio of the available soil strength (stabilising forces) over the driving forces or moments, where summation is made over all slices within the failure surface:

$$FoS = \frac{\sum M_{stabilising}}{\sum M_{driving}} \qquad (8\text{-}3)$$

In the method of slices basically circular failure surfaces are considered. However some slice methods have also been developed to analyse non-circular failure surfaces and arbitrarily defined multi-planar failure surfaces. In Table 8.12 an overview is given of different slice methods, the type of failure surface considered and the equilibrium equations that taken into account.

### Block stability

Limit Equilibrium Methods (LEM's) that verify the stability by means of soil blocks within the failure surface require more complex calculations of all forces acting in between the blocks. These methods have also been implemented in software for a quick analysis.

The block stability analysis is a more appropriate method to verify the stability in case soil conditions and geometries lead to non-circular failure sur-

Table 8.12 *Different methods of slice assumptions (Duncan and Wright, 2005)*

| Procedure | Assumptions | Equilibrium equations |
|---|---|---|
| Infinite slope | The slip surface is parallel to the slope face | Σ forces parallel to slope<br>Σ forces perpendicular to slope |
| Logarithmic spiral | The slip surface is a logarithmic spiral | Σ moments about centre of spiral |
| Swedish Circle ($\varphi' = 0$) | The slip surface is circular, the friction angle is zero | Σ moments about centre of circle |
| Ordinary method of slices (also Fellenius method or Swedish method of slices) | The slip circle is circular; the forces on the sides of the slices are neglected | Σ moments about centre of circle |
| Simplified Bishop | The slip surface is circular; the forces on the sides of the slices are horizontal (i.e. there is no shear force between the slices) | Σ moments about centre of circle<br>Σ forces in vertical direction |
| Force equilibrium (Lowe and Karafiath, Simplified Janbu, Corps of Engineers, modified Swedish), Janbu's GPS procedure | The inclination of the interslice forces are assumed; assumptions vary with procedure | Σ forces in the horizontal direction<br>Σ forces in the vertical direction |
| Spencer | Interslice forces are parallel (i.e. all have the same inclination); the normal force acts at the centre of the base of the slice (typically). | Σ moments about any selected point<br>Σ forces in horizontal direction<br>Σ forces in vertical direction |
| Chen and Morgenstern | Interslice shear force is related to interslice normal force by means of a function to be defined; the normal force acts at the center of the base of the slice (typically) | Σ moments about any selected point<br>Σ forces in horizontal direction<br>Σ forces in vertical direction |
| Sarma | Interslice shear force is related to interslice shear strength $S_u$ by means of a function to be defined; interslice shear strength depends on shear strength parameters, pore water pressures and horizontal component of interslice force; the normal force acts at the center of the base of the slice (typically) | Σ moments about any selected point<br>Σ forces in horizontal direction<br>Σ forces in vertical direction |

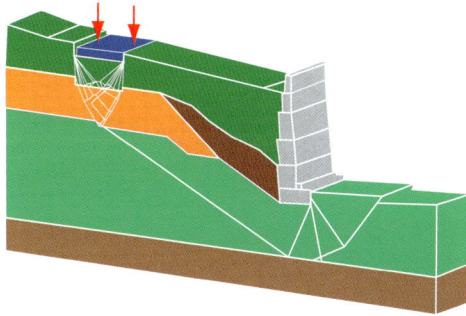

Figure 8.22  *Block stability, combined bearing-retaining wall failure mechanism (from Limitstate Geo brochure).*

faces, e.g. when structural elements are present in the geometry to be analysed. A combined bearing capacity and slope stability analysis can be realised with such methods (Figure 8.22).

**Limitations of Limit Equilibrium Methods**

In those events that a combined failure of soil and structural elements can be expected, due consideration must be given to the soil-structure interaction in terms of their relative stiffness. These cases include, for instance, failure surfaces intersecting structural members such as piles, flexible walls or armouring (e.g. geotextiles). Such an approach is, however, not always feasible in Limit Equilibrium Methods. The Finite Element Method is a more appropriate tool for the analysis of soil-structure interaction.

In LEMs, it is assumed that the maximum shear resistance is simultaneously mobilised in all soil layers over the full length of the failure surface. This assumption is not always correct since the development of the shear stress very much depends on the type of soil and its behaviour (e.g. stress-strain relationship). In Figure 8.23 this is schematically shown. The top two drawings depict the situation at a certain level of shear strain ($\gamma$). For the soil in point A, the mobilised shear resistance is at maximum shear strength while in all other points the shear strain is lower and thus the mobilised shear resistance has not reached the maximum yet. The mobilised shear resistance is different from point to point; in the drawing it is assumed that points B have experienced the same strain as wel as points C. The lower two drawings depict a situation where larger shear strains have occurred. The mobilised shear resistance of the soil in Point A now has reached the residual shear strength while the mobilised shear resistance of the soil in points B is at maximum shear strength and the shear strength of the soil in points C still has not been fully mobilised.

The selection of soil parameters and safety factor should consider this basic modelling assumption.

241

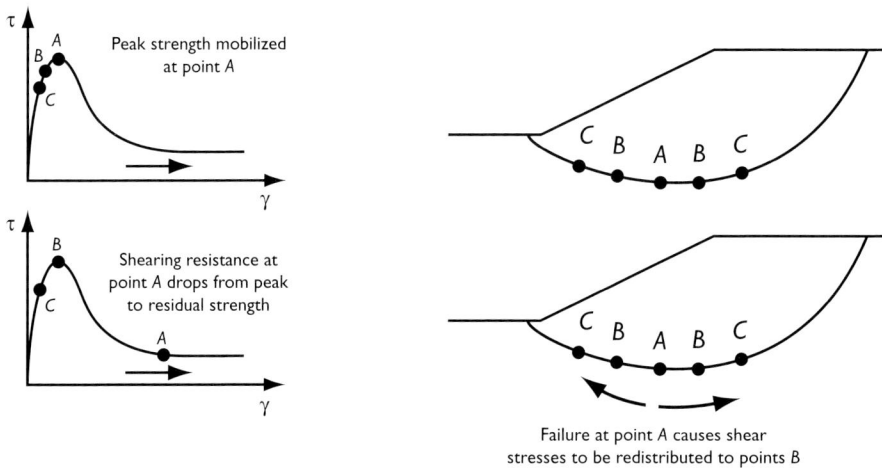

Figure 8.23    *Principle of mobilising shear resistance along a failure surface.*

The governing failure surface found by geotechnical software is sometimes physically incorrect. The designer must therefore always check the feasibility (kinematically) of the resulting failure surface.

### 8.4.3.7.3    Finite Element Method

The main advantage of using a Finite Element Method to verify slope stability failure is that a potential failure mechanism does not have to be predefined. The critical failure surface automatically results from the FEM calculations, provided that soil stratification and soil characteristics are correctly modelled. A disadvantage of the software is that sometimes non-relevant local failure mechanisms are found which break off the calculation process. The designer has to avoid these local effects by adjusting the soil characteristics in some of the elements of the FE model.

The Finite Element Method is essentially a method in which the stress-strain behaviour of soil is respected and, in the first instance, deformations are calculated. The Factor of Safety is normally defined by performing a $\varphi/c$-reduction: the shear strength of the soil defined by tan $(\varphi)$ and $c'$ is gradually decreased until 'failure' occurs in the model. Failure is reached when the deformations become very large under a very small further reduction of the shear strength.

Contrary to Limit Equilibrium models, the effect of load spreading is also taken into account.

In general 2D models are used although 3D models exist as well. In almost all cases found in literature the stability factor for a 3D model exceeds that of the 2D geometry (although the difference also depends on specific conditions and the geometry).

### 8.4.3.8   Construction of a slope on soft soil

When a reclamation has to be realised on a natural soft soil (e.g. recent marine sediments), a stability analysis of the containment bunds and the fill front itself has to be made. The loading introduced by the bunds and the fill is considered to be a quick loading condition, compared to the rate of consolidation of the natural subsoil. As such, the soft cohesive material behaves undrained and the most critical situation is the undrained condition as illustrated in Figure 8.24. This figure demonstrates the effect of excess pore water pressure (positive or negative), dissipation and the increase/decrease of the shear strength with consolidation.

Figure 8.24 (left) demonstrates that when soft soil is loaded, the most critical situation is immediately after loading. At this moment, the excess pore water pressure is at its highest level. During the subsequent consolidation period, the excess pore water pressure gradually diminishes, resulting in an increasing resisting shear strength and higher safety factor.

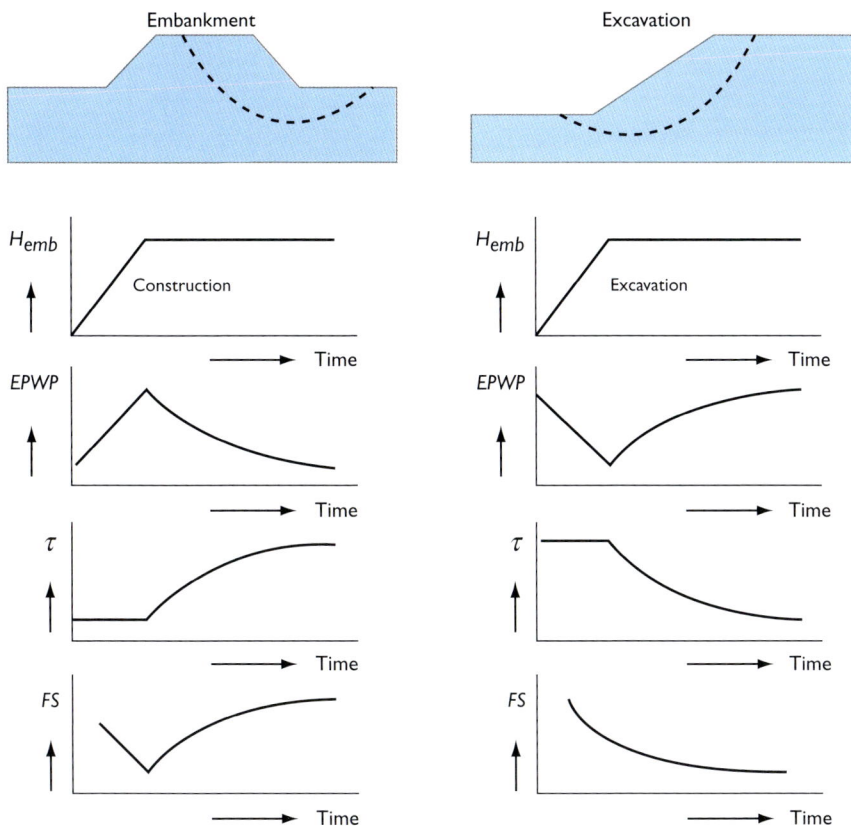

Figure 8.24   *Evolution of stability of a slope in a loading situation (embankment) or unloading situation (excavation).*

On the other hand, when realising an excavation (Figure 8.24, right), the most critical condition occurs after full equilibration of porewater pressures due to unloading (which is also a consolidation phenomenon) of the soil while the immediate condition (during and) after the excavation can be sufficiently safe.

If the construction of an embankment on soft soil is not stable under the assumption that the entire structure is instantly constructed, then the loading schedule should be examined in more detail. In principle, the construction of an embankment will always take some time. During this time period, the load due to the weight of the fill only gradually increases while the subsoil already consolidates to a certain extent. As a result of this consolidation, the shear strength of the subsoil increases as discussed in section 8.4.2.1. The construction of an embankment in phases, with the purpose of allowing the soil to improve, is called "staged construction". This staged construction process can even further be improved by using vertical drains to accelerate the consolidation rate of the cohesive soil.

Figure 8.25 presents the principle of "staged construction". In the top figure, the loading stages and the changes in total stress and effective stress in the soil is shown. The middle figure represents the settlements and the lower figure shows the safety factor as a function of the different construction stages. As a result of a careful study of a staged construction process different loading steps can be defined to guarantee a minimum safety factor per phase as required.

Time is an important aspect for the analysis of the stability. A clear distinction between the stability during construction/excavation and the final stage is necessary. For each of those stages different partial/overall safety factors can be considered. In the example of Figure 8.25, a temporary safety factor of 1.1 is required, while the required long term final safety is 1.3.

### 8.4.3.9 Effect of groundwater flow on slope stability

For each slope stability analysis, the groundwater flow regime is to be studied thoroughly since pore water pressures reduces the effective stress and thus the shear strength. Moreover, in case of seepage, the danger exists for erosion and/or piping. Both phenomena can initiate failure of the slope.

The most critical water levels both inside and outside the soil mass need to be selected with care. In general the (upstream) groundwater in the soil is taken at a high level because this influences seepage while outside water levels at the downstream face are chosen at lowest level for slope stability analysis. Both selected water levels, upstream and downstream, need to be compatible and time effects may have to be taken into account. A typical example is a water reservoir for which

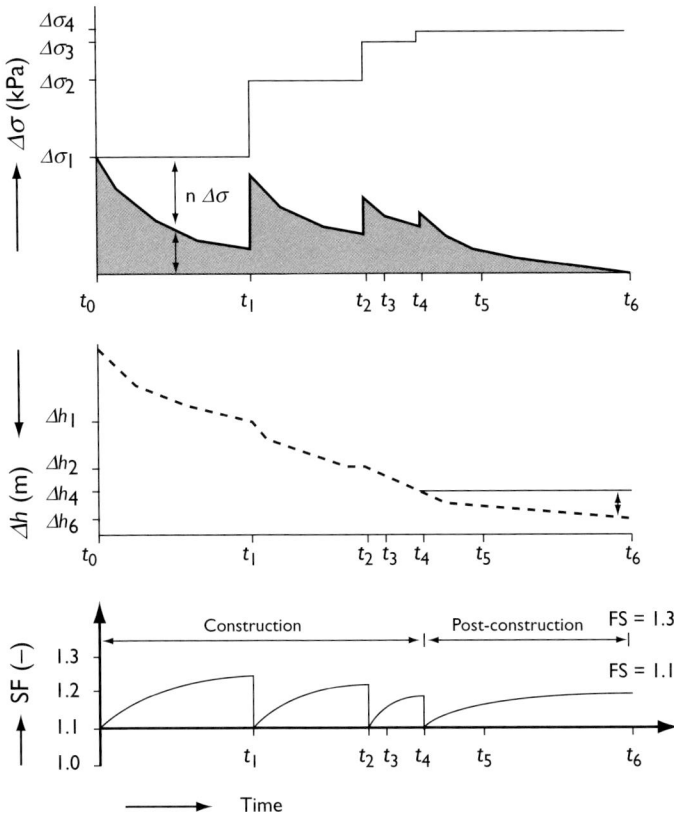

Figure 8.25 *The development of excess porewater pressure, settlement and safety factor as a function of time during a staged construction.*

a quick drawdown analysis needs to be performed in order to determine the maximum possible difference between both levels.

As a general rule, a seepage face on a slope should be avoided by taking appropriate measures in order to keep the phreatic line as far as possible from the slope face. This can be realised by ditches, (toe) drains, impermeable layers or other means such as demonstrated in Figure 8.26. It is noted that a ditch such as shown in Figure 8.26a may affect the stability of the structure in a negative manner. Its efficiency will depend on the local soil conditions and the water level in the ditch.

Water management is typically an issue in case hydraulic fills are deposited above the groundwater table. Verification of the stability of retaining bunds is often a requirement for hydraulic fill works.

Retaining bunds are generally built using local soil, at least for the first lift of 2 m to 3 m. For the next lifts, the fill material itself is used in a system comparable to

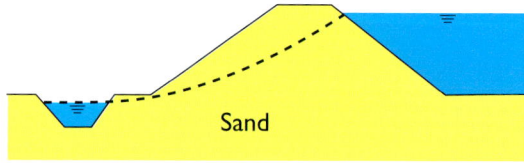

(a) Ditch at the toe of the slope

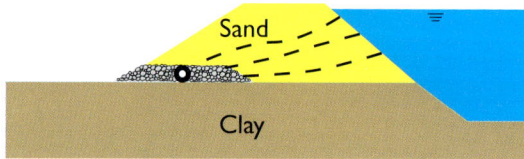

(b) Drain at the toe of the slope

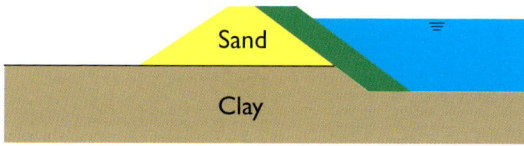

(c) Impermeable layer at the upstream face

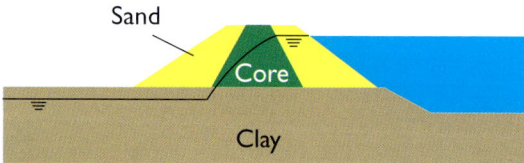

(d) Impermeable core

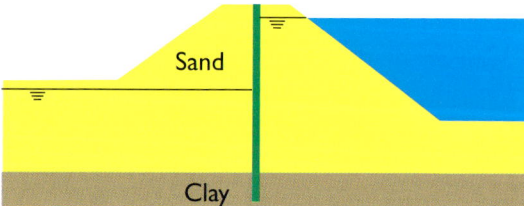

(e) Impermeable screen in the embankment

Figure 8.26    *Measures to avoid water seeping out of a slope face.*

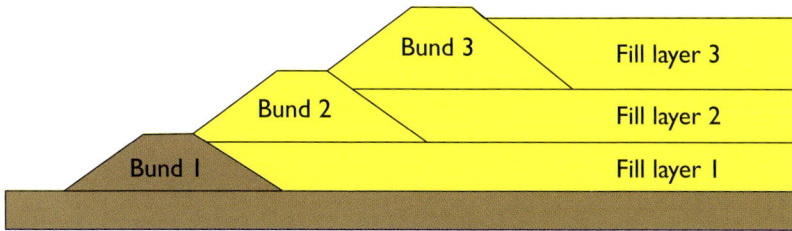

Figure 8.27  *Tailing dam system for multi-lift hydraulic fills.*

tailing dam construction (Figure 8.27). When the natural soil layer is of low permeability and if the fill material consists of sand, the results are a side slope of the reclamation which mainly consists of sand with a free water table on top of the reclaimed mass during further hydraulic filling. This problem specifically needs attention when the fill consists of 3 or more lifts (e.g. for a sand stockpile). The outside slope needs to be sufficiently gentle and/or berms should be introduced. This also means that the slopes of such a reclamation can take quite some surface area which must be physically available.

For the first lift in the construction of containment bunds, frequently local soil is used. Moderate compaction (e.g. by a bulldozer) might be required to prevent the creation of more permeable zones which may soften during subsequent fill lifts.

Hydraulic fill material is normally used for the next lifts although it could be optional to use less permeable material as well if this is locally available. It has to be realised that recently deposited hydraulic fill remains saturated for a period of time. Such a hydraulic boundary condition needs to be considered in the analysis of the slope stability.

The height and slope of the retaining bunds basically depend on the stability requirements, but generally inside and outside slopes of $3_H$ over $2_V$ are realised. The crest width is about 4 m to allow track-type equipment to drive on top. The outside slope may need to be more gentle than the inside slope in case a high water level can be expected in the hydraulic fill reservoir.

In some cases drains have to be installed during the construction of subsequent levels of the containment bund. Even other measures may be necessary to control the groundwater table such as for instance dewatering by means of vacuum pumping.

Ignoring such measures introduces a risk for slope instability. Such a slope failure can result in a large volume of process water that is released through this breach in the bund, with possible catastrophic consequences for people and structures in the area next to the hydraulic fill area (Figure 8.28). Even when the slope remains

Figure 8.28   *Failure of a hydraulic fill bund.*

stable, water seeping out of the slope face near the toe can cause non-cohesive material to be transported with the water. As a consequence, the lower part of the slope may deform to a more gentle slope (more or less parallel to $\varphi'/2$) while becoming inaccessible because of the low density and saturation of the material.

During the execution, strong currents occur in the hydraulic fill reservoir. If the bunds are constructed from non-cohesive material, the inner slopes may easily erode with possibly failure of the bund as a consequence. To avoid this phenomenon erosion-resistant material should be used for the construction of the inner slopes of the bund. An alternative is to place a protective layer (e.g. geotextile or PVC sheet) over the slope. Such a PVC sheet will also prevent or reduce the flow of process water through the bund. However, such sheets are installed with simple overlaps and they get easily damaged. Consequently the positive effect of such sheets in the long term is not reliable.

Retaining bunds have to be at least 0.5 m higher than the highest water level in the fill area in order to avoid overtopping. Generally no special measures for protection of the outer slopes against water action are taken. Such a water flow would introduce immediate erosion and a potential risk for failure of the bund.

**Remark**

As a result of the hydraulic fill process, the fill material is saturated and might remain saturated for some time after the filling works, depending on the permeability and water storage capacity of the subsoil and the containment bunds. Sometimes even a higher permanent water table is encountered in reclaimed land compared to the original situation. This effect not only has to be taken into account for slope stability studies, but also for other design aspects such as land drainage, consolidation and settlement.

Other potential failure mechanisms involving (ground)water are uplift, piping and (internal) erosion. These mechanisms are illustrated in Figure 8.29 and in Figure 8.30. Especially uplift is a typical failure mechanism studied in dike stability in The Netherlands.

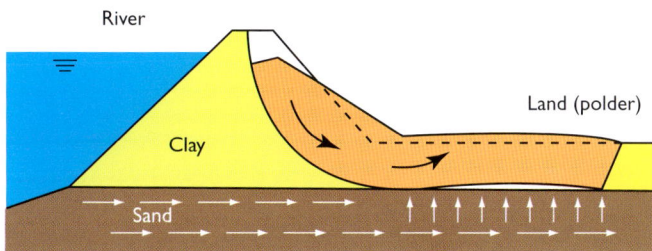

Figure 8.29   *Uplift mechanism leading to dike instability (TAW, 2001).*

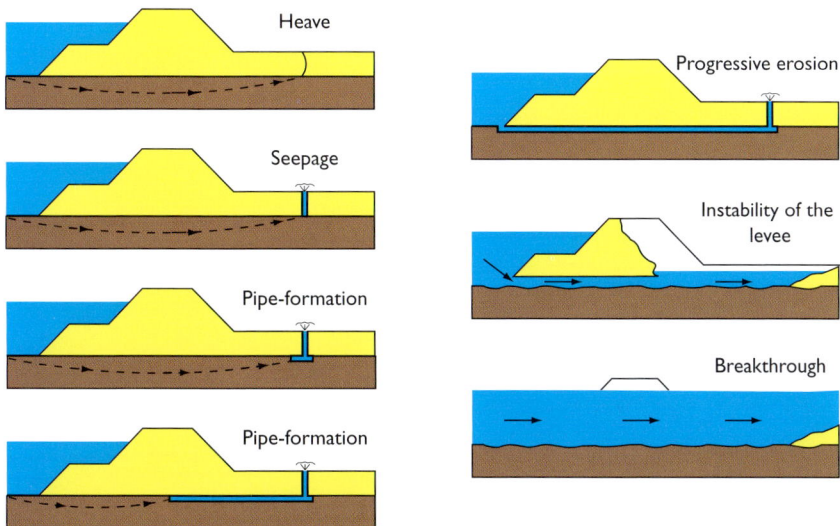

Figure 8.30   *Piping mechanism leading to dike instability.*

249

8.4.3.10  Earthquakes and slope stability

A practical approach for assessing the effects of earthquakes on port structures can be found in a report prepared by Working Group No. 34 of the Maritime Navigation Commission (PIANC 2001) and in Seismic Design Guidelines for Port Structures (2001). In the Technical Commentary No. 7 of the latter reference, following statement concerning breakwaters, embankments and slopes can be found in section T7.1.4:

"In most cases involving soils that do not exhibit considerable strength loss due to shaking, common pseudo-static rigid body methods assuming slip surfaces will generally suffice for evaluating the stability of slopes in Ports. The seismic stability of slopes is evaluated by modifying static limit equilibrium analyses with the addition of a lateral seismic body force that is the product of a seismic coefficient $k_h$ and the mass of the soil bounded by the potential slip surface".

"The seismic coefficient $k_h$ is specified as a fraction, roughly 1/2, of the peak horizontal acceleration ratio $a_{max}/g$ at the crest of an earth dam because the lateral inertial force is applied for only a short interval during transient earthquake loading."

A similar kind of approach is followed within the Eurocode 8, where the horizontal ground acceleration is taken as half of the Peak Ground Acceleration. In addition, however, the Eurocode also considers the effect of a vertical ground acceleration, which equals 1/2 to 1/3 of the horizontal ground acceleration.

8.4.3.11  Stabilising measures for slope stability

The design should ensure that all construction activities on site can be planned and executed such that the occurrence of an Ultimate Limit State is unlikely.

If the stability of a slope cannot be guaranteed or when the deformations are considered unacceptable for the intended use of the reclamation, then stabilising methods are required.

**Slope formation with hydraulic fill**

Very gentle slopes can be realised by hydraulic placement of fill. This way, stability problems when adding fill on very soft material will generally be limited, compared to dry fill procedures.

The slopes formed during hydraulic fill operations mainly depend on the particle size of the sand and whether the fill is deposited above or below the water table (or in the tidal zone).

Reference is made to Chapter 6, Construction methods reclamation area for more information on this subject.

Stabilising methods which have been developed to make the safe execution of earth structures possible are:

– Slope geometry: Adapt the slope geometry to obtain a sufficient stable slope or use structural elements to guarantee slope stability.
– Staged construction: Gradually increase the height of the fill (or carefully raise the stress level by phased filling) and allow for consolidation of the soft subsoil in between loading steps; the increase in shear strength that comes with consolidation is counted upon for the stability of the next filling phase.
– Soil improvement: Increase the shear strength in situ by soil replacement/improvement Accelerated consolidation by means of the installation of vertical drains.
– Use of geosynthetics.
– In the following paragraphs, some measures are discussed in more detail.

8.4.3.11.1   Optimizing the slope geometry by using counterweight berms

A commonly used solution, applicable in case sufficient space is available during the construction period, is the use of a counterweight berm at the toe of the slope. By using a berm, the stability of a slope is favourably influenced because its weight increases the counteracting moment considerably. Its presence can be temporary, just for increasing the stability during the most critical construction phases. The counterweight berm is considered to be a relatively cheap and easy-to-apply method. This is especially the case when ditches need to be excavated before the construction is started. The soil from the ditches can directly be used for the berm construction.

8.4.3.11.2   Staged construction

This technique has already been discussed in section 6.4. The construction is realised in small lifts with allowance for some consolidation time in between. The effect of (partial) consolidation is taken into account in order to define the increase in shear strength of the soft subsoil.

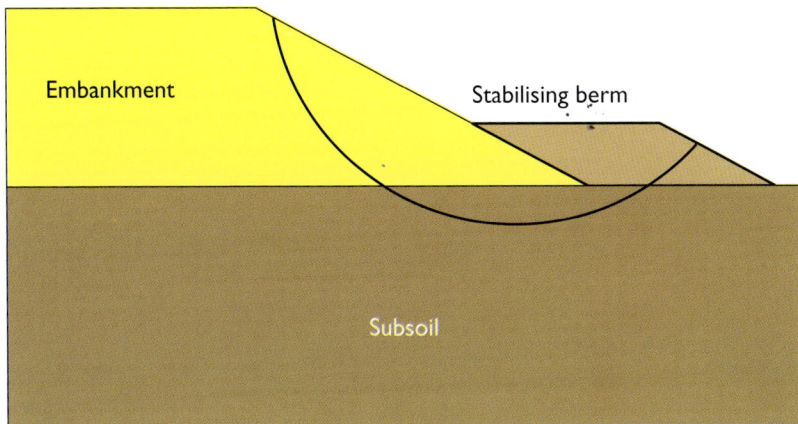

Figure 8.31   *Stability of a slope is favourably influenced by a counterweight berm.*

Drainage of the subsoil to reduce waiting times in between subsequent fill lifts, and thus minimising the construction period, is a commonly used construction method. Dissipation of the excess pore pressures can be achieved by the installation of vertical drains. In addition, vacuum pressure can be applied to the system. The vacuum pressure also acts as an additional preload.

8.4.3.11.3   Soil replacement (sand key)

With this method, a part of the soft subsoil is excavated and often replaced by sand. This can be realised below as well as above the water table.

Soil replacement can be realised just at the toe of a slope in order to obtain a larger soil (counter) weight (Figure 8.32). The effect of increased shear strength at this location is, however, limited because of the relatively small effective vertical stresses.

Alternatively (or in combination with the above), soil replacement is realised underneath the entire slope and the crest. The effect of the increase in shear strength is, in this case, more significant and potential slip surfaces will be "pushed" backwards resulting in a deeper and less critical failure mechanism.

8.4.3.11.4   Stone columns, sand compaction piles

In order to improve the shear strength, the bearing capacity and consolidation behaviour of soft subsoil layers, stone columns or sand compaction piles can be used. If necessary, they can be installed from floating equipment.

Reference is made to sections 7.6.3 and 7.6.4 for more information on these techniques.

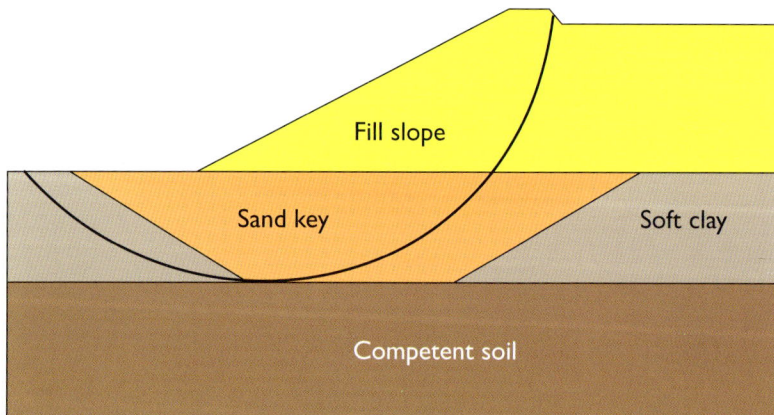

Figure 8.32   *Soil improvement at the toe of the slope (Sand Key).*

### 8.4.3.11.5  Geosynthetics

The use of reinforcing geosynthetics in slopes is a commonly used technique in road embankment construction and can be used to stabilise slopes of bunds or dikes. The geotextile types used for soil reinforcement are woven high strength materials or geogrid type of materials. The design with these elements is outside the scope of this Manual.

Geosynthetics are produced in many forms (i.e., geotextiles, geomembranes, geogrids and geocomposites), each with their own characteristic properties regarding strength, stiffness, permeability, puncture-resistance and interaction with the surrounding soil (see Figure 8.33). Consulting with the producers of the fabrics is the most advisable way to determine which type meets the intended function best.

There is a large variety in reinforcement techniques with geosynthetics, from a simple horizontally placed reinforcement underneath embankments to steep geotextile/geogrid reinforced slopes.

Figure 8.34 demonstrates the principle of a reinforcement of an embankment as a result of the presence of a (strong) geogrid intersecting a potential shear plane.

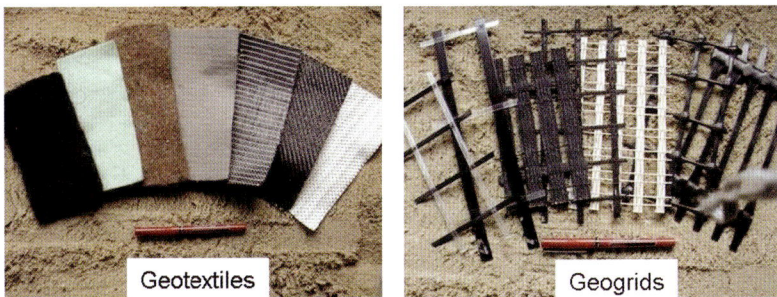

Geotextiles          Geogrids

Figure 8.33    *Geotextile and geogrid.*

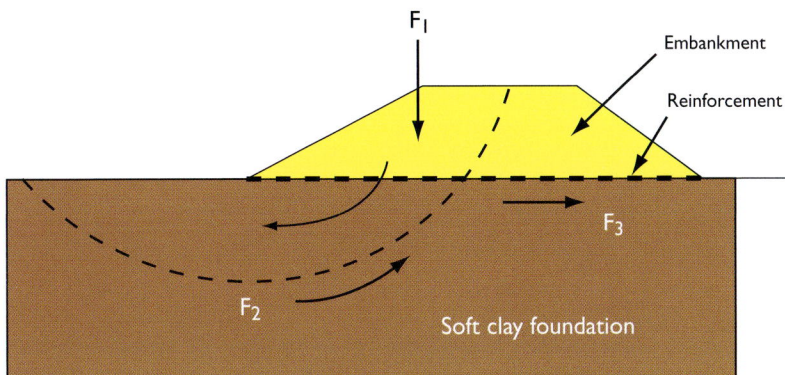

Figure 8.34    *Soil reinforcement with geotextile.*

The weight of the fill increases the friction between the geogrid and the soil, preventing the geogrid from being pulled out by the sliding fill mass.

For the design of geosynthetic elements which need to function in the long term, very high reduction factors have to be accounted for as a consequence of creep and potential damage.

## 8.5 Stiffness and deformation

### 8.5.1 *Introduction*

The stiffness of subsoil and fill material will define the deformation (or strain) to occur during and after construction of the reclamation. The deformation and consolidation behaviour of a reclamation must be estimated correctly for several reasons:

– Prediction of the volume of fill material needed to guarantee a reclamation level at a certain milestone during the construction and/or in the long term as defined by the client.
– Prediction of the settlement that the reclaimed area will undergo after handover and start of the construction of superstructures (residual settlements).
– Prediction of differential settlements that may occur between several zones of the reclaimed area; differential settlements might be more problematic than absolute settlements.
– Prediction of the strength increase of compressible (natural) soil layers due to consolidation.

The stress-strain behaviour of a soil is an important issue in soil mechanics. The "strength" of a soil is linked to its "deformation" (e.g. peak and residual shear strength). The definition of "failure" is the occurrence of large and uncontrolled deformations (also called Ultimate Limit State, ULS).

When designing a fill, the stiffness of a soil is a means to calculate the deformation (Serviceability Limit State, SLS). The stress-strain relation is important to select the correct stiffness at a certain mobilised stress level.

### 8.5.2 *Stiffness*

#### 8.5.2.1 General considerations

In soil mechanics, stiffness is a difficult subject because of the non-linear nature of the stress-strain behaviour of soil. The reason for this non-linear behaviour is the change of the soil's skeleton when being loaded (i.e. the porosity changes). The

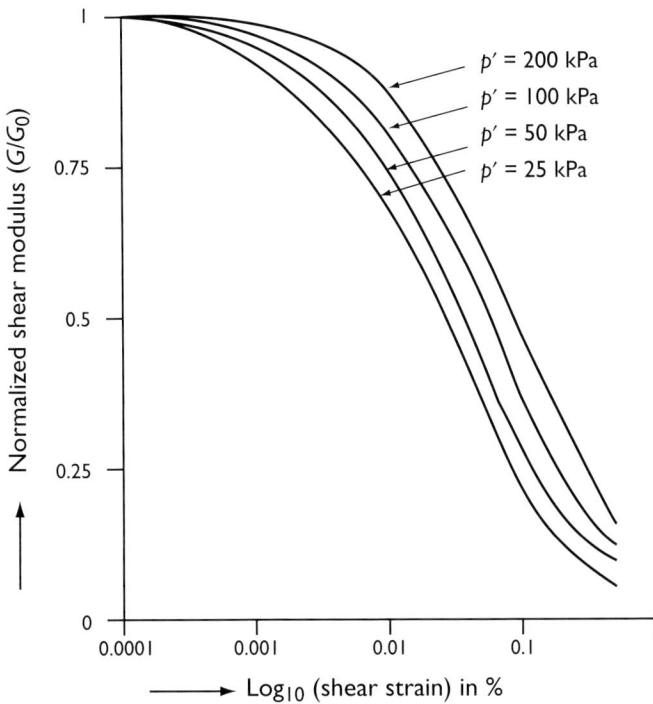

Figure 8.35   *G as a function of strain and stress (Fahey, 1992).*

majority of the strain is irreversible when the soil is unloaded. The stiffness of a soil is strain dependent as shown in Figure 8.35. This figure illustrates the relation between the shear modulus ($G$) and the shear strain ($\gamma$). The stiffness moduli at very low strain are nominated with the suffix 0 ($G_0$ or $E_0$) and are defined by very small deformations.

For cohesive soils a decrease in porosity requires pore water to dissipate. Since the permeability of cohesive soils is very low, the dissipation may take a long time (consolidation). Consequently, the short term (undrained) stiffness of saturated fine-grained soils will be higher than the long-term stiffness. Excess pore pressures will initially carry the load without large deformations.

Stiffness also depends on the stress history or over-consolidation ratio (OCR). The stiffness of soil at virgin loading is lower than the stiffness at unloading/reloading conditions.

The stiffness of soil at cyclic loading conditions is even more specific and very often a hysteresis loop can be found during cyclic testing (e.g. cyclic triaxial testing). Multiple loading cycles can cause degradation of the soil strength. This has to be taken into account for the design of foundations in some specific situations such as earthquake regions. This aspect of soil behaviour is outside the scope of this Manual and reference is made to specialised literature on this topic.

Stiffness is normally not a parameter that is defined as a project requirement. Only for specific tests a stiffness requirement can be defined, such as for the Plate Load Test.

However, specifying a stiffness without referring to a detailed description of the test method, can lead to a misunderstanding of the project specifications. Reference is for instance made to the stiffness used in the field of road construction which is often empirically derived from a CBR value and relates to a dynamic stiffness which is much higher than the stiffness under static loading conditions (e.g. the Plate Load Test see section 11.3.3.1).

### 8.5.2.2 Stiffness of subsoil

Stiffness is defined in terms of modulus of elasticity or shear modulus. As defined in elasticity theory, the modulus of elasticity links normal stress to strain ($\tau = \varepsilon \times E$) and the shear modulus links shear stress to shear strain ($\tau = \gamma \times G$).

Many different elasticity moduli are referred to in Soil Mechanics:

- Young's modulus $E_y$;
- Young's modulus defined from the line between origin and a certain point of the stress-strain curve $E_{sec}$ (secant modulus): e.g. $E_{50}$ is the Young's modulus at 50% of the failure stress;
- Young's modulus defined from the tangent line in a certain point of the stress-strain curve $E_{tan}$ (tangent modulus);
- Young's modulus at very small deformations $E_0$ (is a tangent modulus);
- Constrained modulus denoted by various symbols: $M$, $E_s$ or $E_{oed}$;
- Bulk modulus B;
- Undrained modulus $E_u$;
- Dynamic modulus $E_{dyn}$;
- Plate Load Test modulus $E_{PLT}$ or $E_{v,1}$ $E_{v,2}$;
- Dilatometer modulus $E_{DMT}$;
- Pressuremeter Modulus $E_{PMT}$.

In literature, 2 types of shear moduli are generally used: $G$ and $G_0$. $G_0$ is the shear modulus at very small deformations ($\gamma < 1 \ast 10^{-3}\%$). A correct definition of the shear modulus G should include the corresponding strain.

In the literature, theoretical and empirical formulas exist expressing the relationship between several of these moduli (references). A clear definition of which modulus is meant when discussing stiffness moduli is therefore considered essential.

Testing of the stiffness is in many cases linked to the above defined type of moduli:

- Triaxial test: $E_{50}$ or $E_{tan}$;
- Oedometer test: $E_{oed}$ constrained modulus;
- The stiffness at very small deformations is generally measured by means of seismic methods in the field or in the laboratory (bender elements), using $P$ and $S$ waves (Pressure and Shear waves);
- Plate Load Test: $E_{PLT}$.

Stiffness can also be referred to in terms of compression index, compression ratio or compression coefficient. These parameters are used for settlement analysis and will be discussed in section 8.5.3. Similar parameters exist for the unloading/reloading cycle.

Many correlations exist in the literature between different stiffness moduli and results of field tests and/or other soil parameters. Some examples can be found in Appendix C, Correlations and Correction Methods.

8.5.2.3    Stiffness of fill material

Estimating the stiffness of a hydraulic fill material is a different issue. In fact the fill material can be considered to be a very young, non-aged soil deposit. In order to estimate the stiffness of a granular fill during the design phase, the following sequential procedure can be followed:

- Predict the relative density of the fill material, taking into account the deposition method and the expected required compaction techniques. Reference is made to section 7, see also Table 8.5 in section 8.3, density.
- establish the cone resistance profile to be expected based on this relative density and the future stress condition according to the empirical relationships between relative density and cone resistance (reference is made to Appendix C, Correlations and Correction Methods).
- based on the cone resistance the stiffness can be estimated (Appendix C, Correlations and Correction Methods).

Although this approach is based on several empirical correlations and experience, a reliable estimation of the stiffness is possible.

Once the fill is placed, testing can be performed to verify the initial assumptions. If necessary compaction needs to be performed in order to obtain the required density/stiffness. A CPT is the most common tool to test the fill mass over its entire thickness, however the derivation of stiffness is indirect and may require calibration (for instance Section 9.2.5.2, Cone Penetration and Standard Penetration testing in carbonate sands and see Appendix C). Additionally, with a seismic CPT, also $G_0$ can be measured.

Generally the fill is tested shortly after the reclamation has been completed. This means that the above described approaches neglect the ageing effects. Neglecting these effects is a safe but conservative approach. Observations demonstrate that the cone resistance in granular fill may increase in time after the filling works and compaction of the material.

When using fine grained soil as a fill material, the initial stiffness of the fill is much more difficult to predict. The difficulty lies in the prediction of the initial void ratio and stress state. Reference is made to section 9.1, where the use of fine grained material is discussed.

Two examples are:

− The use of stiff clay as fill material. This will result in a fill consisting of clay lumps or clay balls, with potential large inter lump voids.
− The use of a hydraulically pumped soft clay or silt slurry as a fill material. The void ratio of the slurry depends largely on the mineralogy (clay or silt) and particle size distribution. The void ratio of the fill is the result of the sedimentation process followed by a self-weight consolidation process, and will therefore vary in time.

In both cases, it is strongly advised to preload the fill in order to control the stress history of the fill. The amount of preload should be related to the future design load of the subject area. When the soft subsoil has been pre-loaded and consolidated, the stiffness characteristics of the material can be predicted from laboratory testing. Where necessary, soil improvement techniques should be used to improve the consolidation rate of the soil (see Section 7).

In practice initial volume changes related to fine-grained soil used as reclamation material will often be defined by bulking considerations, without trying to estimate the stiffness of such very soft material. Only when the soil has consolidated under its self-weight then deformations under external loads can be calculated adopting the (predicted) stiffness of the fill mass.

### 8.5.3   Deformations

#### 8.5.3.1   General considerations

The magnitude of deformation depends on the stiffness of the soil and the loads imposed on the reclamation. In order to predict deformations the loading scheme needs to be known in advance.

The following types of deformations are relevant for the design of a reclamation area:

− vertical deformation, or settlement, of the subsoil due to the weight of the fill and future loading conditions;

- vertical deformation, or settlement and (auto)compaction, of the fill material itself;
- horizontal deformations of fill and subsoil caused by future loading conditions.

During the construction of the fill, monitoring of the deformations is required to verify the assumptions made during the design phase. The monitoring results can be used to:

- determine the actually required volume of fill material;
- control settlements and horizontal deformations near existing structures;
- control reclamation level by forecast of the final settlement;
- control and forecast residual and differential settlements;
- control the fill schedule.

The methods to monitor deformations during construction are described in section 11.3. Based on the results, measures can be taken to accelerate or limit the deformations.

Whilst the full monitoring of deformations is desirable for many reasons, this is not always possible from the start of a project. When fill starts under water the installation of traditional settlement measurement equipment (settlement beacons) is not possible. A grid of settlement beacons normally is installed as soon as the fill is above water level.

The settlement (both total and differential) that occurs in the natural soil underneath the fill and within the fill material itself, as well as the rate of settlement/ excess pore-water dissipation is often one of the most critical elements of the project. It may affect many aspects of project development such as cut/fill balance, ground improvement requirements, construction schedule, pavement design, maintenance, foundation and structural design. Prediction of the magnitude and rate of settlements during the design phase is needed to ensure that hydraulic fill requirements, such as a certain percentage of the total (differential) settlements, are achieved within the time frame allocated for construction in the project schedule. More information about settlement requirements is described in section 8.5.3.4.

Typical values of allowable settlements at time of handover of a reclamation area depend strongly on the final use of the area and the type of structures to be realised on top of the fill. Economic considerations might also play a role by balancing capital investment costs versus maintenance cost. Therefore typical settlement specifications cannot generally be defined and "performance based" design is therefore of utmost importance.

For container terminal areas, indicative values for allowable long-term (e.g. 20–30 years) residual settlement range in between 150 mm and 300 mm. For structural elements built on the fill, such as shallow foundations of limited size, settlement

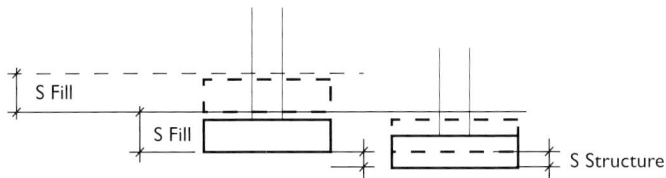

Figure 8.36   *Settlement fill and settlement structure versus fill.*

requirements can be more stringent. For such foundations, settlements have to be considered separately from the settlements induced by the fill itself (Figure 8.36). For shallow foundations, differential settlement between the structural element and the surrounding area should be the main concern, rather than the global settlement of the fill itself. This implies that in case a zone load test is adopted, the measured vertical elevations should not be related to a deep datum beacon, but should be related to the level of the surrounding fill.

For piled foundations, a consideration could be that settlements are not relevant as long as negative skin friction and relative deformations are taken into account. The need for limiting residual settlements to small values in the range of cm's or even mm's should always be verified thoroughly since such requirements can simply not always be met or would require excessive cost for soil stabilisation or other soil improvement techniques.

A consolidation analysis is also related to a strength analysis (section 8.4.2), since the consolidation degree is directly related to the gain in strength. Each stability analysis should accurately model the geometry and characteristics of the fill at the corresponding point in time.

### 8.5.3.2   Settlement calculation methods

The most common approaches to estimate the settlement in a soil stratum are based on stress strain measurements during laboratory tests performed on undisturbed soil samples. Whenever possible, experience gained from prior construction under similar site conditions should be applied. If there is no earlier experience, the construction of a trial embankment for monitoring the settlement/consolidation behaviour may be considered.

Many different methods exist to calculate the settlement of the subsoil. Reference is made to the literature for such theories.

The total settlement consists of the following constituents:

– Primary settlement: This behaviour is a result of dissipation of excess porewater pressure. For granular soils the settlements occur almost immediately during/after loading. However for cohesive soils with relatively low permeability, the overstressed pore water is driven out slowly. The time required for the excess pore water to dissipate is called the consolidation or hydrodynamic period. The settlement during this period is referred to as primary settlement.

– Secondary compression: In addition to the primary settlement there is secondary compression which is related to creep behaviour and which is, in principle, infinite.

**Instantaneous settlement**

In literature, the term Instantaneous (or immediate) settlement is often used, but generally not well defined. This settlement is sometimes also referred to with Elastic settlement. This may lead to confusion.

The settlement of a soil can be described as instantaneous in case:

– The soil is not fully saturated.
– The soil is saturated, but very permeable (sand). The settlement is actually not instantaneous, but rather very quick (the excess pore water is expelled very quickly).
– The instantaneous settlement is a result of a horizontal deformation. The volume of the soil remains unchanged. This kind of deformation is often modelled using the elastic theory. For large reclamation areas, this definition of instantaneous settlement can be neglected since there will be no horizontal deformation.
– Pure elastic deformation is normally neglected.
– Collapse mechanisms can occur in very loose soil which result in a quasi-instantaneous deformation.
– of collapse of the soil skeleton upon wetting (e.g. Sabkha, see Section 9.4.4).

Sometimes, an instantaneous settlement is measured during a consolidation test on saturated samples. This can be attributed to imperfections in the testing method, such as the presence of air bubbles.

Generally, a settlement calculation consists of three distinct steps:

– Calculation of the elastic deformation (constant volume);
– Calculation of the final primary settlement (which occurs after infinite long time);
– Calculation of how the primary settlement develops with time (consolidation rate);
– Calculation of the time dependent secondary settlement.

The total time dependant settlement is considered to be the sum of the primary settlement and the secondary compression. The most common approach assumes that the secondary compression only starts when the primary consolidation is completed. Often, as a practical approach, it is assumed to start at 90% primary consolidation. Two subsequent time-dependent processes must be considered when calculating the total settlement at a certain time after loading.

Other theories have been developed assuming that creep and consolidation are two phenomena that occur simultaneously (e.g. Isotachen theory, Den Haan and Sellmeijer, 2000).

In order to demonstrate the importance of the influencing parameters, the following most common settlement theories and corresponding formulas are briefly discussed in appendix D Geotechnical Principles:

– Final primary settlement: Terzaghi's formula;
– Vertical (Primary) Consolidation: Terzaghi's one-dimensional consolidation theory;
– Horizontal Consolidation (vertical drains):
  o theory of Barron
  o smear effects: Hansbo
– combination of Horizontal Consolidation and Vertical Consolidation—Carillo;
– secondary compression (Mesri).

Since the secondary compression formula applies to naturally occurring creep deformation, it gives a false high value when using prefabricated vertical drains (PVDs) because the primary consolidation period is reduced. This results in a calculated larger secondary compression, because the ratio $t_f/t_p$ becomes larger (see Appendix D). When surcharge is used to (over-) consolidate the soil in order to limit settlements due to future service loads, very often secondary compression is neglected since the magnitude thereof is small and only occurs over a much longer time scale (Mesri and Feng, 1991).

### 8.5.3.3 Additional considerations

**Bjerrum—isochrone theory**
A basic analysis of settlement behaviour of clays is given by Bjerrum (1967) which is summarised in Figure 8.37. This figure illustrates that under a constant load "delayed compression" occurs in time. This phenomenon is called creep. However, when a soil is over-consolidated (up to a stress $p_c$ in the figure) and unloaded to $p_0$, it will demonstrate much less "delayed compression" in time as the realised compression in the over-consolidated state is equivalent to a loading of $p_0$ during a very long time period. In that case, and in an engineering time frame, secondary compression can be neglected. Upon reloading the soil will behave stiffer. Once the pre-consolidation stress is reached, the soil will behave as a soil under virgin compression.

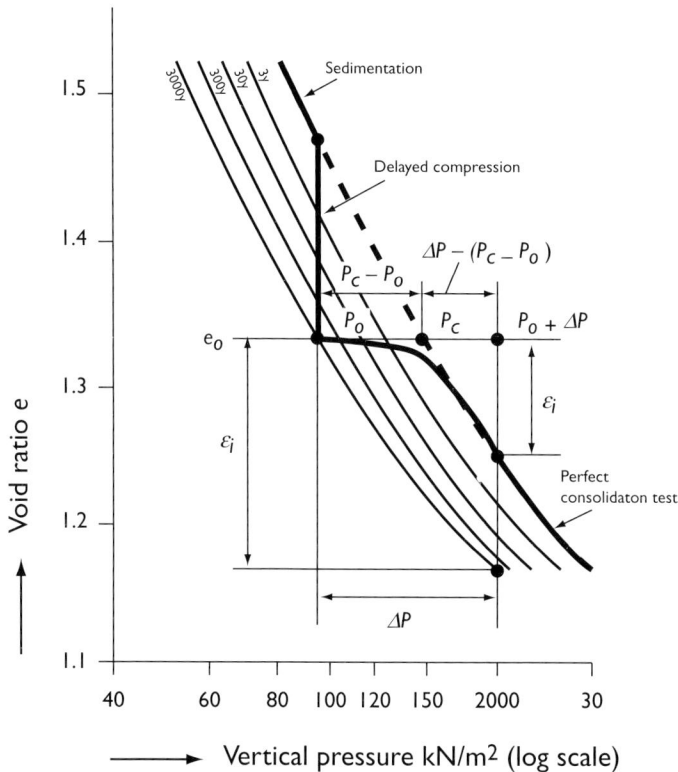

Figure 8.37 *Load-settlement graph taking into account time effect (Bjerrum, 1967).*

Even if a soil is not over-consolidated and ageing occurs under a constant load, it shifts along the vertical undergoing "delayed compression". Upon further loading the soil will demonstrate behaviour similar to an over-consolidated soil even when it, in reality, was never loaded to a higher stress level.

**Elastic deformation calculation**

Sometimes, an elastic deformation calculation can be made using the oedometer modulus or constrained modulus $E_{oed}$. However, this simplified approach is only valid when the load increment is limited compared to the existing stress condition and provided that an appropriate modulus is selected. The modulus should be selected taking into account the current stress conditions and the stress history. It is a method that can be used for a quick settlement estimate, but it should not be used for the final design since it leads to an over-estimation of the settlement. For reclamation projects, the stress increase will generally be much higher than the initial effective vertical stresses, especially near the top of the existing soil layers. The elastic method is therefore generally not suitable.

**Large strain theories**

When analysing the deformation of highly compressible soils (e.g. sludge, tailings), the assumption of constant soil parameters such as stiffness and permeability is not correct. The change of density and permeability as a function of stress level (constitutive e-$\sigma'_v$ and e-k relationships) should be considered. Once these constitutive relationships are known an estimation of the load-settlement behaviour as a function of time based "large strain" theories such as Gibson's theory (Gibson *et al.*, 1967) may be adopted. Using this method, however, requires detailed laboratory testing to define the relevant constitutive relationships. Secondary compression is not taken into account in these methods. In these theories 'degree of consolidation', either defined by settlements or by excess porewater pressures, will be different. In the classical Terzaghi consolidation theory, both definitions are identical.

The theories discussed above are all based on the analysis of one-dimensional compression (or oedometer) tests. The calculation of deformation requires the interpretation of these compression tests according to the theoretical assumptions of the calculation model. Different calculation methods therefore require a different interpretation of the compression test.

**Calculating settlements with the Finite Element Method (FEM)**

The discussion in the paragraph on strength already mentioned that the FEM is a suitable alternative for the analytical method, providing the constitutive soil models and their related parameters are well chosen and—when possible—calibrated.

The FEM is in the first place a method to calculate deformations (horizontally and vertically). The analysis of time-dependent settlements is possible as well, although the calculation of the degree of consolidation will require the input of the horizontal and vertical permeability of the soil and the computation of groundwater flow (dissipation of excess pore-water pressures) in the FE mesh.

For basic calculations, the FEM is not often used for settlement analysis. The use of Terzaghi's formula is often more straightforward. When using the FEM, the simple Mohr-Coulomb model (linear elastic-plastic) should not be used for a reliable prediction of the deformations.

8.5.3.4   Vertical deformation of a reclamation surface

**General**

One of the most important requirements for a reclamation project is the final design level. This level is to be guaranteed at time of hand-over time or after a certain specified design life of the structure. An accurate prediction of the settlement behaviour is therefore essential. This becomes more and more challenging when the subsoil consists of a soft compressible layer.

When specifications are given in terms of settlement, in general 2 types of requirements apply:

- Allowed residual settlement after construction: The residual settlement can be caused by the weight of the fill, future permanent and temporary service loads or a combination thereof.
- Allowed differential settlement after construction: This should be defined as the allowed settlement difference over a certain distance (angular distortion), possibly also caused by a specified loading condition.

It is the responsibility of the designer to accurately predict the settlement in time as a function of the loading conditions and to account for this behaviour in the planning of the construction works.

At the time of handover, the (sub)soil does not necessarily have to be completely consolidated. If the allowed residual settlement is larger than the expected secondary compression, then part of the residual settlement may comprise some primary settlement as well. In that case there is no need to wait for full consolidation of the subsoil during the construction period. It is, in that respect, noted that a much longer time is required to obtain an increase of the consolidation percentage from 95% to 99% than for instance from 80% to 95%.

On the other hand, the secondary compression may possibly exceed the allowable residual settlement. In that case a preload by means of a temporary surcharge will be needed to reduce the secondary compression to an allowable limit.

The magnitude of differential settlements is difficult to predict because they strongly depend on the variability in the subsoil. Only a very rough estimate based on available borehole information and CPTs is possible. Sometimes a certain percentage (e.g. 50%) of the absolute total settlement is assumed to represent the differential settlement. Another option is a statistical analysis based on the variability of the soil parameters. Because of these uncertainties, requirements related to differential settlements are often omitted.

Preloading a reclamation area is an effective measure to reduce differential settlements since after removal of an adequate preload only (more or less) elastic deformations occur which are much smaller than the primary deformations.

**Deformation of subsoil**
The distinction should always be made between settlements caused by the weight of the reclamation itself and settlements as a consequence of future service loads. In most cases the settlement caused by the weight of the fill is governing since this gives the first stress increase for the subsoil at a very low initial stress.

When the settlement requirement also applies for future service loads, then preloading is the most appropriate soil improvement method. Compaction of fine-grained soils by means of mechanical compaction is not feasible. Other soil improvement techniques (such as stone columns, soil mixing, etc.) are feasible, but can be expensive. Soil replacement could be an option as long as the extent of the soft soil layer is limited and severe environmental constraints do not exist. Alternative solutions are only selected if a preload is not feasible because of stability reasons, timing or when very stringent residual deformation requirements have to be met, see also Chapter 7, Ground improvement.

A preload is often combined with the installation of prefabricated vertical drains.

In some cases a preload higher than the future service load is adopted in order to ensure that future settlements already occur during the construction period. The design of a reclamation including a preload in combination with PVDs offers three parameters to vary:

- the magnitude of the preload;
- the grid spacing of the PVDs (related to consolidation time);
- the consolidation period (related to the project and contract milestones).

Depending on local site conditions and the specifications set by the Client, different situations may occur. Some possibilities are illustrated in Figure 8.38 and Figure 8.39.

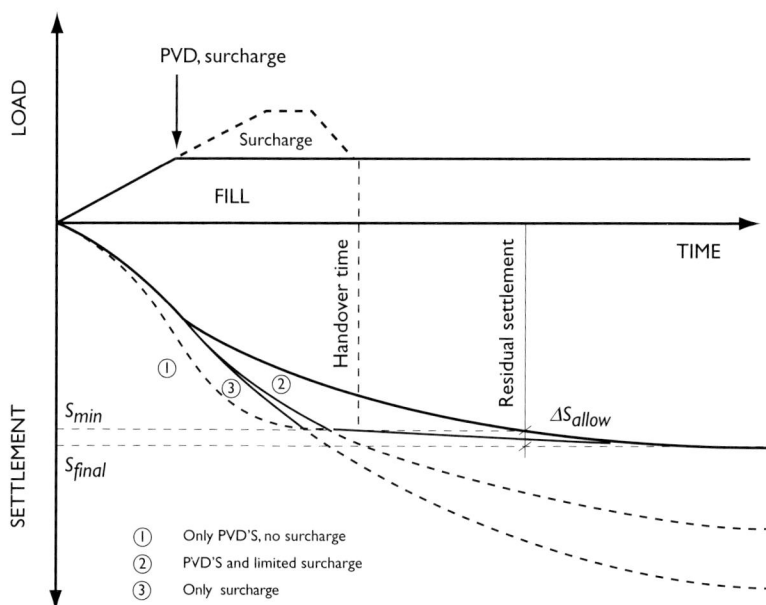

Figure 8.38    *Time-load-settlement graph for loading with fill alone.*

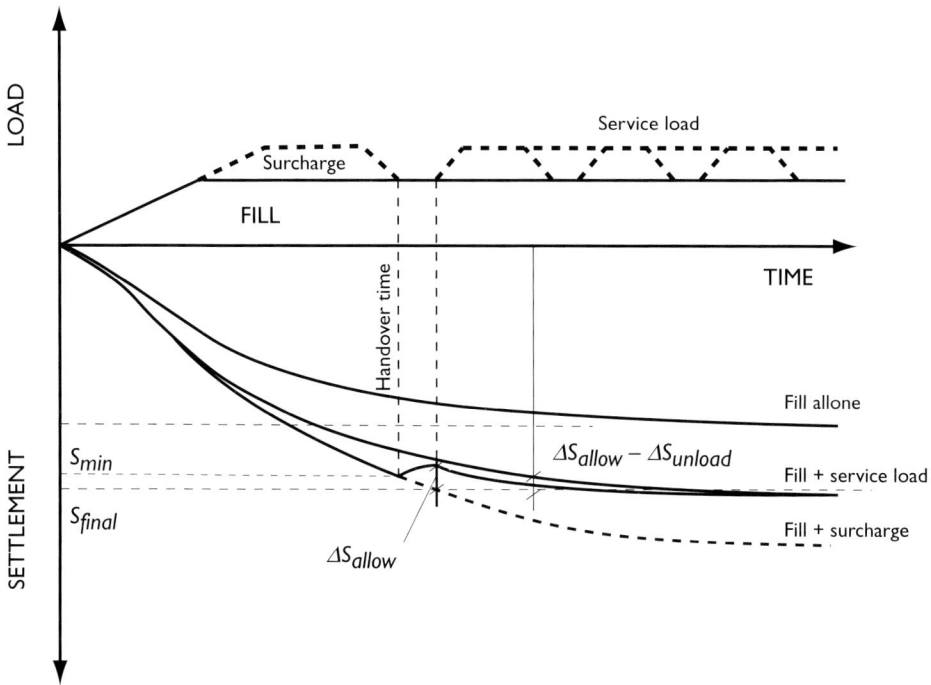

Figure 8.39  *Time-load settlement graph for loading with fill and service load.*

In Figure 8.38 a situation is presented where the weight of the fill is the only long-term loading condition. The specified allowable residual settlement at the time of hand-over is $\Delta S_{allow}$. This means that the settlement $S_{min} = S_{final} - \Delta S_{allow}$ should be reached at time of hand-over. The consolidation of the subsoil should progress fast enough to achieve this.

When this is not the case (as shown in the figure), soil improvement techniques are required.

Three different remedial options using surcharge and/or PVDs are illustrated in Figure 8.38:

1. Installation of PVDs to accelerate the consolidation process in order to achieve $S_{min}$ within the available time period (curve 1 in the figure).
2. Application of a limited preload (resulting in a larger final settlement if removal of the surcharge would not take place) in combination with PVDs to accelerate the consolidation process. Compared with scenario 1, a larger drainspacing can be adopted (curve 2 in the figure).
3. Application of a preload (resulting in the largest settlement if removal of the surcharge would not take place) in order to achieve $S_{min}$ within the available time period (curve 3 in the figure).

In Figure 8.39, a situation is presented in which a service load is defined on top of the fill (as specified in the contract). The specified allowable residual settlement after the construction period at time of hand-over is $\Delta S_{allow}$. In the figure $S_{min}$ is larger than $S_{final,\ fill\ alon}$ which will result in additional unloading/reloading deformations when the service load is applied. This means that the settlement $S_{min} = S_{final} - (\Delta S_{allow} - \Delta S_{unload})$ should be reached at time of hand-over. Clearly, in this situation, consolidation (with or without ground improvement) under the weight of the fill alone will never lead to sufficient settlement during the construction period. As illustrated in the figure, the solution is to use a temporary preload in combination with PVDs in order to achieve the settlement $S_{min}$. The higher the preload, the less the number of drains needed as discussed above. Different solutions can be defined based on practical and theoretical limitations.

**Deformation of fill material**
The fill material itself will experience deformations as well. After hydraulic placement of the fill, the fresh deposit will have a density which differs above and below the water table. A change in volume can occur, which may be caused by:

- Auto-compaction caused by rearrangement of the grain skeleton as a consequence of small stress changes, fluctuations of the water table, vibrations, etc. Deformations caused by the fill's own weight during its construction can also be expected.
- Surface and/or deep compaction in order to obtain a higher density of the fill may result in an important reduction of the volume. Depending on the difference between the initial (as placed) and final (after compaction), density a volume change of the order of 5% to 10% of the layer thickness may be expected.
- The weight of structures built on top of the reclamation and temporary or permanent service loads will introduce instantaneous settlements.
- Finally a long-term secondary compression or creep (ageing) can occur. This is explained by crushing effects at the grain contact points resulting in a further rearrangement of the grains. This effect is very often considered to be negligible for granular material. It will, however, strongly depend on the particle size distribution and mineralogy of the fill material.

During hydraulic placement of fill, and depending on its particle size distribution, fines can get segregated from coarser material. Special measures might have to be taken in order to avoid sedimentation of fines within the reclaimed area. Despite these measures, however, sometimes pockets or veneers of fine material get entrapped in the fill. This may influence the deformation behaviour and bearing capacity of the reclamation. Should segregation of fines occur, then the effect of these inclusions should be analysed. It can very often be demonstrated (e.g. by means of a Zone Load Test) that the presence of fines has limited effects provided that the inclusions do not occur in the top 2 to 3 m of the fill layer. The settlements may be somewhat larger, but the consolidation of thin silt layers goes very fast and when inclusions occur at a larger depth, they only have little effect on the bearing capacity and settlement of the reclamation. Moreover, some silt type materials (e.g. in regions with calcareous sediments) have the same mineralogy as the fill material itself and exhibit a similar effective shear strength under static loading as the coarser material. This silt, material may, however, behave undrained under dynamic loading (e.g. earthquake loading) which needs proper attention when relevant.

Similar to the occurrence of inclusions of fine material, sometimes lenses of stiff clay or "clay balls" can be encountered within the reclamation. Depending on the number of "clay balls" and their distribution within the fill, similar considerations apply as for the fine material discussed above. The clay of a "clay ball" is normally stiff which means that the material exhibits a reasonable strength and a limited deformation upon loading.

In many situations, the application of a temporary preload will be very effective to solve (differential) deformation problems related to the inclusions of fines.

### 8.5.3.5  Vertical deformations of structures

In the preceding sections, settlements were considered with a constant stress increase $\Delta\sigma'$ over the entire depth of the compressible subsoil layer. This is a correct assumption when the load occupies a very large area which is normally the case for the fill material in reclamation projects.

However, for foundations (footings) with limited dimensions at or near the surface of a reclaimed area, the spreading effect of the stress increase as a function of the depth must be accounted for. This is illustrated in Figure 8.40 where the vertical stress distribution is given on two horizontal sections at depth $d_1$ and $d_2$.

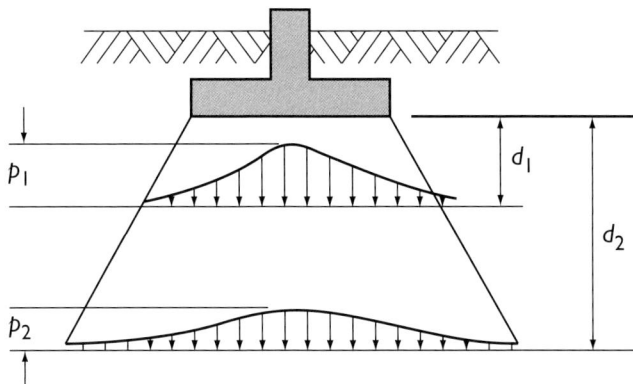

Figure 8.40  *Vertical stress distribution at different levels in the soil.*

$$q = \frac{Q}{(B + z)(L + z)}$$

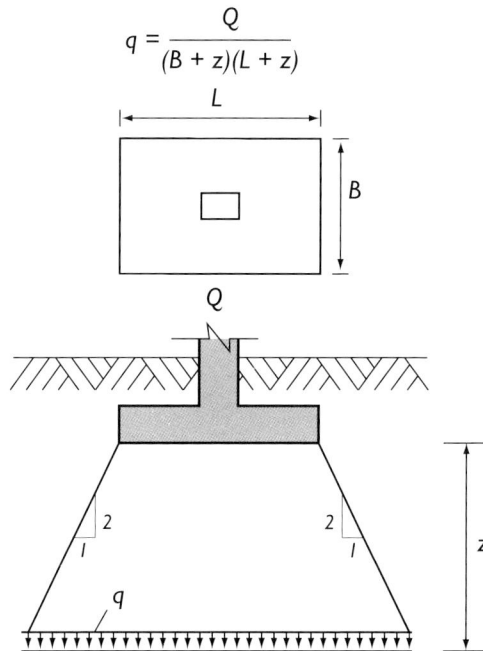

Figure 8.41 *Approximation of the stress increase with depth (uniform load distribution).*

The theoretical approach is generally based on the theory of Boussinesq giving the stress increase at an arbitrary point in an elastic half space. Approximations may also be used based on the assumption of a uniform stress distribution underneath the foundation. The stress increase is then calculated according to simplified spreading rules. In Figure 8.41 a uniform stress increase at depth z under a footing (total load Q) is calculated assuming load spreading according to $2_V$ on $1_H$. Reference is made to various handbooks on soil mechanics for this subject.

### 8.5.3.6   Horizontal deformations

Horizontal deformations of the reclamation fill are seldom or never specified as a project requirement. It is generally accepted that a stable structure, which is designed with an adequate safety factor, will experience limited horizontal deformations.

However, in the design of quay walls or sheet-pile structures, horizontal deformations can be an issue and the stresses introduced by fill material deposited behind the structure should be taken into account. This also applies for temporary conditions, for instance, during staged filling operations (Figure 8.25).

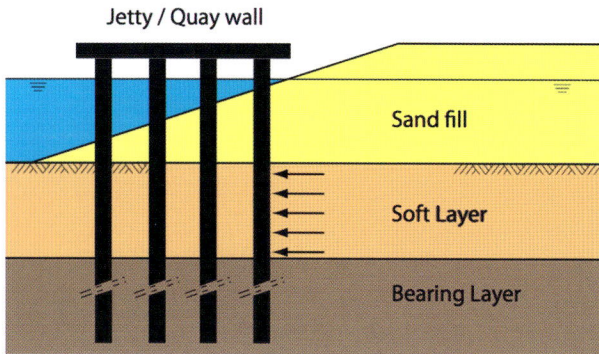

Figure 8.42    *Horizontal loading on pile foundations by hydraulic fill.*

Simple methods to calculate horizontal deformations do not exist. Finite Element Methods (FEM) or the theory of a beam on an elastic subgrade (e.g. sheet-pile design), are normally used for an analysis of horizontal displacements.

Some specific situations need special attention when realising a hydraulic fill project for land reclamation. Horizontal deformations may have a detrimental effect on existing neighbouring structures, specifically if they are founded on deep foundations which are quite vulnerable for such type of loading conditions (Figure 8.42).

Measurement of horizontal deformations can be realised by means of inclinometers (see Section 11.3.3.2 on monitoring) which are generally used to verify the stability of a slope or a retaining structure.

### 8.5.4    *Techniques for limiting settlement*

Apart from the technique of a surcharge with accelerated consolidation (PVDs) as discussed above, various other methods are available to reduce the settlement. Normally a distinction is made between techniques to reduce the settlement of the existing subsoil and techniques to limit the settlement of the fill material. Besides limiting the settlements, most techniques will also improve the stability.

Typical techniques to improve a soft compressible subsoil are:

– soil replacement;
– use of stone columns or related systems such as sand compaction piles;
– deep mixing or soil mixing.

In special situations another option may be to use lightweight fill material or to design structural elements transferring the loads to deeper more competent layers (e.g. relieve platforms).

Techniques to improve granular fill material are all based on compaction by means of the introduction of vibrations or heavy impacts (surface and deep compaction).

Reference is made to Chapter 7, Ground improvement, for details about these techniques.

The application of ground improvement techniques will improve the behaviour of a hydraulic fill and reduce the maintenance costs. Such activities should, however, be taken into account during the design (including soil investigation) and planning phase of a project. The execution of the reclamation may become more complex because of the need for different dredging techniques and other types of equipment to be on site simultaneously.

## 8.6 Liquefaction and earthquakes

### 8.6.1 *Overview*

Soil liquefaction is a phenomenon in which soil loses much of its strength or stiffness, generally for no more than a few minutes, but nevertheless for long enough to make liquefaction the cause of many failures, deaths and major financial losses.

The nature of the liquefaction is easy to understand: soil particles re-arranging to denser state under a stress perturbation and causing excess pore water pressure to develop because of insufficient time for the displaced water to escape. The stress perturbations causing liquefaction are often earthquakes, but these are by no means the only cause. Storm wave loadings may be important with coastal/offshore projects and liquefaction has been induced by vibration of machinery. Liquefaction can also be triggered statically, either by slope steepening or by increasing pore water pressures within the fill mass.

Static liquefaction slides have also developed in quite dense sands (e.g. Fort Peck and Nerlerk) when those sands have experienced reducing mean stress in part of the slopes as the underlying foundation was displaced (yielded) horizontally because of the load static imposed by the fill; sand behaviour under reducing mean stress is a current research interest (e.g. Chu *et al.*, 2012) but the practical implication is straightforward: consider possible foundation movement as well as the issues around the hydraulic fill itself.

Depending on the geometry of the slope or foundation and soil density, the consequences of liquefaction can range from a rapid, large-scale movement of the fill (Figure 8.43, left) through to a general softening of the soil with substantial settlement/shear of the foundation bringing the supported structure to failure (Figure 8.43, right). Movements resulting from liquefaction do not necessarily

Figure 8.43   (l) and (r): Failure of the Fort Peck Dam, Montana (USA) as a result of flow liquefaction (left, Olson, 2001); waterfront damage from cyclic softening at Kobe (Japan).

arise at the same instant as the loading, with movement sometimes occurring minutes after an earthquake as the induced pore pressure redistribute and trigger the failure. Sand boils may develop after liquefaction as the excess water migrates to the surface.

A wide range of soils are susceptible to liquefaction, certainly from clean sands through to non-plastic silts. Practically, this includes all soils likely to be used for hydraulic fills.

Liquefaction studies in the framework of land reclamation are required to decide whether or not ground improvement will be necessary and, if so, to which minimum level. In general, this will be improvement of the fill material, but sometimes (static) liquefaction of the natural underlying soil also needs to be prevented.

The assessment of liquefaction is usually sub-divided by its consequences, in particular into *flow liquefaction* (with rapid, large deformations) and *cyclic softening* (where deformations accumulate in a fatigue-like manner).

- Flow liquefaction can only develop if the soil loses strength as it is loaded, a situation that only arises if the soil is 'loose' (loose is defined and quantified later in this section).
- Cyclic softening can develop with even quite dense soils.

The underlying soil behaviour giving the differing flow liquefaction and cyclic softening responses is considered later in this section, as it is necessary to introduce some more background about soil mechanics before these aspects of liquefaction can be clarified.

Practically, as a result of their apparently spontaneous initiation, fast propagation and large deformations, flow slides are important geo-hazards. Flow slides can be

**Breaching**

When a local steep is constructed at the toe of a slope, it initiates a retrogressive failure mechanism. As the local steep moves upward along the slope, the sand that flows down erodes the slope below. As long as the sand is removed at the bottom, this process will continue until the slope reaches equilibrium. This process is denoted by the term breaching.

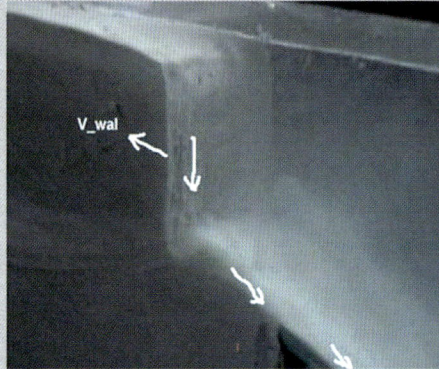

Figure 8.44    *Breaching mechanism as observed in laboratory.*

Breaching is well known in the dredging industry as an effective method to optimize productions during sand mining. However, recently breaching has been related to slope failure as well (e.g. De Groot *et al.*, 2009). An unstable breach may be initiated during construction or by a liquefaction mechanism, thereby leading to an overall retrogressive mechanism. Internationally, the term breaching is not generally recognized as a separate failure mechanism. Retrogressive mechanisms are observed after the occurrence of slope failures. They are considered to be the result of rather than the initiation for slope failures. Breaching is therefore not separately discussed as a failure mechanism in the Hydraulic Fill Manual.

Breaching as a failure mechanism can only occur in dense sands where the local steep remains stable for a short time period, depending on the suction caused by dilation and the hydraulic permeability of the soil. In loose sand flow liquefaction will immediately be initiated when a local steep is realized and thus the breaching mechanism cannot develop.

triggered by adverse soil conditions below the fill, but dredging or soil erosion at the toe of a slope is a more common cause. See breaching in box above.

Because of the potential for rapid and seemingly spontaneous development, the assessment that a fill is not susceptible to flow liquefaction must precede any consideration of cyclic softening. If a susceptibility to flow liquefaction is indicated, then ground improvement is required; the only question is: how much improvement is needed? If flow liquefaction is not indicated, then cyclic softening must still be considered.

Possibly the key question for many involved with hydraulic fills will be: what testing is needed to support a liquefaction assessment? This section of the manual responds to that question by providing a briefing or guide to all who may be involved in a hydraulic fill project. Most of the calculation procedures are now found in commercial software and some widely used software is listed later in this section. However, actual testing needs are poorly covered by Codes of Practice and this section scopes out what would be a usual standard of care.

One caution regarding this standard of care is that it is based on earthquake loadings, but the methods used for that contain many empirical factors. Offshore construction, with its storm wave and ice loads, will need caution as these empirical factors have been established for very few loading cycles (say no more than 20 'significant' cycles in even a very severe earthquake) whereas offshore situations may involve thousands of cycles in a storm. If the fill being engineered will be exposed to storm loading, a greater scope of engineering and possibly supporting testing will be needed than indicated in this Manual.

### 8.6.2  *History of understanding*

Dutch engineers have been engineering against liquefaction for centuries in their efforts to protect their country from the sea, coastal flow slides being a long-standing problem. Although mainly caused by channel erosion, there have been other causes with flow slides in the approach to a railway bridge near Weesp in 1918, being triggered by vibrations from a passing train. A comparable awareness of liquefaction developed in North America on the same timescale, but for different reasons. Today, large and mobile earthmoving equipment is a basic tool of civil engineering; this equipment was not available to 19th century engineers. However those 19th century engineers did have pumps, and hydraulic fill construction became a normal construction method for earth dams. Some of those dams collapsed during construction, notably the Calaveras Dam. By 1918 some civil engineers in the United States were aware of liquefaction and aspects of it had been published in the journals of the day.

The development of understanding in the decade that followed is obscure, but in 1935 the US Corps of Engineers were explicitly considering liquefaction in the design of Franklin Falls dam (and specified explosive compaction to remove liquefaction risk). In some ways, this is a remarkable project as it was the start of putting the 'mechanics' in geotechnical engineering. As part of the engineering work for Franklin Falls, extensive shear box tests were undertaken in the

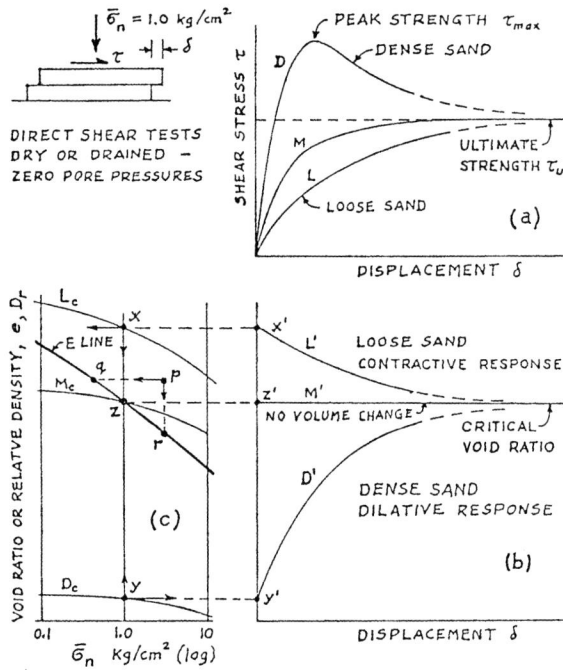

Figure 8.45  *Identification of critical void ratio from direct shear tests (Casagrande, 1936).*

engineering laboratory of Harvard University. These shear box tests showed that loose sands contracted and dense sands dilated until approximately the same void ratio was attained at large strains, Figure 8.45. Note that this figure is "historical", being taken from the original paper published some seventy years ago. Interest here is the void ratio at large strain that distinguishes which mode of behaviour the soil exhibited; Casagrande (1936) termed the distinguishing void ratio the *critical void ratio*. That terminology has continued, with the ideas further developed using the mathematical theory of plasticity into *Critical State Soil Mechanics*—a theoretical framework that now dominates current understanding of soil behaviour (including liquefaction).

Casagrande's work was followed by further testing (just down the road) at MIT during the 1940's, in part because of the Fort Peck failure (Figure 8.43). It was found that the critical void ratio became smaller as the effective confining stress increased (Taylor, 1948); the relationship between critical void ratio and mean effective stress is called the *critical state locus* (CSL), see Figure 8.46. The CSL is often approximated as a straight line on a plot of void ratio versus the logarithm of mean effective stress, the slope of this line being denoted as $\lambda$.

Slightly out of historical sequence, the 1980's saw the realization that the CSL did not just distinguish the mode of soil behaviour but that the difference in void ratio

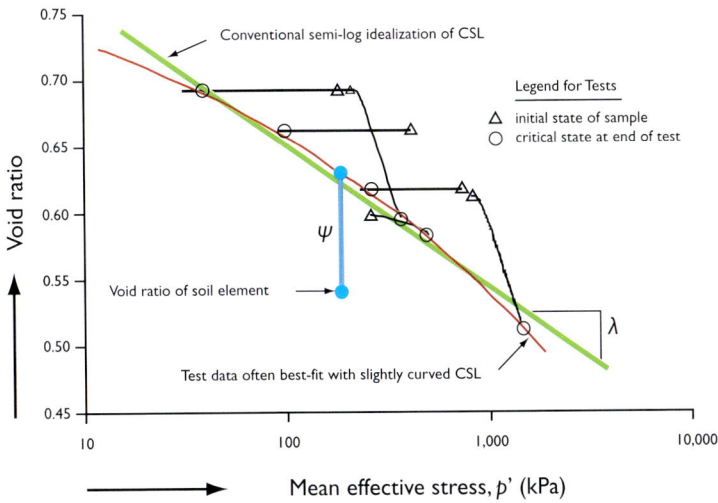

Figure 8.46    *Illustration of the CSL and the state parameter ψ (The CSL of sands to silts is often slightly curved, as seen with this data, but the simple semi-log idealization highlighted in green is sufficient for most practical engineering).*

between a soil's current condition and the critical void ratio at the same stress, termed the state parameter ψ, characterized the strength of soils. Today, the CSL and ψ are the basis of every advanced constitutive model for soil; broadly, the slope of the CSL defines the intrinsic compressibility or plastic hardening of the soil while ψ determines its dilatancy aspects that are fundamental to any proper constitutive model. However, although these modern models readily compute the stress-strain-pore pressure behaviour of soils during all forms of liquefaction, practical engineering only uses the mechanics underlying these models to characterize the resistance to liquefaction of soil—in effect, to just use the strength rather than the entire stress-strain curve. Figure 8.46 illustrates the framework; the CSL is a soil-specific behaviour being affected by gradation and mineralogy of the soil's constituents.

The range for the state parameter in the laboratory is about −0.3 to +0.1 for sands to sandy silts with predominantly quartz particles with negative values being the dilatant or the "strong and stiff" situation (the negative term arises because of the "compression positive" convention of soil mechanics). Calcareous sands, or soil with high mica content, will exhibit greater compressibility (i.e. a greater slope to the CSL) but the state parameter continues to characterize the dilatancy.

Returning to the evolution of understanding, a large earthquake in Japan demonstrated the damage caused by earthquake-induced liquefaction; and, lead to a different method of liquefaction assessment to that just described. The city of Niigata lies on the west coast of Japan and is underlain by about 30 m of fine alluvial sand. The 16th June 1964 earthquake liquefied loose sand deposits in low-lying areas. An apartment building tilted almost on its side because of bearing capacity failure in the liquefied ground. Underground structures such as septic tanks, storage tanks, sewage conduits and manholes floated upward out of the ground. Five spans of the Showa Bridge across the river fell when pier foundation piles deflected because of lateral support loss from liquefaction.

Laboratory testing in the 1960's soon showed that being denser than the critical state was not sufficient to prevent cyclic softening. The framework then emerged that a measure of the stresses imposed on the ground by the earthquake—called the *cyclic stress ratio* (CSR)—should be compared to an equivalent available strength of the ground, the *cyclic resistance ratio* (CRR).

Development of methods to determine the CSR and the CRR have diverged in terms of the representation of soil behaviour. In the case of the CSR, it was recognized that site response affected the CSR and that the site response depended on the soil behaviour in cyclic loading. Rather simple approximations of soil behaviour became standard and widely used. Kramer (1996) provides a good text and guide to this aspect of liquefaction evaluation.

In the 1960's, constitutive modelling of liquefaction was a distant dream and there was no obviously applicable method/protocol to determine the CRR; what amounts to a geological 'classification' approach had to be followed. Various instances of liquefaction were identified (e.g. from visual evidence such as sand boils) with the sites characterized in terms of their penetration resistance; the cyclic stresses imposed by the earthquake were estimated. When adding in nearby sites that had not liquefied, allowed a chart to be developed with a line distinguishing between occurrences of liquefaction and no-liquefaction for a range of soil penetration resistance and a range of cyclic stresses. The method was pioneered by Seed and co-workers at Berkeley (University of California), but has seen contributions since then by workers in Japan, Canada, and the US. This approach is today generally referred to as the NCEER method (National Center for Earthquake Engineering Research, Youd *et al.*, 2001).

The hazard liquefaction presents was emphasized by a near-catastrophe in 1971. The 43 m high Lower San Fernando Dam was the terminus of the main aqueduct system for Los Angeles. It was built by hydraulic filling. Shaking during the 1971 San Fernando earthquake caused a slide of the top thirty feet of the dam, illustrated on Figure 8.47. About 80,000 people lived in a 6 mile long area down the valley from the dam and were threatened by the very real possibility that the dam would fail completely, inundating the area by a catastrophic flood wave. Disaster was narrowly averted by drawing down the reservoir before the remnant of the crest gave way. An interesting aspect of this failure is that it did not occur during

Figure 8.47    *The effect of an earthquake on a hydraulic fill, the 1971 failure of Lower San Fernando Dam. Note the paved dam crest road that is mostly now under the reservoir following a liquefaction slide in the upstream shell. (Photograph from Karl V. Steinbrugge Collection, Earthquake Engineering Research Center, University of California, Berkeley).*

the earthquake itself, but developed a few minutes after the shaking stopped as the earthquake-induced pore water pressures moved in the dam body.

An important drive to understand liquefaction has come from the US under the auspices of The National Earthquake Hazards Reduction Program established by the U.S. Congress in 1977. Soil liquefaction was recognized as a hazard under this program. A number of agencies implement various activities under the program, of which the National Science Foundation has been the most important in advancing the understanding of liquefactions (and remediation to prevent it). This program has had wider impact than the US, as the NSF has involved engineers from other countries and, by doing so, influenced the research within those countries as well as ensuring dissemination of findings. The text by Jefferies & Been (2006) provides a good overview of the physical processes in liquefaction and how they fit within the modern, and computable, understanding of soil behaviour that has come out of this work. Today, the understanding of liquefaction as instances of whether or not sand boils were observed—the 'geological' approach—has been supplemented by an 'engineering' approach anchored in fundamental understanding of stress-strain behaviour of particulate media.

Liquefaction engineering continues to evolve and the 'state of practice' is changing, and likely to continue changing for another decade. However, in essence the principles are now set and evolution is about the coefficients or form of the equations in the design/assessment methods. With this in mind, it is now appropriate to return to flow liquefaction and cyclic softening.

### 8.6.3 *Flow slides versus Cyclic softening*

The consequences of soil liquefaction, and the way soil liquefies, depend on whether or not the soil is looser or denser than the CSL, recall Figure. 8.45. There are quite fundamental differences between liquefaction of loose soils and liquefaction of dense soils.

In the case of "loose" soil ($\psi > 0$), the intrinsic behaviour is readily observed in an undrained triaxial compression test and Figure 8.48 illustrates 'stress path' followed by the sample to the left and the stress-strain behaviour to the right. The important behaviour is the rapid loss of soil strength following the peak: static liquefaction—if such behaviour develops where the loads cannot reduce, which can be the situation in a slope, then there will be large scale movement. This is generally referred to as a flow slide as it develops rapidly and usually with little warning by way of tension cracks.

The soil behaviour illustrated on Figure 8.48 is not restricted to the triaxial test, and the understanding generalizes for all stress conditions. A line drawn from the origin through the peak strength in the test is called the Instability Line as

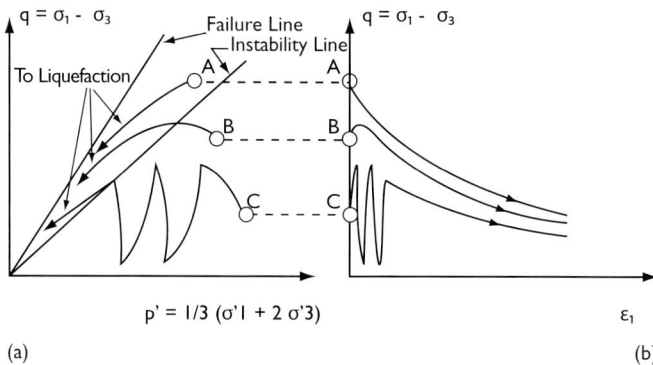

Figure 8.48   *Loose sand behaviour showing (a) stress paths and (b) stress–strain relations for initiation of static (A, B) and cyclic (C) instability at small strains and subsequent liquefaction at large strains.*

illustrated on Figure 8.48. And, as also illustrated on the figure, the instability line may be markedly lower than the soil's critical state frictional strength. The region above the instability line is generally referred to as a 'zone of potential instability' (e.g. Lade & Yamamuro, 2011) because whether flow sliding or other collapse develops depends on whether it is a 'dead load' driving force or if the loads can re-distribute away from the strength-loss zone.

It does not matter how the instability line is approached. It can be approached statically, that is drained loading, by excavation at the toe of a slope (case B on Figure 8.48). Once the instability line is reached, the soil behaviour can immediately transition to undrained response regardless of how slowly further excavation is done. The instability line can also be approached by the cyclic loading of an earthquake increasing the pore pressures in a slope (case C on Figure 8.48), with the same consequences. It can be approached by seepage raising the pore pressures in a slope (case A on Figure 8.48), also with the same consequences.

The instability line is not a "property" of the soil, but depends directly on the state parameter. The slope of the instability line becomes smaller as the state parameter becomes more positive. There is a literature on this functional behaviour, but that literature is of no practical consequence. If the situation in the slope is $\psi > 0$, then the questions becomes this: would slope failure be a hazard or otherwise unacceptable? If a rapid slope failure, developing without warning, is unacceptable then remedial work is needed; an "observational" approach will not provide safety (for example, the slope in the Aberfan disaster was inspected, and deemed safe, shortly before it failed killing 144 people).

Dense soils can still 'liquefy', with most of the experience of "liquefaction" worldwide actually being on soils denser than the critical state. But, the entire character of the soil behaviour changes if $\psi < 0$. For $\psi < 0$ there is no brittle strength reduction with a gentle softening of the soil shear stiffness developing instead. This changed behaviour is illustrated on Figure 8.49 which shows the behaviour of a dense soil in cyclic simple shear as it experiences multiple cycles of uniform shear loading (analogous to an earthquake). The lower part of the figure shows the excess pore pressure ratio versus the cyclic strain, with the upper part showing the corresponding stress strain behaviour. As can be seen, even though the sample is dense excess pore pressure accumulates with each loading cycle. The soil response remains comparable to its initial stiffness until the excess pore pressure ratio exceeds 0.5, whereupon a progressive softening of the shear modulus develops. This form of stress-strain behaviour can manifest itself as an accumulation of plastic strain (displacement) with each additional loading cycle; there is no "run-away" failure, but extensive damage can still develop.

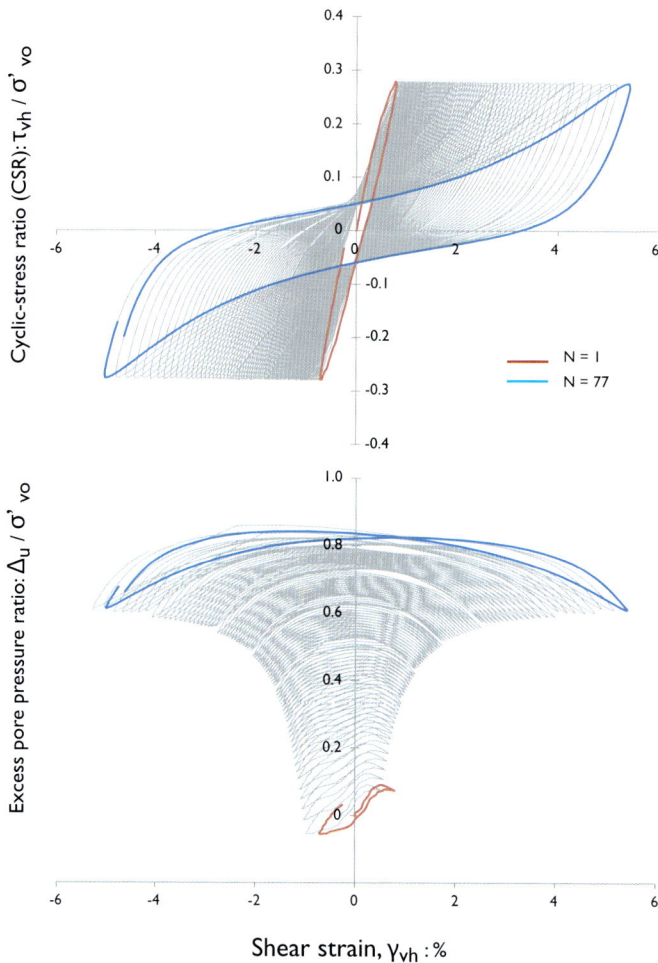

Figure 8.49   *Dense soil behaviour in cyclic simple shear.*

## 8.6.4   *Assessing liquefaction susceptibility*

*Preamble: Design/assessment methods use the ratio of imposed cyclic shear stress to the pre-event vertical effective stress—what is called the cyclic stress ratio (CSR)—as the "load" in a liquefaction analysis or "ACTION" as it is referred to in EC7/EC8. The "resistance" of the soil to liquefaction is expressed as a similar ratio termed the cyclic resistance ratio (CRR). The margin to resist liquefaction can be expressed as a safety factor, $FS_L = CRR/CSR$.*

*The CSR depends on both the earthquake scenarios adopted as the appropriate hazard for the site and how the site responds to the ground motions imposed by earthquakes. There are standard methods to assess the CSR imposed by earthquakes, ranging from an approximate calculation that can be done on a spread sheet to commercial numerical analyses. These methods have widespread use and can be viewed as standard practice. The situation for other loadings, for example storm waves or aero-elastic vibrations from wind turbines, is less standardized. This manual only discusses calculation of the CSR caused by earthquakes.*

*Estimating the soil resistance to imposed load, the CRR, requires two steps:*

*– Evaluate susceptibility to flow-slide movement;*
*– Provided not vulnerable to movement, then evaluate the CRR.*

*Both steps involve assessing "liquefaction". The first step is a check that the fill is suf-ficiently strong (dense) to make calculation of a CRR meaningful. The second step then provides the strength for assessing the vulnerability of the works to cyclically induced softening; there may also be associated settlements and these are considered in Section 8.6.5.2.*

*There are options for the methods used to calculate the CRR, and these options are presented with their limitations and other considerations. The necessary test-ing depends on the choice of design approach and is discussed subsequently in Section 8.6.6.*

### 8.6.4.1   Codes & Standards

European Standards EN 1997 and EN 1998 ("Eurocodes" EC7 and EC8) are widely used, both within their jurisdiction and in other parts of the world where a formal code may not have been adopted. In North America, the comparable docu-ments are the International Building Code (which is actually a US code despite its name) and the National Building Code of Canada.

These codes have a bias to buildings, not civil infrastructure where hydraulic fills are most likely to be used. These codes also distinguish the importance of buildings, which might be viewed as 'routine commercial', 'lifeline', and 'unusually important'. The same logic of design standards depending upon use should be applied across to hydraulic fills.

EC8 does not require seismic design in countries or areas where there is very low seismic activity. Low seismic activity is defined as either a design ground accelera-tion on Type A ground (see Table 8.13) $a_g < 0.04$ g or those where the product of S and $a_g$ is less than 0.05 g. The selection of whether to use the value of $a_g$ or the product of S and $a_g$ follows from the country's national annex.

EC8 (2004) provides for a two-level seismic design. The two performance objectives can be roughly compared to the Ultimate Limit State (no-collapse)

283

and the Serviceability Limit State (damage limitation). For structures of ordinary importance Eurocode 8 recommends:

- A design seismic action (for local collapse prevention) with 10% exceedance probability in 50 years (mean return period: 475 years)
- A serviceability seismic action (for damage limitation) with 10% exceedance probability in 10 years (mean return period: 95 years). For buildings with

Table 8.13   *Ground types according to EC8–1 (2004).*

| Ground type | Description of stratigraphic profile | $V_s$ (m/s) | $N_{SPT}$ (blows/30 cm) | $C_u$ (kPa) |
|---|---|---|---|---|
| | | **Parameters** | | |
| A | Rock or other rock-like geological formation, including at most 5 m of weaker material at the surface | >800 | – | – |
| B | Deposits of very dense sand, gravel, or very stiff clay, at least several tens of metres in thickness, characterised by a gradual increase of mechanical properties with depth | 360–800 | >50 | >250 |
| C | Deep deposits of dense or medium dense sand, gravel or stiff clay with thickness from several tens to many hundreds of metres | 180–360 | 15–50 | 70–250 |
| D | Deposits of loose-to-medium cohesionless soil (with or without some soft cohesive layers), or of predominantly soft-to-firm cohesive soil | <180 | <15 | <70 |
| E | A soil profile consisting of a surface alluvium layer with $V_s$ values of type C or D and thickness varying between about 5 m and 20 m, underlain by stiffer material with $V_s$ > 800 m/s | | | |
| $S_1$ | Deposits consisting, or containing a layer at least 10 m thick, of soft clay/silts with a high plasticity index (PI > 40) and high water content | <100 (indicative) | | 10–20 |
| $S_2$ | Deposits of liquefiable soil, of sensitive clays, or any other soil profile not included in types A–E or $S_1$ | | | |

different importance class, the seismic action is determined by multiplying the reference seismic action for structures of ordinary importance with an importance factor $\gamma_I$ (refer to Eurocode 8 for further elaboration on this).

The design motions for earthquake-induced liquefaction are trending to more stringent criteria than found in the present Eurocode 8, with North American codes now looking to a 2% probability of exceedance in 50 year design life for 'routine commercial' work. The 'lifeline' structures will require a higher standard.

A 'factor of safety' ($FS_L$) against liquefaction, analogous to use in other geotechnical stability calculations, may be used to express the results of a liquefaction assessment; EC8–5 (2004) requires $FS_L > 1.25$. This $FS_L$ requirement is consistent with the excess pore pressure ratio being less than <0.5 (e.g. Tokimatsu & Yoshimini, 1983), a ratio that will keep cyclic strains small (say <1%, depending on the imposed CSR) as illustrated earlier on Figure 8.49.

Alternatively, the various uncertainties in the liquefaction analysis can be combined into an overall probability that the site will perform better than needed during the design life. This approach appears to be gaining favour in research studies but does not appear to have been adopted by codes of practice (or similar regulatory requirements).

8.6.4.2   Loading: Estimating CSR by site response analysis

The CSR is the imposed ACTION in the context of EC7/EC8. In more common engineering language it is the "load" on the soil.

The minimum description of a design earthquake scenario for a site is the peak ground acceleration (PGA) at bedrock level in combination with the governing earthquake magnitude (M); these values need to conform to the code-required return period and exceedance probability. The PGA scales the induced cyclic shear stresses; the magnitude M is a proxy for the number of loading cycles that are imposed (larger magnitude earthquakes last longer and cause more loading cycles). Both aspects of earthquake loading are needed. In general, several design scenarios are considered, covering both near but lower magnitude and more distant and larger magnitude events. Design scenarios may be a few chosen situations or a more formal probabilistic assessment of 'hazard'.

Most hydraulic fill projects will require a site-specific ground motion study in any area where the ground motions might exceed (say) 0.05 g in a 500 year period. These studies may range from a desk study for a few thousand $ to a full study including geological mapping, seismic source identification and advanced probabilistic

**Earthquake magnitude**

Several sources and projects define the earthquake magnitude in different ways. Day (2002) and Youd *et al.* (2001) differentiate between the local magnitude ($M_L$), the surface wave magnitude ($M_S$) and the moment magnitude ($M_W$). The latter ($M_W$) is defined as $M_S$ in Eurocode, and best represents the energy associated with the earthquake. However it is concluded that below a value of 7 the magnitude are reasonably close to one another and there is no need to identify which scale is meant in design specifications.

When no input is available the relation between the epicentral acceleration and the magnitude from a relation provided by Day (2002) can be used as a very rough initial assumption (Table 8.14).

Table 8.14  *Relation between earthquake-magnitude and (epicentral) acceleration (Day 2002).*

| Local Magnitude ML | Typical peak ground acceleration $a_{max}$ near the vicinity of the fault rupture | Typical duration of ground shaking near the vicinity of the fault rupture | Modified Mercalli intensity level near the vicinity of the fault rupture |
|---|---|---|---|
| ≤2 | – | – | I–II |
| 3 | – | – | III |
| 4 | – | – | IV–V |
| 5 | 0.09 g | 2 s | VI–VII |
| 6 | 0.22 g | 12s | VII–VIII |
| 7 | 0.37 g | 24s | IX–X |
| ≥8 | ≥0.50 g | ≥34s | XI–XII |

modelling. In the absence of a site-specific study the following sources can be used to develop an initial estimate for design ground motions:

- The Global Seismic Hazard Assessment Program provides a world map in digital format, including raw data, with PGA (http://www.seismo.ethz.ch/static/GSHAP/).
- The Uniform Building Code (UBC), a code developed in the United States, is perhaps one of the most advanced seismic codes worldwide; the code is readily available (e.g. from www.amazon.com). Many engineers using this code outside the United States, in part because Division III, Section 1653, of the code zones earthquake hazard for practically the entire world. For each of these zones a reference peak ground acceleration at bedrock is defined. The 1997 edition of the UBC provided the PGA at bedrock with 10% exceedance probability in 50 years (mean return period: 475 years); this design hazard would generally be regarded as insufficient today. The UBC is routinely updated, and is

now known as the International Building Code although, despite this name, it remains a United States, not an International, code. Current revisions look to a 2% exceedance in 50 year standard. The design peak ground acceleration at bedrock for return periods other than the reference period is obtained by multiplying with an importance factor.

Earthquake ground motions in bedrock tend to be amplified as they propagate vertically upwards at a site. This amplification must be included in the liquefaction assessment as it can substantially increase the CSR imposed on the soil. There are four levels of assessment:

i) simplified;
ii) 1D 'equivalent linear';
iii) 1D 'effective stress';
iv) 2D methods.

All are briefly discussed, but only (i) and (ii) are widely used in engineering practice.

The 'simplified' procedure was proposed by Seed & Idriss (1971). In this procedure the 'load' is computed using the expression:

$$CSR = 0.65 \frac{a_{max}}{g} \frac{\sigma_v}{\overline{\sigma}_v} r_d$$

where the '0.65' is an empirical factor to account for the somewhat variable nature of each cycle in the earthquake and reduce it to an equivalent number of uniform cycles; $a_{max}$ is the single greatest value in the ground motion (PGA or peak ground acceleration). The "site effects" are encapsulated in the dimensionless response coefficient $r_d$.

Seed & Idriss evaluated $r_d$ by simulating numerous sites and suggested an average trend for $r_d$ with depth be used in "simplified" site assessment. This average trend is commonly found embedded in software and spread sheets used for site assessment. More recently, Idriss (1999) has suggested that there is a systematic bias with earthquake magnitude ( = duration of shaking) as illustrated in Figure 8.50.

Although the simplified method is easy to use, and imposes no demand for site-specific data beyond the PGA, Seed et al. (2003) have shown a rather large variation in $r_d$ from one site to another. There is on-going discussion about its biases and uncertainties and the simplified method is not necessarily "conservative". Because of these concerns over the factors influencing $r_d$, the simplified method is most appropriate for initial stages of design; detailed design will usually be based on a site-specific calculation, and the 'equivalent linear' method dominates that engineering.

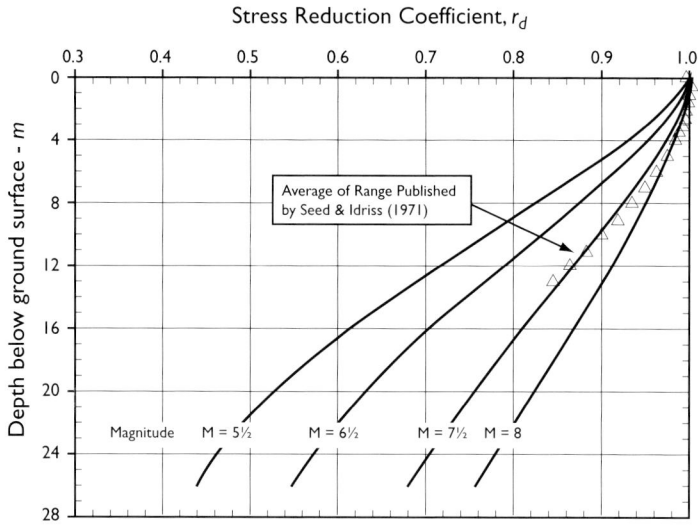

Figure 8.50  *Response coefficient versus depth (Idriss, 1999).*

The equivalent linear method idealizes soil as a linear visco-elastic material with the non-linearity of soil behaviour captured by "iterating" to find an effective shear strain which gives equivalent secant shear modulus and equivalent linear damping ratio that best approximates the actual nonlinear hysteretic stress—strain behaviour of cyclically loaded soils. This is a total stress approach and requires that the soil must not liquefy (albeit that some excess pore pressure is embedded in the modulus degradation). This calculation method can only be done in a computer; the original program for this purpose was SHAKE (Schnabel *et al.* 1972). This program has been public-domain since its inception (and is still available at http:// nisee.berkeley.edu/elibrary/getpkg?id = SHAKE91), but, as might be expected from the era of its development, it is not user-friendly. Today, the underlying code has been configured for use on a Windows-based PC and is available as the program PROSHAKE at http://www.proshake.com/. This program is an accepted standard to assess the CSR for liquefaction assessment.

A SHAKE analysis requires a profile of elastic shear modulus in all soil layers, as well as a curve, or a set of curves, that shows how the shear modulus degrades with increasing shear strain. The model does not explicitly distinguish between plastic hysteretic damping and viscous damping, but instead simply considers all damping as viscous in nature. Hence, the analysis also requires a curve, or a set of curves, that characterizes how the damping ratio increases with shear strain. These modulus degradation and damping curves are usually simply adopted as "average" values from the literature (and are included in the SHAKE manual), but damping in particular can markedly affect the results.

The SHAKE analysis operates on a ground motion history so a set of digitized acceleration versus time 'strong motion' records are needed. These ground motion records have to be matched for frequency content, causative fault mechanism and depth, and distance from the site to the design earthquakes identified in the seismic hazard assessment; the selected records are then scaled to the design PGA. Engineering companies carrying out liquefaction analysis will normally have a library of strong motion records available.

From the perspective of validation requirements embedded in EC7 and EC8, there is a wide experience base with SHAKE and with calibration of its response to measured ground behaviour (e.g. Borja, 2002).

Soil behaves according to effective stress, and a number of workers have proposed similar 1-D analyses to SHAKE but basing the stress-strain behaviour on a computed excess pore pressure. These models claim to better represent soil behaviour than SHAKE, and certainly in principle should be able to properly simulate cyclic softening. But, regardless of technical merit, engineering practice has not yet embraced any of the effective stress methods. There is also the issue that none of the effective stress methods appear validated to meet EC7/EC8 standards, potentially placing a large burden on engineering practice.

Everything so far has treated the CSR as being caused by vertically propagating shear waves, a 1-D situation. This is obviously of uncertain relevance for soil-structure response situations such as quay walls. A reasonable view is that analysis where 2-D "effects" may be important will require finite element simulations with simple to advanced soil models. Discussion of these models is beyond the scope of this manual. However, the recommended site characterization presented in Section 8.6.6 does include the data needs for these types of analysis.

8.6.4.3   Resistance, Step 1: Susceptibility to large deformations

The starting point for a liquefaction resistance is the adequacy in overall stability of the proposed works; that is, susceptibility to mass-movement of the hydraulic fill (flow slides). Two approaches are followed in assessing vulnerability to large deformations: i) to evaluate the type of stress-strain response of the fill to check that it is not brittle (i.e. to avoid post-peak strength loss in undrained loading); and/or, ii) to use post-liquefaction (residual) strengths in a limit equilibrium analysis to check that FS > 1 for overall stability with these 'minimum assured' undrained strengths.

i) *Soil Behavior Approach.* The principle behind this approach is to evaluate whether the soil is dilatant or not. However, while the understanding is based on laboratory test data the practical application of the approach has been calibrated by back-analysis of static liquefaction slides. This methodology means the strength includes some degree of internal pore water redistribution and thus the strengths are closer to the minimum strength in 'limited liquefaction' than the true critical state value seen in small laboratory tests. The approach has evolved over perhaps twenty years with the consensus becoming apparent in a discussion of Shuttle & Cunning (2007) by Robertson (2008) and Shuttle & Cunning (2008), as well as in Robertson (2010).

The input to the assessment is piezocone (CPTu) data. In a design situation, experience with other fills (i.e. CPTu data in the fills considered similar to what is expected) is used and the design parameters will need subsequent 'verification' by tests after fill placement. Each CPTu test needs processing into two data streams with tip data aligned and averaged over the centre of the friction sleeve, giving a depth aligned data stream of normalized penetration resistance $Q_u$ (where $Q_u = Q(1 - Bq) + 1$) and friction ratio F.

The consensus takes the form of a chart where $Q_u$ versus F data from the CPT is compared to a decision criterion, Figure 8.51. If the data plots consistently below the decision line, then a vulnerability to flow liquefaction is indicated and engineering should consider ground improvement without any further discussion about cyclic softening. Conversely, if the data plots above the decision line then the assumptions in the methods used to estimate CRR are reasonably met and the engineering moves to assessing the extent of cyclic softening that may develop and the consequences of that.

Strictly, the decision criterion on Figure 8.51 is affected by both the elastic shear modulus of the soil "$G_{max}$" and the drained friction angle $\varphi'_{crit}$ of the soil at critical state. Present practice is to neglect these effects as secondary to the check that the soil will have the required stress-strain behaviour under large strains. However, it is essential to consider the effect of the variability in soil density. Because the loosest 10% or so of the soil mass controls its behaviour, the CPT data compared to the criterion on the figure must be that for the loosest 10%. Assessing CPT data could use statistical methods "in principle", but practically it is usually done by "eye" with simple judgment as to whether or not 90% of the data lies above the decision criterion or not (the difficulty being that there is presently no commercial software for statistical processing of CPT data—this area of work is limited to academic studies of liquefaction, for example Hicks & Onisiphorou, 2005).

ii) *Residual Strength Approach.* This approach uses a "minimum assured" undrained strength $s_r$ in a conventional limit equilibrium analysis and, provided that FS > 1, it follows that large scale movements will not develop. This "minimum assured" strength is sometimes referred to as the 'steady state' strength. The important step is the strength that is used in the analysis. This strength is primarily based on calibration to past failures and uses penetration resistance as the index of soil state. The idea of using case-history data to estimate the operational residual strength is more than 50 years old; the idea gained momentum in the 1980's and several groups have contributed to it.

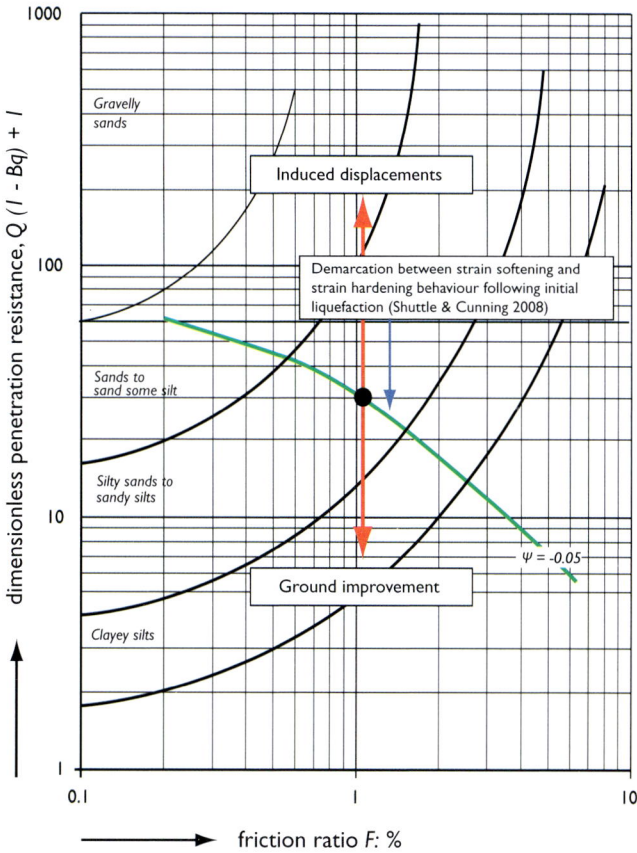

Figure 8.51   *CPTu based liquefaction screening chart (adapted, after Shuttle and Cunning, 2008).*

Although the concept of using case history data is widely accepted (e.g. Stark & Mesri, 2003), and most workers calculate comparable strengths for any case history, there is less clarity on what penetration resistance corresponds to that strength. And, there is no consensus on how the experience should be synthesized. The guidance from the constitutive behaviour of soils is never-theless simple and unambiguous: the ratio $s_r/\sigma'_{v0}$ must depend on $\psi$ with that dependence being a function of the soil properties $M$ and $\lambda$. Since $\psi$ controls the CPT penetration resistance, there is an expectation that the ratio $s_r/\sigma'_{v0}$ will be directly related to the stress-normalized CPT resistance, which turns out to be the case as illustrated on Figure 8.52.

There is some scatter on this figure, but that is not surprising as the soil proper-ties are regarded as the same across all case histories—a more or less necessary assumption as few of the case histories have measured supporting data to do better than this.

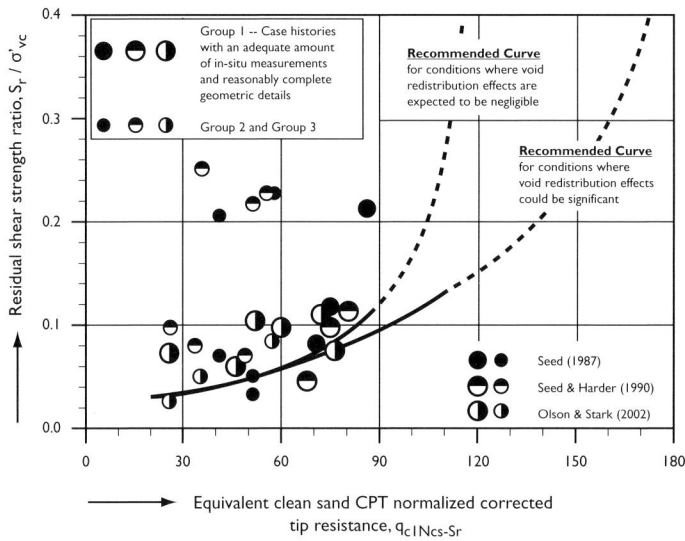

Figure 8.52    *Mobilized steady state strength from case history data (Idriss & Boulanger, 2008).*

Truly dilatant soils will produce a very rapid increase in the post-liquefaction strength, so the trend drawn as the 'proposed relationship' in Figure 8.52 does not continue in a straight line. The Idriss and Boulanger (2008) work illustrated on Figure 8.52, suggests that a very rapid increase in strength develops at about $q_{c1} > 100$ with sands; this is consistent with soil states above the 'green line' in Figure 8.51.

### 8.6.4.4    Resistance, Step 2: Evaluation of CRR

There are two general and standard methods for evaluating the CRR and one new method for finer grained soils (sandy silts to clayey silts). The two standard methods are based on processing penetration test data; the new method requires laboratory testing of undisturbed samples.

The two standard methods suit the assessment of cyclic mobility for level to slightly sloping ground. The extension to soil structure situations, where 2D (or even 3D) effects may be important requires further consideration—something that may well arise in hydraulic fills associated with new marine facilities. In that situation, finite element analysis to assess the imposed cyclic stress paths is important; this analysis will likely need further substantiation by laboratory tests to assess the fill behaviour under specific load paths. In effect, the simpler 1-D methods have to be superseded. However, if the 1-D response has a sufficient margin of safety then engineering judgment may lead to the conclusion that no further testing or analysis is warranted.

## Method 1: The NCEER approach

Correlating observed occurrence-non-occurrence of liquefaction to site penetration resistance (Figure 8.53 below) suggested by Seed & Idriss (1971) has seen a progressive accumulation of experience and updating. A particularly important updating was a workshop in 1996, sponsored by the National Center for Earthquake Engineering Research (NCEER) with support from the National Science Foundation. This workshop systematically explored and discussed the overall approach and resulted in a "definitive" paper on the methodology: Youd *et al.* (2001). As a consequence of the workshop the approach originated by Seed & Idriss has become known as the NCEER method consistent with the much wider group of workers involved in developing it.

The only measures of soil properties used in the NCEER method are geological classifications, "fines content" (the fraction passing the # 200 sieve) and the Atterberg limits. In one version of this method, these geological classification parameters have been correlated to the "soil behaviour type" indicated by the CPT; this allows direct use of the method based on CPT data alone.

The NCEER method involves several mathematical transformations of the measured data (called "corrections") to compute the $CRR_{75}$ where the subscript '75' refers to the strength available under the loading of a M7.5 earthquake (taken to be equivalent to 15 cycles of uniform load). This standard strength is then further adjusted to the CRR involved in the particular earthquake scenario under consideration through a so called "magnitude scaling factor".

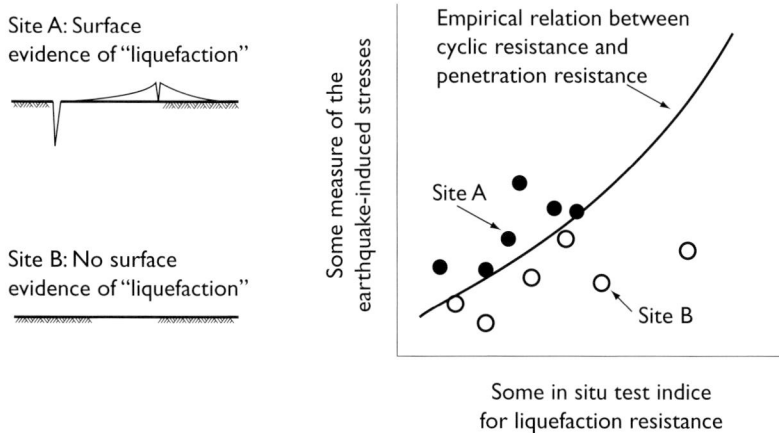

Figure 8.53 *Principle of the semi-empirical cyclic stress method (Idriss and Boulanger, 2008).*

The NCEER CPT method is based on Robertson & Wride (1998), with minor variations. The calculation procedure is documented in a flowchart, and can be easily implemented as a 'user defined function' in a spreadsheet. However, commercial CPT processing software now includes this method and it is becoming normal engineering practice to use this commercial software (in part because it makes Quality Assurance straightforward). Examples of software include: http://www.geologismiki.gr/Products/CLiq.html and http://www.datgel.com/Datgel-CPTTool.aspx.

A convenient public domain implementation of the NCEER method (also including Method 2, below) in a spread sheet environment can be downloaded from www.golder.com/liq".

The NCEER method is widely used in practice. It continues to evolve with further suggested changes to some of the coefficients in the method, recent contributions including Cetin *et al.* (2004), Idriss & Boulanger (2008), and Boulanger & Idriss (2011). Youd (2011) compared the proposed changes and found that the computed factors of safety against liquefaction differed by as much as ±30%. At the moment, these new contributions do not have widespread acceptance and engineering practice tends to remain with the Youd *et al.* (2001) variant of the method.

Further refinements of the NCEER approach may not matter to practical engineering as all variants of the methodology are unclear on how natural site variability is to be represented; that is, should average properties be used or something akin to a characteristic value as conceptualized in the Eurocodes. This site variability issue likely dominates further refinement of the method. We return to this point on site characterization below.

Method 2: State parameter approach

Concerns with the NCEER method are that it is neither based on a modern understanding of soil behaviour nor does it properly include the effect of soil properties on soil behaviour (something that is certainly an issue with calcareous or micaceous sands). And, there are now numerous publications showing that liquefaction, in all its forms, is a readily computed behaviour. However, a purely-mechanics based approach might be too idealized and not include aspects like the effects of multi-directional shaking. An approach that has developed is to use the more reliable data underlying the NCEER method and to put that data in a consistent framework.

The state parameter framework continues to be based on the CPT but separates the data processing into two steps. First, the CPT data at the site is assessed to determining the characteristic in situ state parameter $\psi_k$. Second, the CRR for that characteristic state parameter is determined. Two steps are used because the soil properties, particularly $G_{max}$ and compressibility, influence both penetration resistance and CRR but to possibly different amounts. A convenient single-source reference for the methodology is Jefferies & Been (2006); supporting data and programs can be downloaded from www.golder.com/liq.

Practically, the CRR can be evaluated from CPT data directly using the state parameter method with no further soil properties—exactly as with the NCEER approach. The difference is that the equations are simpler in the state parameter and the extension to finer grained soils is based on a large body of test data rather than poorly-constrained correlations at a few sites. Robertson (2010) suggests that the two approaches are essentially equivalent; Jefferies & Shuttle (2011) suggest that the NCEER method is intrinsically flawed when dealing with finer grained soil.

Where the state parameter approach substantially improves on the NCEER method is if the effects of soil properties are included by measurement/calibration rather than correlation. Extension to directly include the effect of soil properties requires measurement of $G_{max}$ in situ (which is frequently carried out anyway for the site response analysis) and measurement of the soil compressibility $\lambda$ (defined on Figure 8.46, and typically requiring a program of 5–10 triaxial tests).

To date, the state parameter approach has seen greatest use with tailings dam and projects where the soil properties are markedly different from clean sands.

Method 3: Laboratory cyclic simple shear tests
Both of the previous methods developed because it is effectively impossible to obtain undisturbed samples of sands. Reliance on in situ tests was the only practical approach. However, it became apparent over the past decade or so that finer grained soils are more prone to liquefaction than thought to be the case. Some of these soils can be recovered as near-undisturbed samples. Further, the widespread commercial availability of computer-controlled cyclic simple shear testing, with its close simulation of the imposed earthquake loading on soils, has made testing more attractive than it used to be. These cyclic simple shear tests provide data on the entire pattern of soil stiffness degradation during loading as well as the number of cycles to liquefaction for each sample.

Laboratory testing requires a suite of tests, in part to cover the range of CSR of the design scenarios. However, even the best sample handling procedures will cause some sample densification between in situ and as tested conditions with the effect that the as-tested behaviour is likely stronger than in situ. Commonly, this disturbance effect can be estimated, for samples recovered from below the water table, by comparing the void ratio of the soil recovered in the field (i.e. by measuring the water content of trimmings from the base of sample tube) with the 'as-sheared' void ratio of the test specimen. They will usually be different, and further laboratory tests, using re-constituted samples at several densities, will be needed to develop a correction factor for "lab to insitu". This correction factor could also be established by formal constitutive modelling of the effect of changed void ratio on the test data using one of the modern state-parameter based models.

Laboratory testing to assess CRR may be needed for the foundation soils underlying the hydraulic fill as well as the fill itself.

### 8.6.5 Movements caused by liquefaction

Assuming that flow liquefaction does not occur, earthquake induced pore pressures can result in soil movements that need to be considered in design. These movements can be classified as:

- slope movements, where the ground is not level and the result is downslope displacement of soils;
- lateral spreads, where near level ground is extended laterally as movement occurs towards an exposed face, such as a river bank;
- settlements.

Each is discussed briefly below.

### 8.6.5.1 Slope deformations

The starting point of assessing slope behaviour under earthquakes is the presumption that the slope has been adequately designed for normal static loads. The effect of an earthquake is to impose additional transient loads.

Historically, earthquake loading of slopes was evaluated using the 'pseudo-static' method earthquake-induced horizontal and vertical forces are applied to every portion of the soil mass and to any gravity loads acting on top of the slope. This method was already introduced in section 8.4.3.10. For guidance in the selection of the vertical and horizontal accelerations, reference can be made to EC8–5 (2004). The pseudo-static method approximates the cyclic nature of the earthquake as if the earthquake only applies an additional static force upon the slope, with that additional force including the effects of the earthquakes's horizontal and vertical accelerations; these accelerations will usually be smaller than the site's design "$a_{max}$" value because of plastic deformations and EC8–5 provides the 'partial factors' for this aspect. The pseudo-static method also needs an assessment of any earthquake induced strength reduction of the soil because of excess pore pressures.

A substantial divergence in practice is developing over the use of the pseudo-static method. The EC8–5 standard includes various partial factors for use with the pseudo-static method, implying acceptance of that method within the EU. But, North American practice would generally not accept the pseudo-static method because: (i) there is a lack of clarity in the meaning of the factor of safety under transient loads - a FS < 1 simply means the slope may move but that movement may not be very large as the earthquake is short duration; and, (ii) the 'partial factors' to scale $a_{max}$ to a pseudo-static value are not constant but depend very much on the fundamental period of the structure/site and the frequency content of the ground motion. Modern practice is to move directly to calculation of potential slope displacements rather than to accept the idealizations/approximations implicit in the pseudo-static method.

The original concept to gain an indication of the expected permanent deformation was by Newmark (Newark 1965) and assumed that the sliding mass is rigid and will deform only during those portions of the earthquake where the acceleration is above a "yield" threshold (generally determined from a pseudo-static slope stability analysis). Figure 8.54 illustrates the principle of the method. Application the Newmark method requires as input the acceleration versus time record(s) of the design earthquake(s), something that should be available from the site response studies discussed earlier. The main limitation of the Newmark method is that it requires that the soil not be brittle; this means there must be a reasonable $FS_L$ against liquefaction. Within its limitations, the Newmark method is widely accepted. The recent development by Bray & Travasarou (2007) extends the method and with very substantive validation.

Bray & Travasarou used a nonlinear fully coupled stick-slip sliding block model, essentially the Newmark idealization with a deformable slope mass, to capture the dynamic performance of an earth dam, natural slope, or earth fill during earthquake loading. The nonlinear coupled stick-slip deformable sliding model is characterized by: i) its

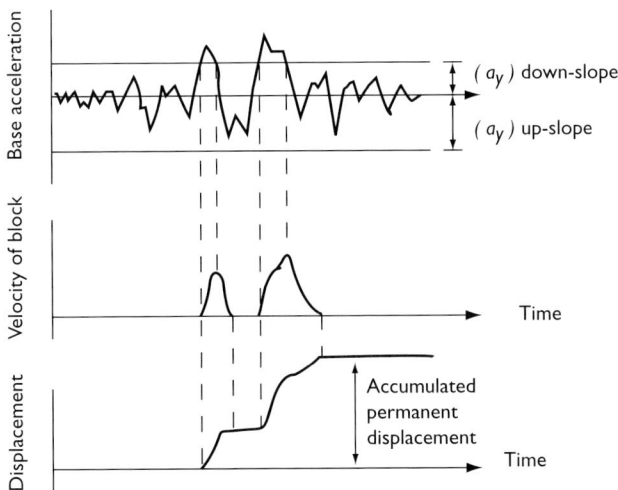

Figure 8.54 *Diagram illustrating the Newmark method, $a_y$ is the yield acceleration corresponding to FSpseudo-static = 1.0.*

strength as represented by its yield coefficient ($k_y$), and, ii) its dynamic stiffness as represented by its initial fundamental period ($T_s$). A comprehensive database containing 688 recorded ground motions was used to compute seismic displacements as a "performance function" given probabilities of deformations. A very simple framework emerged that allows straightforward evaluation of the seismic displacement potential.

The key inputs to the Bray & Travasarou method are the representation of the slope in terms of a yield coefficient and its fundamental period. This yield acceleration is the uniform horizontal acceleration required to produce a FS = 1; exactly analogous to the long usage of a pseudo-static coefficient, but now using the analysis as an input to estimating displacements. Practically, this means the effect of the earthquake in degrading the strength needs assessing (and the method only remains valid provided that brittle failure does not develop, that is, the site must lie above the green line of Figure 8.51). This assessment of strength degradation uses the computed factor of safety against liquefaction, $FS_L$. Figure 8.55 illustrates how the excess pore pressures induced by earthquake loading are related to $FS_L$ and this figure is one guide; The Designer's Guide to EC8–1 (Fardis *et al.*, 2005) also discusses this point. However, actual testing of the soil involved, using cyclic simple shear, is always an option to give a soil-specific design basis.

If $FS_L > 1.5$ then excess pore pressure will be negligible and the soil does not experience an appreciable reduction in shear strength—the drained shear strength is used directly in the limit equilibrium analysis to estimate $k_y$.

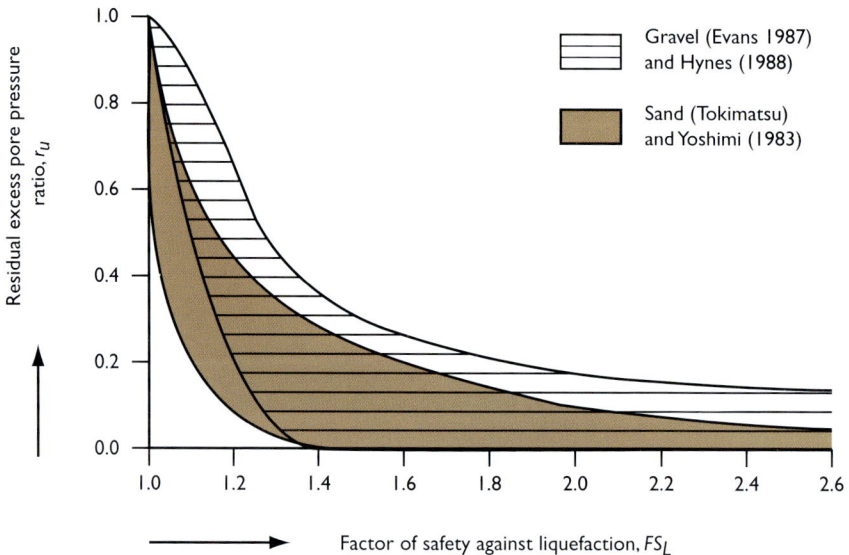

Figure 8.55 *Relationship between residual excess pore pressure and factor of safety against liquefaction according to Marcuson, Hynes and Franklin, from Day (2009).*

If $1.25 < FS_L < 1.5$, then the estimated excess pore pressure is commonly used to reduce the static strength through the Mohr-Coulomb shear strength equation for non-cohesive soils. The reduced, large-strain shear strength is used in computing $k_y$ and this strength can be expressed by:

$$\tan \varphi'_{equivalent} = (1 - r_u) \tan \varphi'_{static}$$

If $FS_L < 1.25$ then the analysis no longer meets the requirements of EC8. In this case, $k_y$ should be estimated using post-liquefaction strengths (these strengths were discussed earlier).

The site fundamental period, the other input to the Bray & Travasarou method, will have been determined during the site response analysis.

### 8.6.5.2 Lateral spreads

Lateral spreading is the displacement of near-horizontal ground that arises over liquefied soil in an earthquake. During lateral spreading, mostly intact blocks move on an underlying and liquefied layer with opening of cracks and buckling of the ground surface. Displacements can be fractions of a meter to several meters, with the ground movement imposing substantial loads on quay walls and bridge piers that extend downward below the liquefied layer. Lateral spreading has been the cause of much damage during earthquakes.

In the case of hydraulic fills, the principal vulnerability is likely to be in the underlying (natural) strata since if the fill is loose enough to cause a lateral spread it will have been compacted for serviceability of the built infrastructure over it. The exception to this consideration is very large fills, when it could be envisaged that ground improvement might be limited to only those parts of the fill where settlement was an issue. In either case, this may be a reason to estimate displacements in a lateral spread as it is those displacements that will load the other structures.

Estimating the displacement during a lateral spread requires a design ground motion scenario (since the duration of shaking partly controls how far the ground will move) and the penetration resistance (or state parameter) of the ground. There are three classes of method to estimate displacement:

i) Those that use the earthquake scenario (magnitude and epicentral distance from site) as the input and use correlations to prior observations as the basis of the performance model. Youd *et al.* (2002) is the most recent version of this class of analysis, extending work by various groups and now comprising a substantial data base. However, the method does not appear particularly accurate in validation studies reported by Valsamis *et al.* (2007).

ii) Methods that take the site acceleration and duration of that shaking as the input, but with a similar regression performance model much as class (i). The site acceleration and duration are the same information developed for site response analysis. The methodology of Hamada (1999) did particularly well in the validation study of Valsamis *et al.* (2007).

iii) Numerical methods explicitly simulating the soil response. Valsamis *et al.* (2007) report on the performance of a modern model implemented in FLAC; the model validated very well. It requires determination of a few soil properties and the in situ state parameter.

Both the empirical classes of methods (i) and (ii) relate horizontal displacement of the lateral spread to the cumulative thickness (cumulative because several layers may be involved) of the soil with penetration resistance less than a particular value. Although the methods have their origin in the standard penetration test (SPT), functionally the criterion used is the same as the green line shown on Figure 8.51. That is, lateral spreads only develop if the soils are vulnerable to flow sliding; this criterion must be extended into the foundation beneath the fill.

Method (iii), numerical analysis, is interesting in that it allows evaluation of detail in site response that is absent from the empirical methods. But, it also involves far more effort than using a regression equation. The merit in method (iii) is if the soils involved are "unusual", for example carbonate sands, and as such not present in the data base use to develop the regression equation.

In choosing between these methods to assess deformations in a lateral spread, the first question must be: why is lateral spreading a concern? The usual engineering approach for a fill will be to densify that fill so that spreading is not an issue. If a decision is made to leave a vulnerable stratum unimproved, possibly because of the scale of the works, then likely deformations can be estimated using Hamada (1999) as straightforward supplement to the site response analysis. If that is regarded as insufficiently reliable for a particular project, then specialist advice will be needed and almost certainly detailed numerical analysis.

### 8.6.5.3 Settlements

Restricting attention to saturated ground (i.e. below the water table), settlement can only develop as pore water flows out of the ground—settlement is a 'consolidation' process. This is most readily seen during explosive compaction, a ground improvement method where timed sequences of explosive charges are used to liquefy the ground with the subsequent settlement producing the desired densification. Observation during numerous explosive compaction projects is that settlement never develops during the vibration shocks, with no settlement evident for many minutes afterwards; then a consolidation becomes apparent with water flowing out of the ground as the surface settles.

The fact that settlement follows sometime after an earthquake, not concurrently with ground motion, is not controversial. Indeed, this was the basis of an interesting research program in Japan where sand was loaded by simulated earthquakes while preventing drainage, with the drainage then being subsequently allowed and the volumetric strain (= scaled settlement in the field) measured. A range of earthquake simulations were used, with different shaking intensities, and imposed on a range of sand densities. The work is reported by Nagase & Ishihara (1988); this data was further processed by Ishihara & Yoshimine (1991) into simplified trends. More recently, Sanin and Wijewickreme (2006) have reported similar findings for Fraser River Sand.

Figure 8.56 below (from Gohl *et al.*, 2000) compares the laboratory data on Fuji River sand, the line labelled as 'after Ishihara & Yoshimine, 1991' with the measured results achieved using explosive compaction. The Fuji River data is a lower bound to the explosive compaction case-histories, likely because the Fuji River sand is the least compressible soil in the figure. This figure is for the maximum strains following full liquefaction, which is the limiting case for consolidation.

Current practice for estimating liquefaction-induced settlements has grown from the methodology for liquefaction susceptibility described in Section 8.6.4 and the Fuji River sand data set discussed above. The first attempt to relate post-earthquake settlements to ground conditions was based on SPT data as the input (Tokimatsu & Seed, 1987) but that has been superseded by the CPT-based method of Zhang *et al.* (2002) concurrent with the move to CPTu from the SPT.

The Zhang *et al.* method infers the potential vertical settlement as being influenced by both the initial soil density (as indexed by the normalized CPT resistance) and the factor of safety against liquefaction. Zhang *et al.* compared (favourably) the settlements computed as above to observed surface settlements at 5 locations within the Marina District and Treasure Island following the 1989 Loma Prieta earthquake. The Zhang *et al.* method is particularly attractive as it is entirely based the CPT and is straightforward to implement in a spread sheet or commercial CPT processing software packages. As a result, the method appears to be widely used.

There are several limitations of the Zhang Method that need to be recognized: it is based on data from one incompressible sand; it approximates the effect of soil

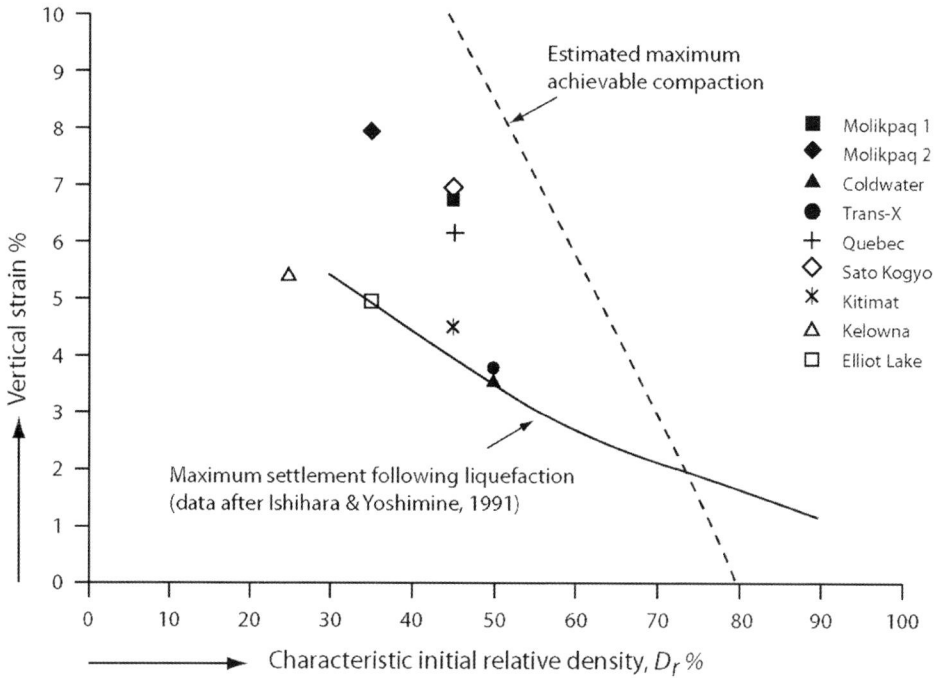

Figure 8.56    *Comparison of maximum settlements following liquefaction ( Points are explosive-induced liquefaction case histories; solid-line is laboratory test trend on Fuji River sand.)*

compressibility uniquely related to fines content; the validations themselves have inconsistent changes in penetration resistance with reported settlements.

Because of concerns with the assumptions, idealizations and mechanics of the Zhang *et al.* method, a better approach is to compute the maximum settlement assuming that full liquefaction occurs, which sets the maximum possible effective stress change (since the excess pore pressure ratio, $r_u$, cannot exceed unity as that is the physical limit of liquefaction). Settlement is then a straightforward relationship between compressibility (or modulus) and the effective stress change. The soil property that needs to be measured is therefore compressibility, which can be done several ways:

- A series of cyclic simple shear (CDSS) tests on typical sand materials at a range of initial density and stress level, with the post liquefaction consolidation settlement monitored. This will give a combined elastic and plastic compressibility.
- In situ elastic modulus can be measured using a seismic cone. Elastic shear modulus is directly related to the shear wave velocity through elasticity theory.
- Plastic compressibility is a function of elastic compressibility, providing a straightforward way to estimate plastic compressibility from the CPT, but standard laboratory oedometer tests can also be carried out to measure compressibility directly.

### 8.6.6   Fill characterization for liquefaction assessment

The most important part of a liquefaction assessment is to characterize the in situ state of the soil, whether the parameter of interest is $\psi$, relative density, or anything else. For design, before a hydraulic fill is placed, the in situ state has to be estimated from past experience and case histories in which the in situ state has been measured, see section 8.3.5. However, during construction it is essential that the design estimate is confirmed by measurement. Design for structures to be placed on the hydraulic fill will also need these measurements of in situ state of the hydraulic fill.

### 8.6.6.1   Necessity for in situ tests

It has long been known that measurement if the in situ density of sands and silts by sampling is unreliable because sampling substantially disturbs the soil. Loose samples will tend to densify during sampling whereas dense samples will dilate and loosen during sampling. The net result is most samples tend to be "medium dense" when the density (or unit weight) is measured, and there is insufficient precision in this method for liquefaction assessment. This densification problem does not exist for clays, partly because of the much lower permeability of clays and partly as the disturbance zone around the cutting edge of the sampler is smaller than it would be for sands. Silts fall somewhere between sands and clays, and in a few cases workers have reported successful sampling and water content (i.e. density if the material is saturated) measurement of silts and silty sands.

It is possible to freeze saturated sand before coring a sample (e.g. Sego *et al.*, 1994), which then needs to be kept frozen until the unit weight is determined. This is a rather sophisticated, tricky and expensive. If freezing is not done at the appropriate speed for silts, there is a possibility of both ice lensing (where water is drawn to the freezing front) or expansion of the water in the pore space as it freezes. While freezing is considered the "gold standard" for undisturbed sampling of soils that are susceptible to liquefaction, it is not very practical or economical on most engineering projects. It may be carried out for very high value or high risk projects where the expense and time available allow for it—Wride *et al.* (2000) report a cost of $50,000 for a single cored hole in the late 1990s.

Borehole logging using nuclear density tools appears to be an attractive option to determine in situ density, but is not as attractive as it seems because soil behaviour is controlled by the state parameter. If the *in situ* density (or void ratio) is determined, then the corresponding CSL must also be determined. While this is not a difficult task if a uniform soil is being investigated, the reality is that few real soils, even well-controlled hydraulic fills, are uniform and that seemingly minor changes in gradation may have a marked influence on the CSL. This then presents a situation where the associated laboratory testing effort to support the assessment of *in situ* state from direct measurements of density may be unrealistic and impractical.

Liquefaction assessment methods have therefore developed with a heavy reliance on in situ testing. Initially this was the SPT, but that test has poor repeatability and the variable contact area with the soil makes precise evaluation of the data difficult. It also does not measure excess pore water pressure, and important factor with finer grained soils. The modern electronic piezocone or CPTu test, with its accuracy and resolution of even thin layers in the soil, has effectively displaced the SPT.

The CPT is an acceptable test under every methodology for liquefaction assessment. The CPT is a necessary test for any of the advanced methods of liquefaction assessment, particularly when moving into 2D cases of soil structure interaction or if looking to offshore situations. The CPT will even be necessary if taking undisturbed samples for cyclic simple shear testing, both as a means to extrapolate the test data obtained across the area of the site (only a few of these laboratory tests will normally be carried out) as well as provide data to estimate the sample disturbance from *in situ* to *as-tested* conditions. For a description of the CPT, see Appendix B-2.3.2.

A further development, the seismic or SCPT, is also relevant. A seismic cone is simply a modified cone that contains one or more geophones at fixed spacings behind the cone tip. At each rod change during the test, a shear wave signal is generated at the ground surface and the first arrival time of the signal at each geophone is measured. In this way a profile of shear wave velocity with depth is developed; it is almost "free" data as an add-on to a conventional CPT test.

If there is no access to a seismic CPT, shear wave velocity can also be measured by lowering geophones down a PVC cased borehole and measuring shear wave arrival times at regular depth intervals down the hole. The shear wave source can either be at the ground surface (vertical shear wave velocity profiling) or in an adjacent borehole (cross hole shear wave velocity measurement).

More recently (in the last 25 years or so), surface wave methods have become popular for measurement of shear wave velocity. These include SASW (spectral analysis of surface waves, e.g. Stokoe *et al.,* 1994), see Appendix B 5.3.6 and MASW (or multichannel analysis of surface waves, e.g. www.masw.com by Park Seismic LLC). The basis of the technique is the dispersive property of Rayleigh-type surface waves when propagating in a layered system. No subsurface drilling or equipment is needed, but more sophisticated data processing is required to determine the shear wave velocity.

Although the CPTu, SCPT, and geophysical methods are standard and necessary for characterizing hydraulic fills, they are not the only applicable tests. In some circumstances the self-bored pressure meter can provide data on the degradation of soil modulus with strain (useful for evaluation of lateral pile capacity in fills, especially under earthquake loading). The 'cone pressure meter', a combination of CPT and full-displacement pressure meter may also be useful for the same purpose.

### 8.6.6.2 Required number of CPT soundings

How many CPTu/SCPT tests are needed? This is an important question, but with little guidance in the various codes of practice. There are two issues: i) developing a statistically reliable characterization of the site; and, ii) identifying what value from that statistical characterization is "characteristic". The underlying concept is that all soils are spatially variable in their properties but that design calculations treat the formation as having uniform (or uniformly trending) properties; the concept of a 'characteristic' property is that it is the particular value from a distribution of properties that gives the same predicted performance when treating the soil as uniform as the performance that actually develops in the real soil with all its variability. In Eurocodes, characteristic values are subscripted by 'k' and apply to all materials (i.e. steel, concrete, timber, as well as soil).

Figure 8.57 shows data from an underwater hydraulic fill where some twenty individual CPT sounds have been processed to illustrate the distribution of tip resistance at various depths in the fill (this was done as a simple 'bin count' after slicing the site horizontally into 1 m thick layers). Broadly, the distributions are consistent with void ratio being normally distributed.

In the context of Eurocode, liquefaction assessments should use the penetration resistance at approximately the "mean minus 1.5 sigma" as characteristic; that is, it is the looser parts of a fill that control the mass behaviour. The characteristic $q_{c,k}$ will also be a profile with depth, not a single value (all else equal, penetration resistance increases with depth because of confining stress).

How many CPT's are needed to assess the characteristic resistance? Provided conditions are "stationary", about the same number as used in any other branch of civil engineering—say 6 to 8. However the catch is the concept of "stationary" which means there are no underlying spatial trends—a situation unlikely to be true with hydraulic fills where the distance from spigot point(s) matters as do any machine operations in developing beaches and so forth. It is necessary to take a general view of the site and classify it into representative areas, each of which will need their own 6–8 CPT soundings. This may add up to a lot of testing, and more than 50 CPTs are not uncommon on large projects; even small fills may need 20 soundings.

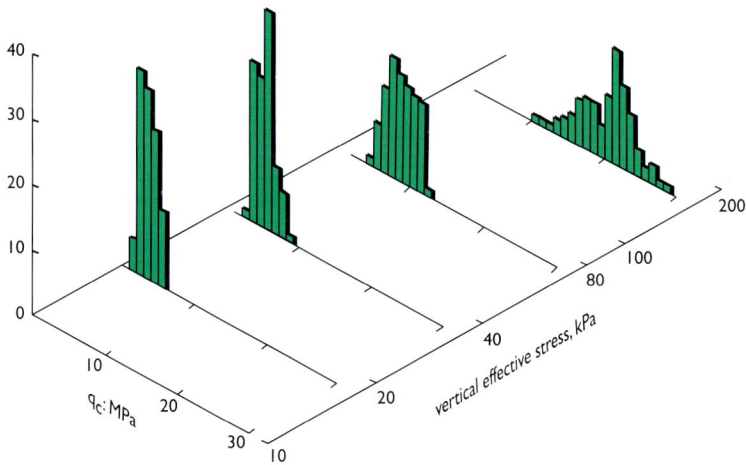

Figure 8.57  *Example of distribution of measured CPT resistance in a hydraulic fill.*

Data processing to determine the characteristic penetration resistance can be done formally with statistical processing or using engineering judgment. The latter is often sufficient as the eye is quite good at bisection, so a little less (a "smidgen') than halfway between the mean and the minimum trend is a reasonable judgment of the characteristic value—and within the precision of current knowledge. If formal processing is used, then it needs to be carried out assuming that $q_c$ values are log-normally distributed; the process can be set up in a spread sheet (there is no commercial software for statistical processing of CPT data at present.

### 8.6.6.3   CPT calibration

The CPT is often used empirically as in the NCEER method, but can do much more. If thoroughly analysed, the CPT is capable of determining the state parameter to a precision of about ± 0.03. The matching precision in void ratio is less accurate because the CPT responds to the soil's state parameter, not its void ratio, and calculating void ratio from the state parameters need the CSL—a further set of data with its own uncertainty caused by natural variation in fines content.

Thorough analysis of the CPT is based on calibration in a large cell, involving as much as 2 tonnes of sand for a single calibration point. Such calibrations have been carried out mainly for research sands, but also including some sands used in large scale hydraulic fill projects. The challenge these calibration data present is that there is not a unique relationship between tip resistance, void ratio and confining stress; the relationship is unique for any sand, but differs from one sand to another. However, the calibration is now computable with advanced finite element simulations and repeated simulations have provided simple approximations in terms of the soils critical friction ratio, compressibility, and elastic modulus. If this soil property data is available from laboratory testing of the fill, the coefficients for reasonably precise evaluation of the CPT can be quickly calculated. The basis, supporting calibration data, procedures, and details to develop soil-specific calibrations for the CPT are summarized in Jefferies & Been (2006).

### 8.6.6.4   Supporting laboratory data

A consequence of the NCEER approach relying on the CPT alone is that the approach gives up precision. Sole reliance on the CPT also limits any possibility of more advanced analysis. In general, the scale of most hydraulic fill projects will warrant a program of laboratory tests to determine the engineering parameters and support the CPT characterization of in situ state of the fill.

An appropriate laboratory testing program is quite small in scope—say 8 to 12 triaxial tests, possibly supplemented by some cyclic simple shear tests—on a single representative gradation of the fill. In the scale of a hydraulic fill project, this level of testing is minimal cost (although obtaining samples of the proposed fill at design stage may be a real difficulty). The sample of 'representative' fill should be sufficiently large (~20 kg) so that all tests can be completed using identical soil (using different soil gradations for each triaxial test, even if from the same fill will mislead the assessed properties). For larger projects, the laboratory testing might be extended to consider several gradations to cover the range (or expected range) of fill insitu. Details on the appropriate testing procedures and data reduction to engineering properties can be found in Jefferies & Been (2006).

### 8.6.7   *Note on soil type ( Calcareous and other non-standard sands)*

Liquefaction practice in North America is essentially captured in the NCEER method (Youd *et al.*, 2001), but that method includes the following statement:

> *Data were collected mostly from sites on level to gently sloping terrain, underlain by Holocene alluvial or fluvial sediments at shallow depths (<15m). The original procedure was verified for, and is applicable only to, these site conditions. Similar restrictions apply to the implementation of the updated procedures recommended in this report.*

From this statement it is not apparent that several important soils other than siliceous sands were considered such as: carbonate sands, micaceous or platy sands and volcanic materials (e.g. sands derived from tuffs).

In particular carbonate sands are often used as a fill for land reclamation. Origin and composition, typical properties and mechanical behavior of these materials are described in section 9.2.

When studying the liquefaction susceptibility of carbonate sands two key factors need to be considered:

1. Penetration resistance for the SPT and CPT is considered to be reduced for calcareous sands or materials that are more crushable and compressible than typical soils (e.g. Robertson & Campanella, 1983). Wehr (2005) and Lunne (2006) indicate how these difference may be quantified.
2. Pore pressures generated during cyclic loading depend on the compressibility of the soil, however the trend is potentially counter-intuitive. Smaller pore pressures develop in more compressible soils under cyclic loading, because the pore pressures are a response to a potential volume change caused by the loading. A smaller change in effective stress (i.e. pore pressure) results in a larger volume change in a more compressible soil.

Liquefaction assessment, whether using the NCEER or the state parameter framework, are anchored to penetration test data. As noted earlier, calibrations of penetration tests to soil properties and void ratio depend on calibration chamber studies. There are no such reported calibrations for the SPT in carbonate sand fill; there is one reported study for the CPT (Nutt, 1993) although this used a small calibration chamber (and which cautions against direct use of the data). Accordingly, liquefaction assessments for calcareous or other non-silica sands will require more engineering than needed for fills with predominantly silica (quartz) particles; and that engineering may be further complicated by grain crushing. However, the state parameter methodology is applicable to carbonate sands, based on the detailed laboratory test data and trends reported by Coop (1990).

Good practice dictates that the mechanics based methods (i.e. critical state methods and the state parameter $\psi$) described in this section must be used for calcareous, micaceous or volcanic sands. These methods require that the engineering properties of the soils are measured in the laboratory (in particular $\varphi'_c$, $\lambda$ and plastic hardening modulus as well as the cyclic resistance to liquefaction, CRR). A few more laboratory tests will be needed for carbonate sands because of the potential "curved" rather than "semi-log" CSL, see Fig. 8.45. In addition, the shear modulus must be measured in situ, and the CPT is interpreted considering these measured engineering properties. What is required is to develop a sand-specific calibration to the proposed/existing fill? Typically this would require numerical analysis, such as that described in Shuttle and Cunning (2007) supported by the measured engineering properties. The liquefaction analysis will also involve advanced numerical analysis, depending on the problem, rather than simplified analyses which are common practice for less important projects on typical siliceous sands.

# SPECIAL FILL MATERIALS AND PROBLEMATIC SUBSOILS

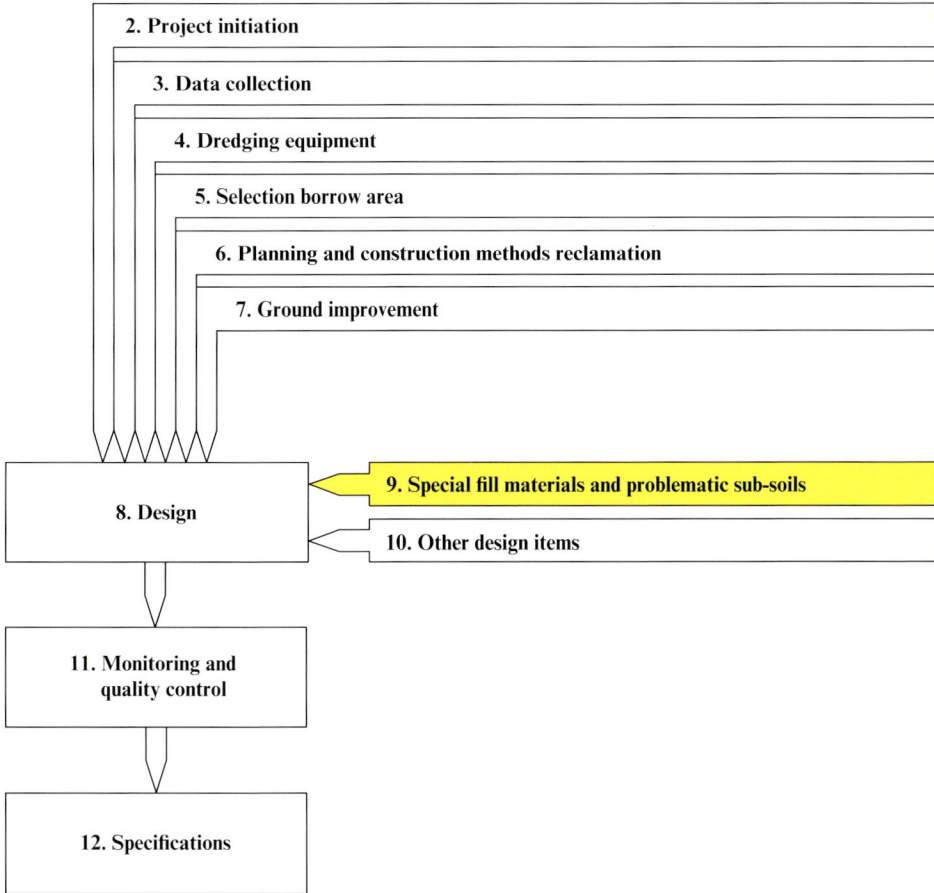

| 2. Project initiation |
| 3. Data collection |
| 4. Dredging equipment |
| 5. Selection borrow area |
| 6. Planning and construction methods reclamation |
| 7. Ground improvement |

| 8. Design | 9. Special fill materials and problematic sub-soils |
| | 10. Other design items |

| 11. Monitoring and quality control |

| 12. Specifications |

## 9 Special Fill Materials and Problematic Subsoils

### 9.1 Cohesive or fine grained materials

| | | | |
|---|---|---|---|
| **9.1.1** Introduction | **9.1.2** Segregation of fines | **9.1.3** Soft clay or soft silt | **9.1.4** Stiff clay or silt |

### 9.2 Carbonate sand fill material

| | | | |
|---|---|---|---|
| **9.2.1** Introduction | **9.2.2** Origin and composition of carbonate sands | **9.2.3** Typical properties of carbonate sands | **9.2.4** Mechanical behaviour of carbonate sands |
| **9.2.5** The use of carbonate sand as fill | | | |

### 9.3 Hydraulic rock fill

| | | | |
|---|---|---|---|
| **9.3.1** Introduction | **9.3.2** Lump size | **9.3.3** Compaction and measurement of compaction result | **9.3.4** Grading |
| **9.3.5** Fines | **9.3.6** Wear and tear | **9.3.7** Pumping distance during rock dredging | **9.3.8** Specification rock fill |

### 9.4 Problematic subsoils

| | | | |
|---|---|---|---|
| **9.4.1** Sensitive clays | **9.4.2** Peat | **9.4.3** Glacial soils | **9.4.4** Sabkha |
| **9.4.5** Karst | **9.4.6** Laterite | | |

## 9.1  Cohesive or fine-grained fill materials

### 9.1.1  *Introduction*

In some cases, the use of cohesive or fine-grained soils like silt and clay can be an economical alternative for the use of sand as fill material, for instance, when only cohesive soils are (economically) available in the vicinity of the reclamation area or when large volumes of cohesive soils become available from excavation or dredging at the project area itself. As an alternative to granular material like sand, some of these cohesive soils can be (made) suitable for use as hydraulic fill.

The advantages of using these soft soils as fill material are:

–  the reuse of available material reduces the volume that has to be extracted from the borrow area;
–  cohesive soils may be less expensive than sand;
–  the disposal of (contaminated) soft soils from the project area is avoided, thereby reducing both cost and environmental impact.

Although the use of cohesive soils as fill material is possible, specific constraints concerning the design and construction of hydraulic fills must be taken into account. Some problems associated with the use of cohesive soils are:

–  The low shear strength and high compressibility of soft soils may result in:
   o  excessive deformations of the fill after loading;
   o  slope failure;
   o  bearing capacity failure of foundations.
–  Uncertainty about the design parameters to be used for the (remoulded) cohesive fill: Excavation or dredging of cohesive material can result in a drastic change of its geotechnical characteristics.
–  Dredged cohesive soil may require a long ripening or consolidation period:
   o  Ripening: when disposing soft clays/silts above the waterline, one or two years in a ripening field is required before the material is suitable for construction purposes. Desiccation and chemical processes will improve the mechanical properties of the clay.
   o  Consolidation: should a surcharge system be adopted, a sufficiently long consolidation period will be required. The larger the layer thickness of the cohesive fill, the larger the required consolidation period.
–  If stiff clays are used as fill material, clay balls are likely to be formed during the dredging process:
   o  Production: clay lumps might cause significant hydraulic friction losses in the discharge pipeline. The maximum achievable pumping distance of such a clay lump mixture will therefore be shorter compared to a fine to medium sand.
   o  Segregation: during dredging and transport the clay lumps will (partly) erode producing rounded 'balls' and slurry. At the point of discharge this

will cause segregation as the clay balls will be deposited in the immediate vicinity of the discharge pipe while the slurry will settle at larger distances.

Other compaction techniques are required compared to sandy soils. Compaction methods based on vibration are not applicable in the case of cohesive or fine-grained soils, see Chapter 7. Table 9.1 gives an overview of some frequently encountered cohesive fill materials.

The geotechnical behaviour of a fine-grained soil is strongly influenced by its clay content. In the next sections, an overview of some typical cohesive soils used as fill material is given, along with a discussion of the major geotechnical and practical challenges resulting from the use of such materials.

The main characteristic properties of a clay are its cohesive behaviour and the ability to hold water. Clay particles consist of 'metal silicates' (Figure 9.1). Their structure is plate-like. These particles have strong cohesive properties, largely attributable to chemical forces acting between the grains on a small scale. The mineral structure of the clay particles enhances the absorption of organic material, which strongly affects the suitability of the soil in structural applications.

Table 9.1 *Overview of cohesive soils used as fill material.*

| Dredged soil | Soil characteristics in the borrow area | Resulting cohesive material within the reclamation area | Ref. § |
|---|---|---|---|
| Sands or gravels containing high percentage of fines | % fines >15% | The silt or clay fraction will segregate during the discharge process and may form lenses in the fill | 9.1.2 |
| Degradable soils such as weak rock and calcareous sand | | The silt or clay fraction will segregate during the discharge process and may form lenses in the fill | 9.1.2 |
| Soft clay or silt | | Low density slurry | 9.1.3 |
| Stiff clay or silt (pumped) | Very low plasticity $I_p < 20$ | Significant disintegration, occasionally clay ball | 9.1.4 |
| | Low plasticity $I_p = 20$–$30$ | Some clay balls in slurry | 9.1.4 |
| | High plasticity $I_p > 30$ | Clay balls and slurry | 9.1.4 |
| Stiff clay or silt (mechanically excavated) | | Clay lumps | |

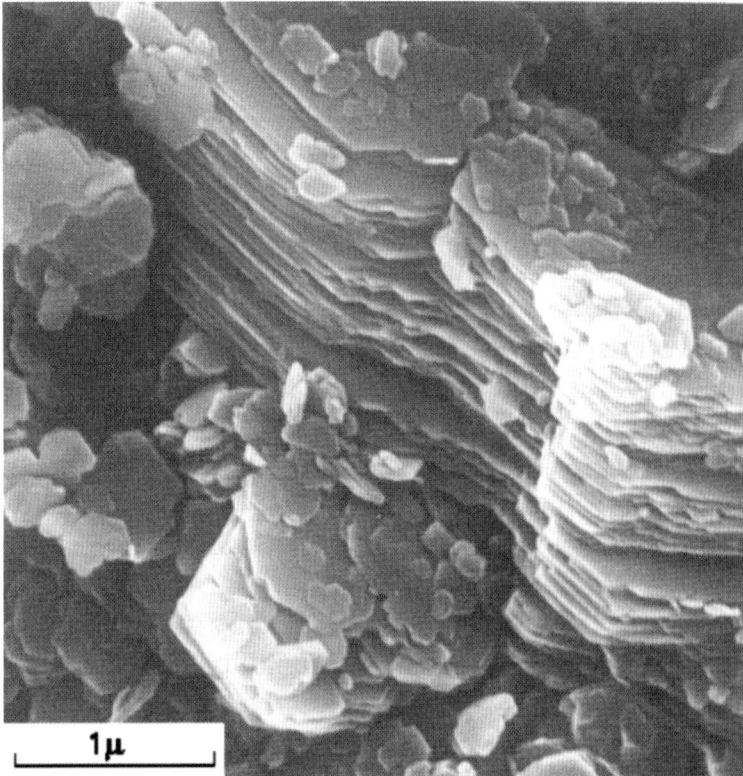

Figure 9.1    *Microscopic image of clay particles (J. van de Graaff, 2006).*

## 9.1.2    *Segregation of fines*

Although a silty/clayey sand or gravel is not necessarily a cohesive soil, the segregation of fines during the dredging and discharging process might result in a soft cohesive soil at the reclamation area. Similarly dredging and discharging of soils which are sensitive to degradation and/or disintegration (like carbonate sands) might result in a significant volume of soft cohesive materials at the reclamation area.

The segregated fines may accumulate, forming layers and lenses in the fill, or may even dominate whole parts of the reclamation area. Fines typically settle in still water areas. The finer the grain size, the further away from the discharge point it will settle.

Some considerations with regards to the management of the reclamation area can be made:

− Good planning of the reclamation area with respect to process water management is required. Well-defined drainage paths and settling ponds can drastically reduce the volume of fines captured in the reclamation area. Still water areas should be avoided as much as possible. It should be anticipated, however, that an occasional clay or silt lens in the fill is almost unavoidable. This will especially be the case if coarser silt fractions are involved.

– In most reclamation projects with soils having a significant fine content, the resulting fill will show a relatively thin bottom layer of soft cohesive material. The classic solution to this problem is to specify in the contract that fines in the reclaimed fill are not allowed. A more realistic approach would be to take the presence of fines into account during the design phase of the project. From a geotechnical point of view the presence of a thin layer of fine material is not necessarily a major problem:

o Fines are not always highly compressible. For instance, experience in the Middle East suggests that the fines (mainly silts) originating from carbonate sands may have a compressibility equal to that of the carbonate sand itself. In addition, as a result of their relatively high coefficient of consolidation, settlements develop rapidly. Laboratory testing (oedometer tests) on prepared samples should give a decisive answer. Note that the shear strength of such deposits is generally significantly lower than that of the carbonate sand.

o If the fines are situated at the bottom of the fill and the thickness of the overlying granular fill is sufficient, they will not have much influence on the bearing capacity of shallow foundations Bearing capacity analyses (including punch through calculations) must confirm whether the thickness of the overlying granular fill is sufficient (see Section 8.4.3.5).

o Excessive settlements resulting from a uniform distributed load can generally be remediated by applying a surcharge (if necessary, in combination with vertical drains). Settlement analyses should be undertaken to determine the required surcharge load and the corresponding drain spacing. Reference is made to Section 8.5.3.2.

o The stress increase as a result of a loaded foundation of limited dimensions may cause settlements. However, as a result of stress distribution the stress increase will reduce with depth: at a depth of 3 times the width of the foundation, the stress increase is usually less than 5% of the original load. If the compressible fines occur below such a depth, settlements of shallow foundations will be negligible. Further information on settlement analyses of foundations with limited dimensions can be found in Section 8.5.3.5.

o Since fines layers are generally of a limited thickness, the required consolidation period will be relatively short.

o The overall stability of the fill should be checked, since the soft layers can form a preferential slip surface. In general, avoiding soft layers near the perimeter of the reclamation area will prevent most slope stability issues.

o In addition, it may be important to determine the maximum thickness of the lifts of granular fill that can be placed on the fines layer without jeopardising the stability of the slope of the fill itself.

For further information on stability analyses reference is made to Section 8.4.3.

It should be noted that during the design stage, the subject clay or silt layers cannot be sampled, since they result from segregation and degradation during the hydraulic reclamation. The resulting silt or clay can be simulated by artificially washing out or degrading samples taken in the borrow area. Eventually samples can be taken at nearby silt ponds from previous reclamation activities.

Often it will be necessary to make a prognosis of the characteristics of the soft clay or silt layers. (shear strength, density, consolidation characteristics). This prognosis should be checked during the construction phase. A well-designed monitoring plan is a vital tool to verify the design assumptions and check the behaviour of a fill containing silt layers.

### 9.1.3   Soft clay or soft silt

When dredging soft clays or silts, it is obvious that this will result in a soft slurry on the reclamation area. Since water is added during dredging and discharging, significant bulking can occur, depending on the initial density and dredging method (see Section 5.4.1).

After disposal on the reclamation area, the slurry will consolidate under its own weight. The resulting density on the reclamation area is strongly related to the clay content of the soil. Various authors (Skempton, 1970, Buchan et al., 1990, Tsuchida et al., 1995) suggest that the water content of clays and silts immediately after settling from a slurry, will equal 1.5–2.0 times their liquid limit. With time the water content will reduce and the shear strength will increase as a result of self-weight consolidation and desiccation (only above water).

In case soft clays or silts are disposed above the waterline, these layers can be left exposed in order to ripen and make them suitable for construction purposes. The ripening of clay is a combination of desiccation and chemical processes ultimately resulting in a workable clay.

Alternatively the strength of soft clay or silt can be improved by using consolidation techniques such as surcharging, vacuum consolidation, etc.. These techniques can be applied above and below the waterline.

The disposed material will behave as any natural soil deposit, although it might be more homogeneous than a natural clay or silt deposit.

### 9.1.3.1 Suitability of soft (organic) clay or silt as fill material

Because of their high compressibility, materials with a large organic content are not suitable for construction applications. Moreover, the oxidation (decomposition) of organic compounds may cause a volume decrease of up to 50%, resulting in relatively large vertical deformations. For these reasons untreated organic soils, such as peat, should preferably not be applied in constructions. Peat and soft clay can be used for non-constructive applications, provided that there are no strict requirements for height, bearing capacity and stability of the structure, for example, as a cover layer or for a nature development.

The distinction between cohesive and organic soils is not always clear. Often clay contains some degree of organic material. For construction applications, generally a maximum organic content of 5 to 10% is allowed.

Compared to granular material such as sand, soft soils show some typical engineering properties. These properties can have strong implications for the design of structures containing these materials:

High compressibility – potentially large vertical and horizontal displacements can cause damage to structures and must be taken into account in the structural design (see also Section 8.5.3.5 and 8.5.3.6).

Relatively low shear strength – often relatively large foundation structures are necessary to spread the imposed loads. There is an increased risk of slope failure (see Section 8.4.3)

Low permeability – dewatering and (rain)water run-off may become problematic. Special measures must be taken to ensure adequate drainage of the structure. Reference is made to Section 10.2.

### 9.1.3.2 Workability of clay

The workability of clay refers to whether the clay can be handled by, driven on and compacted by construction equipment. This is a crucial aspect when using clay as a structural fill above the waterline. The workability of clay depends on its moisture content and Atterberg limits and can be described by the consistency index, $I_c$, (see box Soil consistency). Clay, which is suitable for use in landfills above the waterline, should have a consistency index larger than 0.6. Clay with a consistency below 0.6 is soft, has a low accessibility and is difficult to compact.

**Soil consistency**

The consistency and liquidity index of a soil sample can be determined from its Atterberg limits and water content:

Consistency index: $\quad I_C = \dfrac{w_1 - w}{w_L - w_P} = \dfrac{w_I - w}{I_P}$

Liquidity Index: $\quad I_L = \dfrac{w - w_P}{w_L - w_P} = \dfrac{w - w_P}{I_P}$

Where:

$I_P$ = plasticity index

$w$ = water content

$w_L$ = liquid limit

$w_P$ = plastic limit

**Pitfall**

The Atterberg limits are known to reflect the behaviour of clay. However, the use of salt water or fresh water can have an important effect on the Atterberg limits.

Ref ASTM D4318 - 1.6

"The composition and concentration of soluble salts in a soil affect the values of the liquid and plastic limits as the water content values of soils (see Method D 2216). Special consideration should therefore be given to soils from a marine environment or other sources where high soluble salt concentrations may be present. The degree to which the salts present in these soils are diluted or concentrated must be given careful consideration".

### 9.1.3.3   Effects of winning method

Clay can be obtained from mechanical excavation (by means of backhoe dredger, grab dredger or bucket dredger) or from hydraulic dredging (i.e., cutting suction dredger, trailing suction hopper dredger). The most important difference between the two methods is the resulting water content of the clay after deposition on the reclamation area. Clay that is mechanically excavated will have a water content that is usually equal to or only slightly higher than the natural water content. Clay that is dredged by hydraulic means will arrive on the reclamation area with a water content that will be (significantly) higher than its original water content as a result of mixing with water during the hydraulic dredging operation and hydraulic transport. The water content of clay is an important factor that may determine the time needed for dewatering and/or desiccation of the clay. Moreover, increasing the water content will result in a volume increase of the dredged soil, necessitating larger storage areas.

Essentially, clay from dredging is not any different than 'natural' clay from excavation. However, there are some aspects specific to dredged clay that can be relevant for design and construction:

– As a result of the extra mixing of the material during the dredging process, clay from dredging is often more homogeneous than clay from mechanical excavation.
– Dredged material may form an environmental threat in case of contamination. Dangerous contaminants, such as heavy metals and PCBs (Poly Chlorinated Biphenyls) may have to be accounted for (J. van de Graaff, 2006). Highly contaminated dredged material has to be treated as chemical waste. This can have significant consequences for the cost of a dredging project in an existing harbour.

9.1.3.4   Measures to improve the fill properties after disposal

There are several methods available to improve the properties of soft clay or silt. These include:

1. ripening of the clay (dewatering) which is only feasible above the waterline;
2. use of additives (for instance dry powder clay, quicklime, cement, fly-ash);
3. application of a surcharge;
4. use of sandwich structure;
5. vacuum consolidation;
6. accelerating the consolidation process of the soft clay or silt by means of vertical drains or drainage layers.

Without these measures the fill properties will improve slowly over time.

*Ad 1)   Ripening of the clay*

When soft clay is dredged and disposed of, its initial water content is very high and the consistency index is below zero. When the clay is disposed of above the waterline, the mechanical properties and workability of clay can be improved by placing the clay in a field in layers with a maximum thickness of 1.0 to 1.5 m and leaving them to ripen for a period of one or two years, depending on the climatological conditions. This is called the ripening process. During the ripening process the properties of the clay change such that its workability improves.

**Effects of ripening of clay**
Dredged soil can be ripened in ripening fields during which the following processes take place:

o   the water content of the soil decreases by evaporation;
o   the density of the clay increases by consolidation;
o   a chemical bonding is created between the clay particles.

As a result of these processes the consistency index of the material increases. Also the concentration of some (organic) contaminants may decrease. Unripened soil as a result of dredging usually has a consistency index below zero. The target consistency for workable clay is between 0.6 to 1.0. Table 9.2 shows the change of several relevant parameters during the ripening process (DWW 2005).

As an example, some relevant material properties for clayey material and clays suitable for structural fill above the waterline are listed in Table 9.3. This Table has been setup by CROW based on research in The Netherlands.

Table 9.2 *Changes in the properties of clay during the ripening process.*

| Parameter | Symbol | Change during the ripening process |
|---|---|---|
| Consistency index | $I_c$ | Increases of up to 0.6–0.8 |
| Water content | $w$ | Decreases |
| Volume | $V$ | Decreases |
| Organic (humus) content | – | Slightly decreases |
| Contaminations | – | Decreases |
| Liquid limit | $w_L$ | Slightly decreases |
| Plastic limit | $w_P$ | Remains approximately equal |
| Plasticity index | $I_P$ | Slightly decreases |

Table 9.3 *Characteristic properties of clay suitable for constructional landfill based on experience in The Netherlands-(CROW publication 281).*

| Material property | Unit | Clay |
|---|---|---|
| *Classification properties* | | |
| Grain(size) distribution | % <2 µm | 8–100 |
| Grading | % 2–63 µm | 0–75 |
| Plasticity index ($I_P$) | % | <20–90 |
| Organic content | % | <5 |
| *Mechanical properties* | | |
| Density (wet) | kg/m³ | 1,600–2,000 |
| Density (dry) | kg/m³ | 1,200–1,600 |
| Maximum density | kg/m³ | 1,350–1,700 |
| Undrained shear strength | kPa | 25–250 |
| $E_{dyn}$ | MPa | >25 |
| Effective cohesion | kPa | 0–50 |
| Effective internal friction angle | ° | 15–35 |
| Compression index | – | 0.05–0.3 |
| Consolidation coefficient | m²/s | $1.0 \times 10^{-9}$–$1.0 \times 10^{-6}$ |
| Permeability (hydraulic conductivity) | m/s | $1.5 \times 10^{-12}$–$1.0 \times 10^{-9}$ |
| *Workability properties* | | |
| Relative compaction $R_c$ | % | 85–95 |
| Consistency index | – | 0.6–1.0 |

**Acceleration of the ripening process**

Depending on the climatological conditions, it takes approximately one to two years for dredged material to ripen to workable clay. The speed of the ripening process mainly depends on temperature, precipitation and wind. The evaporation rate and thus the ripening process will be slower in humid or cold seasons or climates. The natural dewatering of material dredged on a given project can be improved in several ways (Herbich 1992, DWW 2005):

o Optimise the dredging method to minimise the water content of the dredged clay; During the dredging process the mixing of clay and process water should be avoided as much as possible. For excavation and/or transport, preferably mechanical techniques (backhoe, grab or clamshell dredger) should be used instead of hydraulic techniques.
o Plan the dredging operations to maximise the time the surface of the discharged dredged material is exposed to the atmosphere during the period when evaporation rates are high.
o Place dredged material under a mild slope to improve water run-off (approximately 1:50).
o Place dredged material in thin layers which can undergo at least partial drying before being covered by subsequent lifts. This can be accomplished by compartmentalising the bunded area to allow a drying period in between lifts of about 30 days (depending on the climate at the particular site).
o Consider the use of vegetation with a high transpiration ratio to remove moisture from dredged material. This measure may not be applicable in land disposal areas until the dredging process is completed.
o Stimulate drainage by making ditches with special equipment on the reclaimed land.
o Periodically redistribute the clay to expose underlying layers.

**Deceleration of the ripening process**

In some cases slowing down the ripening process is necessary. This applies for instance when the target consistency of the dredged material has been reached but the clay cannot yet be used in a landfill. Further ripening may lead to clay that is too dry for processing. The ripening process can be slowed down by redistributing the clay in high mounds with a relatively small exposed surface area.

**Prediction of properties after ripening**

An accelerated ripening test can be setup in order to gain insight into the consistency index that can be achieved after one or two years of ripening. Standard laboratory tests such as triaxial testing, consolidation tests, etc. are conducted on a small volume of dredged soil. These tests give a reasonable prediction of the engineering and environmental properties at the end of the ripening period. Clay that has ripened this way has approximately the same water content as clay that has ripened 'naturally'. However, processes such as chemical reactions and moulding of the clay cannot be accelerated in the lab. Therefore the parameters that are obtained from accelerated ripening should only be used as an indication of future properties. The accelerated ripening of clays takes about two months.

*Ad 2)   Additives*

The properties of clay can be altered by the use of additives. Examples of additives are dry powder clay, quicklime, fly ash or cement. Resulting effects are (DWW 2005):

- improved workability
- increase of bearing capacity and shear strength
- decrease of compressibility
- decrease of plasticity index ($P_I$)
- decrease of liquidity index $I_L$ and increase of shear strength
- increase or decrease in permeability (depending on the type of additive); an increase in permeability can be beneficial for accelerating the consolidation process, or improve the drainage characteristics of an area. On the other hand, a reduction of the permeability can be desirable when dealing with contaminated materials.

The use of additives will usually require extensive laboratory testing in order to determine the most economic (mixture of) additive(s), the preferred working

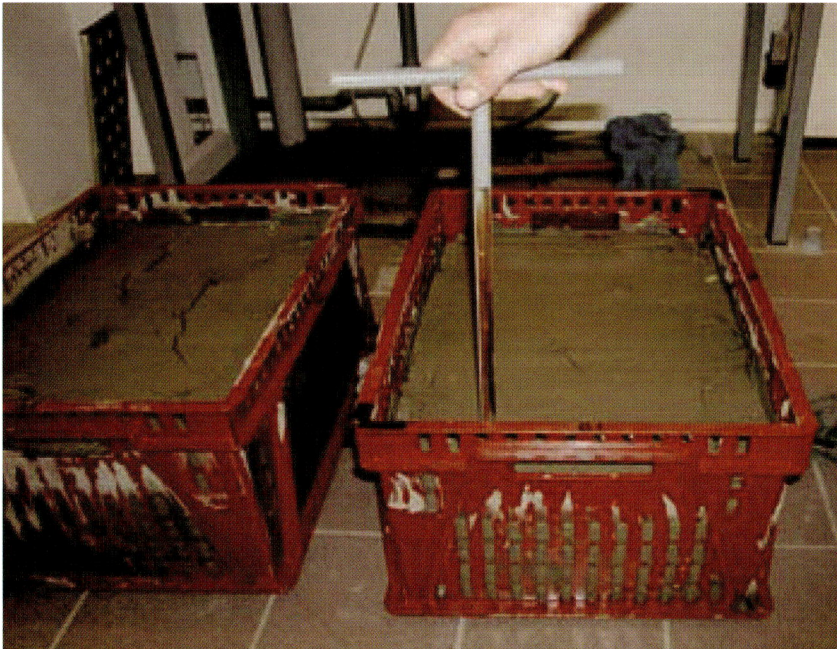

Figure 9.2   *Sampling of material during accelerated ripening test (Rijkswaterstaat 2005).*

method, the final results in terms of shear strength increase and settlement reduction, etc.

It should be noted that using additives is generally a rather expensive soil improvement solutions.

Two main execution methods exist for mixing additives with clay:

– Mixing in situ (after placement): an example of this method is the treatment of 5 ha of dredged material in the port of Valencia (see Figure 9.3).
– Process mixing: mixing after dredging, but before placement in the reclamation area.

*Ad 3) Application of a surcharge*

When adding surcharge to a layer of soft clay or silt, the total vertical stress on the soil is increased. After consolidation, this will result in an increase of the effective vertical stress. The shear strength of a cohesive soil can be correlated to the effective vertical stress (ref. 8.5.2.2).

Figure 9.3    *In situ treatment of dredged material, Port of Valencia Spain (picture ALLU-Finland).*

Figure 9.4    *Process stabilisation – Turku project, Finland.*

*Ad 4)    Use of sandwich structure*

Alternatively, or in addition to other types of soil improvement, the properties of the landfill as a whole can be improved by applying a sandwich structure. In a sandwich structure, layers of clay are alternated with layers of sand. Generally, clay layers with a thickness of 1.0 to 2.0 m are alternated with a layer of sand with a minimum thickness of 0.2 to 0.5 m. A sandwich structure has several advantages compared to monolithic clay elevations (DWW 2005):

– In the case of application above the waterline, the clay layers can be distributed and compacted more effectively because of the stiff and permeable sand base.
– The consolidation of the clay takes place in a shorter time period, because of the limited thickness of these layers.
– The alternating sand layers increase the stiffness of the structure as a whole.
– Clay with a lower consistency index than 0.6 can be used.
– The sand layers provide a suitable sub grade for a temporary construction road.

When applying a sandwich structure above the waterline, the mixing of sand and clay should be avoided where possible. Sand inclusions can collect water, which in turn can lead to local weaknesses in the structure (CROW 2009).

*Ad 5)   Vacuum consolidation*

In some cases, the application of surcharge might be impossible for stability reasons or because of a shortage of suitable surcharge material. In such cases the use of vacuum consolidation techniques to improve the shear strength of the cohesive soil might be a suitable alternative. Reference is made to the case study on the use of alternative material in Bremenhaven, see box below.

The basic working principles of a vacuum consolidation system are sketched in Figure 9.5.

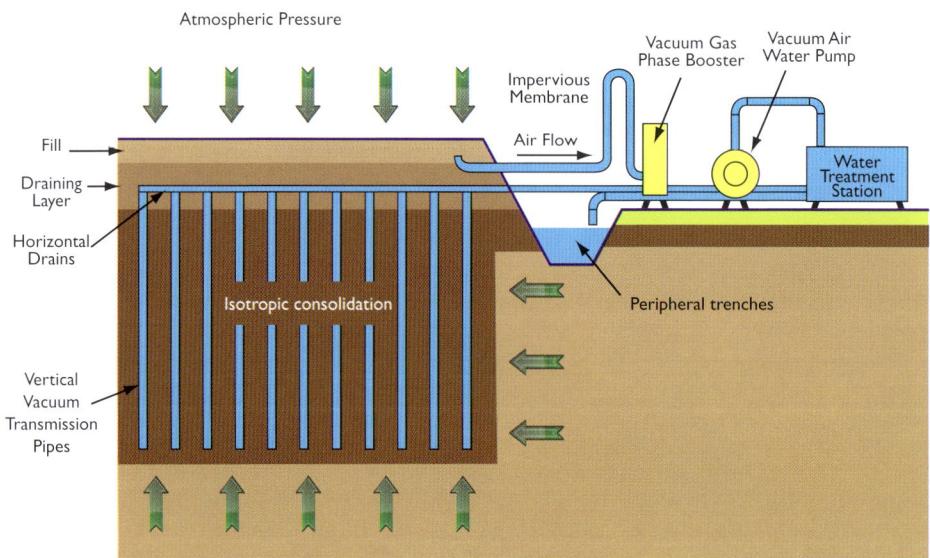

Figure 9.5   *Sketch working principle vacuum consolidation (Menard).*

*Ad 6)   Accelerated drainage*

As an alternative to the sandwich structure, the drainage of dredged material can be improved by methods such as:

– creating trenches through or around the disposal area. The trenches can be filled with sand and gravel to form bottom drains for the next lift;
– installation of vertical drains;
– application of bottom drainage.

**Example of the use of an alternative fill material in Bremerhaven (Nooy van der Kolff et al, 2010)**

Expansion of Bremerhaven's Osthafen (Germany), and in particular its car handling facilities, required reclamation of approximately 6.1 hectares of the existing harbour basin and deepening of the remainder. Because federal environmental regulations prohibited dumping of the dredged, often contaminated sludge at sea it was decided to use the dredged materials as a permanent fill. These materials consisted of soft organic silts and were to be used in the lower part of the fill.

The dredging and reclamation operations as planned comprised placement of a sheet piled wall closing off part of the harbour basin, dredging of the w of the basin, deposition of the dredged mud behind the sheet piled wall (see Figure 9.6), placement of a geotextile and a sand capping on top of the sludge and installation of a system of vertical drains to accelerate the consolidation process of the mud. The design of the heavy superstructures provide for piled foundations transferring the bearing loads to deeper, more competent strata.

Figure 9.6    *Overview Osthafen project during dredging operations: Dredging pontoon in the middle of the basin, floating pipeline and diffuser pontoon discharging the mud in the reclamation area (dark gray area) behind the sheet piled wall.*

Despite all efforts to minimise the increase of the water content of the dredged sludge during dredging, transport and placement (the use of a grab dredger, a positive displacement pump and a diffuser) stability problems occurred when spraying the initial lifts of the sand capping after installation of the geotextile (sand mat) on top of the sludge.

A study of the University of Bremen indicated that the problems had to be attributed to the generation of gas as a result of the organic nature of the sludge and the subsequent reduction of the shear strength of the mud.

In order to overcome the problems it was decided to install a vacuum drainage system to remove the gas, increase the shear strength and accelerate the consolidation process of the sludge. After 3–6 months pumping the first lifts of sand could be placed without stability problems.

This project demonstrates that dredged sludge can successfully be used as a structural fill provided that the problems associated with its typical properties are addressed adequately. The use of the vacuum consolidation system to increase the shear strength of the sludge has played a key role during construction.

**Example of the use of soft clay as fill material at Pulau Tekong (Singapore)**

This Project involved the reclamation of approximately 1,480 ha. of land at Area Y, A,B & D1 off Pulau Ubin and Pulau Tekong. The scope of works included the stabilization of seabed by dredge and sand fill a trench along the reclamation profile, reclamation sand filling, construction of sloping stone revetment/sandy beaches, ground improvement works, extension of rivers and other drainage improvement works.

In order to create a stable reclamation area, a sand key solution was chosen. (ref chapter strength).

A trench was dredged around the perimeter of the future reclamation area. The soft and firm clays at this location were dredged mechanically by means of grab dredgers and were placed inside the reclamation area behind a temporary sand bund (with slope 1/6). A sand cover of approximately 3 m was installed on top of this dumped material by means of a spraying pontoon. A typical cross section of the reclamation area can be seen in Figure 9.10.

**Example of use of soft soil as fill material**

Wenzhou, China (China Communications Construction Company – CCCC)

Due to the lack of suitable granular fill material, mud and very soft clays, dredged in waterways and harbours, were used for land reclamation projects in China during the first decade of the 21st century. Satisfying the requirements for the bearing capacity of the new land is, however, often problematic when using soft cohesive fill material. This example presents an alternative for improvement of the mud by means of vacuum consolidation without the application of a granular drainage blanket.

Project description

The land reclamation project is located in the Dingshan Development Zone in Wenzhou, China. Due to the lack of sand resources near the project site, mud and soft clay were used as fill material. Some project details are listed in Table 9.4.

Table 9.4. *Details of a reclamation project, Wenzhou, China.*

| Location | Reclamation area | Volume fill material | Fill material | Required bearing capacity | Transport distance |
|---|---|---|---|---|---|
| Wenzhou, China | 4,360,000 m$^2$ | 12.1 million m$^3$ | Mud and soft clay | ≥50 kPa (0–1.5 m) (Plate Load Test) | 3 km |

The fill material, originating from dredging activities in a waterway located at a distance of some 3 km from the project site, was excavated by a cutter suction dredge and directly pumped to the reclamation area through a 5 km long pipeline. The reclamation was divided into 6 compartments surrounded by dikes, each having a surface area of approximate 700,000 m$^2$. Ground improvement was subsequently carried out according to the following steps:

– placement of a geotextile on the (mud) surface;
– installation of vertical drains using dual-hull amphibian equipment (Figure 9.7);
– installation of an osmosis hose network, field instruments and a vacuum pump system;
– connection of the osmosis hose network to the vertical drains;
– link to the vacuum pump and run a system test: air and/or water tightness;
– cover the system with non-woven geofabrics;
– cover the system with a sealing membrane;
– application of a preload by filling the compartment with water.

Figure 9.7   *Installation of vertical drains using dual-hull amphibian equipment.*

Ground improvement technique

Consolidation of the hydraulically placed soft mud was realized by applying a traditional ground improvement method (Figure 9.8) which is composed of the following elements:

1. vertical drains, installed in a square grid of 0.8 m × 0.8 m;
2. the horizontal drainage system consisted of an osmosis hose network only. The normally applied drainage blanket (approx. 0.5 m sand) was absent due to the lack of granular material;
3. sealing of the vacuum system was carried out by means of a traditional cut-off wall and a PE membrane on top of the fill;
4. the preload consisted of a water column of 2 m instead of a more traditional sand layer;
5. jetting vacuum pump.

In principle all common components of a traditional vacuum preloading system were used, except for the drainage (sand) blanket.

Because of the absence of the drainage blanket at the top of the system the following deficiencies were anticipated during the design phase compared to a more traditional vacuum transfer system:

– higher vacuum losses;
– lower discharge capacity in the horizontal drainage system.

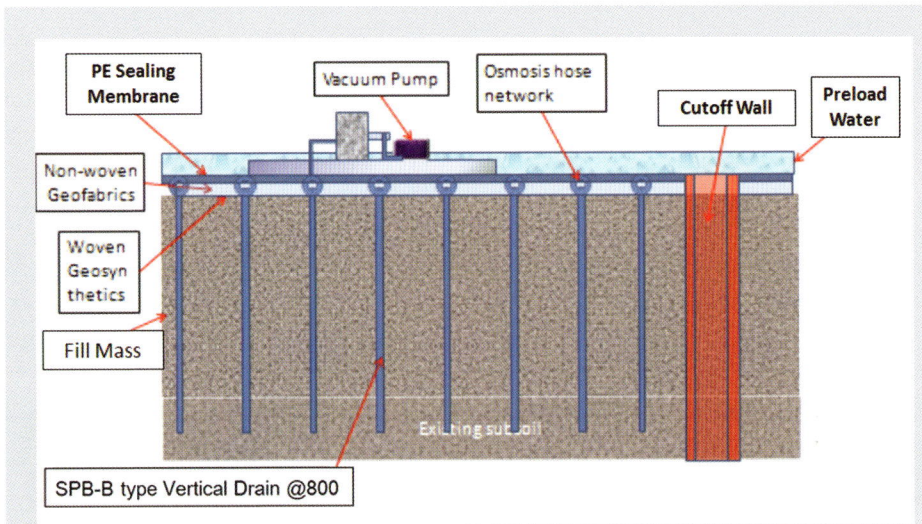

Figure 9.8  *Scheme of the vacuum preloading technique without a drainage blanket.*

Having no sand sources in the Wenzhou area, it was a great challenge to make this alternative ground improvement method feasible, both technically and financial. In order to achieve these goals, the following measures were taken:

– the application of a denser grid for the osmosis hoses in order to apply as much vacuum as possible to the surface of the fill (to make it more comparable with that of a drainage blanket) and to increase the horizontal discharge capacity of the system;
– the application of a newly invented elbow joint in order to create a watertight connection between the osmosis hoses and the vertical drains. In this way vacuum leakage is prevented in the connection, which partly compensates for vacuum losses on the surface and within the treated soil.

The vacuum was gradually increased from 40 kPa to 90 kPa. The vacuum is maintained at $85 \pm 5$ kPa underneath the membrane and at $80 \pm 5$ kPa in the vertical drains. Air and water tightness of the hoses and the individual joints is carefully monitored during the entire duration of the preload.

After a preloading period of 3 to 5 months, the fill mass was finally consolidated from a soft (fluid) mud to a clayey soil capable to withstand a pressure of 50 kPa during an in-situ Plate Load Test. Based on the tests results the required specification for acceptance was realized. The results of the field tests are listed in Table 9.5. For comparison, the initial status of the mud just after filling is listed as well.

Figure 9.9 shows the state of the fill material during the different construction phases.

Table 9.5    *Properties of the fill before and after improvement.*

|  | Mud before improvement | | | Mud after improvement | | | | | |
|---|---|---|---|---|---|---|---|---|---|
|  | Water content % | Density g/cm³ | Porosity – | Water content % | Density g/cm³ | Porosity – | Cohesion kN/m² | Friction angle (°) | Bearing capacity (PLT) kPa |
| Average | 120–130 | – | >3.0 | 49.3 | 1.70 | 1.4 | 13.3 | 9.6 | >50 |
| St. dev. | – | – | – | 3.9 | 0 | 0.1 | 1.4 | 0.2 | |

Hydraulic fill          Fill before improvement          Fill after improvement

Figure 9.9    *Fill material during different construction phases.*

The success of the project demonstrates that the adopted ground improvement method, without the application of a filter (sand) blanket, is a feasible and cost-effective alternative in areas with a desperate lack of sand sources such as the Wenzhou district in China.

Other observations

Although the vacuum was carefully controlled, the consolidation process appeared to be effective within a distance of approximate 40 cm around the vertical drain only which is 50% of the radius predicted during the design phase. Beyond this distance, consolidation hardly occurred in the first 1 to 2 months after the start of the preloading phase. Consequently, the fill did not achieve the required bearing capacity after the anticipated 3 months pre-loading period. Possible reasons for this phenomenon are that very fine particles in the mud may have:

- blocked the filter material in the vertical drains;
- blocked the filter membrane (geofabrics);
- formed a kind of "cut-off" peripheral membrane around the vertical drains reducing seepage.

Field and laboratory tests conducted later did not show blockage of the filter membrane. It, however, appeared that the properties of the mud (in terms of density and permeability) had changed as a function of the distance to the vertical drains. The permeability of the mud near the drains was 1/2 to 1/3 of the permeability of the material located in the marginal zone. Although further study is needed for confirmation, this could be a plausible explanation for the observed phenomenon.

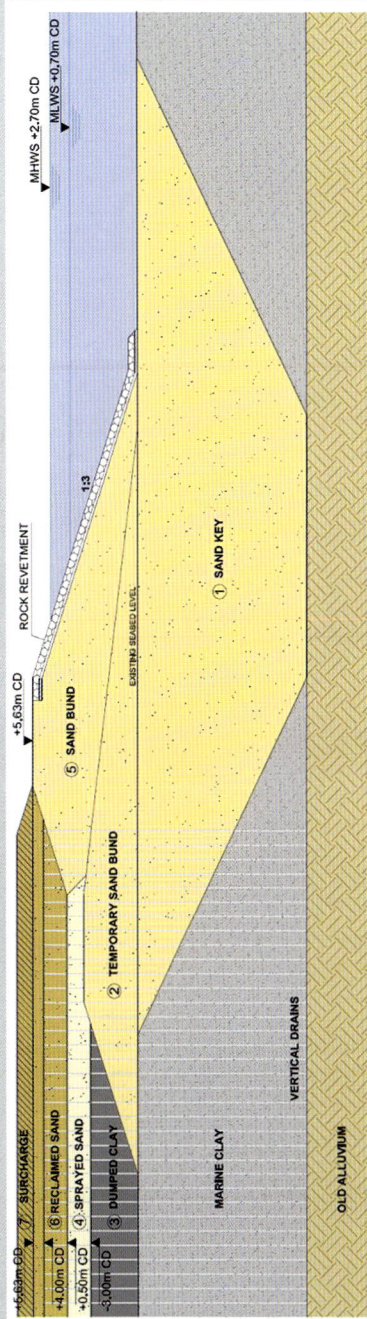

Figure 9.10   Example of re-use of dredged soft clay resulting from a sand key trench.

331

The construction sequence of the above example is shown below.

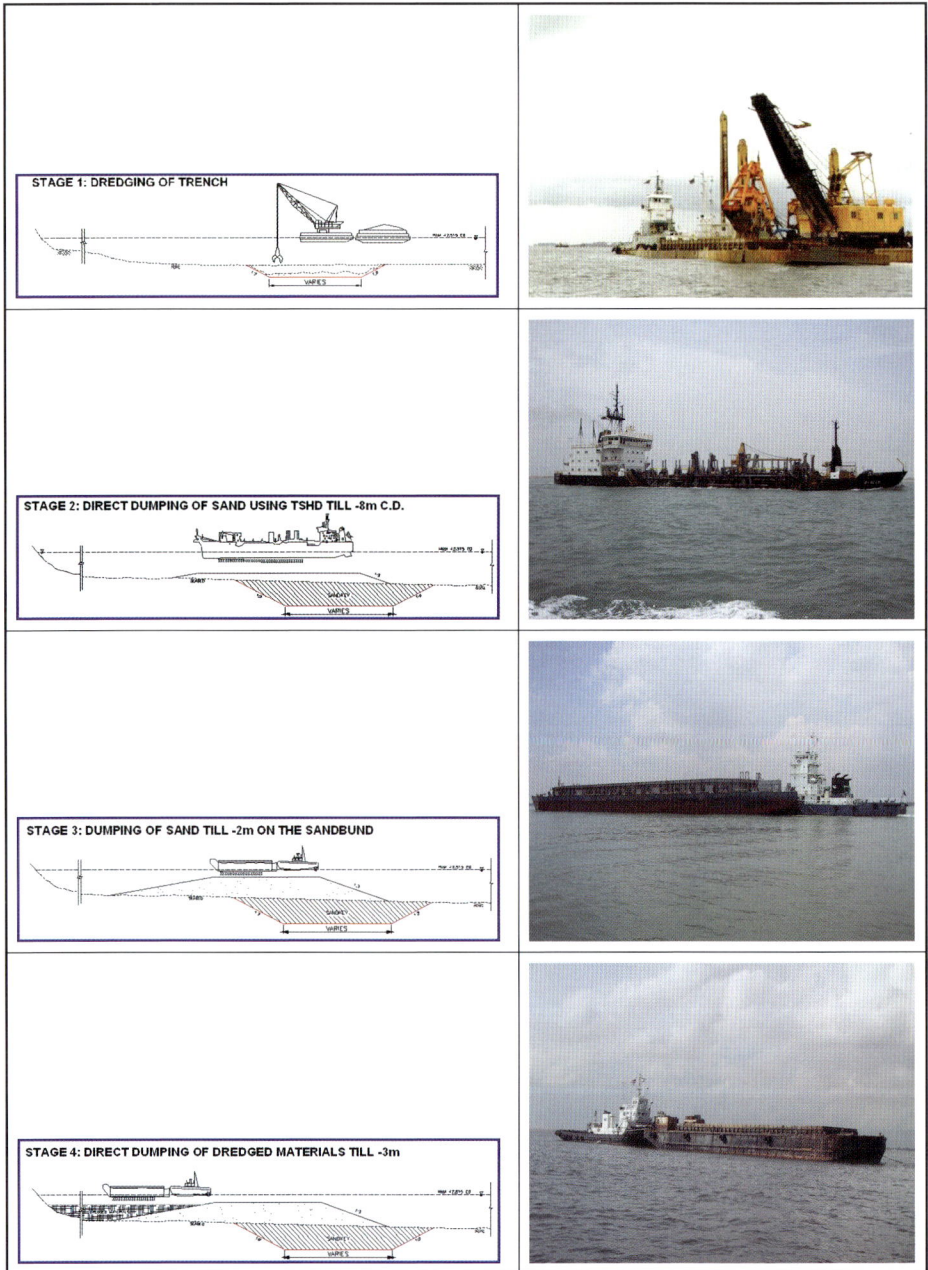

Figure 9.11  (*Continued*).

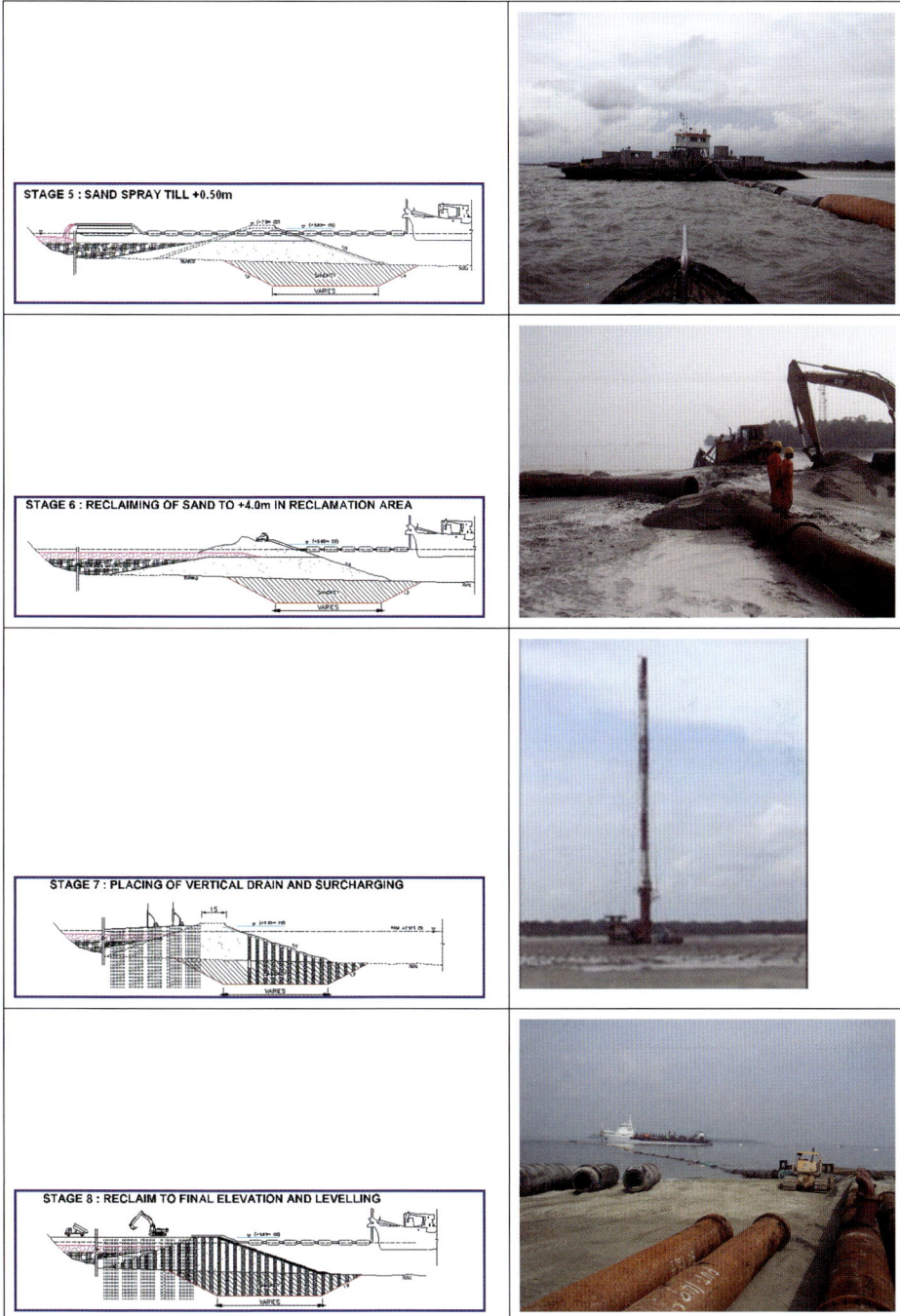

Figure 9.11  (*Continued*).

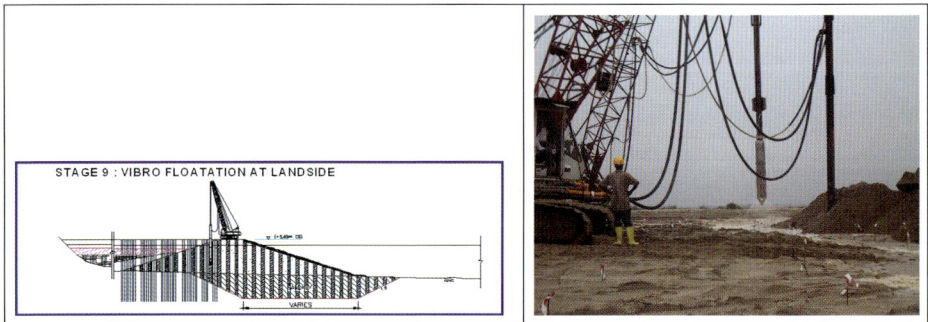

Figure 9.11 *Typical working sequence involving sandkey construction, installation of vertical drains, spraying on top of soft soil, compaction (Tekong island – Singapore).*

### 9.1.3.5 Construction aspects of soft soils in case of application above the waterline

An example of the use of ripened clay for construction above the waterline is presented in Figure 9.12 showing the construction of a highway exit.

The characteristic behaviour of cohesive soils requires some additional considerations during the construction phase. Weather conditions such as drought or precipitation influence the consistency index of cohesive soil (see Figure 9.13). Since the consistency is a direct measure for the workability of the soil, anticipating unfavourable weather conditions prior to and during construction is important. The possible effect of weather conditions is demonstrated in Figure 9.13. Furthermore the following aspects of clay should be taken into account:

- cohesive soils have a minimum strength during construction;
- mixing with other soil types should be avoided;
- compaction should take place in relatively thin layers (0.20 to 0.25 m);
- drainage should be adequate in order to prevent mixing with water from precipitation;
- during periods of rain or frost, cohesive soils cannot be processed.

In order to improve water run-off, it is generally sufficient to apply the clay under a gentle slope (approximately 1:50). The driving direction of the roller should be in the direction of the slope (see Figure 9.14). This is the most favourable configuration for water run-off and prevents the formation of puddles on the surface.

For the use of clay in landfills above the waterline, the following considerations apply (J. van de Graaff, 2006):

- Landfills that are loaded from the top must be prepared with a permeable cover layer, which is capable of distributing loads to the underlying soft clay. Clay

Figure 9.12 *Construction of highway exit with dredged (ripened) clay.*

Figure 9.13 *Two extremes: Wet clay with puddle formation after rainfall and dried out clay after an extended period of drought (DWW 2005).*

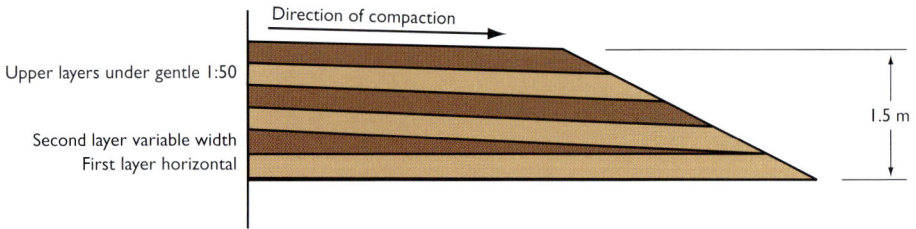

Figure 9.14    *The usage of layers with a variable thickness in order to create a slope to improve water run-off.*

itself cannot be used for this purpose. Higher loads and/or soft clay require a thicker load distributing layer.

– The load distributing layer should include drainage measures in order to prevent water from accumulating on top of the underlying clay. Accumulated water results in softening of the clay.

– For protection against erosive forces from weather, flora and fauna, etc. the permeable cover layer should have a thickness of at least 1.5 m. The protective layer can be prepared from sand or from cohesive soil. If cohesive soil is used as a protective layer, a foil or gravel layer must be used to separate the cover layer from the underlying clay in order to prevent water accumulation.

– Cables and pipelines should preferably be constructed within the protective cover layer.

### 9.1.3.6    Construction aspects of soft soils in case of application below the waterline

In case soft soil is placed below the waterline, a containment bund has to be constructed prior to the placement of the soft soil. This containment bund can be a sand bund, sheet pile wall, rubble bund or combination thereof.

An example of such a containment bund is presented in Figure 9.15. This example shows a containment bund which consists of a (submerged) sand bund in combination with a sheet pile wall. Some openings in the bund are foreseen in order to allow for the entrance of barges.

### 9.1.4    *Stiff clay or silt*

Stiff clay or silt can be dredged by backhoe dredgers, trailer hopper suction dredgers or cutter suction dredgers. The dredging technique has a significant influence on the resulting properties of the soil in the reclamation area.

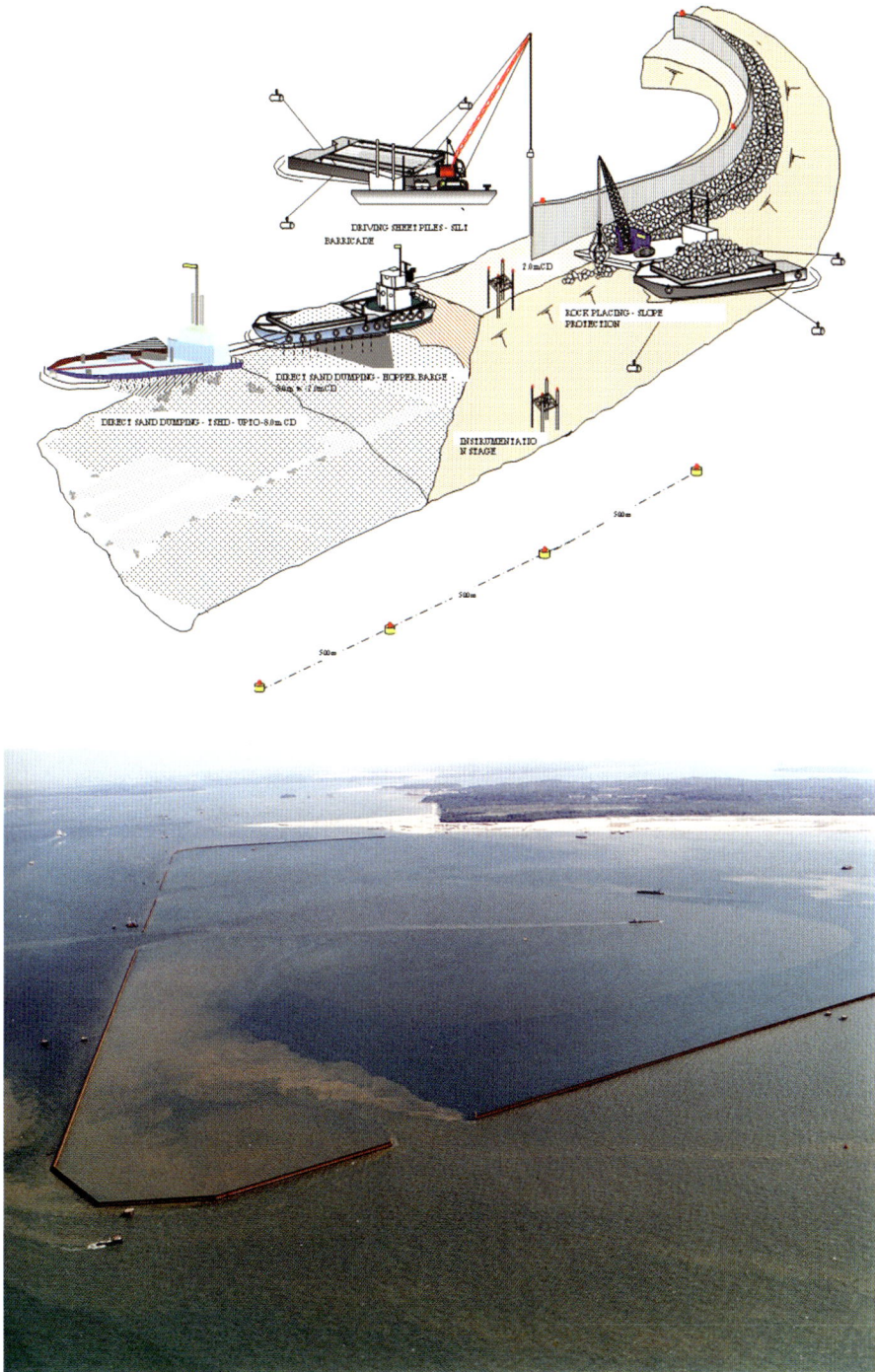

Figure 9.15    *Construction of perimeter bund near Tekong island – Singapore.*

When stiff clay/silt is dredged with a trailing suction hopper dredger, the dredge productions will generally be low. An important fraction of the stiff clay/silt will disintegrate into fines. These fines will not settle well in the hopper, but will be discharged back to the dredge area via the overflow system of the dredger. The remaining clay lumps will settle in the hopper. Especially when dealing with a clay of high plasticity, discharging the load may prove to be difficult because the material will stick to the walls of the hopper, causing bridge formation (Figure 9.16). Additional disintegration will occur when discharging the clay.

Figure 9.16    *Sticky stiff clay inside a hopper – difficult to discharge.*

Figure 9.17    *Dredging silt with a backhoe dredger.*

When dredging stiff clay with a backhoe dredger, large lumps will be formed. Generally the lumps are loaded into a split hopper barge and discharged in the reclamation (or disposal) area. Almost no disintegration will occur and no slurry is generated.

When dredging stiff clay/silt with a cutter suction dredger, the clay/silt will initially be cut into pieces by the cutter head. The material will then be loaded into barges or hydraulically transported via pipelines to the discharge area. During transport in the discharge pipeline, the clay/silt lumps will erode into clay balls. The eroded particles will mix with the process water and form a low density slurry.

The resulting geotechnical properties will strongly depend on the volume of clay balls in relation to the volume of slurry. This relationship is very difficult to predict. If the stiff clay is dredged in combination with more sandy layers, the clay balls can be encapsulated in a sand matrix.

The research of hydraulically transported clay balls dates back to 1949 when Casagrande studied the hydraulic reclamation of 30 million $m^3$ of clay balls for the construction of the Boston airport (Casagrande, 1949).

Figure 9.18  *Formation of clay balls on the reclamation area.*

The results were re-evaluated by Carrier & Bromwell in 1984. Their conclusion was that when the initial voids between the clay balls are large, most of the settlement is probably caused by a rearrangement of the slippery clay balls into a denser matrix (Carrier & Bromwell, 1984).

Generally speaking, the consolidation process of a fill consisting of clay lumps or clay balls has two separate consolidation processes:

–  Initial settlement resulting from the rearrangement of the clay lumps. This is a rather quick process and will generate the most settlement. This consolidation process is often completed during the construction period.
–  Settlement of the clay lump material itself.

Various researchers have studied the behaviour of lumpy clay fills in Singapore (dredged by backhoe dredgers). The conclusions from laboratory experiments were (Lueng *et al.*, 2001):

–  The higher the initial shear strength of the clay, the larger the initial settlement that will occur. This can be explained by the fact that very stiff clay lumps will have larger inter-lump voids and therefore show a higher bulking ratio after excavation.
–  The higher the initial shear strength, the lower the settlement that will occur after the initial settlement. This could be explained by the fact that the intact clay ball will have a higher pre-consolidation pressure compared to a clay ball with lower shear strength.
–  Round lumps show less initial settlement.
–  The larger the lump, the larger the initial settlement.

Some research with regard to the disintegration of clay balls during the pumping process has been done by D. Leshchinsky *et al.* (1994). Results of this research are shown in Figure 9.19, indicating the relationship between mixture velocity and rate of disintegration. Please note that in this figure, the definition of soft clay, medium clay and stiff clay was based on the degree of compaction (% of maximum density). The conclusions were:

- The rate of degradation of a clay ball is strongly related to the plasticity index ($I_p$). If the plasticity index of a stiff clay ball is above 35, the disintegration of the clay ball is negligible. If the clay balls have a plasticity index of 25 to 35, the disintegration is still rather low. The disintegration of clay balls with a plasticity index below 25 is significant.
- Higher transport velocity will cause higher disintegration of clay balls.

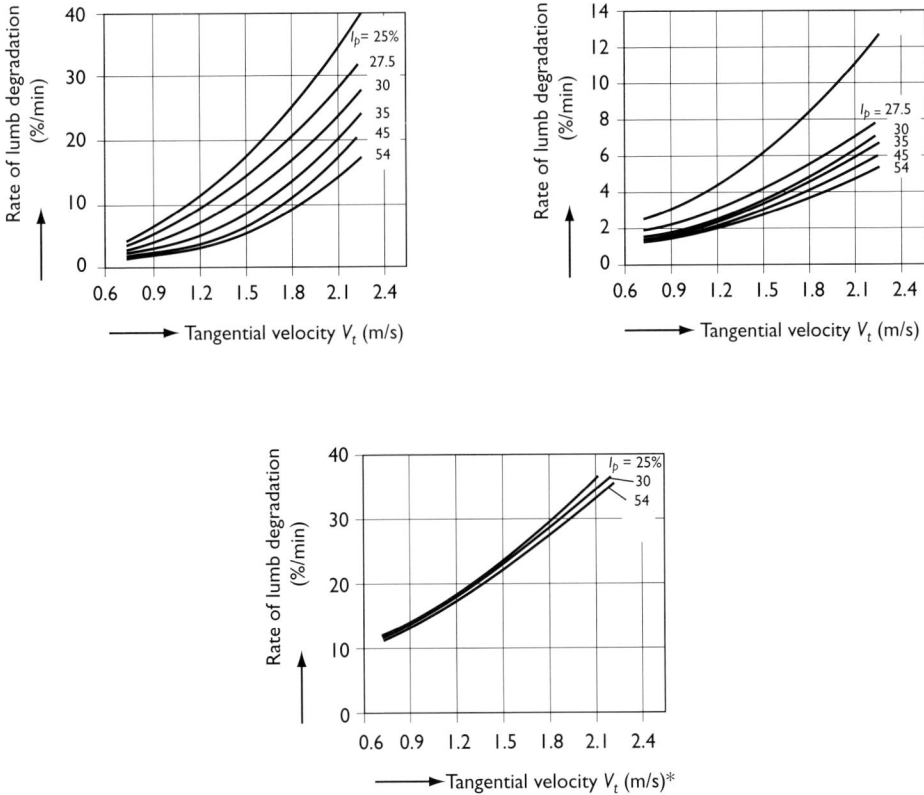

a) Soft clay (density = 80% of maximum value)
b) Medium clay (density = 90% of maximum value)
c) Stiff clay (density = 100% of maximum value)

Figure 9.19   *Design chart for the estimation of clay ball desintegration (after Leshchinsky et al. 1994)*
*(\*) Tangential velocity refers to the rotational velocity of the testing drum.*

The plasticity index clearly cannot be used as unique parameter to indicate the formation of clay balls. Consider a clay with a high plasticity index, but with a water content above the liquid limit. It is clear that this clay will behave as a liquid and will not form clay balls. Intuitively it can be said that the water content of the clay should be somewhere in between the liquid and plastic limit and not too close to the liquid limit. As an additional parameter, the consistency index could be used. As a rule of thumb, the consistency index should be larger than 0.3. This corresponds with a liquidity index smaller than 0.7, since $I_c = 1 - I_L$. Based on the correlation between the liquidity index and the shear strength (see Section 8.4.2.2), the minimum undrained shear strength cu of natural soil samples may be equal to 15 to 20 kPa (Figure 8.11). The lower the plasticity index and shear strength of the clay balls, the larger the disintegration during the hydraulic transport will be. The longer the pumping distance, the higher the disintegration of the clay balls.

It should be noted that the formation of clay balls and the rate of disintegration remains an issue which is often difficult to predict.

In view of the rapid initial consolidation of a clay ball structure, the application of a surcharge is a very effective way of anticipating future settlement during the design lifetime of the reclamation area.

## 9.2 Carbonate sand fill material

### 9.2.1 Introduction

In certain parts of the world the only fill material available may consist of carbonate sand. Their properties may deviate significantly from those of the more commonly used quartz sands, therefore these materials may display a different behaviour during and after the reclamation process. Site investigations and laboratory testing, design and subsequent technical specifications of a reclamation area filled with carbonate sand have to account for this deviating behaviour.

The typical performance of a carbonate sand fill is mostly a result of the crushability and angularity of the carbonate grains, the high initial void ratio of the carbonate fill mass after deposition and the cementation between the particles. Issues may include:

− a lower cone resistance and N-value (SPT) than encountered in quartz sands at comparable stress levels and (relative) densities;
− degradation of the grain size distribution (increase fines content) during dredging, transport and deposition;
− results of laboratory tests that may fall outside the common range of test results of quartz sands;

- a higher compressibility than quartz sands (at high stress levels);
- a higher peak friction angle than quartz sands;
- a higher (or at least different) resistance against cyclic loading than quartz sands at comparable stress levels and (relative) densities;
- possible lower productions of the dredger in the borrow area as a result of in situ cementation;
- increasing strength and stiffness of fill mass over time after deposition of the fill.

These typical issues and their consequences will be discussed in more detail in the following sections.


9.2.2   *Origin and composition of carbonate sands*

Carbonate soils are defined as soils in which carbonate minerals predominate. They are widely distributed in the warm and shallow seas and oceans of the world's tropical and subtropical regions (latitude 30 degrees north to 30 degrees south) covering almost 40% of the ocean floor (Holmes, 1978). Carbonate deposits are usually formed by accumulation of skeletal remains of small marine organisms. from the upper waters of the ocean (bioclastic deposits), but may also have a non-organic origin, for instance, as a result of chemical precipitation from carbonate-rich water (oolites) (Lunne, 1997; Mitchell, 2005; Coop and Airey, 2003).

The typical engineering issues as discussed in this section mainly concern the bioclastic, skeletal sediments (see Figure 9.20 and Figure 9.21) and not the pure oolite sands. Unique post-depositional processes of cementation, dissolution, re-crystallisation and other diagenetic changes have produced various types of carbonate deposits with varying mechanical properties (Celestino and Mitchell, 1983) that may deviate significantly from those of the commonly known quartz sands. It is therefore important to properly classify the nature of the carbonate deposits in the borrow area and in the reclamation area (when applicable) in an early stage of the design.

The adjective "carbonate" is generally being used as a generic description of all sediments containing calcium or magnesium carbonate. According to the classification of carbonate rocks as proposed by Clark and Walker (1977) and Meigh (1987):

- a sand (clay/mud, silt, gravel) classifies as a '*carbonate*' (clay/mud, silt, gravel) sand when its carbonate content is in excess of 90%;
- a '*siliceous carbonate*' (clay/mud, silt, gravel) sand has a carbonate content of 50–90%;
- the adjective '*calcareous*' is used for deposits with a carbonate content of 10–50%.

In practice, however, a less strict nomenclature is often used and both calcareous sand and carbonate sand are encountered in soil classification as a general description.

Figure 9.20    *Typical bioclastic carbonate sand mainly consisting of (fragments of) mollusks and corals.*

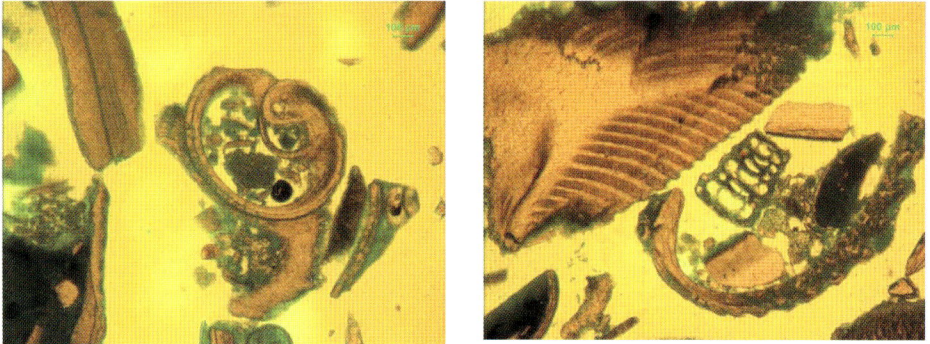

Figure 9.21    *Electron photomicrographs of thin sections of bioclastic carbonate sand originating from the Arabian Gulf showing some complete shells (bivalvia and gastropods) and smaller porous fragments believed to represent bryozoa, algae and corals.*

### 9.2.3    *Typical properties of carbonate sands*

**Main material and mass features**

The main material and mass features of carbonate sands that dominate their engineering properties are:

– Hardness: on Moh's hardness scale, calcium carbonate has a value of 3, compared to a hardness of 7 for quartz, indicating lower intermolecular bonds and, hence, a lower material strength.

- Particle shape: grains of bioclastic carbonate sand may be curved (shells), platy, thin-walled, flaky and hollow or porous (intra-particle porosity) and often have a high angularity, while oolites generally are rounded and massive (see Figure 9.21).
- Void ratio: as a result of their angularity, poor grading and intra-particle porosity, bioclastic carbonate sands usually have higher initial void ratios than silicate sands. Their relatively low minimum and maximum (dry) densities are attributed to the same grain properties.

**Crushability**
The material and mass features mentioned in the previous paragraph make carbonate sands more susceptible to particle crushing and, hence, more compressible than silica sands, especially at elevated stress levels. This will not only affect their mechanical behaviour after deposition in the reclamation area, but will also lead to the generation of more fines during the process of dredging and hydraulic transport and will influence the results of in-situ and laboratory testing. To avoid premature degradation, recovery and subsequent handling of samples of carbonate deposits during site investigations and laboratory testing will require more care and attention compared to quartz sands.

Various authors have suggested methods to quantify the crushability. Hardin (1985) proposed the Relative Breakage (Br) expressed as the ratio of the actual breakage and the breakage potential that can be derived from the difference between the particle size distributions of the sand before and after crushing (see Figure 9.22). It quantifies the amount of crushing that a sand has undergone based on the overall change of the grading.

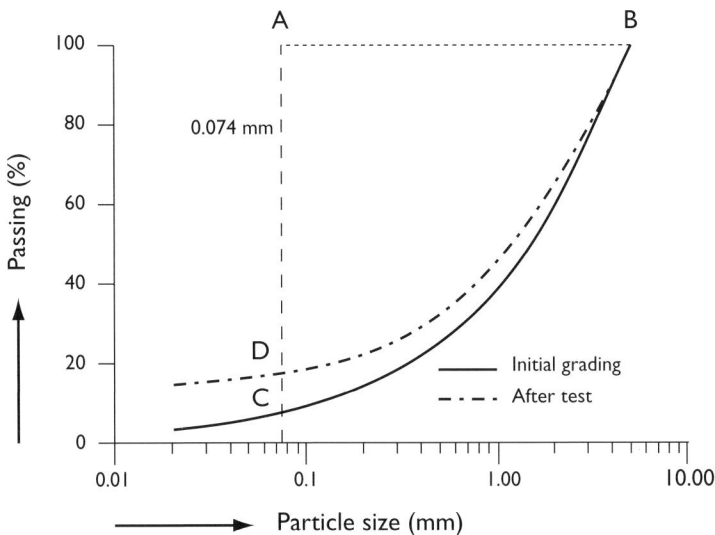

Figure 9.22. *Definition of Relative Breakage (Hardin, 1985).*

Kwag *et al.* (1999) found that crushability is a function of the relative density and the corresponding yield stress (i.e., the stress at which crushing starts). The yield point was defined by De Souza (1958) as the point of maximum curvature of the void ratio versus the logarithm of the effective stress curve. The stress at this yield point during one-dimensional compression marks the onset of particle crushing (Nakata *et al.*, 2001) and can be graphically determined by the method of Casagrande (1936). Subsequently they presented a crushability index that determines the sensitivity of a sand to crushing and allows for an estimate of the yield stress at a given relative density:

$$K = \ln p_c - 2.27 \times R_e$$

where:

$K$ = crushability index
$R_e$ = relative density [%]
$p_c$ = yield stress [kPa] corresponding to $R_e$

Table 9.6 presents some index properties, the crushability index and yield stresses (at different relative densities) of a number of well-known sands (Kwag *et al.*, 1999). The results clearly demonstrate the influence of the mineral composition of the particles on the crushability and yield stress. Note that the larger the crushability index $K$ is, the smaller the crushability of the sand will be.

Table 9.6.  *Some index properties, crushability indices and yield stresses of well-known sands (Kwag et al., 1999).*

|  | Description (Clark and Walker, 1977) | $e_{max}$ | $e_{min}$ | $CaCO_3$ | K-value | $p_c$ [kPa] at $R_e = 60\%$ | $p_c$ [kPa] at $R_e = 30\%$ |
|---|---|---|---|---|---|---|---|
| Toyoura sand | Silica sand | 0.985 | 0.606 | 0% | 8.15 | 13 521 | 6 843 |
| Amami Sand | Calcareous silica sand | 1.114 | 0.711 | 50% | 7.4 | 6 387 | 3 232 |
| Quiou Sand | Bioclastic siliceous carbonate sand | 1.303 | 0.786 | 81% | 5.8 | 1 289 | 653 |
| Dogs Bay Sand | Bioclastic carbonate sand | 1.720 | 1.080 | 93% | 5.9 | 1 425 | 721 |

**Angularity**

The angular nature of the bioclastic particles may result in relatively high void ratios (see Table 9.6) and high angles of internal friction at stress levels relevant to engineering practice. At higher stress levels this angularity, however, may also induce high stress concentrations at the grain contacts leading to crushing of the grains.

Angularity is difficult to quantify. Powers (1953) has proposed a classification system based on a visual examination of the grains and a comparison with a roundness scale (see Appendix B).

**Cementation**
The grains of carbonate sands are often cemented as a result of precipitation of carbonates (like aragonite, calcite and/or dolomite) or other salts (like halites) from saturated pore water (see Figure 9.23). Such cementation may not only be encountered in-situ in the borrow area, but may also develop at or soon after placement in the reclamation area (Fookes, 1988).

As the precipitation of carbonates from the pore water is very sensitive to variables like porosity, permeability, texture and composition of the fill mass, the water depth, temperature and chemistry of the pore water and the burial depth, the resulting cementation may be highly variable as well, both in terms of strength and in terms of spatial distribution (vertical and lateral extent).

This tendency of dissolved salts to precipitate from the pore water has its consequences for the laboratory testing procedures: particle size analyses should preferably

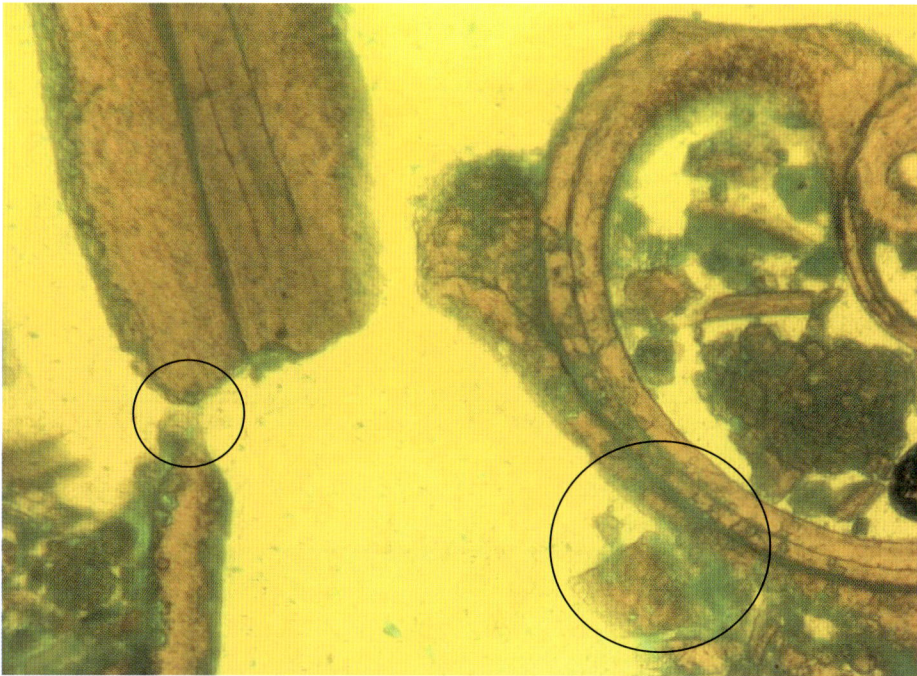

Figure 9.23 *Enlargement of electron microphotograph of thin section showing fragile calcite bonds (cementation) between particles (micrite, green to blue on photograph).*

be carried out using the wet sieving technique, drying of samples must be undertaken at moderate temperatures, etc.

The bonds resulting from cementation of the particles may range from strong to extremely weak and tend to break up during sampling. Identification of cementation is therefore very difficult by conventional site investigation techniques (sampling and visual inspection) and will generally require microscopic examination and/or chemical analyses. The Alizarin Red Test is a simple quick field test to distinguish between calcium carbonate and other minerals. It will stain the calcium carbonates (calcite, aragonite, etc.), but will leave all other minerals (dolomite, siderite, halite, etc.) uncoloured. If the grains and cement material have a different mineralogy a subsequent visual inspection (preferably with a magnifying glass or microscope) will give a qualitative indication of the presence of cementation.

Cementation, even if the bonds between the particles are weak, may adversely affect the production of the dredger in the borrow area, but may also increase the shear strength and stiffness of a fill mass after deposition in the reclamation area.

Occasionally it may be possible to observe the development of cementation in a fill mass after deposition by recording increasing cone resistances during consecutive cone penetration tests at the same location. However, given the variable nature of the cementation it is generally not possible to predict this effect with sufficient accuracy during design or construction stage.

Figure 9.24 presents the cone resistances measured immediately after deposition and compaction and after approximately 3 months. The CPTs were undertaken at (almost) the same location at a carbonate sand fill in the Middle East (from top CPT profile down to +1.75 m QNHD). A clear increase in cone resistance due to cementation can be observed in this zone above the groundwater level. However, below sea level and in other comparable areas of the same site, no increase of the cone resistance with time has been observed.

**Suitability as a fill material**
It must be stressed that these specific properties of carbonate sands do not imply that they are not suitable as a fill material (crushing, cementitious bonding). Ample experience, in particular in the Middle East, indicates that carbonate sands perform satisfactorily as a bulk fill provided that during design and construction their typical properties are taken into account.

Well-established, empirically based interpretation methods and correlations developed for quartz sands may not be applicable or may need to be adapted to account for the specific properties of carbonate sands. Conventional design procedures based on the concept of the relative density and related parameters have been

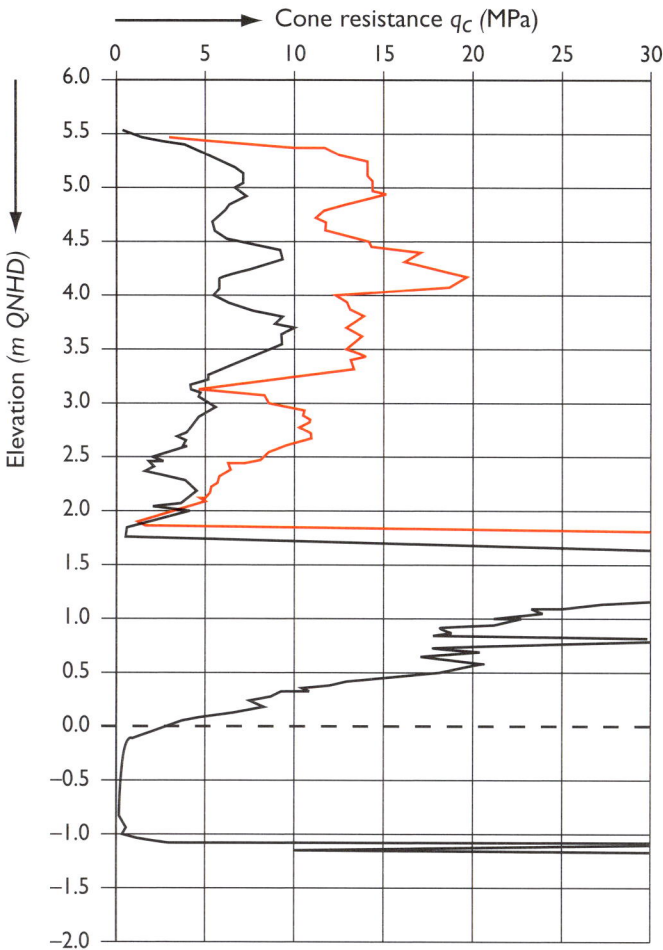

Figure 9.24  *Increase of the cone resistance as a result of cementation after 3 months (mean sea level = 0.00 m QNHD).*

proven to be unreliable for carbonate sands and other compressible sands (Coop and Airey, 2003) and should therefore not be used.

### 9.2.4  *Mechanical behaviour of carbonate sands*

According to studies undertaken by Coop (1990) and Coop and Airey (2003) the specific properties of carbonate sands result in:

–  behaviour that does not fundamentally deviate from that of sands with other mineralogical compositions;
–  parameters defining the behaviour of carbonate sands that are generally outside the common range of values of other types of sand.

349

Coop and Airey (2003) have proposed a framework to describe the mechanical behaviour of carbonate sands that shows many similarities to the Critical State framework used for clayey soils. Some basic concepts of the critical state soil mechanics like the normal compression line (NCL) and critical state line (CSL) are used in the following sections. Detailed references to geotechnical literature can be found in appendix D.

For more detailed information reference is made to the paper of Coop and Airey. Note that so far only a limited number of papers has been published on this subject. More research in the future may be required to fully understand the mechanical behaviour of carbonate sands.

**Compression**
Volumetric strains as a result of isotropic loading of sand are usually associated with particle crushing. Results of isotropic compression tests on carbonate sands at a wide range of stress levels indicate a clear, straight and unique normal compression line NCL, similar to that which may be expected from quartz sands, see Figure 9.25, (Coop, 1990). The main difference between the compression lines of carbonate and quartz sands is the slope $\lambda$ (i.e., the gradient of the relation between mean effective stress and void ratio $e$ or specific volume $v = 1 + e$, equivalent to compression index $C_c$ in $e/\ln p'$ diagram) that in the case of carbonate sands is generally much steeper. This steep slope points at a higher compressibility of the carbonate deposits, which is attributed to the high initial void ratio and subsequent particle breakage during loading.

Figure 9.25 (Coop, 1990) presents a number of results of isotropic compression data of a carbonate sand that initially shows a relatively stiff behaviour until a stress level of approximately 500–700 kPa has been reached. Further isotropic loading beyond this stress level will gradually set off the crushing of the particles and, hence, increase the compressibility until it reaches a constant value on the NCL. As a result of the gradual start of particle breakage it is difficult to precisely determine the yield stress, i.e., the stress above which significant crushing commences. As mentioned in Section 9.2.3 Kwag *et al.* (1999) suggest to use the graphical method as proposed by Casagrande (1936).

Carefully compacted (i.e., by gentle tamping during pluviation in order to avoid crushing) sample $P$ (see Figure 9.25) displays the same behaviour as the same material isotropically compacted from a higher specific volume once it reached the NCL, but with a higher yield stress. Again, large volumetric strains occur before the NCL has been reached and the onset of crushing is poorly defined.

In contrast, unloading – reloading of the samples (see sample $Q$, Figure 9.25) will result in a rigid, elastic behaviour and a distinct, well-defined yield stress once the NCL has been reached and crushing is resumed.

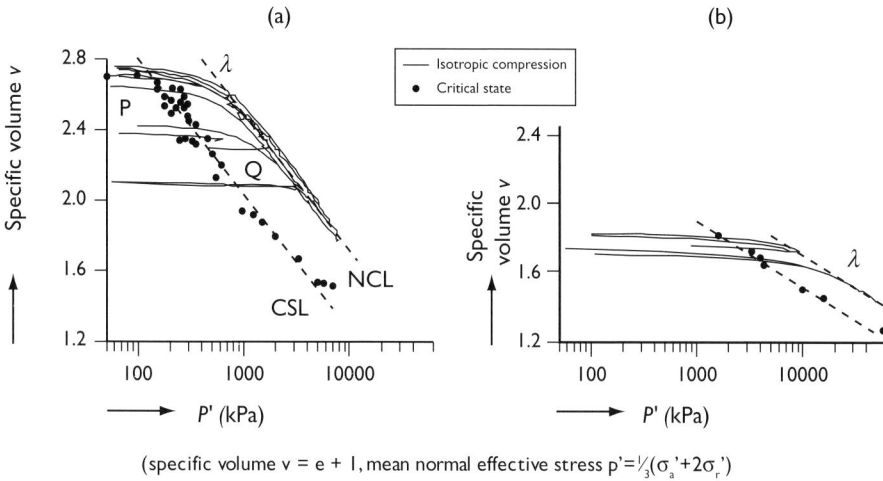

(specific volume v = e + 1, mean normal effective stress $p' = \frac{1}{3}(\sigma_a' + 2\sigma_r')$)

Figure 9.25   *Isotropic compression data for Dogs Bay carbonate sand (a) and Ham River quartz sand (b) at comparable scales (Coop and Airey, 2003).*

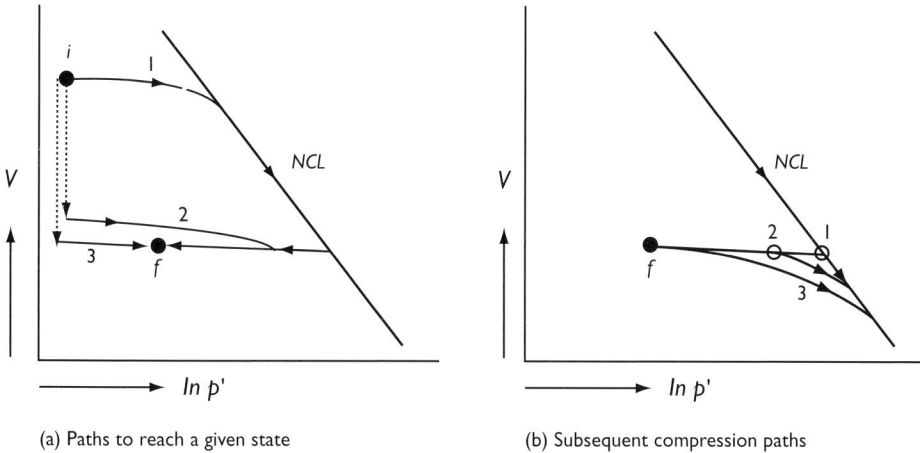

(a) Paths to reach a given state                    (b) Subsequent compression paths

Figure 9.26   *Effect of loading history on compressibility (Coop and Airey, 2003).*

This example demonstrates that not only the (initial) void ratio, but also the stress history may have an important influence on the volumetric behaviour of a carbonate sand. Figure 9.26 illustrates schematically the effect of the stress path followed (i.e. moving from state *i* to state *f*, Figure 9.26a) on subsequent (re)compression (i.e. beyond state *f*, Figure 9.26b): it is clear that stress path 1 (only preloading) will result in less settlement than stress path 2 (partly compaction, partly preloading). Stress path 3 (only compaction) will result in even more settlement. It should be noted that in this example the assumption is made that preloading will be accompanied by crushing, while compaction is solely the result of re-orientation of the particles without crushing. In practice, however, it is unlikely that compaction will not induce any crushing.

Coop and Airey (2003) conclude that the compressibility cannot be described by a unique bulk modulus, but will depend on stress level, initial (depositional) void ratio and previous loading history.

This conclusion has its implications for sampling and testing of carbonate sands originating from both the borrow area and the fill itself. Because a better graded sand usually produces a NCL with a smaller gradient $\lambda$ it is important to test a sample that is representative of the fill in the reclamation area. This sample may have a grading that, as a result of dredging, hydraulic transport and deposition, differs from the fill encountered in the borrow area.

Figure 9.27 (Coop and Airey, 2003) presents the isotropic compression data of a sample that has been loaded well beyond its yield stress. After reaching its maximum load, the test was stopped, the sand retrieved, the sample reconstituted and re-tested. The resulting compression line has a smaller gradient than the initial loading curve as a result of the changed grading after particle breakage during the initial loading phase and the re-arrangement of the grains during reconstituting of the sample (smaller initial void ratio). This example clearly demonstrates the need to start each compression test at an initial void ratio and grading equal to that of the fill mass after deposition in the reclamation area.

The yield stresses determined for carbonate sands are usually well above common engineering stress levels, even for sands with high void ratios. Should these stresses

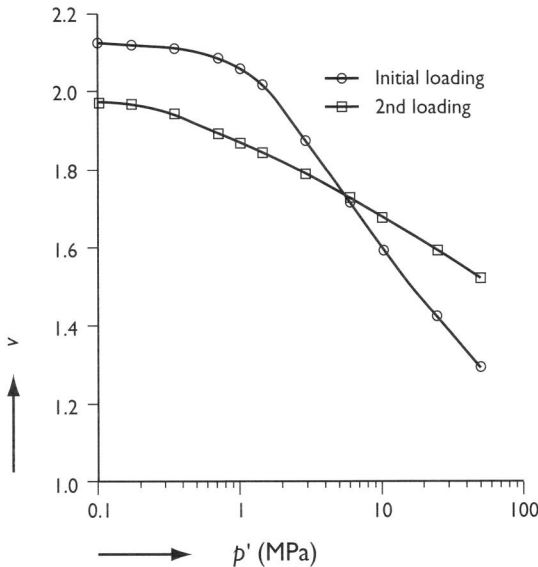

Figure 9.27 *Influence of initial grading on isotropic NCL (Coop and Airey, 2003).*

be of the same order of magnitude, compaction or surcharging of the carbonate fill mass may be required. This will reduce the void ratio of the fill and increase the yield stress, thus preventing excessive future settlements resulting from particle breakage (see Figure 9.26). The density required to control the settlements of a loaded fill mass can be derived from the NCL (or the $K_0$ NCL, see below) of the potential fill material.

If compression testing in the laboratory has to be undertaken to determine the degree of compaction required to limit future settlements at the reclamation area (i.e., future loads are estimated to be in excess of the yield stress at depositional void ratio), Figure 9.27 suggests that the initial grading of the laboratory sample shall be the same as the grading of the fill after deposition. Moreover, subsequent compaction of the sample prior to compression testing should simulate the compaction operations in the field as closely as possible inducing the same degree of crushing. As it is unlikely that the initial gradings and the effects of compaction are known or can be estimated with sufficient accuracy, a series of compression tests with varying initial void ratios, fines contents and compaction efforts are recommended.

Figure 9.26 suggests that pre-loading may be an even better option than compaction to reduce future settlements as after removal of the surcharge and re-loading, particle breakage will start at a higher and better-defined stress level. However, in most cases surcharging will not be feasible as a result of the very high stress levels required to reach and move down the NCL.

One-dimensional compression tests, which may be more representative of compression of a fill mass, result in a $K_0$ NCL that occurs below and parallel to the NCL of isotropic compression. This implies that the yield stress in one-dimensional compression will be less than in isotropic compression and that the difference between surcharging and compaction as explained above may be more pronounced.

The $K_0$-value derived from these one-dimensional compression tests on a number of different carbonate sands is generally significantly higher than when applying the formula of Jaky (1944):

$$K_0 = 1 - \sin \phi'$$

This may have its bearing on the horizontal earth pressure and, hence, on the design of structures retaining a carbonate sand fill.

### Shearing behaviour
Peak friction angles of carbonate sands are generally higher than those found for quartz sands of a similar density (Brandes, 2010). This is thought to be a result of the more angular and elongated nature of the carbonate particles that will result

in more pronounced dilation during shearing. Wide ranges of peak friction angles varying between 35° and 50° are reported by Datta *et al.* (1979), Hull *et al.* (1988) and Brandes *et al.* (2000). Brandes (2010) suggests that the large variation of the peak friction angles reflect the large variation of particle angularity of the different carbonate sands tested.

As illustrated schematically in Figure 9.28 (Coop and Airey, 2003), the magnitude of the peak friction angle of a sand depends on the current state of the sand and its position relative to the NCL and CSL in the $v$:ln $p'$ plane. Sands with a state below (or left of) the CSL (A and C) will display dilation during shearing and the magnitude of the resulting peak friction angle will depend on the distance from the CSL, while shearing of samples with a state plotting between the NCL and the CSL (B and D) will result in compression only without dilation and corresponding peak friction angle.

Figure 9.28 also clearly shows that the (relative) density alone (without stress level) is not a very useful parameter to describe the mechanical behaviour of a sand: states A and B have the same (relative) density, but will display a completely different shearing behaviour.

Although Golightly (1989) suggested that at high stresses crushing of carbonate particles leads to a decrease of the critical state friction angle, Coop (1990) concluded from the results of a number of high stress triaxial tests that the critical state line (CSL) of carbonate sands, just like that of quartz sands, is generally

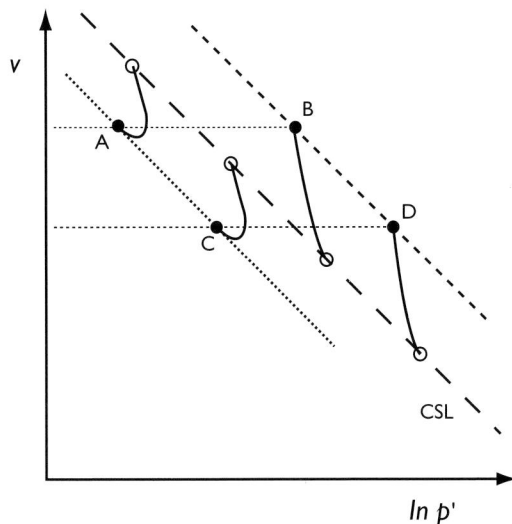

Figure 9.28   *Influence of state on volumetric change (Coop and Airey, 2003).*

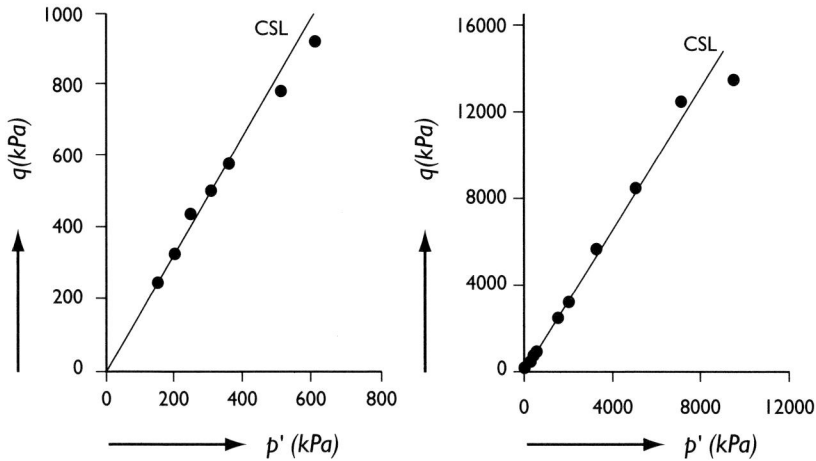

Figure 9.29  *Critical states of Dogs Bay sand over an extended pressure range (Coop, 1990).*

straight over an extended stress range, but that the strains required to reach the CSL are very large (see Figure 9.29). It was suggested that the observation of a decreasing friction angle was probably due to common laboratory triaxial testing procedures that limit the maximum strain to 20% at which the ultimate state may not have been reached. Ring shear tests undertaken by Coop *et al.* (2004) indicate that stable critical state friction angles may require shear strains in the order of 30% or more.

Slopes of the CSL of carbonate sands are usually steeper than those found for quartz sands indicating that critical state friction angles of carbonate sands are higher than those of quartz sands. This is mainly attributed to the angular nature of the carbonate particles. It appears that in terms of critical state carbonate sands are stronger than quartz sands. Engineering problems, however, may arise from the large strains required to mobilise this strength.

**Creep**
Creep in sand is generally considered to be a deformation process occurring at a constant stress and to be closely associated with on-going particle breakage and related stress redistribution. Unfortunately, very little has been published on creep in carbonate sands. Coop and Airey (2003) suggest that the ratio between $C_\alpha$ (rate of secondary compression) and the compression index $C_c$ of carbonate sands on their NCL is approx. 0.013, more or less similar to sands of other minera-logical composition. However, because the state of carbonate deposits is usually closer to the NCL than that of quartz sands at comparable stress levels, carbonate sands are expected to display more creep than quartz sands. This implies that, if

355

deformations of the fill mass as a result of creep are considered to be a problem, compaction and/or surcharging may have to be undertaken.

Tests undertaken by Coop and Airey (2003) indicate that at a same void ratio compacted samples will display more creep than samples of the same carbonate sand that have been over-consolidated. However, as indicated before, surcharging may not always be feasible because of the high loads required to move down the NCL.

### Cyclic behaviour: susceptibility to liquefaction

Limited evidence derived from the results of cyclic triaxial tests seems to suggest that at a given stress level and relative density carbonate sands display a higher resistance against cyclic loading than quartz sands (Coop and Airey, 2003, Olgun *et al.*, 2009). This may partly be attributed to the higher critical state friction angles of carbonate sands.

Olgun *et al.* (2009) compared the susceptibility of a carbonate sand (Playa Santa sand: a poorly graded, angular, bioclastic carbonate sand with large intra-granular voids) at various relative densities with a commercially available silica sand (Monterey sand: a poorly graded, medium to fine silica sand) by plotting the cyclic stress ratio CSR (see Figure 9.30) against the number of loading cycles required to cause liquefaction (i.e., liquefaction defined as 100% excess pore water pressure). Figure 9.30 clearly shows the higher resistance of the carbonate sand at a comparable relative density.

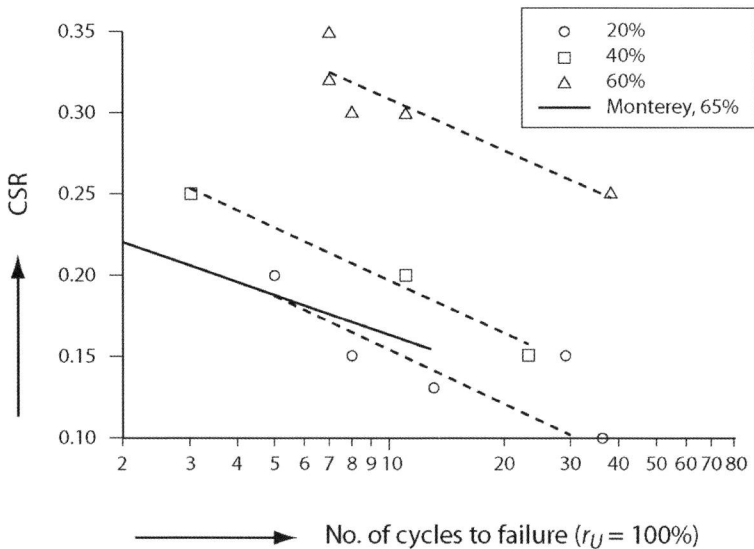

Figure 9.30   *CSR plot of Playa Santa carbonate sand at relative densities of 20%, 40% and 60% and Monterey quartz sand at a relative density of 65% (Olgun et al., 2009).*

Liquefaction evaluations are often undertaken according to empirical methods as proposed by the NCEER (2001) or Idriss and Boulanger (2008) (see also 8.6). These methods are based on historical records (valid for mainly non-carbonate sands) used to establish a relation between the occurrence of liquefaction and the state of the soil mass. The state of the soil mass is generally expressed in terms of cone resistance, N-value (SPT) and/or the shear wave propagation velocity (see Appendix B.2.3.2). When undertaking a liquefaction analysis of a reclamation area consisting of carbonate sand it is important to account for the effects of the specific behaviour of carbonate sand on the results of these tests (see 9.2.5.2, 9.2.5.3 and 9.2.5.4).

Alternatively, susceptibility to liquefaction can be assessed by the evaluation of the cyclic stresses induced by the design earthquake shaking at different levels in the fill mass. These cyclic stresses should be compared to the cyclic stresses, which, at given confining pressures corresponding to the same levels in the fill mass, will cause the carbonate sand to liquefy (100% peak cyclic pore pressure ratio) or will induce a certain degree of cyclic strain in cyclic simple shear tests. Care must be taken to ensure that the tested samples have the same grading (and density) as the fill after deposition in the reclamation area. This may imply that fines have to be added to (or removed from) the fill retrieved from the borrow area to simulate the effects of degradation of the fill during dredging, transport and deposition (see 5.2.1 and 8.6). Further reference is made to section 8.6.

## 9.2.5   The use of carbonate sand as fill

As stated before, despite some of their properties being outside the range of more common soils, carbonate sands can satisfactorily be used as a bulk fill provided that the design takes these deviating parameters into account. Ample experience, in particular in the Middle East, confirms that land reclamation with carbonate sand may result in a fill mass with appropriate mechanical properties.

The following sections describe some of the engineering problems related to the use of carbonate sands as a fill.

### 9.2.5.1   Typical behaviour during dredging and hydraulic transport

**Degradation of particles**
Degradation of carbonate particles during dredging and hydraulic transport may result in a higher fines content of the fill after deposition than determined during the site investigations in the borrow area prior to dredging. The high impact stresses of the cutter teeth, the impellors of the pumps and the abrasive action along the pipeline wall will increase the fines content of the fill changing the original in-situ particle size distribution.

The amount of degradation depends on a number of variables. These may be related to the nature of the fill (carbonate content, original grain size distribution, particle shape, etc.), but also to the operational parameters (type of dredger, number of pumps, length of pipeline, pumping velocity, concentration of the sand-water mixture, etc.).

Although attempts have been made (Ngan *et al.*, 2009) it is still very difficult to quantitatively predict (i.e., in terms of changing particle size distributions) the degradation as a function of the material and process parameters. Better-graded sand will generally result in a smaller gradient $\lambda$ of the NCL (see Figure 9.27) indicating that the fill is less sensitive to crushing. Therefore, testing the samples of the fill material with gradings that closely resemble the particle size distributions after dredging, hydraulic transport and deposition on the reclamation area is important. Given the problems to predict degradation (and segregation during deposition), it is highly recommended to undertake sensitivity analyses by laboratory testing on samples with a range of different grading.

### Hydraulic transport

Carbonate sands often consists of (fragments of) shells that are not spherical, but more disk-shaped. This typical shape will influence the behaviour of the particles during hydraulic transport in a discharge pipe.

Miedema and Ramsdell (2011) demonstrated that the typical disk-shape of shells will reduce the settling velocity of the grains, but increase the erosion velocity required to force settled particles into suspension again. This is caused by the relative orientation of the shells to the direction of the flow during transport and after settlement.

During transport the particles will preferably expose their smallest cross-sectional area perpendicular to the flow and, subsequently, will tend to settle like leaves. This will reduce the settling velocity.

Once settled the shells will be lying flat on the sediment (often with their convex side turned upward), hence protecting themselves and any underlying sediment against erosion. Moving these sediments into suspension again will require a much higher mixture velocity than when pumping more spherical grained materials.

These phenomena may affect hydraulic transport: the critical velocity of the sand-water mixture (see Section 4.5.2) in the pipeline may be less than when pumping normal quartz sands, but when it drops below a certain threshold value the discharge pipe may choke up suddenly and rapidly. This problem may be aggravated by the tendency of the flow velocity to decrease with the development of a bed load in the pipeline.

9.2.5.2   Cone Penetration and Standard Penetration testing in carbonate sands

Cone Penetration Tests (CPTs) are not only widely used to classify soils, but also to establish soil parameters for engineering purposes from well-established and accepted correlations. Although they require boreholes, their procedures are more laborious and their results do not correlate well to the engineering properties of the strata penetrated, Standard Penetration Tests (SPTs) have the advantage that they produce a (disturbed) sample of the soils tested. On reclaimed land cone penetration tests and—to a lesser extent—standard penetration tests are often specified to confirm compliance with the contractual requirements in terms of fill quality (presence of accumulations of fines), relative density, shear strength and stiffness.

Based on experiences with CPTs in carbonate sands, it has been recognised that the cone resistance in these carbonate deposits is less than in quartz (or silica) sands of a comparable relative density at the same depth. During a CPT, the stresses around the tip of the cone are likely to exceed the yield stress of the carbonate sands resulting in significant crushing in front (compression) and in the immediate vicinity (shearing) of the cone. Direct evidence of the crushing has been reported by Nutt and Houlsby (1991), who found fine powdered sand adhering to the cone after extraction of the cone from a calibration chamber, and White (2002), who used a calibration chamber and close-up photography to study the penetration mechanism of a jacked pile (see Figure 9.31).

The same high stresses are expected to occur in the immediate vicinity of the cutting shoe of the split spoon of the standard penetration test when driven into carbonate deposits. Therefore the resulting N-value will be less than the N-value that will be recorded at the same depth in silica sands of comparable density.

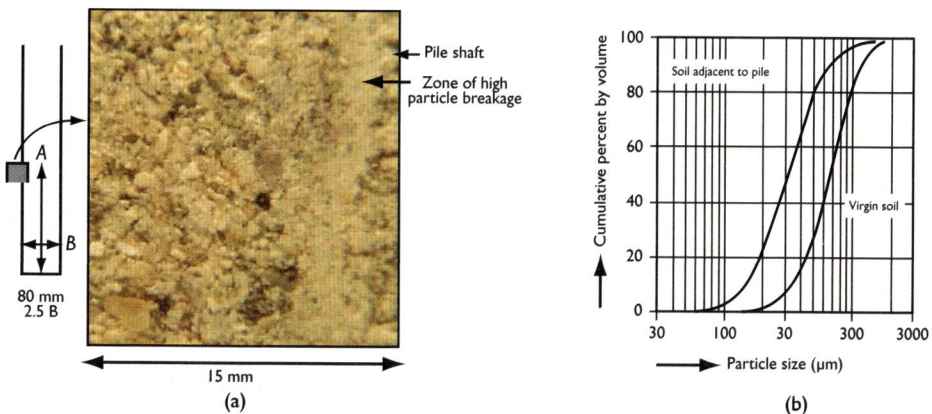

Figure 9.31   *Post-mortem analysis of sand adjacent to pile shaft (White and Bolton, 2002).*

**Soil classification**

When using the CPT for soil classification the following must be taken into consideration (Lunne *et al.*, 1997):

– loose carbonate sands tend to have a low cone resistance and a relatively high friction ratio;
– cemented carbonate sands will have a higher cone resistance, but a very low sleeve friction.

As a result boundaries between various soil classes in generally accepted classification systems (such as, for instance, Robertson, 1990) may have to be reviewed critically. Preferably such a system has to be adapted to the local conditions by comparing the results of boreholes and subsequent sieve analyses with data of cone penetration tests undertaken at the same time in the immediate vicinity of the boreholes.

**Relative density**

Although the relative density as such is not a very useful parameter to describe the volumetric behaviour (and, subsequently, to quantify the friction angle) of a sand, it is still widely used in specifications of land reclamation projects because it can be determined through correlations with the cone resistance (also below the water table). These correlations have been established under controlled conditions in large calibration chambers, but their validity is almost exclusively limited to clean silica sand.

If the relative density is considered to be an important design parameter of a fill mass consisting of compressible or crushable sands and its value has to be determined by cone penetration testing, establishing the unique stress-dependent correlation between cone resistance and relative density of the proposed fill in large calibration chambers is recommended. As follows from the previous sections, testing fill material with a grading similar to the material that will be deposited in the reclamation area and preparing the samples following as much as possible the stress history the fill will experience during construction is essential. Such calibration chamber tests may result in new correlations, but more often correction factors (cone correction factor or shell factor) are derived that can be used to multiply the recorded cone resistance and, thus, allow for the use of existing correlations determined for quartz sands (Wehr, 2005). According to Wehr (2005) these correction factors tend to increase with increasing relative density. Lunne (2006) suggests a less laborious method to establish the cone correction factor: should the compressibility ($Cc/(1 + e_0)$) of the carbonate sand determined by a standardized oedometer test exceed 0.050 a cone correction factor should be applied that depends on the relative density as determined by the Baldi (1986) method. Further reference is made to Section 8.3.5.3 and Appendix C.

Such a correction factor is less commonly used for SPT-N values. Although more difficult to establish, this correction should be applied as well.

**Other correlations (friction angle, stiffness, liquefaction susceptibility, etc.)**
If other parameters have to be derived from the cone resistance or the N-value, correcting for the effects of the crushability of the carbonate sands is important.

As they mostly apply to quartz sands, existing correlations between the cone resistance or N-value and the friction angle will generally underestimate the shear strength of the carbonate fill. The same applies to the correlations between the cone resistance and the stiffness.

In order to provide a timely indication of the material performance, cone penetration testing (or standard penetration testing) is generally undertaken immediately after placement and compaction of the material. It is recognized that as a result of ageing and re-cementation processes the cone resistance or N-value may increase over time. These processes are observed especially in the fill above the water level and will generally result in an increase of strength and stiffness of the carbonate fill mass.

Unfortunately it is not possible to predict the increase of the cone resistance or N-value as a function of time, but occasionally there may be a possibility to repeat cone penetration tests or standard penetration tests at locations of the reclamation area where such tests were carried out previously.

### 9.2.5.3   Laboratory testing

The fragile nature of the particles of carbonate sands and their tendency to cement together may also have its implications for the laboratory testing.

**Sieving and drying**
In order to limit particle breakage and aggregations during sieving the use of the wet sieving method as described in the BS 1377, Part 2 is recommended.

Temperatures at which carbonate sands shall be dried shall preferably not exceed 50°C (air drying or fan-assisted oven drying) in order to avoid aggregation of the particles. If required, aggregations of particles shall be broken down after partial drying without crushing the individual particles (BS 1377, Part 1).

**Determining maximum (dry) density of carbonate sand in laboratory**
Heavy impact stresses as generated during laboratory compaction tests like the BS 1377, Part 4 (Determination of the Dry Density–Moisture Content Relationship, previously known as the Proctor Test), the ASTM D 698 or ASTM D 1557 (Laboratory Compaction Characteristics of Soil Using Standard Effort: 600 kN-m/m$^3$ or Modified Effort: 2700 kN-m/m$^3$) will cause significant

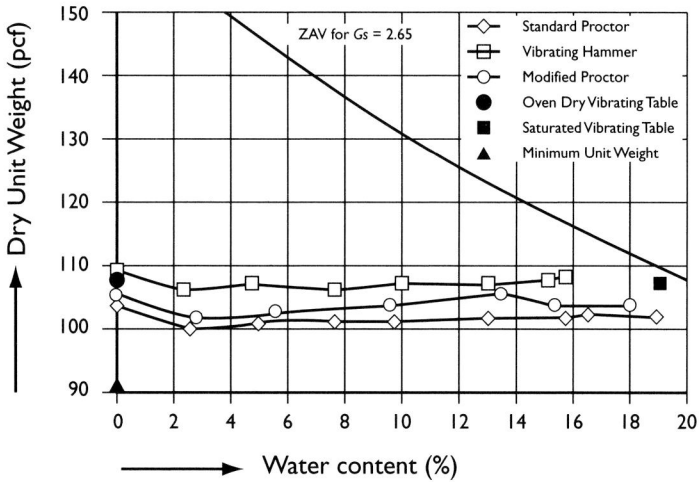

Figure 9.32 *Compaction results for a free draining dune sand. At moisture contents in excess of the tested values water bled from the bottom of the mould (Drnevich et al., 2007).*

particle breakage and, subsequently, in an increasing percentage, fines. As a result of the changed grading the resulting (dry) density will be much higher than the in-situ (dry) density that—even after compaction—can be achieved in the field.

Moreover, the free-draining characteristics of the carbonate sands may result in meaningless values of the optimum moisture content as the moisture content may vary over the height of the compaction mould, water may bleed from the bottom of the mould or the test results indicate more than one optimum moisture content (see Figure 9.32).

Ample experience indicates that as a result of such difficulties the results of these laboratory compaction tests may provide a poor guide for specifications on field compaction.

Should it be required to determine a maximum (dry) density, the replacement of such a test by a less destructive test such as the ASTM D 4253 (Standard Test method for Maximum Index Density and Unit Weight of Soils Using a Vibratory Table) is recommended.

**Triaxial and oedometer testing**
As explained before, the engineering properties of a carbonate fill are closely related to the grading of the sand after dredging, transport and deposition. When

in the design stage of a project samples retrieved from the borrow area are being tested to obtain information on the mechanical behaviour of the fill mass after placement, an estimate has to be made of the grading of the fill after deposition (and compaction, if required). As at present it is not possible to accurately predict the degradation of the fill as a function of material and process parameters, testing a number of samples with varying fines content is recommended. This will allow a better understanding of the influence of the fines on the mechanical behaviour of the fill and may serve as a rational basis for technical specifications with respect to the maximum fines content of the fill.

Oedometer testing on samples of carbonate silts generated during dredging and transport and accumulated in the fill mass as lenses or layers during deposition of the fill has occasionally indicated that the compressibility of the silts is in the same order of magnitude as that of the parent carbonate deposits. This may imply that the presence of accumulations of these silts will not influence the future settlement behaviour of the fill mass. As a result of their relatively high permeability the consolidation period of these fine deposits is generally limited as well.

Their presence could, however, still affect the shear strength of the fill. Whether this is detrimental for the future performance of the reclamation area depends strongly on the shear strength of the fill required at the level of the silt inclusions: if potential shear planes resulting from the superstructures or slopes do not intersect these accumulations, most likely their presence will have no effect on the performance of the fill mass.

### 9.2.5.4   Field compaction

Only limited information is available on the influence of the carbonate content of sand on compaction methods like vibroflotation of dynamic compaction. However, most likely compaction will not just re-arrange the sand particles into a denser state, but will also induce crushing. This crushing will absorb part of the compaction energy. In addition, the angular nature of the carbonate particles will hamper re-arrangement or—in other words—re-arrangement will require more energy.

In line with these considerations experience suggests that in carbonate sands the radius of the column influenced by vibroflotation is smaller than in less angular silica sands of comparable (relative) density.

Particle breakage is also expected as a result of the impacts of dynamic compaction. This is believed to limit the influence depth of the dynamic compaction method.

## 9.3  Hydraulic rock fill

### 9.3.1  Introduction

Capital dredging works often involve the dredging of rocky material. Dredged rock can be used as a high quality fill material.

The use of rock in a reclamation area involves some specific issues and challenges:

– large lump size;
– grading curve is difficult to control and predict;
– compaction is difficult;
– measurement of compaction result is difficult;
– creation of fines is possible;
– high wear of the dredging equipment;
– limited pumping distance;
– reduced liquefaction risk.

### 9.3.2  Lump size

The lump size of the dredged rock is directly related to the dredging technique. In case the rock is very weak to moderately strong it can be dredged by a cutter suction dredge. The rock can be loaded in barges (Figure 9.33), or hydraulically transported to a reclamation area (Figure 9.34).

When dredging rock with a cutter suction dredge, the maximum lump size is generally limited to 30–50 cm. This corresponds with the maximum spherical passage of the dredge pumps and the cutter head.

In case the rock is strong to very strong, pre-treatment by hammering or drilling and blasting might be required. After pre-treatment, the rock can be removed by a Cutter suction dredge or backhoe dredge. In case of removal by a backhoe dredge, the resulting lump size can be considerable (>1 m, Figure 9.35).

### 9.3.3  Compaction and measurement of compaction result

Compaction by means of vibroflotation can generally not be applied, since the vibroflotation probe is not capable to penetrate the rock fill. Dynamic compaction is the only possible compaction technique.

It should be noted that the measurement of the compaction result is difficult. The standard approach used for granular fill, measuring the density as described in

Figure 9.33    *Barge loading with dredged rock.*

Figure 9.34    *Hydraulic fill of rock material (moderately strong lime stone).*

Figure 9.35    *Backhoe: Dredging blasted basalt in the Panama Canal (bucket size = 11 m³).*

Section 8.3.5 and Appendix B) is generally not possible. Often a Zone Load Test is used to monitor the behaviour of rock fill.

### 9.3.4   *Grading*

In case dredged rock is used for construction purposes or as shore protection, a specific grading might be required. It should be noted that it is very difficult (if not impossible) to predict the grading of dredged rock. Since most of the dredge-able rock is of sedimentary origin, it will be rather heterogeneous. Often part of the dredged rock will disintegrate into a sand, silt or even a clay fraction (see paragraph 5 below). Some typical examples of grading curves are presented in Figure 9.36. The rock was dredged with a mega cutter suction dredge and ranged from weak calcerenite to moderately strong limestone.

### 9.3.5   *Fines*

Dredging of rock will often result in the creation of fines. This can be due to the disintegration of the rock (e.g. a weak claystone will partly disintegrate during the process of dredging and hydraulic transport). The top of a rock mass is often weathered. In case of complete weathering this is referred to as a residual soil. Since these layers will frequently be dredged together with sound rock, this results in a considerable volume of smaller fractions.

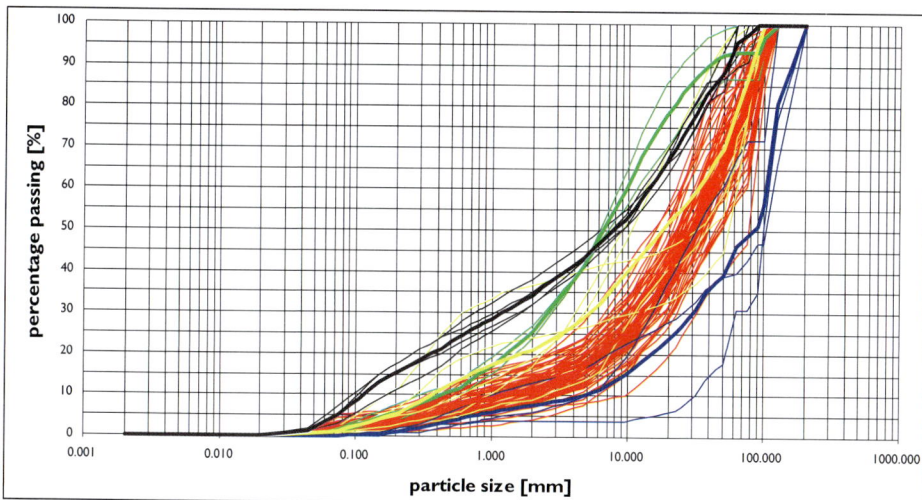

Figure 9.36   *Typical grading curves of dredged rock.*

Figure 9.37    *Wear and tear to cutter head during rock dredging.*

Pockets or voids within a rock mass might be filled with finer material. These fines will be released during the dredging works.

### 9.3.6    *Wear and tear*

Dredging of rock can cause considerable wear and tear to the dredging equipment (Figure 9.37).

### 9.3.7    *Pumping distance during rock dredging*

The maximum pumping distance when dredging rock is rather limited compared to pumping fine to medium sand.

On the 4th graph in figure A-8 in Appendix A, the pumping capacity of some typical vessels for a 10 mm gravel are displayed. The achievable pumping distance when reclaiming rock will be similar to the displayed values on these gravel graphs.

## 9.3.8 *Specifications rock fill*

**Lump size**

Often contract documents contain specifications related to the maximum allowable lump size. This might be useful for the top layers in case future excavations are planned or in case pile foundations will be installed.

As mentioned before, the maximum lump size will generally be limited to 30–50 cm if rock is dredged with a CSD. In case this size is not allowed within the reclamation then additional treatment might be required (e.g. by breaker). This may result in considerable additional cost.

It is therefore recommended to only specify a maximum lump size if that is absolutely necessary. A possible solution could be to specify a maximum lump size for the top layers only, and allow larger lumps in the deeper layers.

**Grading**

In some cases, a specific rock grading is required or specified. In case sand or gravel fractions are not allowed, then the dredged rock will have to be screened. Dredged rock with a broad range in grain sizes is often used as core material for breakwaters.

**Strength**

The unconfined strength of the dredged rock will generally be less than 60 MPa, since this is about the limit for economical dredging with a jumbo cutter suction dredger. Higher strength values generally require the rock to be drilled and blasted followed by excavation with a backhoe (Figure 9.38).

Figure 9.38 *Excavation of drilled & blasted rock and removal of boulders in São Francisco do Sul (Brasil).*

**Density**
Density specifications for rock fill are often useless, since the in situ density within the reclamation is impossible to measure. A Zone Load Test can be used as an alternative. This type of test has the advantage that the actual performance of the fill (bearing capacity and settlement) is measured directly.


## 9.4   Problematic subsoils

Some materials show different behaviour because of their varying mineral content, proportions of constituent minerals, environment, process of formation, and subsequent geological history. Some materials have properties which create special engineering difficulties. Some typical 'problem' soils which are discussed in this section are:

- sensitive clay;
- peat;
- glacial soils;
- sabkha;
- karst;
- laterite.


### 9.4.1   *Sensitive clay*

**Introduction**
The presence of sensitive clay in a reclamation or borrow area involves some specific issues and challenges:

- behaviour is difficult to control and predict;
- high liquefaction risk;
- sudden loss of strength;
- relatively high cost for precautionary and remedial measures;
- extensive laboratory testing inevitable.

Sensitive clays can be very hazardous. Disturbances caused by dredging, construction activities or changes in chemical (e.g. salt) content of the groundwater can, for instance, result in a sudden loss of strength of sensitive clay. This may trigger a soil mass instability responsible for damage to equipment, structures or even loss of life.

**Sensitive and quick clay**
The sensitivity of clays is defined as the ratio of their undisturbed and remoulded strength, and varies from about 1.0 for heavily over-consolidated clays to values

of 50–100 for the so-called extra sensitive or quick clays (Skempton and Northey, 1952).

The more sensitive a clay, the more difficult the engineering problems it imposes. In some cases, and especially in the case of quick clay, the sensitivity is very high. Undisturbed quick clay resembles a water-saturated gel (Wikipedia). When a mass of quick clay undergoes sufficient stress, however, it instantly turns into a flowing ooze, a process known as liquefaction. Quick clay behaves this way because, although it is solid, it has a very high water content, up to 80%. The Plastic Limit as well as the Liquid Limit are relatively high indicating a low Plasticity Index which is not necessarily the case for sensitive clays in general. The quick clay retains a solid structure despite the high water content, because surface tension holds water-coated flocs of clay together in a delicate structure. When the structure is broken by a shock, it reverts to a fluid state. External construction activities, like pile driving or excavation, may trigger mass instability (Price, 1986). Both in field and in laboratory conditions clear evidence is present that slopes in over-consolidated clays may fail at specific static stress conditions as a result of viscous effects. Several case histories of failures of long-term slopes in over-consolidated clays are reported by Bjerrum (1967), while Tavenas et al. (1978), using stress path testing in the laboratory, related the acceleration of viscous effects and subsequent failure to the position of the stress state relative to the failure line.

Quick clay is defined as clay with a sensitivity of 50 or more having a remoulded shear strength of less than 0.4 kPa (Karlsson & Hansbo, 1989). The latter value corresponds to a penetration of 20 mm by the 60 g cone with 60° tip angle in the fall-cone test. Quick clays involve considerable risks, for example, in connection with stability problems where small initial slips may result in large landslides involving the entire quick clay formation. The liquefied material can flow over a large distance causing extensive damage to anything in its path. Quick clays are common in those parts of the world that have geological and climatic conditions similar to those in Sweden, Norway, Russia and Canada.

Quick clay is formed through slow geological processes (Swedish Geotechnical Institute, 2004). Most quick clays have been formed in sediments that were deposited in sea water at the last deglaciation. The land then heaved as the inland ice retreated, leaving the clay deposits located above sea level. The clay deposits have then been subjected to leaching, whereby the ion concentration in the pore water has changed. The leaching has been caused by infiltration of rain water, artesian water pressures in underlying permeable soil or rock, and by diffusion. These processes are very slow and entail that quick clay is found more often in clay deposits with moderate thickness and less frequently in thick deposits, where it only occurs close to permeable layers and the ground surface. However, there are also quick clays that have been deposited in brackish and sweet water. Through contact with organic substances in peat and other humus-rich soils for example, the ion concentration in the pore water may change and the clay becomes a quick clay.

**Precautionary and remedial measures**

Control of sensitive clays is very difficult to achieve, apart from the obvious precaution of avoiding sources of disturbance like pile driving, excavation, etc. Exceeding a certain (peak) shear strength may result in a considerable loss in strength. Most measures focus on damage control rather than measures related to the clay sensitivity itself. Possible measures include:

- reduction of the loading gradient (apparent slope) in these soils;
- buttresses of earth to protect vulnerable areas against clay flow;
- excavation and replacement by more stable soil;
- ground improvement by means of the installation of chalk-cement piles and/or electro-osmosis.

**Consequences**

Knowledge of the presence of sensitive clays at or near borrow and/or reclamation areas in the planning phase of a project is clearly of great importance. The selection of these areas depends on the risks imposed by these sensitive clays and the cost for precautionary or remedial measures. Since the precise combination of factors that determine the sensitivity of clay is still unknown, occurrence of sudden strength loss is difficult to predict. Therefore, laboratory tests are necessary when the presence of sensitive clays is likely. These tests include the determination of the strain rate dependent strength behaviour (Leroueil *et al.,* 1985), mineralogical and chemical content, the moisture content and the particle size distribution.

The presence of sensitive clays may have a significant impact on the design of the reclamation and the foundations of the future structures. Further, due attention shall be paid to an appropriate execution method for the dredging and reclamation works. Engineers should be aware of the consequences preferably already during the initial design stage of the project.

### 9.4.2   *Peat*

**Introduction**

The presence of peat underneath a reclamation area involves some specific issues and challenges:

- highly compressible material;
- need for settlement compensation if left in place;
- long-term creep deformation;
- risks for loss of stability;
- staged construction required;

– ground improvement measures inevitable;
– laboratory tests not reliable.

Peat is a highly compressible material with specific challenges when occurring underneath the reclamation. Although removal of these soils may be regarded as the most convenient solution, both environmental and economic consequences would make this less attractive. Furthermore, at the boundary of the excavation where peat layers are still present, horizontal loading of these layers and stress redistribution may result in progressive failures. Peat is formed during decomposition of dead organic substances like remnants of plants and animals. The formation is stimulated by high temperatures, suitable humidity and access to oxygen from the air. The variety of stems, leaves, biological matter and biochemical circumstances are the main causes for the natural heterogeneity of these soils. Peat formed in early post-glacial times may occur at depth, buried beneath more recent sediments. Potential geotechnical problems are related to the bearing capacity and the compression and consolidation and creep characteristics of this material, which are difficult to define. Moreover, accurate predictions of the strength and settlement are complex because of the heterogeneity and anisotropy of the material. A more detailed description of the material behaviour and constitutive modelling of peat and other organic soils is presented in Mathijssen *et al.* (2009).

**Remedial measures**
Peat can best be removed and replaced before the construction of a landfill, especially along the boundary of a reclamation. However, as indicated above, special attention is required to prevent local failures at the excavation boundaries by keeping the loading gradient on these soils as low as possible.

Alternatively, the drainage capacity, and consequently the strength, can be improved by installing vertical drains. Also stone columns or piles may be installed to transfer foundation loads to more competent layers underneath the peat.

If sufficient time is available for the construction of a land reclamation on peat, the fill may be placed to a temporary higher level than actually required to allow for time-dependent deformations introduced by consolidation and viscous effects of the material resulting in reduction of deformations after construction (Samson and La Rochelle, 1972). With sufficient surcharge, peat may compress to less than 25% of its original volume.

Areas covered with peat even experience settlements without being loaded. This phenomenon of volume reduction is a consequence of oxidation of the material and is called autonomous settlement. In The Netherlands this settlement is estimated to be approximate 0.01 m/year.

**Consequences**

The potential presence of peat in the subsoil at a reclamation site must be investigated in an early phase of the design. Aspects such as the required volume of fill material (not only to reach the specified platform level, but also to compensate for settlements and to be used as a surcharge), the stability of slopes, the influence on adjacent structures, facilities, services, etc., the foundation depth of structures and the necessity of remedial measures have a great influence on the design, planning and cost of the reclamation. Although extensive research has been performed on the topic of deformation of peat and organic soils, autonomous viscous effects, biochemical degradation and temperature effects are not yet incorporated in state-of-the-art isotache models (for calculating settlements) leaving a significant uncertainty.

If peat underneath a site to be reclaimed is not removed then serious geotechnical problems can be expected. Due attention shall be paid to the consequences thereof, for as well the design as for the execution of the reclamation works and the foundations of the structures planned to be constructed on the new land.

### 9.4.3   Glacial soils

**Introduction**

The use of glacial soils in a reclamation area involves some specific issues and challenges:

- presence of voids;
- absence of grain sorting;
- large variations in mechanical properties;
- unpredictable behaviour;
- risks for geotechnical problems.

Because of their unpredictable composition and mechanical properties, glacial soils are difficult to use as construction material. Also their presence in the subsoil underneath a land reclamation project may result in problems with foundation works and unexpected development of voids during or even after construction.

**Till and fluvio-glacial deposits**

There is a very large range of glacial soils. Their nature varies according to their constituents, the way of transport and the environment of deposition. Generally, two types of glacial soils can be discerned (Price, 1986):

- till: deposited directly by the ice (i.e., simply dropped);
- fluvio-glacial soils: deposited by meltwater near the ice front or at greater distance.

Tills are sometimes referred to as 'boulder clay' although this does not always imply that these soils contain boulders or clay. Till is deposited by ice and characterised by the absence of both grain sorting and stratification. Tills are variable in nature and consequently also in their mechanical properties. In general they consist of an assortment of rock debris, ranging in size from rock flour (very fine) to boulders. Gap grading is usual.

Fluvio-glacial soils have been transported and deposited by meltwater and are therefore more sorted and stratified than tills. Many of such deposits consist of gravel and sand characterised by low relative densities. Pockets of very low density can generate geotechnical problems in foundation works. Also voids may appear during or even after construction, possibly caused by melting of ice blocks that have been deposited within the gravel (Price, 1986).

**Remedial measures**
Voids in fluvio-glacial soils are difficult to locate, and this problem is usually anticipated and dealt with by assuming their presence and using compaction equipment before the start of construction.

**Consequences**
Glacial soils should be treated with great caution when used as fill material or when present in the subsoil underneath a reclamation site. Mechanical properties are difficult to predict and should therefore be analysed in the laboratory. The presence of gravel pockets and possible voids in the subsoil must be taken into account during the foundation design.

9.4.4   *Sabkha*

**Introduction**
The presence of sabkha in a reclamation area involves some specific issues and challenges:

- potential for collapse;
- strength decrease due to rainfall;
- impassable for traffic when wetted;
- large variations in mechanical properties;
- risks for geotechnical problems.

There is no consensus on the precise definition of the term sabkha. The expression sabkha (spelled in some publications as sabkhah, subkha, sebkha, etc.) is originally an Arabic name that has been used to describe saline flats that are underlain by sand, silt or clay and often encrusted by salt. From a broader perspective, sabhka flats are distinguished as being large (in dimensions), flat (gentle slopes), salt-encrusted, evaporate terrains (the rate of evaporation is significantly higher than that of pre-cipitation) situated either along the coast or further inland (Abduljauwad and Al-Amoudi, 1995). The vast bulk of the evaporites are sulphates (gypsum and anhydrite), chlorites (halite) and carbonates.

When such deposits are encountered further inland they are generally called salina or salt playa which is schematically shown in Figure 9.39 below.

The deposits can generally be described as follows (Nooy van der Kolff, 1993):

- Sabkha:       Gently dipping salt flats along the coast line, consisting of alter-nating layers of algae matts, clastic materials and evaporites; accumulation by evaporation of groundwater and seawater dur-ing periodical inundation; salts originating from seawater.
- Salina:       Salt pans usually occurring in depressions near the capillary fringe; capillary rise of groundwater is the main accumulating medium; salts are originating from evaporitic deposits included in the bedrock.
- Salt playa:   Inland salt pans usually occurring in depressions; (rain)water and wind are the main accumulating media; salts are originating from evaporitic deposits included in the bedrock: sulphates (gypsum, anhydrite), chlorites (halite), carbonates.

Sabkha's are most often found in hyper-arid to semi-arid areas of the world. Today arid areas form two belts between 15° and 45° north and south of the equa-tor. Climatic, geochemical, hydrological and geomorphological factors play dif-ferent roles in the genesis of the sabkha system. The interaction of these factors

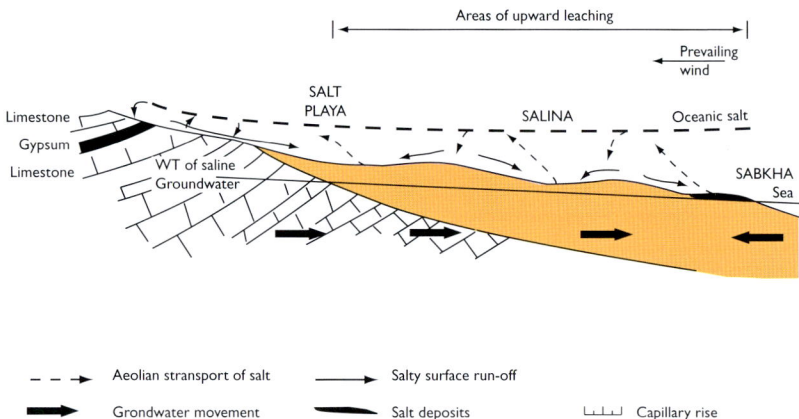

Figure 9.39   *Occurrence of sabkha, salina and salt playa (Fookes, 1985).*

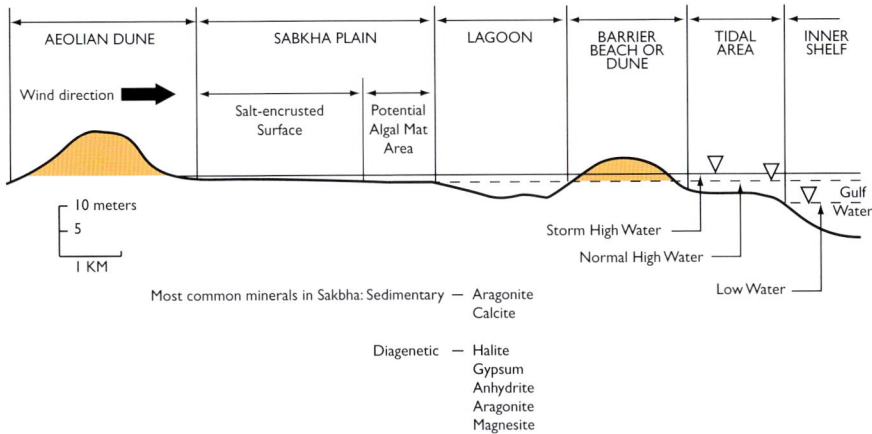

Figure 9.40 *General cross-section of a typical coastal sabkha with surface features.*

will result in processes and reactions, which characterise the sabkha environment (Abduljauwad and Al-Amoudi, 1995). Figure 9.40 which can be considered to be a detailed part of Figure 9.39 above gives a general cross-section of a typical coastal sabkha area.

Sabkha's have generally been considered to be unconsolidated, heterogeneous, layered or un-layered sedimentological frameworks, bathed in highly concentrated subsurface brines. They normally have a loose, rather porous and permeable, gritty structure (Al-Amoudi *et al.* 1991).

The sedimentary features, the mineralogical composition and the chemistry of the interstitial brines in coastal sabkha's vary greatly in both horizontal and vertical directions.

If an area covered with sabkha has to be reclaimed one should be aware of the following hazards:

– if the salt encrusted surface of the sabkha is removed and "fresh" sabkha sediments are exposed to fresh (waste)water then a decrease of strength and bearing capacity may be expected. Collapsing soils can possibly occur caused by the induction of fresh water (see Figure 9.41);
– a strength decrease of the surface crust of the sabkha as a result of rainfall, flash floods, storm tides or the absorption from a humid atmosphere. This decrease in strength can render the normally stable surface crusts impassable for traffic;
– a potential variation of the compressibility characteristics of sabkha sediments;
– hydration and dehydration of gypsum. The associated volume changes and pressures may cause problems;
– attempts to densify the upper loose portion of sabkha material by conventional means could break up cementation bonds resulting in lower bearing capacities;

Figure 9.41 *Collapsed soil as a result of the intrusion of fresh water (C. de Rooij, Azerbaijan).*

– induction of fresh (waste) water into sabkha should be done carefully since it could dissolve some of the cementing agents resulting in a strength reduction of the material while possibly introducing settlements.

**Presence of salts**

A distinguishing feature of a sabkha is the presence of salt. Normally the salinity of groundwater rapidly increases in a landward direction, with the result that the creation of salt crystals, caused by evaporation and desiccation, can form relatively hard crusty surfaces. The salt encrusted surfaces are sufficiently strong and durable. They become impassable upon wetting from rainfall or storm tides. In general a soil with a salt content equal to or more than 0.3% can be defined as a saline soil (Naifeng 1994). Salt can also be transported by wind.

The first mineral to precipitate is aragonite, followed by gypsum and halite, and finally the highly soluble magnesium and potassium salts. Evaporation can be a major source of cementation that appears to hold sand and silt particles together to form cemented layers and cemented zones, particularly in the loose portion of the sabkha near the surface.

**Prediction of collapsibility of sabkha sediments**

Most sabkha sediments will not collapse upon wetting. A prediction can be made by pouring fresh water over sabkha lumps. If after 20 minutes the lumps have completely disintegrated (fallen apart) then the soil might be prone to collapse.

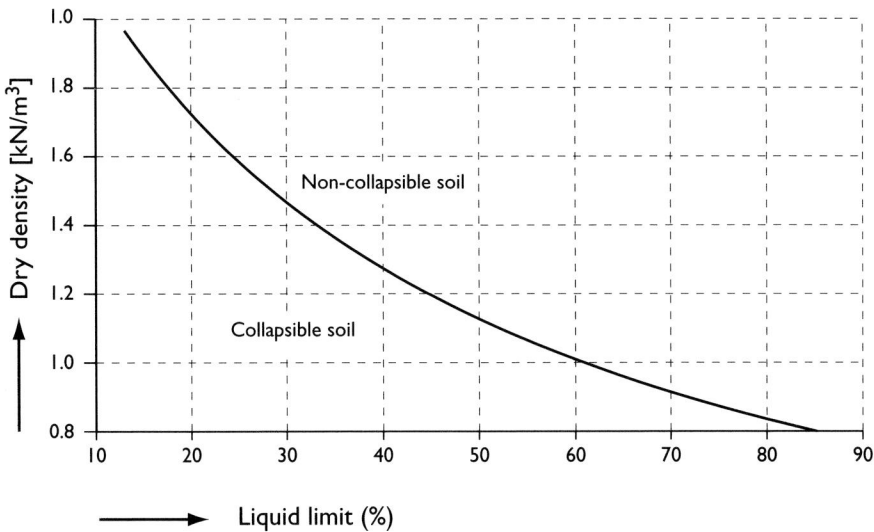

Figure 9.42   *Collapsible and non-collapsible soil as a function of dry unit weight and liquid limit, after Gibbs 1961.*

In the literature several collapse indices are given. Most frequently the collapse index of Gibbs (1961) is used. Gibbs used the dry density and liquid limit as criteria to distinguish between collapsible and non-collapsible soils (see Figure 9.42). His method is based on the premise that a soil, which has a high enough void space to hold its liquid limit moisture content at saturation, is susceptible to collapse on wetting.

### 9.4.5   *Karst*

**Introduction**
The presence of karst in a reclamation area involves some specific issues and challenges:

– dissolution of bedrock;
– underground caves and channels;
– sinkholes;
– potential for collapse;
– need for extensive site investigation;
– risk assessment required.

About 10% of the Earth's land surface consists of limestone, which can easily be dissolved by mildly acidic groundwater. When groundwater containing dissolved $CO_2$ percolates into limestone (see Figure 9.43) it will dissolve the rock limestone

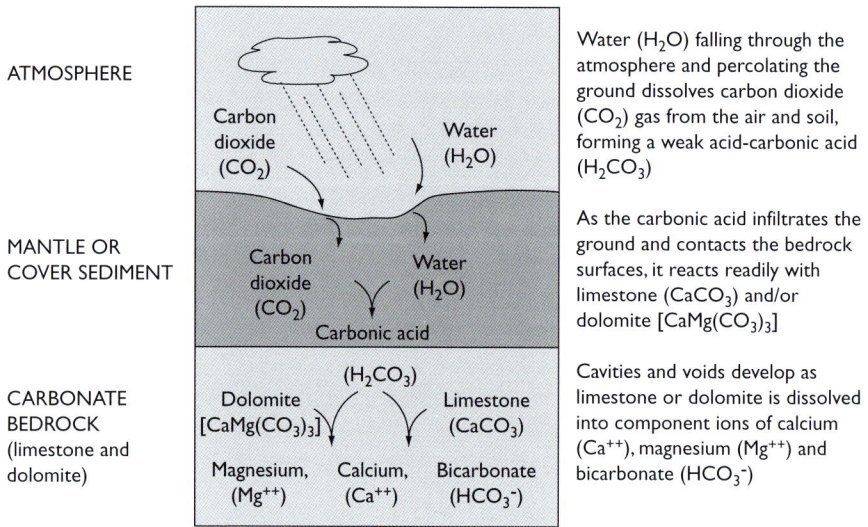

Figure 9.43    *Dissolution process of soluble carbonate rocks by weakly acidic water (Galloway et al., 2005).*

Figure 9.44    *Karst landscape near Minerve, Hérault, France [www.en.wikipedia.org/wiki/ Karst#Morphology].*

along existing joints, fractures and other water bearing discontinuities. This will result in a karst topography comprising caves, underground channels and a rough and bumpy ground surface (see Figure 9.44). Karstification may result in a variety of large and small features that occur at and below the ground surface.

Karst topography is named after the Dinaric Kras Region of western Slovenia and eastern Italy [Ref. www.geography.about.com/od/physicalgeography/a/karst.htm] where the first scientific studies of these phenomena were undertaken. Karst is primarily associated with limestone, but may also be encountered in other carbonate and soluble rocks (gypsum, anhydrite).

Karstic landforms vary enormously in character, shape and size and are formed in wet climatic conditions at present or in the geological past. When developed in the past, surface features may be buried and hidden by more recent sediments deposited in a changed climatic environment: e.g., in the Middle East, currently known for its hyper-arid conditions, widespread karstification has occurred during the wet Middle Pleistocene period (325,000–560,000 years ago) affecting the Eocene limestones (Sadiq *et al.*, 2002). The presence of dissolution cavities and other karst-related hazards may not always be visible at the ground surface, but may have to be identified by site investigations specifically aimed at karstic features.

**Hazards**

Typical hazards associated with karstified landforms are sinkholes (dolines), i.e. depressions that may develop gradually or suddenly as a result of dissolution of the bedrock and subsequent erosion of the unconsolidated overlying deposits into the dissolution cavities. Their occurrence is often related to a change of the groundwater regime. Sinkholes can range in size from a few metres to over 100 metres deep. They have been known to "swallow" cars, houses and other structures. [Ref. www.geography. about.com/od/physicalgeography/a/karst.htm]. A sinkhole can even collapse through the roof of an underground cavern and form what is known as a collapse sinkhole, which can become a portal into a deep underground cavern (see Figure 9.45).

Figure 9.45    *Massive sinkhole in Guatemala.*

Karst may also create pinnacled bedrock surfaces covered by unconsolidated deposits that could cause excessive differential settlements of shallow foundations. When structures are founded on deep foundations, pile driving may result in bending or fracturing of the piles and in strongly varying depths of the pile tips.

Another potential hazard of karst is the presence of solution cavities of various dimensions in the bedrock. Depending on the strength of the rock, the thickness of its roof and the load on top of the roof, such a cavity may collapse when loaded by fill or when penetrated by piles.

### Site investigation, risk and hazard assessment

In order to check whether a future reclamation site may be affected by karst, a systematic investigation is required. This investigation should start with a desk study as in many parts of the world national geological surveys have studied and often mapped karst areas (Galloway *et al.,* 2005). If this study indicates that the site may be located in a karst area, a site investigation should follow to confirm these findings. This site investigation should not only attempt to identify the presence of karstic features, but also determine the locations of any voids or caves in the ground, the properties and character of the relevant soil and rock masses, the rock head configuration and the hydro-geological conditions. The rock structure is important as dissolution voids are normally enhanced along fracture zones and at the intersections of discontinuities, while soil properties can indicate the susceptibility and characteristics of potential subsidence sinkholes (Waltham *et al.,* 2005). In karstic areas both intrusive (i.e., boreholes, probing) and non-intrusive (geophysical surveys) site investigation methods should be used. Meteorological information may also be useful, since sinkholes usually develop after heavy rainfall or prolonged water table decline. The design of the reclamation should be based on the results of site investigations, must include a risk assessment and should propose potential ground improvement schemes or remedial design methods, should this be required.

### Prevention and remediation of sinkholes

If significant karstic features have been identified in an initial stage of the project, relocating the site might be preferable. This, however, may not always be possible as karst usually develops over a larger region.

Several soil treatments and remedial works can be identified for land reclamation projects. Most of these treatments should be accompanied by an appropriate drainage control to prevent (further) development of sinkholes.

For soil treatment to prevent sinkhole development the following measures can be considered (Waltham *et al.* 2005):

- Permeation grouting with bulk grouts which are injected through boreholes into the cavities. Large grout losses may occur.

- Compaction grouting over cavernous karst. Injection of a stiff grout at high pressures will displace and compact the surrounding soils.
- Cap grouting. Constructing a concrete cap over the potential sinkhole by injecting a thick, mortar-like grout. Often carried out in combination with compaction grouting of the soils encountered above the cap.
- Dynamic compaction by a heavy weight. Compaction causes subsurface voids to collapse. This technique does not necessarily eliminate further development of sinkholes.
- Preloading. A surcharge may prevent differential settlement caused by, for instance, infilled sinkholes and pinnacled bedrock.
- Lime treatment or cement stabilisation.

Sinkhole remediation may be undertaken by one of the following methods (Waltham *et al.*, 2005):

- Filling of the sinkhole using a graded-filter technique. This technique allows the infiltration of water into the karst conducts without erosion of the soil cover. This can be achieved by constructing a filter in the outlet of the sinkhole.
- Filling the sinkhole after having completely sealed the outlet at the base of the sinkhole by a concrete plug or a reinforced concrete cap. An adequate control of the surface drainage is essential to avoid water seeping through and creating new outlets.

### 9.4.6 *Laterite*

**Introduction**
The use of laterite in a reclamation area involves some specific issues and challenges:

- creation of fines by "over-handling";
- particle breakdown by "over-compaction";
- dusty during dry weather conditions;
- scour effects during wet weather conditions;
- need for extensive site investigation;
- risk assessment required.

Laterites and lateric soils form a group comprising a wide variety of red, brown, and yellow, fine grained residual soils of light texture as well as nodular gravels and cemented soils. They vary from a loose material to a massive rock and usually contain all size fractions up to the gravel size. Laterites often behave as aggregates rather than soils in the conventional sense, since there is not a severe loss of strength with the addition of water (Price, 1986 and U.S. Army, 1997). The reason for this is that the large particles are not separated or lubricated by moisture.

Laterites and lateric soils are formed under tropical climatic conditions and are characterized by an abundance of iron and aluminium oxides and hydroxides as major constituents. These constituents are the end products of intensive and long lasting weathering of rocks and removal (laterization) of their silicates through hydrolysis and oxidation. The variety in colour depends upon the amount of iron oxides present (Price, 1986 and U.S. Army, 1997).

Lateric surface formations are found in hot and wet tropical areas, seated in the equatorial regions of the world. Laterite soil is usually infertile. Examples of finding places are Brazil, Nigeria, Malaysia and Hawaii (Price, 1986). Lateric covers have mostly a thickness of a few metres but they can, occasionally, be much thicker. Laterite covers are often found on top of acidic rock (i.e. granites, granitic gneiss but also many sediments such as clays and sandstones), whereas the softer lateric soils are formed on rocks that are free of quartz (i.e. basalt, serpentinite and such) (Schellmann, 2007).

The usual methods of soil classification, involving grain-size distribution and Atterberg limits, should be performed on laterites or suspected laterites that are anticipated for use as fill material. Because some particles of laterite crush easily, the more the material is handled, the finer the aggregates become.

Figure 9.46    *Characteristic laterite formation on top of basaltic rock.*

When laterite is exposed to air or dried out (i.e. by lowering the ground water table) irreversible hardening often occurs, producing a material suitable for use as road aggregate (U.S. Army, 1997).

Lateric soils behave more like fine-grained sands, gravels, and soft rocks. Some particles of laterite tend to crush easily under impact, disintegrating into a soil material that may be plastic. Lateric soils may be self-hardening when exposed to drying, or if they are not self-hardening, they may contain appreciable amounts of hardened laterite rock or laterite gravel. It is difficult to recognize the material during the execution of a soil investigation.

**Softening and clogging**
The following engineering problems can be expected when constructing with laterite soils:

- Softening:    The laterites can, because of their structural strength, be very suitable building materials for subgrades and (sub) base course. Care should be taken to avoid particle breakdown from "over-compaction". Although laterites are resistant to the effects of moisture, there is a need for good drainage to prevent softening and breakdown of the structure under repeated loadings.
- Clogging:    Laterite can provide a suitable low-grade surfacing for roads when it can be compacted to give a dense, mechanically stable material. However, it tends to corrugate (formation of parallel ridges) under road traffic and becomes dusty during dry weather. In wet weather conditions, it scours and tends to clog the drainage system. Fine fragments are trapped and fill the open pores of the material, resulting in reduction of permeability. The lateritic soils, being weaker than the laterites, are not suitable for use in wearing courses (the top layer of roads) unless for limited use.

385

# OTHER DESIGN ITEMS

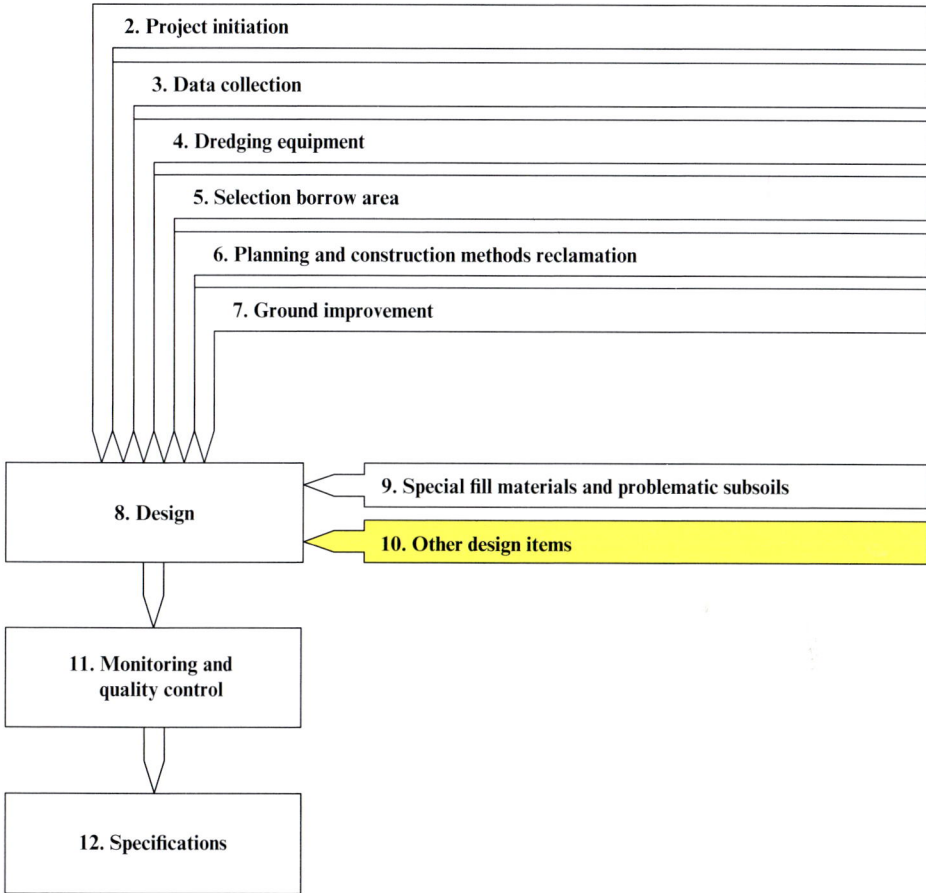

```
┌─────────────────────────────────────────────────────────────────┐
│ 2. Project initiation                                            │
│  ┌────────────────────────────────────────────────────────────┐ │
│  │ 3. Data collection                                         │ │
│  │  ┌─────────────────────────────────────────────────────┐   │ │
│  │  │ 4. Dredging equipment                               │   │ │
│  │  │  ┌──────────────────────────────────────────────┐   │   │ │
│  │  │  │ 5. Selection borrow area                     │   │   │ │
│  │  │  │  ┌────────────────────────────────────────┐  │   │   │ │
│  │  │  │  │ 6. Planning and construction methods   │  │   │   │ │
│  │  │  │  │    reclamation                         │  │   │   │ │
│  │  │  │  │  ┌──────────────────────────────────┐  │  │   │   │ │
│  │  │  │  │  │ 7. Ground improvement            │  │  │   │   │ │
```

| 8. Design | 9. Special fill materials and problematic subsoils |
| | **10. Other design items** |

| 11. Monitoring and quality control |

| 12. Specifications |

Chapter 10 Other Design Items

10.1 Introduction

10.2 Drainage

| 10.2.1 Infiltration | 10.2.2 Surface runoff | 10.2.3 Artificial drainage systems |

10.3 Wind erosion

10.4 Slope, bank and bed protection

10.5 Interaction between reclamation and civil works

| 10.5.1 General | 10.5.2 Foundations | 10.5.3 Construction sequence | 10.5.4 Impact on existing structures |

10.6 Earthquakes

10.7 Tsunamis

## 10.1 Introduction

The main design aspects of hydraulic fills are discussed in the previous chapters which mainly focus on the geomechanical properties of the fill mass and subsoil. There are, however, more design aspects such as drainage, wind erosion and slope protection that require attention as well. Although this Manual will not discuss these aspects in detail, an overview and relevant references will be presented.

## 10.2 Drainage

Inadequate drainage may seriously hamper the use of a reclamation as water from natural precipitation (rain, melting snow and ice), overtopping or other sources may cause problems such as ponding, flooding, erosion, and inaccessibility. Also during the construction phase process water may adversely affect the mechanical properties of the fill mass (in particular the accessibility for construction equipment) should the drainage capacity of the placed fill be insufficient.

During the design stage of a reclamation, the need for a drainage system must be determined as it could have an important impact on the final lay-out and costs of the facility. Such a need strongly depends on the anticipated intensity and amount of precipitation and/or overtopping, the nature of the development (housing area, nature development, industrial estate, etc.) and on the natural drainage capacity of the fill mass itself.

Natural drainage is defined as the infiltration of water from the surface into the fill mass (subsurface drainage) in combination with (uncontrolled) surface runoff (surface drainage) (see Figure 10.1). If the intensity of the precipitation (and/or overtopping) exceeds the infiltration capacity of the fill mass, ponding and, subsequently, natural surface runoff will occur.

Should the anticipated amount of ponding and/or (uncontrolled) surface runoff be considered unacceptable, then an artificial drainage system may be necessary. This usually comprises a number of special measures that can be taken to improve the infiltration and/or to control the ponding and surface runoff in order to transport the water to a suitable location where it can be discharged into open water (river, sea) without affecting surrounding facilities.

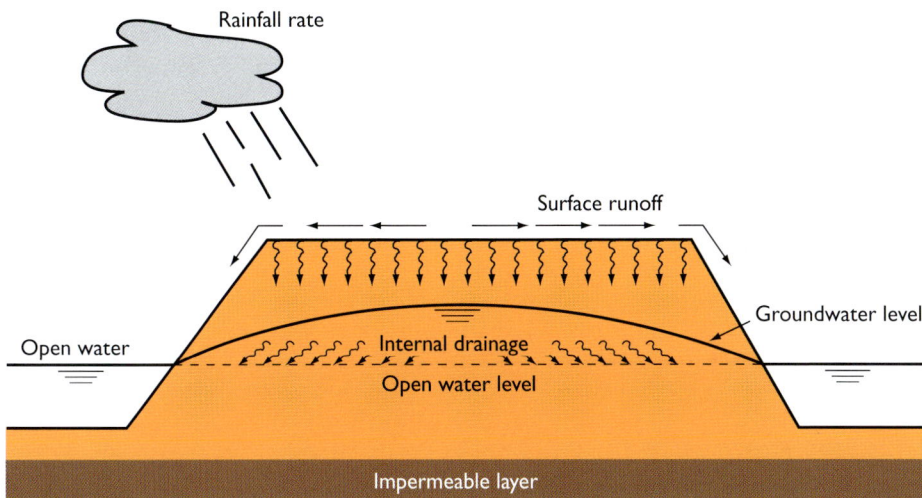

Figure 10.1   *Natural drainage.*

### 10.2.1 Infiltration

Infiltration is the process by which water at ground surface enters the subsoil. The infiltration capacity depends on various fill mass properties such as permeability and storage capacity.

**Permeability**

The permeability or hydraulic conductivity of the fill must be large enough not only to allow for percolation of water through the unsaturated zone, but also to ensure sufficient groundwater flow (internal drainage) to prevent the water from rising to an unacceptable level.

Compaction of the fill will generally result in a reduction of the permeability and, hence, of the infiltration capacity. In moderate and (sub)arctic climates the upper zone of the fill may be frozen during certain periods of the year. This will cause a strong reduction of the permeability near the surface which will hamper the infiltration capacity. In particular at the end of a frost period (melting of snow and ice), this often results in ponding and an increase of the surface runoff.

As a result of the hydraulic placement method a thin crust of fine-grained, aggregated soil can develop at the fill's surface that may act as a low-permeable seal, hindering natural infiltration and promoting ponding and surface runoff.

The most important parameters that control the permeability are the grain size distribution and the density of the fill. A first estimate of the permeability can be obtained from the particle size distribution by empirical formulae proposed by Hazen (1911), Kozeny-Carman (Carman, 1956) and others. In particular the fines content has a strong influence on the permeability.

**Storage capacity**

The storage capacity is defined as the volume of pores in the unsaturated zone that is available to store infiltrated water. It mainly depends on the grain size distribution, the density and water content of the soil and the depth to the groundwater table.

A rising groundwater table or compaction of the fill will reduce the storage capacity and, hence, the infiltration capacity. A high natural water content of the fill mass will also adversely affect the storage capacity.

There are various methods to calculate the infiltration capacity. Reference is made to Mays (2005).

### 10.2.2 Surface runoff

When the water supply as a result of precipitation, overtopping, etc. exceeds the infiltration capacity of the fill then initially ponding will occur in depressions at the surface of the reclamation. Once the depression storage is filled, surface run-off will start, i.e., the excess water will flow over the surface to lower ground levels, eventually draining into streams, rivers or the sea.

The surface runoff rate depends on many variables like the intensity of the precipitation, the infiltration capacity of the fill, evaporation and transpiration, surface cover (vegetation, pavement, buildings, etc.) of the reclamation and the gradient and roughness of the surface.

To estimate the amount of runoff, predictive computer models of varying complexity are used. Reference is made to TR-55, U.S. Department of Agriculture (1986) and Chow *et al.* (1988). Surface runoff is usually associated with flooding and erosion and should therefore be limited or, at least, controlled.

### 10.2.3 *Artificial drainage systems*

An artificial drainage system generally consists of:

- A field drainage system draining the water away from the fill area. It can be divided into surface and subsurface drainage systems intended to increase the infiltration capacity and/or to control the excess surface runoff.
- An external (main) drainage system transporting the water from the fill area to a point of discharge where the water is being disposed of into the surrounding natural water system through an outfall or outlet.

Depending on the groundwater table and the level of the open water in which the drainage water has to be discharged, pumping facilities may be needed.

**Design process**
The design process of an artificial field drainage system of a reclamation area may comprise the following steps:

- data collection: precipitation and overtopping (intensity, duration), fill mass properties after deposition, temperature, lay-out of planned developments, etc.;
- assessment of the capacity of the natural drainage system of the future fill mass (i.e., the capacity of natural infiltration and allowable surface runoff) and, subsequently, the need for an artificial drainage system;
- design of an artificial field drainage system to improve the infiltration capacity and/or to control the groundwater table and surface runoff taking into account the lay-out of the future development which may include pumping facilities;
- design of an artificial external drainage system of sufficient capacity to transport the water from the reclamation area to the discharge location which may also include pumping facilities.

Special features are wadi's. A wadi is a river valley present in dry areas that is dehydrated throughout most of the year. However, during wet periods and heavy rainfall water flows through the wadi. This may happen unexpectedly and these events are called "flash floods".

Note that land reclamations are often constructed along existing shores and banks and they may therefore block existing outlets/outfalls of systems draining the hinterland.

### Artificial field drainage system
An artificial field drainage system is generally designed to improve the infiltration capacity and/or to control the surface runoff.

The infiltration capacity of a fill can be improved by:

- the construction of a surface drainage system including (deep) open drains (ditches, channels, etc.) to control the groundwater table and, hence, the storage capacity (see Figure 10.2);
- the installation of a subsurface drainage system consisting of prefabricated horizontal drains (pipe drains) to control the groundwater table and, hence, the storage capacity (see Figure 10.3);
- growing of vegetation to increase the permeability of the upper zone of the fill mass;
- the construction of retention basins.

It is important to realise that future developments of a reclamation (urbanisation, industrialisation, etc.) often reduce the infiltration capacity of the fill mass as the presence of buildings, pavements and other constructions will seal (part of) the fill surface. In addition, surface compaction to provide sufficient bearing capacity or to restrict residual settlements may lower the infiltration capacity as well.

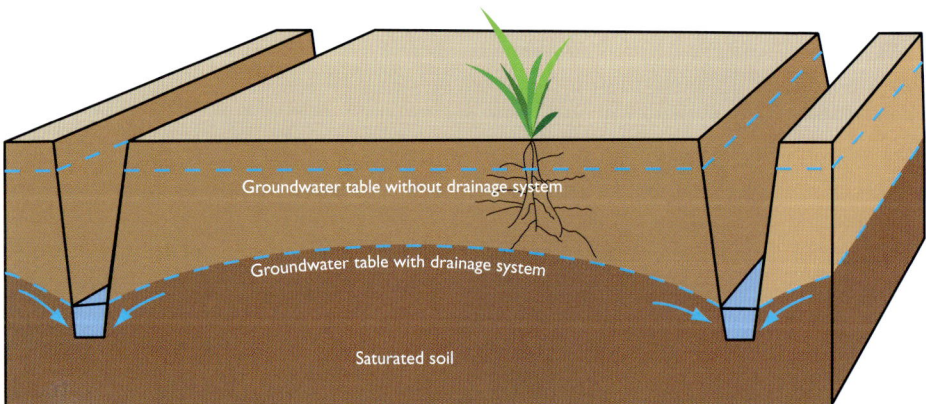

Figure 10.2 *Control of the groundwater table by means of deep open drains (ref.: www.fao.org).*

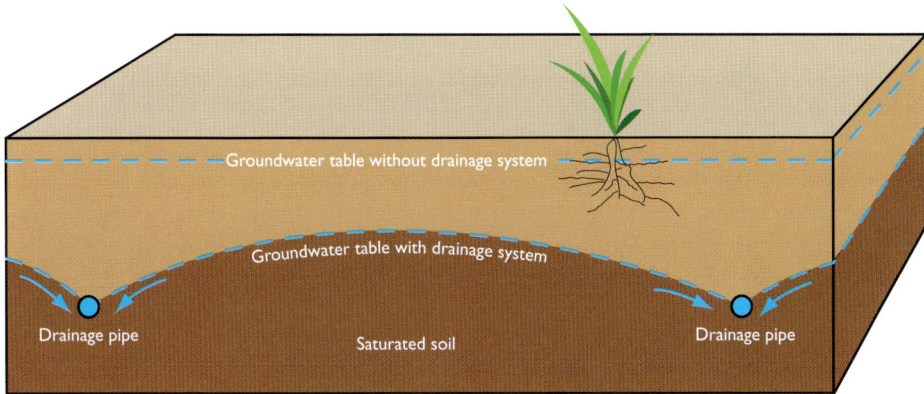

Figure 10.3 *Control of the groundwater table by means of buried pipes (ref.: www.fao.org).*

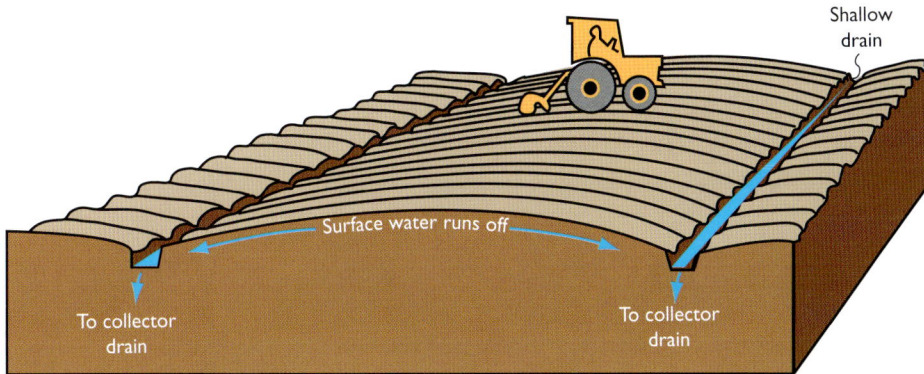

Figure 10.4 *Network of shallow open drains and sloping surface to control surface runoff (ref.: www. fao.org).*

This implies that in these areas surface runoff will prevail which, in principle, needs to be controlled. Should, however, the area be developed as a nature reserve then future vegetation can be expected reducing the surface runoff such that additional measures may not be required.

An artificial field drainage system to control the surface runoff may include:

- a network of open shallow surface drains (such as gutters, shallow ditches, ducts, overflows) of sufficient capacity to intercept and carry off water;
- specified gradients of the fill surface to create watersheds of limited size that direct water towards drains, retention basins and other facilities constructed as part of an artificial drainage system (see Figure 10.4);
- other measures such as vegetation or increasing the surface roughness of the reclamation area.

The capacity (dimensions, slopes, spacing of drains, gutters, ditches, etc.) of the artificial drainage system must be sufficient to remove the water from the surface without causing flooding and erosion. The open drains themselves may also have to be lined or protected against erosion by water flow.

The design of an artificial drainage system must take into account the lay-out of the future development of the reclamation area: locations of retention basins, ditches, drains and so on should not interfere with future structures. Further reference is made to Chow *et al.* (1988).

**Artificial external (main) drainage system**
The main drainage system must carry off the water to the discharge location(s) where it drains into natural open water. The system generally consists of shallow and/or deep collector drains that collect water of respectively the field surface drains or the field subsurface drains. Main drains transport the drainage water from the collector drains to the locations where the water is discharged into the river, sea or other open water. Outfalls (if required) are usually also considered to be part of the main drainage system. Shallow collector drains can be channels or ditches, while the deep collector drains include pipe drains and (deep) ditches.

## 10.3   Wind erosion

Erosion of a landfill by wind is generally not significant in terms of volume loss of material. However, effects like dust and sand drift can be annoying, detrimental to equipment, structures, industrial processes and crop (abrasiveness) or even hazardous to safety and health (low visibility, respiratory distress) of construction workers on the reclaimed area and drivers, residents and other people downwind of the fill area (see Figure 10.5).

Figure 10.5   *Typical effects of wind erosion (reference: WERU, United States Department of Agriculture, www.weru.ksu.ed).*

Soil movement as a result of wind forces can be divided into three categories (see Figure 10.6):

- suspension (typically associated with fine grained materials < 0.1 mm);
- saltation (generally related to grain sizes between 0.05 mm and 0.5 mm);
- surface creep (grain sizes in excess of 0.5 mm).

While soil can be blown to virtually any height, over 90% of the soil movement takes place within one metre above the ground surface.

The occurrence of wind erosion mainly depends on the wind velocity, the grain size distribution of the fill and the structure of the soil preventing particle detachment. The minimum wind velocity required to cause entrainment in the erosional agent of wind is known as the threshold velocity. The threshold velocity is strongly related to the size of the grains, but also to the vegetation, surface roughness, soil structure (surface crust), precipitation and humidity, surface soil moisture content, etc.

A minimum static threshold velocity of approx. 5–10 m/s (measured at a height of approx. 0.15 m above the ground) has been reported for a loose fine sand/silt with grain sizes between 0.1 mm and 0.15 mm (Bagnold, 1941; Stout and Aritmo, 2010). This value increases with increasing grain sizes. When smaller clay-sized particles are involved, the threshold velocity may also increase as a result of aggregation of the particles (structure of the soil). It is important to note that the threshold velocity depends on the maximum momentary velocity of turbulent flow and not on

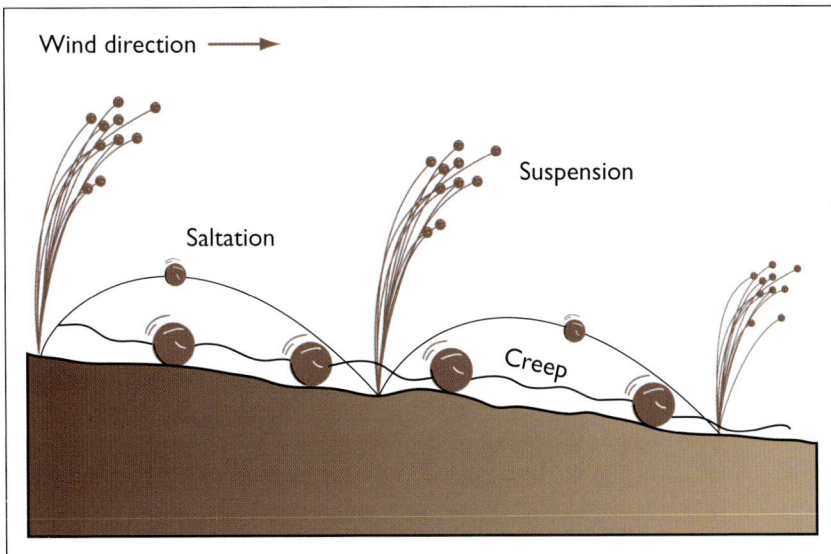

Figure 10.6   *The wind erosion process (based on Wind Erosion Research Unit, USDA, www.weru.ksu.edu/weps.html).*

the average forward velocity (i.e., eddies and currents are more important in lifting and transporting of soil than the average wind velocity).

Measures to prevent the effects of dust formation are generally based on:

− Reduction of wind velocity at the fill surface and protection of facilities by strategic placement of wind screens. Given the fact that the majority of the soil movement as a result of wind erosion occurs within one metre of the ground, artificial screens, rows of natural willow twigs and/or other natural materials of limited height may be particularly useful to mitigate the saltation and creep types of soil transport.
− Preservation of the vulnerable fill surface. As a result of the hydraulic placing method the surface of the fill may be covered by a thin layer of fine-grained, aggregated soil that could be critical to resist wind erosion. This thin surface crust can readily be destroyed by physical disturbance as a result of, for instance, vehicle traffic. Concentrating the traffic on the reclamation within certain designated corridors will keep the destruction of the vulnerable surface of the fill mass and subsequent generation of sand and dust drifts to a minimum.
− Stabilisation of the fill surface will prevent wind erosion and the subsequent sand and dust drift. Typical stabilising measures include:
  o Spraying of water during dry conditions until surface finishing works like paving start;
  o Spraying of stabilising agents (starch-based or oil-based stabilisers or other chemicals like polymers, sodium silicate and synthetic latex) are occasionally used, although these substances are generally considered to be expensive and often unacceptable from an environmental point of view;
  o Placement of a top layer of gravelly material not containing fine-grained particles below say 2 mm for a more permanent solution;
  o Placement of a top layer of fertile soil seeded with grass varieties or other vegetation has proved to flourish in local (possibly salty) conditions, see Figure 10.7.

The final choice for any of these options is very much dependent on local circumstances. Further reference is made to the website of the Wind Erosion Research Unit (WERU) of the United States Department of Agriculture (http://www.weru. ksu.edu).

**Grass test fields at Maasvlakte 2**

In order to accommodate the growth of the Port of Rotterdam, The Netherlands, the existing harbour, the Maasvlakte, has been extended by reclaiming approximately 2000 ha from the North Sea. As part of this project an extensive study is undertaken to enable a selection of the best variety of grasses to stabilise the surface of the reclaimed land and the inner slopes of the sea defence. This will not only prevent sand and dust drifts reducing the nuisance in the vicinity, but it is also meant to maintain the final platform level of the newly created land while attracting insects and birds to improve the ecological value of the area. The grasses have to survive in a harsh environment characterised by a poor subsoil, strong winds, sand drifts and significant salt spray. Two areas of approximately one ha have each been seeded with different seed mixtures, while a third area has not been treated at all in order to serve as a reference. Regular measurements using gauging rods in all three areas are carried out to determine the sand loss or accretion as a result of the wind erosion. Figure 10.7 shows the extensive test fields.

Figure 10.7    *Test fields (on the right) of grass varieties for surface stabilisation at Maasvlakte 2.*

## 10.4  Slope, bank and bed protection

Reclamations directly adjacent to open water (marine and coastal waters) require carefully designed and constructed shore protection measures to defend the new land against hydraulic attack (currents, waves). Banks and beds of canals, channels, ditches, outfalls and other types of open surface drains (inland waters) constructed on or adjacent to the reclamation may have to be lined to prevent scour. Depending on the type of structure, the future development, the magnitude and intensity of hydraulic loading and the level of damage that is considered to be acceptable, various options exist to protect the reclamation area:

– if the anticipated damage is less than the acceptable damage, no measures are required;
– regular replenishment of eroded fill (for instance, beach nourishment);
– protection of reclaimed land (for instance, armour stone, gabions, concrete lining, asphalt revetment, mattresses, grass cover);
– elimination or reduction of the hydraulic loads by protecting the reclamation area behind natural or artificial barriers (for instance, coral reefs, vegetation, groynes, breakwaters).

Information on the physical site conditions required to design an appropriate protection include, see Chapter 3 and Appendix B:

– bathymetry at the reclamation site and layout of the future development including levels, slopes, etc.;
– marine and coastal waters: wind data, tidal water levels, storm surges, waves, sea level rise, design water levels;
– inland waters: discharge and flow, turbulence, flood waves, translation waves, ship-induced waves, currents;
– geotechnical data;
– ice conditions.

The most appropriate type of protection depends on many variables including hydraulic, economic, environmental and climatic conditions. In general, the protection has to be more robust if the severity of the hydraulic loads increases.

If no slope protection is applied then natural gentle slopes develop which may be subject to the erosive action of currents and waves. Depending on the exposure to wave action and its grain size, beach slopes in sand may vary between 1:20 and 1:50 (Wiegel 1964, CUR 130). Slopes not exposed to waves may be stable at 1:4 to 1:6 for longer periods of time. Undesirable erosion should be compensated with beach nourishment works.

If hydraulic loads are not too high then natural protection may be an attractive option. Vegetation like mangroves along coastal waters, willows, reed and other aquatic plants along river banks and beds, grass on dikes, and such may offer protection against waves and currents and may be desirable from an ecological point of view. Roots of the vegetation can increase the strength of the protection by reinforcing the structure of the subsoil and stabilising the unconsolidated sediments of coasts, banks and beds.

Preservation of existing coral reefs or sand bars near the reclamation area may reduce the wave energy and/or deflect (tidal) currents and can therefore be an attractive solution both from an environmental and aesthetic point of view.

More artificial structures may include:

– Bed protection of channels, ditches and outfalls consisting of rock, concrete blocks, mattresses and concrete lining. Loose-grained bed protection may require a filter to prevent loss of bottom material.
– Bank protection along channels, ditches and open drains including revetments of loose rock, concrete blocks, asphalt, but also rigid structures such as sheet pile walls (often with an earth retaining function). Groynes are often used to protect river banks from erosion. If loss of loose-grained bank material through a permeable structure (loose rock, concrete blocks) is expected then filters have to be applied.
– Shore protection comprising revetments of rock, concrete blocks, asphalt, groynes and breakwaters with or without filter structures to stabilise the underlying cohesionless materials.

Failure mechanisms can be divided into various categories:

– Failure of the protection itself: failure of the armour layer, filter construction, asphalt layer, and so on.
– Failure of the structure underlying the protection: generally some kind of a slope stability or settlement failure as a result of, for instance, the load of the protection, toe erosion, lack of compaction.
– Damage to existing shores, beds, banks and structures in the vicinity of a newly constructed, protected reclamation area as a result of changed morphological conditions of the region: for instance, unwanted erosion and/or sedimentation elsewhere along the existing shoreline or damage to existing coastal or bank protection works related to changed wave conditions, (tidal) currents.

Further reference is made to handbooks on this subject like the Rock Manual (2007) and Schiereck (2001) in which the theoretical backgrounds and design of these protective structures are described.

## 10.5 Interaction between reclamation and civil works

### 10.5.1 *General*

Land reclamation is generally undertaken for a specific purpose: for example, the extension or construction of a port facility, the construction of an industrial estate or the development of a housing area. The type of usage and, hence, the structures to be built on the reclamation dictate, to a great extent, the functional and technical requirements to be considered during the design and construction of the fill mass.

The civil works constructed on reclaimed land generally include structures (e.g. buildings, storage tanks, bridges), infrastructure (e.g. runways for airports, roads, railways, pipelines/cables), and structures located along the boundary (e.g. quay walls). Each type of structure will have its own interaction with the reclaimed land, usually in terms of required behaviour of the fill mass.

Attaining symbiosis between the reclamation and civil works is often important. The required fill mass properties derived from an optimised structural design may, however, not always be technically and economically feasible given the boundary conditions (e.g., nature of existing subsoil, quality of the fill material available) and the dredging equipment available (see Figure 2.3).

### 10.5.2 *Foundations*

All structures are subject to foundation design. In some cases piled foundations are preferable. Heavy structures with very strict specifications regarding (differential) settlement will normally require a piled foundation when constructed on reclaimed land overlying soft subsoil. Small sized structures of limited weight can generally be supported by shallow foundations. Based on a technical and economical evaluation of the structural and reclamation design, the decision whether a piled foundation is required should be made in an early phase of the project.

When piled foundations are foreseen for structures, compaction of the fill could be limited. A too high level of compaction would yield problems when driving piles to their design depth. In case of cast-in-situ piles, very high energy levels may be required to reach the design tip elevation.

When shallow foundations are foreseen in the design, then an adequate compaction of the fill may be required to guarantee the bearing capacity with sufficient safety.

Loads from infrastructure such as roads and railways are relatively small compared to those of civil structures but they might have a more dynamic character. This makes it logical to consider specific (e.g. compaction) requirements for the top few metres of the fill at the dedicated locations only. However, additional measures to limit the magnitude of differential settlements may be required should the fill mass be placed on soft subsoil.

### 10.5.3 *Construction sequence*

Although often land reclamation works precede the civil works, occasionally another construction sequence may be preferable. This may, in particular, relate to structures such as quay walls that have to retain the fill or structures that have to be buried in the fill (e.g., submerged tunnels). The construction sequence may have significant effects on the planning and costs.

Benefits and drawbacks of three different construction sequences can be described as follows:

– First reclaim land and subsequently construct the civil structures in the dry after which any excess fill material is excavated. A significant advantage of this sequence is that the civil works can be executed from land which may be very

Figure 10.8 *Simultaneous dredging and construction works at Maasvlakte 2, Rotterdam, The Netherlands.*

cost effective. On the other hand, removal ('second handling') of the fill placed outside the footprint of the reclamation area, idle time of the dredging plant and loss of fill as a result of erosion may increase costs considerably.

- In certain circumstances, partially completing the reclamation works, carrying out the civil works and then completing the project by raising the fill to the required platform level may be preferable.
- First construct the civil structures in the wet followed by hydraulic placement of the fill material. This construction sequence usually requires less space and may therefore be an attractive alternative within confined areas. Another advantage of this sequence is that rock fill can be placed in the active wedge of a quay wall which may optimize the design. Hydraulic filling of relatively small areas behind structures is, however, often less attractive compared to dry filling methods using earth-moving equipment.

Preference for the sequence to be adopted during the execution cannot be deduced from a straightforward procedure. The sequence depends on a diversity of aspects, varying from specific project conditions to contract philosophy, and should, on an individual basis, be evaluated during the design stage of each project.

### 10.5.4  *Impact on existing structures*

The construction of a new reclamation or an extension close to existing structures and/or infrastructure may have a serious impact on these elements. The weight of the new fill introduces additional vertical and/or horizontal stresses in the existing soil body and the underlying soil strata, resulting in settlements and/or lateral displacements. Consequently, existing structures may experience deformations and could be damaged, even if they are located at some distance from the new fill. Examples of structures vulnerable to the effects of an extension in, for instance, port areas are:

- infrastructure (roads, pavements, railways, tunnels, cables and pipelines);
- structures (piled or shallow foundations);
- revetments (shore protection, breakwaters, etc.);
- special structures (e.g. quay walls).

In an early phase of the design an inventory should be made of existing structures and infrastructure close to the planned reclamation so that, if necessary, measures can be taken to prevent or to limit the impact.

Figure 10.9 shows two common landfill scenarios: a port extension where the fill influences existing structures in both vertical and horizontal direction, and the construction of a reclamation in open water crossing infrastructure (i.e. pipelines or cables).

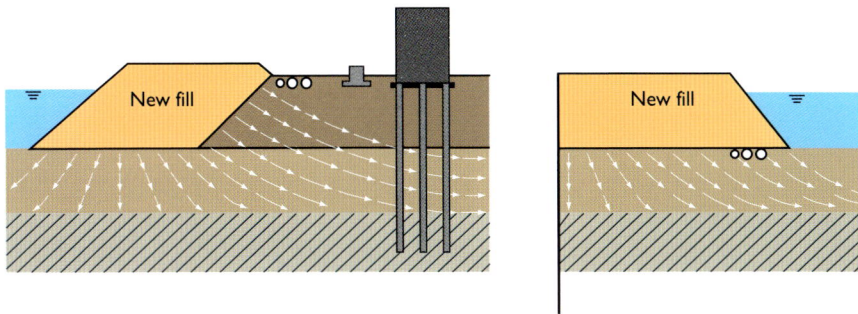

Figure 10.9 *Impact of new fill on existing structures.*

Differential displacements in vertical and/or horizontal direction are associated with a variety of problems related to existing structures. Consequently, prior to the execution of a landfill project an assessment should be made with respect to the expected soil deformations and their effects on existing structures. These effects can be evaluated according to two principle states:

– Serviceability Limit State (SLS) – loss of function or discomfort during daily use, sections 8.4.1 and 8.5.1;
– Ultimate Limit State (ULS) – damage to structure (breakage, cracks), sections 8.4.1 and 8.5.1.

Some interactional problems and the potential measures to prevent or limit the damage are described below.

### a. Infrastructure (roads, pavements, railways, tunnels, cables and pipelines)

Roads, pavements and railways often have a stiff foundation layer or contain rigid elements which are vulnerable to deformations. Differential settlements may even cause damage. The choice for taking measures is often a trade-off between cost and accepted damage level. Some examples of possible measures are:

– accept deformation or damage and repair if necessary;
– reduce deformations of the subsoil by using lightweight material in the new landfill;
– limit propagation of deformations by the installation of, for instance, a sheet pile wall.

Cables and pipelines are frequently present in coastal areas. They are used for transport of data, electricity, liquids or gasses. The weight of a landfill can result in significant settlements for existing underground structures, especially in case of soft subsoil conditions. These deformations become a problem when the associated stresses cause leakage or collapse. Possible damage to underground infrastructure is, however, not the only reason for taking measures. Pipelines and cables will become inaccessible after being covered with large volumes of fill material and maintenance will become practically impossible. Removal or diversion of the underground infrastructure is often the only reliable option.

### b. Structures (piled or shallow foundations)

Differential settlements give rise to problems for all foundation types. Shallow foundations experience similar kind of problems as discussed above with respect to the infrastructure. Vertical and horizontal soil displacements near piles result in negative skin friction and bending moments. Damage to foundations is difficult and costly to repair, so measures are based on prevention rather than on mitigation. Some examples of possible measures are:

- reduce deformations of the subsoil by locally using lightweight fill material;
- retain a minimum distance to existing structures to keep the foundation out of the influence zone of the landfill;
- limit propagation of deformations by the installation of, for instance, an anchored sheet pile wall.

### c. Revetments (shore protection, breakwaters, etc.)

Boundaries of coastal areas are frequently provided with a shore protection which becomes ineffective if fill material is placed in case of an extension. Instead of burying the existing shore protection underneath the new landfill, it is an option to remove this material. Shore protection consisting of loose elements such as rocks or blocks can easily be removed and possibly reused. Asphalt or concrete revetments can be broken down and recycled.

Leaving an existing shore protection in place can save some costs. However, the presence of hard elements in between the land masses, can give rise to the following problems:

- local deformations in the fill due to loss of fine fragments into the pores of the protection layer;
- hard elements form an obstruction for future foundations and underground infrastructure.

To prevent these problems it is best to consider the removal of existing revetments prior to or during the landfill operation.

### d. Special structures (e.g. quay walls)

A quay wall is a typical example of a special structure present in port areas. The quay wall may, for instance, consist of a vertical structure used for handling vessels. Scenario's as presented in Figure 10.10 may occur in case of a port extension.

If landfill is placed to extend the coastline in front of the quay wall, then it will become enclosed by the two sand bodies and consequently loose its function.

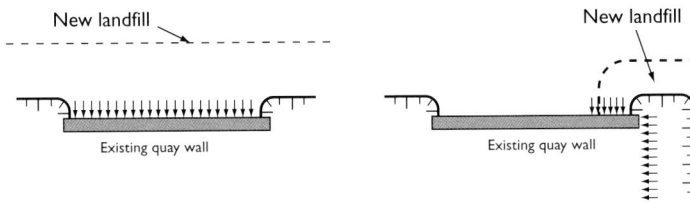

Figure 10.10    *Impact of port extension on existing quay wall.*

If not removed, the quay wall forms an obstruction for future underground infra-structure such as cables and pipelines. Best measure is to (partially) remove the quay wall.

If the new landfill is located next to or only partially in front of the existing quay wall then other problems can be expected. Any local changes in soil pressure can cause an unevenly distributed loading on the wall, resulting in additional bending moments and deformations of the structure.

As quay walls are used for loading and unloading, they are often equipped with rail mounted gantry cranes for ship-to-shore transport of cargo. The deformation tolerances of this type of rails are often very limited. Some possible measures to limit quay wall deformations are:

– reduce the soil stresses by locally using lightweight fill material;
– reduce the loads by lowering the groundwater level behind the quay wall.

### e. Buried structures
Piled underground foundations which are not removed before the placement of fill material will behave like hard spots that may cause problems in view of differential settlements. The decision to remove old foundations depends on the requirements with respect to the future use of the area. Should the foundations maintain their function after completion of the filling works then it is to be investigated which measures are required to satisfy that requirement.

## 10.6    Earthquakes

Although the Earth's surface may look stable, it is in fact a dynamic system consisting of large rigid tectonic plates (lithosphere) moving relative to each other over the underlying, ductilely deforming, visco-elastic asthenosphere. The occurrence of earthquakes is mainly associated with the plate boundaries, where the plates will either slide along (transform boundaries) or towards each other (convergent boundaries causing subduction or continental collision). Frictional resistance along a fault plane will prevent movement and will increase stresses until the fault reaches its point of rupture and strain energy is suddenly released. Resulting movements will create seismic waves that will cause shaking and sometimes displacements at the Earth's surface. Earthquakes may trigger tsunamis, landslides and occasionally volcanic activity. They may induce liquefaction and can cause structural damage to buildings, bridges, cables and other infrastructural facilities.

When plotting past earthquakes on a map, earthquake zones become visible. Figure 10.11 shows areas of seismic activity, represented by the magnitude of the peak ground acceleration, on a global seismic hazard map.

Local seismic hazard maps for different countries worldwide can be found at the following link: www.geology.about.com/library/bl/maps/blworldindex.htm. Many small earthquakes occur daily around the world. Most of them are only noticed by seismographs and not by people, as they are very small or they take place in remote areas.

Earthquakes may affect a site in two ways:

- directly if a fault crosses the site;
- indirectly by seismic waves hitting the site.

The first mechanism causes differential displacements between both sides of a fault. The differential displacement can be quite large: values up to 3 m have been observed.

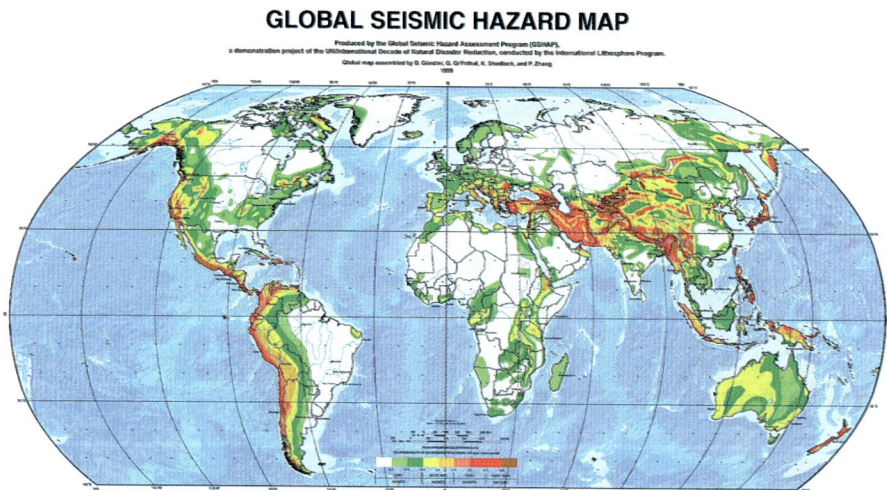

Figure 10.11   *Global seismic hazard map (ref. http://geology.about.com/library/bl/maps/blworldindex.htm).*

The second mechanism causes cyclic loading of the soil which in turn results in the generation of excess pore water pressures. These excess pore water pressures may cause liquefaction, densification and deformation of slopes and structures, see section 8.6. An example of liquefaction which occurred in a housing area is presented in Figure 10.12.

Other sources of earthquakes are "induced" earthquakes, caused by reservoir impoundment, pumping into deep wells, fluid extraction, volcanic eruption, gas bursts, geysers, mining, underground nuclear blasts, depletion of oil and gas reservoirs, the collapse of underground mines and nuclear bomb testing. The hypocenters of these earthquakes are at a relatively shallow depth.

While seismic magnitude is interesting, seismic intensity is more important. Seismic intensity is the motion that actually affects people and buildings. Intensity maps are very helpful for practical issues such as city planning, building codes and emergency response. Many seismic intensity scales have been developed over the years. One of the most well-known scales is the Scale of Richter that quantifies the energy contained in an earthquake by calculating the logarithm of the amplitude of the seismic waves measured by a seismograph. The Modified Mercalli (Intensity) Scale is used in the United States to classify earthquakes by their effects. The Omori Scale

Figure 10.12    *Liquefaction in a housing area, Nigata, Japan, 1964.*

Table 10.1   *The European Macroseismic Scale (EMS) of Earthquake Intensity.*

| Intensity | Definition | Typical observed effects |
|---|---|---|
| I | Not felt | Not felt, even under the most favourable circumstances. |
| II | Scarcely felt | Vibration is felt only by individual people at rest in houses, especially on upper floors of buildings. |
| III | Weak | The vibration is weak and is felt indoors by a few people. People at rest feel a swaying or light trembling. |
| IV | Largely observed | The earthquake is felt indoors by many people, outdoors by very few. A few people are awakened. The level of vibration is not frightening. Windows, doors and dishes rattle. Hanging objects swing. |
| V | Strong | The earthquake is felt indoors by most, outdoors by few. Many sleeping people awake. A few run outdoors. Buildings tremble throughout. Hanging objects swing considerably. China and glasses clatter together. The vibration is strong. Top-heavy objects topple over. Doors and windows swing open or shut. |
| VI | Slightly damaging | Felt by most indoors and by many outdoors. Many people in buildings are frightened and run outdoors. Small objects fall. Slight damage to many ordinary buildings; for example, fine cracks in plaster and small pieces of plaster fall. |
| VII | Damaging | Most people are frightened and run outdoors. Furniture is shifted and objects fall from shelves in large numbers. Many ordinary buildings suffer moderate damage: small cracks in walls; partial collapse of chimneys. |
| VIII | Heavily damaging | Furniture may be overturned. Many ordinary buildings suffer damage: chimneys fall; large cracks appear in walls and a few buildings may partially collapse |
| IX | Destructive | Monuments and columns fall or are twisted. Many ordinary buildings partially collapse and a few collapse completely. |
| X | Very destructive | Many ordinary buildings collapse. |
| XI | Devastating | Most ordinary buildings collapse. |
| XII | Completely devastating | Practically all structures above and below ground are heavily damaged or destroyed. |

is used throughout Japan, and the European Macroseismic Scale is the current standard in Europe. Many other countries use local versions of these scales for purposes of emergency planning and earthquake-resistant construction guidelines.

The European Macroseismic Scale (ref. www.geology.about.com/od/quakemags/a/ European-Macroseismic-Scale-EMS98.htm), updated in 1998, is the basis for evaluation of seismic intensities in European countries. Despite the name (which reflects the fact that it was developed at the instigation of the European Seismological Commission), the Scale is equally suitable for use outside Europe and has been successfully used for assessing earthquakes in many parts of the world. Unlike earthquake magnitude, which indicates the energy a quake expends, EMS98 intensity denotes how strongly an earthquake affects a specific place. The European Macroseismic Scale has a 12 level division (see Table 10.1).

Earthquake waves cause a two-way movement of the ground in horizontal (parallel and perpendicular to the direction of wave propagation) and vertical direction. How big and how fast (acceleration and velocity) this deviation is depends on various factors such as the length of the fault, the distance between the site and the fault and the geology of the subsurface. The last can have a large influence on the ground acceleration and wave form, which makes it very important to know about even slight changes in the subsoil underneath the site.

It follows that the values of ground acceleration show a high variation even within small regions. This is especially true for medium- to large-size earthquakes and interpolating the values is thus difficult. Generally the acceleration will decrease the further away from the fault. The calculation of the Peak Ground Acceleration (PGA) is based on horizontal ground movements. The acceleration is expressed as a decimal or percentage of the Earth's gravitational acceleration, $g = 9.81$ m/s$^2$. The PGA is often used in the design specifications of reclamation projects and usually based on (local) Building Codes and experience. The analysis of the liquefaction potential and stability of earth structures is generally based on the PGA (see also section 8.6.4.2).

## 10.7   Tsunamis

A tsunami (Japanese: 'harbour wave') is generally defined as a series of water waves caused by a (sudden) displacement of a large volume of water. This displacement can be a result of an earthquake, a volcanic eruption, an underwater explosion, a landslide, calving of glaciers or a more unlikely event like the impact of a meteorite. The waves generated by these events could be high and could consequently have a devastating effect on the coastal region affected. Although tsunamis are often related to oceans they do also occur in large lakes and fjords.

Figure 10.13   *Tsunami wave in Japan 2011, (www.jagansindia.co.in/2011/03/japan-tsunami-devasting-live-footage.html).*

The risks of hydraulic hazards such as tsunamis should be analysed at the initial (feasibility) stage of a reclamation project. The frequency of occurrence and the expected impact of extreme events should be properly evaluated and related to the expected costs for protection against such events. Such a risk-cost assessment should be the basis for the design of reclamation areas in terms of being 'hazard-proof' or not.

The risk for tsunamis is in some parts of the world higher than in others. In general, the frequency and intensity of tsunamis in the Indian and Pacific Oceans and the Mediterranean Sea is much higher than in other regions of the world. A global overview of tsunami events and risks is for example presented in the following link: http://www.nerc-bas.ac.uk/tsunami-risks/. At present, however, no accurate site-specific information is available with respect to the chance of occurrence and intensities of tsunamis.

The propagation of a tsunami wave towards the coast can be predicted by means of numerical modelling, i.e., by linking a seismic fault model with a wave propagation model. The modelling can predict, with a reasonable degree of accuracy, the time of arrival and height of a tsunami wave at the shoreline. The modelling results can be used as input for risk, design, evacuation or emergency assessments. Data on sub-sea faults can be derived from seismic observations and/or earthquake recurrence models, which are referenced in various literature sources.

# CHAPTER 11

## MONITORING AND QUALITY CONTROL

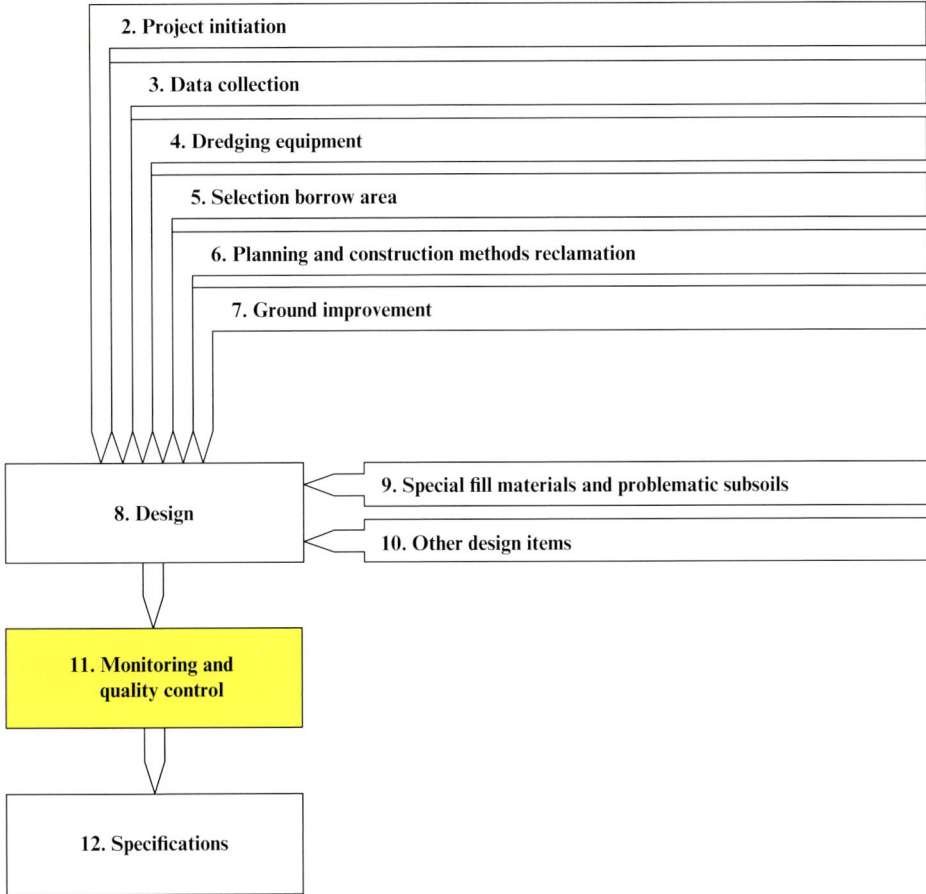

```
┌─────────────────────────────────────────────────────────────┐
│  2. Project initiation                                       │
│   ┌─────────────────────────────────────────────────────────┤
│   │  3. Data collection                                      │
│   │   ┌─────────────────────────────────────────────────────┤
│   │   │  4. Dredging equipment                               │
│   │   │   ┌─────────────────────────────────────────────────┤
│   │   │   │  5. Selection borrow area                        │
│   │   │   │   ┌─────────────────────────────────────────────┤
│   │   │   │   │  6. Planning and construction methods reclamation │
│   │   │   │   │   ┌─────────────────────────────────────────┤
│   │   │   │   │   │  7. Ground improvement                   │
```

| 8. Design | ← | 9. Special fill materials and problematic subsoils |
| | ← | 10. Other design items |

**11. Monitoring and quality control**

12. Specifications

The flow chart below shows the structure of this chapter and where to find the information corresponding to the various items.

| Chapter 11 Monitoring and Quality Control |
|---|

| 11.1 Introduction |
|---|

| 11.2 Quality Control Plan |
|---|

| 11.3 Monitoring and testing | | | |
|---|---|---|---|
| 11.3.1 Geometry | 11.3.2 Fill material properties | 11.3.3 Fill mass properties | 11.3.4 Environmental monitoring |

## 11.1   Introduction

The main purpose of a quality control and monitoring programme for a reclamation project is to ensure that the fill mass and subsoil exhibit the behaviour as intended by the designer. This behaviour has to be translated into performance requirements of the fill mass that are often determined by the design of the super-structures (foundations, live loads, etc.) which the reclamation will accommodate in the future. In addition, performance requirements may be subject not only to the loads of superstructures, but also to other external loads such as ground accelerations (as a result of earthquakes) and influences like rainfall. These performance requirements are generally expressed as technical specifications related to:

- The safety against loss of bearing capacity and slope failure. This is mainly defined by the shear strength of the fill mass (and subsoil) and the loads imposed by the geometry (slope angle), foundations and other external influences.
- Allowable residual settlements and horizontal deformations. These are defined by the stiffness of the fill mass (and subsoil) and the design of the superstructures and infrastructure.
- Resistance against liquefaction. This is determined by the density of the fill mass (and subsoil) and the characteristics of the dynamic shear load introduced by earthquakes.
- Drainage characteristics. These are a function of the permeability of the fill mass and the expected rainfall intensity, wave overtopping, etc.

In addition to the quality and performance of the fill mass, quality control also includes verification of the specified geometry.

Table 11.1   *Overview of dimensional and mechanical properties to be monitored and tested.*

| Fill geometry | Bathymetry and topography |
|---|---|
| Fill material properties | – particle size distribution and angularity<br>– minimum and maximum dry density<br>– mineralogy |
| Fill mass properties<br>– shear strength<br>– stiffness<br>– relative density<br>– permeability | – stability and bearing capacity (fill/subsoil)<br>– deformation/settlement (fill/subsoil)<br>– resistance against liquefaction (fill/subsoil)<br>– drainage characteristics |
| Environmental aspects | |

An overview of the most relevant dimensional and technical aspects to be monitored and tested for land reclamation projects is presented in Table 11.1. A subdivision is made with respect to the fill geometry, the fill material properties and the fill mass properties.

Different approaches can be followed for the process of monitoring and quality control:

## 1. Material testing

Technical specifications for hydraulic fill projects often specify criteria for the materials to be used within the reclamation. Verification of these criteria can be done by means of material testing. A physical sample taken from the fill area can be checked in a laboratory with respect to, for example, the specified maximum percentage of fines. Cone Penetration Tests executed on newly reclaimed land can further demonstrate whether a specified cone resistance is satisfied or not.

## 2. Behaviour monitoring

Quality control can also be executed by monitoring the behaviour of the reclamation. The settlement of the fill can, for example, be measured on a regular basis after placement of the material. Interpretation of this time-settlement behaviour will demonstrate whether a specified residual settlement after construction will be satisfied or not. If necessary it can be decided to timely undertake remedial measures. Other aspects of monitoring are related to horizontal soil deformations near a slope or existing structures and the dissipation of excess pore pressures in soft subsoil underneath a fill.

## 3. Performance testing

The performance of the fill mass under future loading conditions can be checked by means of a Zone Load Test. This is a full-scale performance test on site simulating the specified future loading conditions.

## 4. Process monitoring

A completely different approach is process monitoring. Based on fill classification systems, the behaviour of a fill mass may be assured by following a specified work method. For instance, a certain required compaction of a certain class of fill will be achieved in a layer of say 2 m thickness after 10 passes of a certain specified roller. In this case the process is monitored: in other words, the requirements are fulfilled when the roller has indeed executed the prescribed number of passes. The relation between classification, work method and fill mass behaviour can be obtained either from (national) codes of practise or by trial testing. The process monitoring method assumes a delivery of constant fill quality and will provide information to support the objectives of a method specification.

**Comments and remarks**

Monitoring and material testing only makes sense in case realistic specifications result in the desired behaviour of the fill mass. It is recommended to undertake quality control by a well balanced combination of the above described methods. Such a combination requires carefully defined specifications that must be in line with each other dictated by the future land use.

Properties of the fill mass such as shear strength, stiffness, and permeability are not only correlated to in-situ density, cone resistance (CPT) or N-value (SPT), but also to material properties such as minimum and maximum density, particle size distribution, shape and angularity of the grains, mineralogy of the grains, cementation, and so on. Numerous correlations combining material and mass properties are available to enable the assessment of the performance of the fill mass.

Occasionally it may be preferable to directly measure the required fill mass properties. Zone Load Tests and, to a lesser extent, Plate Loading Tests give reliable information about the stiffness and the strength of the fill mass.

Laboratory tests on samples retrieved from the reclamation to determine the shear strength and/or compressibility are often laborious and less reliable as these require reconstituting cohesionless material to its estimated in-situ density.

The specified geometry is verified by regular bathymetrical and topographical surveys.

Monitoring the behaviour of the fill mass after completion of the reclamation includes the observation of time-dependent processes by measuring the deformations of the fill mass (and subsoil). This is done by using instruments such as settlement plates, extensometers, inclinometers, and other methods. Monitoring also includes the interpretation of the measurements and the prediction of the future behaviour of the reclamation by extrapolation of the data.

The quality of the works is ensured by the implementation of an appropriate Quality Control Programme. A well planned and executed Quality Control Programme benefits all parties. The Client, the Designer and the Engineer are ensured that the works are carried out in accordance with the Contract requirements whilst the Contractor can efficiently optimise the planning. In addition, monitoring may enable all parties to identify unforeseen inappropriate fill mass behaviour in an early stage of the construction. Immediate action can then be taken to hold technical and financial consequences to a minimum.

## 11.2   Quality Control Plan

A Quality Control Program includes a Quality Control Plan and the actual testing during and after construction as well as the subsequent interpretation of the results. The Quality Control Plan should include:

– an organisation chart identifying all personnel responsible for quality control;
– procedures for reviewing drawings, samples, certificates, and other submittals necessary for contract compliance;
– description of the services provided by outside organisations such as testing laboratories and consulting engineers;
– document control procedures;
– procedures for non-compliance;
– an overview of all relevant levels and parameters to be checked;
– the types of tests to be performed;
– sampling and test procedures including:
    o   test locations, time and frequency of measurements;
    o   proposed techniques and equipment;
    o   the extent of data to be collected;
    o   accept/reject criteria for test results.

In general, dredging and filling operations progress rapidly and are frequently subject to changes. It is essential that daily aspects and problems related to quality control can be handled by the site organisation since time for consultation might not always be available. Off-site backup support should be available in case of special problems or test results that are in non-compliance with Contract specifications.

All works specified for the project should be monitored and tested and records of such inspections and tests should be documented and maintained throughout the construction period. It is recommended to utilise approved and certified laboratories for testing and to specify Standards (such as BS or ASTM) for all inspections, sampling and testing of materials. Relevant factual information on each non-conformance should be collected and prompt action taken to identify the causes of such an event. Corrective actions can subsequently be taken to prevent recurrence.

As part of the Quality Control Program the Client/Engineer often requires the Contractor to submit a Method Statement for all activities related to monitoring and testing. It is important that all parties agree on such a document before the start of the dredging and reclamation works.

It is not realistic to specify a standard format for a Quality Control Program prescribing a fixed number of monitoring devices such as Settlement Markers, Cone Penetration Tests, Plate Load Tests, Zone Load Tests, etc. for land reclamation projects. For the definition of such a program it should be realised what has to be monitored and, in particular, why it is to be monitored. The main goal of quality control is to verify whether the specified dimensions of the reclamation and the quality as well as the performance requirements of the fill mass are satisfied. Examples of aspects that need monitoring in land reclamation projects using hydraulic fill material are presented below:

– boundaries and levels;
– quality of the fill material;
– consolidation and settlements;
– strength in terms of stability and bearing capacity;
– density in terms of resistance against liquefaction;
– environmental impact.

The contents and structure of a Quality Control Program depends on many different situations including site specific conditions and locations together with the particular project requirements. Below is a range of key aspects which require consideration, assessment and an implementation plan.

**a. Project dimensions and boundary conditions**
Large projects, reclamations with an irregular geometry and fills constructed near existing infrastructure require a more intense monitoring program during their execution than isolated simple land reclamation projects.

**b. Borrow area and quality of fill material**
Small borrow areas and areas with a complex soil stratification (e.g. intermediate unsuitable layers) require a higher intensity of monitoring than large borrow areas covered with a thick layer of suitable fill material with no fines present.

**c. Future land use**
A land reclamation for heavy industry will have more stringent requirements for the performance of the fill mass than, for instance, a reclamation project for the

development of housing or the storage of mass products. Especially the future land use is governing for the performance criteria of the fill mass and thus also governing for the extent of the Quality Control Program.

### d. Earthquake region

Land reclamation projects constructed in earthquake regions require a minimum density for the fill material in order to avoid liquefaction. This is normally achieved by means of deep compaction. It is clear that a Quality Control program for such sensitive areas is more intense than for reclamation projects in non-tectonic regions where generally no density requirements are specified.

### e. Subsoil conditions at the project site

Subsoil containing weak compressible layers requires more monitoring effort in view of consolidation, settlement and strength increase than subsoil composed of hard layers only. Inhomogeneous weak compressible subsoil with a strongly varying stratification in lateral direction requires, on its turn, a higher monitoring intensity than homogeneous weak compressible subsoil having a constant thickness within the entire project area.

### f. Ground improvement

Performance requirements for reclaimed land may imply that ground improvement has to be performed for subsoil and/or fill material. It is in that case required to monitor and verify the specified effect thereof. Each of the ground improvement techniques listed below requires its own particular method in view of monitoring and quality control:

- vertical drains (consolidation, settlement, strength increase);
- deep and surface compaction (density, bearing capacity);
- stone columns (strength, deformation);
- etc.

### g. Available construction time

A short construction period in combination with the presence of soft compressible soil layers at the reclamation site will result in a "sharp" design with an increased risk profile in terms of stability and settlements. A higher frequency in monitoring effort is then obvious in order to avoid failures and excessive settlements after completion. One could, on the other hand, relax on the monitoring program in case of an available long construction period in combination with a design having a low risk profile.

### h. Environmental impact

Some regions in the world are more sensitive with respect to the environment than others. Hydraulic fill projects executed in such areas (e.g. a rich underwater habitat, the presence of coral, etc.) have frequently more severe requirements than projects constructed in an existing industrial environment. More stringent requirements require a more intensive monitoring program.

It may be clear from the above that a Quality Control Program should be "fit for purpose" accounting for the quality and performance criteria of the fill mass and considering the boundary conditions of the project. The following section describes monitoring and testing techniques in a random sequence.

## 11.3  Monitoring and testing

### 11.3.1  *Geometry*

The dimensions of a landfill are checked by means of a geodetic survey. Before the start of construction a so-called "in survey" is performed to determine the actual horizontal and vertical limits of the area to be reclaimed. During construction, geodetic surveys are normally carried out with such frequency as to allow for a controlled execution of the work. After completion a so-called "out survey" is performed in order to check whether the horizontal and vertical limits of the reclamation meet the contractual requirements. Frequently checking the water depth in the borrow area is also important since dredging is often specified to a certain approved depth or because suitable fill material has a limited thickness only. Information on the type of geodetic surveys is presented in Appendix B.

---

**Comments and remarks**

As a result of the limitations of construction and survey equipment, the realised fill level will always show some deviation from the specified height. Achieving accuracies less than 0.05 m with standard land reclamation equipment is very difficult. Deviations should be acceptable to a certain level, also because of the uncertainties in settlement predictions. Claiming millimetre-precision for earthworks is unrealistic. Technical Specifications should therefore allow for pre-defined construction tolerances. If necessary, a distinction in scale can be made by defining macro- and micro tolerances.

---

### 11.3.2  *Fill material properties*

#### 11.3.2.1  Grain size distribution

Specifications are normally designed to ensure a well-mixed fill without unacceptable segregation of fines. This allows the fulfilment of density requirements whilst reducing the risk of liquefaction and/or intolerable deformations and settlements. Specifications for the granular composition of hydraulic fill generally comprise:

– a prescribed maximum percentage of fines;
– a grading envelope which allows for feasibility of compaction;

– a prescribed maximum size of rock fragments or boulders within the upper part of the reclamation.

Methods and tools available for monitoring and testing the granular composition of fill material are presented in Appendix B.

**Comments and remarks**

– Representative samples for checking the quality of the fill material can be taken:
  o onboard the dredging vessel;
  o at the end of the discharge pipeline;
  o at the reclamation site.
– Samples taken onboard the dredging vessel can be visually examined and/or rapidly analysed offering the opportunity to take immediate action (e.g. adjustment of the working method or dredging process) if non-conformance with the specified grading is observed.
– Should samples be collected onboard of a trailer suction hopper dredger, then special care shall be taken with respect to the selection of the sampling location. Typically, the coarse material will accumulate near the loading boxes whilst the finer material will be encountered in the vicinity of the overflow locations.
– Segregation of fines cannot always be avoided in practice and this may also occur during the deposition process on site. It is therefore necessary to also take samples on the reclamation itself.
– It is possible to reduce the percentage of fines by means of overflow onboard trailing suction hopper dredgers. This, however, may not always be feasible because of environmental restrictions. Cutter suction dredgers do not have the possibility to make use of overflow. Fill pumped into the reclamation area then consists of the in-situ material as encountered within the borrow area.
– Samples taken from boreholes, drilled from the reclamation platform, only become available in a later stage of the project execution. Remedial actions in case of non-conformance are then more difficult, especially if too many fines are detected at some depth. It is, on the other hand, not necessarily disastrous if silt inclusions are locally present within the reclamation if it does not affect the intended future land use. The actual behaviour of the fill mass with silt inclusions could, in that case, be checked by means of a Zone Load Test (see section 11.3.3.1) or by means of geotechnical calculations.
– The number of samples to be taken and tests to be performed depends, amongst other things, on the quality of the fill material encountered within the borrow area. For instance, this number could be either one per hopper load or two per day or one per 10,000 $m^2$ of each layer.

### 11.3.2.2 Minimum and maximum dry densities

The minimum and maximum dry densities are material properties, which together with the in-situ density, determine the relative density. Since the relative density is related to the in-situ density it is a mass property. The minimum and maximum dry densities can be determined by various standardised laboratory tests (see Appendix B).

### 11.3.2.3 Mineralogy

The mineralogy (chemical composition of the solid particles) of the fill material and subsoil may significantly affect the construction and subsequent behaviour of the land reclamation.

Fill predominantly consisting of quartz may result in considerable wear and tear on the dredging equipment. Calcareous fill material may display a different mechanical fill mass behaviour as a result of its tendency for crushing. Subsoil consisting of organic clay or peat may cause considerably more settlement than mineral clay or sand.

Certain salts (sulphates like anhydrite and/or gypsum, halite, etc.) may result in unwanted mechanical or chemical behaviour as a result of dissolution, swell, concrete corrosion, etc.

Relevant laboratory tests may further include the determination of the carbonate, the organic and the sulphate content. Reference is made to Chapter 3.

### 11.3.3  *Fill mass properties*

### 11.3.3.1 Shear strength

The shear strength of the fill mass and the subsoil determines the resistance against shear failure. This is relevant for bearing capacity and slope stability. Sufficient resistance against shear failure must be guaranteed during and after the construction. Tests available for determination of the shear strength are referenced in Appendix B.

---

**Comments and remarks**

- The Vane Shear Test measures the actual value of the undrained shear strength in soft cohesive subsoil. The proportional strength increase due

---

to consolidation under the weight of the fill mass (for instance, during a staged construction) can be measured in time. The increased values for the undrained shear strength can be used as input in a geotechnical computer programme which analyzes the bearing capacity or slope stability.
- According to Bjerrum (1972) results of Vane Shear Tests should be corrected to obtain the actual in-situ undrained shear strength. The correction factor depends on the plasticity of the cohesive material encountered.

**Shear strength in relation to bearing capacity**

Specifications for land reclamation projects sometimes include requirements with respect to the bearing capacity of the fill mass in view of the construction of shallow foundations. It is important to realise that the bearing capacity not only depends on the shear strength of the fill mass but also on the geometry, dimensions and depth of the foundation. The bearing capacity can be checked by direct and indirect methods. The following direct methods are available to measure the bearing capacity of the fill:

- Plate Loading Test;
- Zone Load Test.

These tests are in more detail described in Appendix B.

**Comments and remarks**

- Often the contract only specifies the bearing capacity without detailing the geometry and dimensions of the future foundations. It is noted that besides the shear strength of the subsoil, the geometry (shape, size and embedment depth) of a foundation has an effect on the bearing capacity.
- The Plate Loading Test is a more appropriate test method for determining the bearing capacity and load-deflection relationship of the materials used for roads, airport runways, pavements, etc.
- For testing the actual behaviour of future foundations, Zone Load Tests are more meaningful than Plate Loading Tests. They are, however, more time consuming and more expensive than Plate Loading Tests.
- A Zone Load Test should be performed at a location where a Cone Penetration Test has been executed. The elastic settlement measured during the test can then be used to correlate and calibrate the stiffness modulus E with the measured Cone Resistance. This can be done using a finite element programme. Following this approach the number of Zone Load Tests can be reduced and replaced by much cheaper and less time-consuming Cone Penetration Tests.

- The bearing capacity can also be assessed based on indirect measurements such as CPTs and SPT's. The resistance measured in the fill correlates to a friction angle, which is used in a theoretical bearing capacity formula. Another option is to perform triaxial or direct shear tests on the fill material at its in-situ density in order to measure the actual friction angle.

### Shear strength in relation to slope stability

Slope failure can be a consequence of constructing side slopes which are too steep along the periphery of the reclamation area. Precipitous placing of fill on soft subsoil, without allowance for consolidation and subsequent increase of shear strength, may also result in geotechnical instability. In both cases circular or non-circular failure mechanisms can be developed.

Technical Specifications normally prescribe requirements applicable for hydraulic fill structures after completion of the construction. A Quality Control Plan should, however, also pay attention to the above-mentioned failure mechanisms during the actual execution of the works.

The following tools can be used to monitor the geotechnical stability:

- Electrical piezometer;
- Inclinometer.

These tools are in more detail described in Appendix B.

### Comments and remarks

- The electrical piezometer measures the actual value of the excess pore water pressures in (soft) cohesive subsoil. The results are used as input in geotechnical computer programmes to analyse the slope stability. Using excess pore pressures in a geotechnical analysis requires that sufficient information must be available on the drained shear strength characteristics ($\varphi'$, $c'$) of the cohesive soil. Alternatively the undrained shear strength $c_u$ can be used in the analysis but excess pore pressures are then not relevant.
- Piezometers installed in between vertical drains may provide erroneous results. It is impossible to guarantee that the device is centrically installed in between the drains. There is also no guarantee that the drains are perfectly vertically installed. If a piezometer is located close to a drain then lower excess pore pressures are measured than actually present on an average. One should be aware of this phenomenon when evaluating the results of piezometers installed in a soil mass containing vertical drains.

11.3.3.2   Stiffness

Each land reclamation project will be subject to deformations and settlements. Their magnitude depends on the nature of the existing subsoil on which the hydraulic fill is constructed and the stiffness of the fill mass itself. Settlements are relatively small in granular material but can be significant if the subsoil (or the fill) consists of cohesive material such as very soft clay. The majority of the deformation in granular fill material generally occurs immediately during the construction of the reclamation, but some deformation may still occur at a later stage as a result of time-dependent creep. In soft cohesive material most deformations and settlements are time dependent because of consolidation and creep effects. Consequently, unacceptable residual settlements may occur after completion of the construction. Acceleration of the consolidation settlement in cohesive materials is therefore often enforced by special measures such as the installation of vertical drains in combination with a temporary surcharge. Such a temporary surcharge may be required to meet the specified residual settlement requirements within the time frame of the project execution. The following tools and devices are available for monitoring the deformations of a landfill:

– settlement plate;
– extensometer;
– settlement hose;
– inclinometer;
– trial embankment.

These tools are in more detail described in Appendix B.

**Comments and remarks**

– A trial embankment is preferably executed during the design phase of a land reclamation project. Such a trial is sometimes executed during the construction period in order to optimize the drain spacing and the magnitude of the surcharge. This is, however, only feasible for large reclamation projects when sufficient time is available.
– The initial settlement due to deposition of the fill will only be measured if settlement plates are installed at original seabed or ground level before start of the reclamation works. These plates are, however, often physically displaced by the filling works and consequently the initial settlement is not always correctly interpreted.
– Due consideration must be given to the interpretation of settlement plate readings if those plates are installed after the reclamation arrives at a certain level above the water table. The initial settlement of the reclamation is then not measured. This should, however, not be problematic since vertical drains to accelerate the settlements are normally installed at the same level.

– The time-dependent settlement behaviour of soft cohesive material is schematically presented in Figure 11.1 below which also shows the effects of the application of vertical drains in combination with a temporary surcharge.

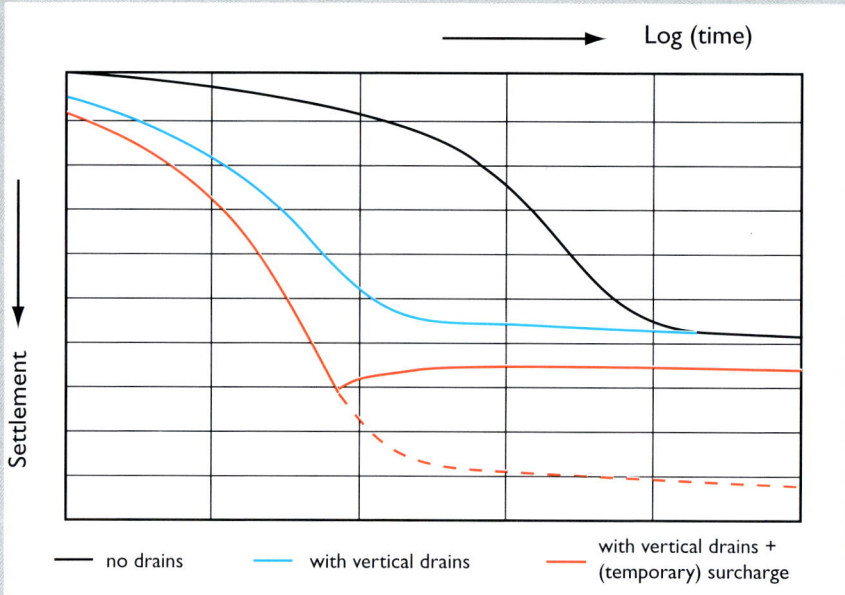

Figure 11.1 *Time dependent settlement behaviour soft soil.*

– The rate of settlement at the start of consolidation is faster than the later stages of the process. The monitoring frequency should be planned accordingly in order to enable an accurate prediction of the residual settlement. Since a land reclamation is normally constructed in phases, the monitoring frequency should be increased in the period just after the placement of the subsequent lifts.
– In addition to the need of frequent measurements it is further important to simultaneously record the progress of the filling process in terms of location, date, height of fill layer, removal of surcharge material, etc.
– It is recommended to frequently inspect and record the condition of the settlement beacons as these instruments are often damaged by other construction activities at the reclamation area. In order to explain possible erroneous readings it is also important to record the date and time of nearby activities (e.g. placement of locally more fill material).

- The settlement of hydraulically placed fill is usually monitored by settlement plates located on a grid of $100 \times 100$ m to $250 \times 250$ m depending on the dimensions of the reclamation site.
- As a result of deformations of the settlement hose it may sometimes not be possible to run the probe through the hose.
- The application of a settlement hose is specifically useful for reclamation areas of limited extent such as embankments.
- The entire system of settlement plates, hoses and inclinometers is designed to enable the Engineer/Contractor to observe the magnitude and rate of deformations and settlements. The monitoring results facilitate, for instance, the decision to proceed with a next construction stage (additional lift), to remove the temporary surcharge, etc. The results can also be used to assess the residual settlement after construction. Remedial measures can be taken during the construction phase if it appears that the specified allowable residual settlement is not satisfied. On the other hand, optimisations can be considered if the total settlement appears to be smaller than expected.
- The progress of consolidation can be defined as the ratio of actual settlement to the ultimate settlement as a result of primary consolidation. The appropriate time for surcharge removal is generally related to the desired degree of consolidation or the allowable residual settlement to be achieved.
- A certain minimum consolidation percentage is frequently prescribed in Specifications. Analysis of the consolidation process in the field is, however, not always that simple, especially if very soft and/or non-homogeneous layers are present. In such cases use could be made of observational methods like the hyperbolic method which assumes that the settlement (s) versus time ($t$) behaviour approaches a straight line after some time in a t/s versus t diagram. This straight line describes a hyperbolic relation, hence the term hyperbolic method. The inverse of the slope of this line would then yield the ultimate settlement. Thus, once sufficient data are available for the settlement to reach the hyperbolic line, the subsequent consolidation can be predicted (Thiam Soon Tan *et al.*, 1991).
- Asaoka's method is another graphical method to assess the ultimate consolidation settlement based on monitoring data of settlement plates. A plot of measured settlements versus time is divided into equal time intervals after which subsequent settlements are plotted against each other. The extrapolation of a straight line finally gives an indication of the ultimate settlement to be expected. It is noted that creep deformation is not included in the graphical method which means that the total settlement may (theoretically) be somewhat underestimated. This method is, however, often used in land reclamation projects. Detailed information

about the procedure followed in Asaoka's method can be found in literature (Asaoka, 1978).
- It was found in the literature (Tan *et al.*, 1996) that settlement data beyond the 60% consolidation stage are needed in both Asaoka's method and the hyperbolic method to enable an accurate prediction of the ultimate primary settlement. Interpretation of the ultimate primary settlement may be obscured due to variations in the ground water level.
- Present computer models have the option to perform a curve fitting analysis. This module allows users to perform automatic parameter fits, based on settlement plate measurements obtained during the construction. More accurate predictions, using adjusted parameters, can then be performed for subsequent construction stages.

11.3.3.3   Density, relative compaction and relative density

**Density**
Because density is a mass property that correlates well with many other geotechnical parameters, it is an important parameter when constructing hydraulic fills. For a definition of the density, reference is made to section 8.3.1. Density of cohesive materials can easily be determined by undistorted sampling. The in-situ density of cohesionless material can only be determined by specific tests (see Appendix B) for fill present above the ground water table. A problem with land reclamation projects, however, is that a considerable part of the fill may be present below the water table. That means that these specific tests which measure the in-situ density in dry circumstances cannot be used, and therefore one has to rely on correlations to determine the in-situ density. Recent developments of the nuclear density probe show promising results to measure the in-situ density directly. The in-situ density is continuously recorded during the penetration of the probe. The principle of the nuclear density probe is described in Appendix B.

**Relative compaction**
Technical Specifications for land reclamation projects frequently prescribe a density for the fill, which is expressed as a certain percentage of the maximum dry density (MDD) determined at optimum moisture content: the relative compaction. The maximum dry density is determined in the laboratory by means of the Proctor compaction test or the more severe modified Proctor compaction test. A sample can be analysed in the laboratory to determine the maximum dry density of the soil, but this gives no information about the actual density in-situ. Indirect tests from which the results are to be correlated with the relative density are then inevitable, especially for fill material below the water table.

## Relative density

The density of underwater granular fill is impossible to measure by sampling, because undisturbed samples cannot be retrieved. Therefore one has to rely on correlations between the Cone Resistance and the Relative Density (as a function of the effective stress). Researchers have established correlations between Cone Resistance (CPT), effective pressure and Relative Density based on results of Calibration Chamber Tests (section 8.3.5.3) performed on typical reference sands. The advantage of the Cone Penetration Test is that the Relative Density can be assessed over the entire depth of the fill mass.

The tests available to measure the in-situ density are described in Appendix B. In the laboratory, Proctor tests can be executed on material taken from the upper part of the fill. The in-situ density of the material is then compared with the maximum Proctor density. Sampling and subsequent testing in the laboratory are usually time-consuming methods. The density of cohesive materials can be determined from undisturbed samples recovered by thin-walled sampling methods.

---

### Comments and remarks

- Crushable material such as carbonate sands will experience breakdown during the Proctor test. This results in a high density in the laboratory, which is not achievable in the field with standard compaction equipment used for land reclamation. It is therefore recommended for such materials to determine the maximum density by means of a Vibrating Table (ASTM-D4253-00).
- For equivalent Relative Densities, the Cone Resistance measured in calcareous sand is less than in quartz sand because of the compressibility and crushing effects of the particles. Depending on the compressibility of the (calcareous) sand, compared with "reference" sand, a shell factor or cone correction factor of the order of 1.2 to 2.3 has to be used. For instance, in the case of an applicable shell factor of for instance 1.5, a specified Cone Resistance of $q_c = 9$ MPa may be reduced to $q_c = 9/1.5 = 6$ MPa.
- In special cases, Calibration Chamber Tests can be considered to be executed in order to establish the actual correlation between Cone Resistance, effective stress and Relative Density for the hydraulic fill material to be used in a reclamation project. It is, however, noted that only a few laboratories in the world have the appropriate equipment to perform this type of test. Availability might be a problem if a Calibration Chamber Test is to be performed within short notice. Performance of such a test in the planning phase of the project is therefore recommended. This will also give the opportunity to specify an appropriate Cone Resistance in Contract Documents avoiding time-consuming discussions during the execution of the project.

- Sometimes a certain minimum degree of compaction is specified for rec-lamation projects. The degree of compaction can, especially for hydraulic fill placed below the water table, not be measured by direct methods. It is more obvious to specify a relative density, which based on correlations, can be assessed by means of a Cone Penetration Test.
- If compaction is required to satisfy a specified density or cone resistance then field trials are normally executed on site to optimise, for instance, the number of passes for compaction rollers or the energy and centre-to-centre distance for vibroflotation equipment. Extensive monitoring and testing as outlined above is then a requirement.
- Hydraulic fill is not a "man-made" material and consequently some hetero-geneity may be encountered, certainly if the material contains an unavoid-able percentage of fines. Consequently, the records of Cone Penetration Tests (for which readings are taken every 20 mm) may show significant

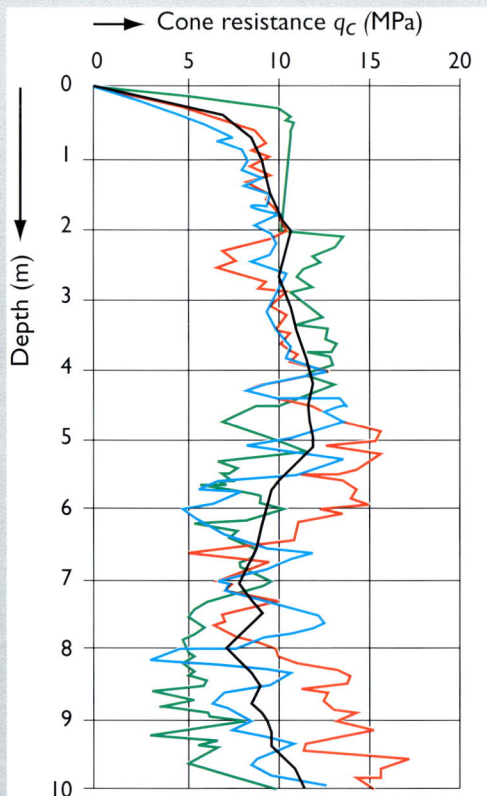

Figure 11.2    *Filtering CPTs by means of running a geometric average over 0.5 m.*

variations in the horizontal and vertical directions of a reclamation site. The peaks and troughs, however, are not representative for the mechanical "overall" behaviour of the fill mass. A realistic interpretation of the CPTs can be obtained by filtering out the peaks and troughs in neighbouring records. This filtering can be made by a running geometric average over a 0.5 m or 1.0 m record length. An example of such a filtering of the Cone Resistance is presented in Figure 11.2. The bold line represents the filtered average values (Massarch *et al.*, 2002). When evaluating the results of compaction by vibroflotation, the CPTs, executed around one compaction point, can be averaged in line with such an approach.

- The Cone Resistance in the upper part of the fill (down to approximate 0.5 m below finished level) is not representative for the assessment of the relative density. Often a low resistance is measured as a consequence of surface failure due to insufficient overburden pressure. One should therefore be cautious when specifying a certain Cone Resistance for the upper part of the fill. Alternatively, for density testing nearer to the surface, the sand replacement or the rubber balloon method could be used (see Appendix B).
- It may be a problem to import nuclear density probes into some countries. This should be checked before such a device is specified in Contract Documents.

### 11.3.4 *Environmental monitoring*

Hydraulic fill operations may have a significant impact on the ecosystem near a reclamation site. Nowadays dredging works are, generally, only authorised after certain environmental requirements have been met. Environmental monitoring provides the required information to demonstrate that the dredging process complies with the requirements, which are either legally or contractually stipulated. Furthermore, monitoring can assist in the communication with regulators, the public and other stakeholders. Environmental monitoring is performed to:

- establish the existing environmental conditions in and around the work areas (baseline monitoring);
- provide environmental data, which can be used to monitor the effects of the works (feedback monitoring);
- provide information, which can be used to demonstrate that the works are in compliance with the environmental requirements (compliance monitoring).

Potential impacts on the environment may be expected in both the borrow area and the fill area. Each of these areas has its own specific environmental impacts which will determine the monitoring strategy. Whilst dredging in a borrow area, suspended sediments (TSS) will be introduced in the water column as a consequence of the excavation, loading and transportation process. At the reclamation site, the environmental impacts are associated with the discharge of residuals materials in the return or decant water. Subject to the nature of local water movement, the introduced sediments may disperse over a large area with the potential for an adverse impact on the natural ecosystem.

Environmental monitoring programmes must specifically be targeted towards environmental parameters relevant to the local ecosystem. The following variables generally indicate the physical condition of water during dredging and reclamation works:

– total Suspended Sediments;
– turbidity;
– dissolved Oxygen;
– additional parameters.

These parameters are in more detail described in Appendix B.

Monitoring of dredging and reclamation works can be done in different ways. Visual inspection is the first step to identifying whether a significant sediment plume is created. During favourable weather and sea conditions, aerial photography or sophisticated remote sensing techniques are very useful for demonstrating the visual extent of the plume at a certain moment. Water sampling, followed by testing in a certified laboratory, is a clear and effective tool to record the state of the water quality. Point or static monitoring provides information about the water quality and/or hydrodynamic parameters of any single location. When environmental information is required from numerous stations or when the installation of offshore buoys, poles or frames is not feasible, environmental data can be collected using a monitoring vessel. Acoustic Doppler current profilers are more frequently used to monitor sediment plumes. The optional monitoring techniques are summarized below and in more detail described in Appendix B.

– visual inspection;
– aerial & satellite imagery;
– water sampling;
– point monitoring;
– mobile vessel-based monitoring;
– acoustic Doppler current profilers (ADCP's).

# CHAPTER 12

## TECHNICAL SPECIFICATIONS

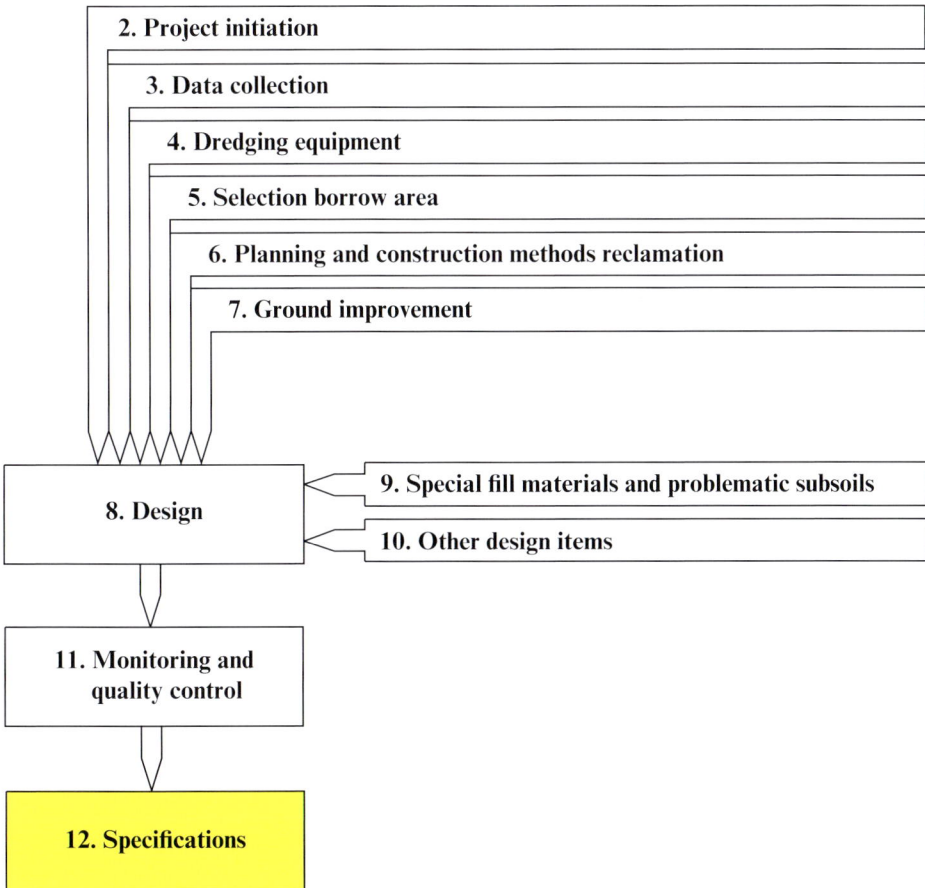

| 2. Project initiation |
|---|

| 3. Data collection |
|---|

| 4. Dredging equipment |
|---|

| 5. Selection borrow area |
|---|

| 6. Planning and construction methods reclamation |
|---|

| 7. Ground improvement |
|---|

**8. Design**

**9. Special fill materials and problematic subsoils**

**10. Other design items**

**11. Monitoring and quality control**

**12. Specifications**

Chapter 12 Technical Specifications

12.1 Introduction

12.2 Roles and responsibilities

12.3 Checklist project requirements

12.4 Commented examples of Technical Specifications

## 12.1 Introduction

This chapter gives an overview of the most common subjects to be addressed in the Technical Specifications for a dredging and land reclamation project. Figure 12.1 presents the logical sequence of a hydraulic fill design as laid down in this Manual starting with the functional requirements. Following the functional requirements, various performance requirements can be identified. Technical Specifications must finally ensure that the system performs according to the performance requirements.

## 12.2 Roles and responsibilities

In the case of traditional land reclamation projects, the translation of functional requirements into performance requirements and Technical Specifications is done by the Client and his Consultant.

In a design and construct type of contract only the project definition and functional requirements are defined by the Client and his Consultant. The performance requirements and resulting Technical Specifications are then deliberated by the Contractor.

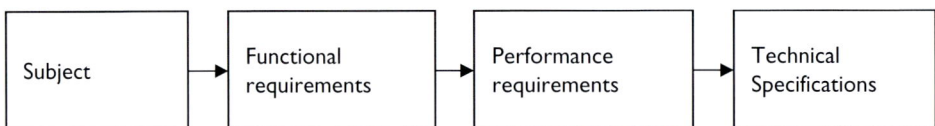

| Subject | → | Functional requirements | → | Performance requirements | → | Technical Specifications |

Figure 12.1  *Sequence of defining requirements.*

An intermediate option is a form of contract in which the performance require-ments (and possibly also the design methods) are specified. The definition of the required fill and fill mass properties (including possibly required soil improvement techniques) is done by the Contractor.

The implications of these 3 contract types on the level of detailing in the Specifica-tions is shown in the flowchart presented in Figure 12.2 and Figure 12.3.

To avoid confusing during the tender and construction phase of a project, it is important to ensure that the technical specifications are in line with the respon-sibilities as defined by the type of contract. Mixing functional requirements with performance requirements and Technical Specifications is a common pitfall for design and construct projects.

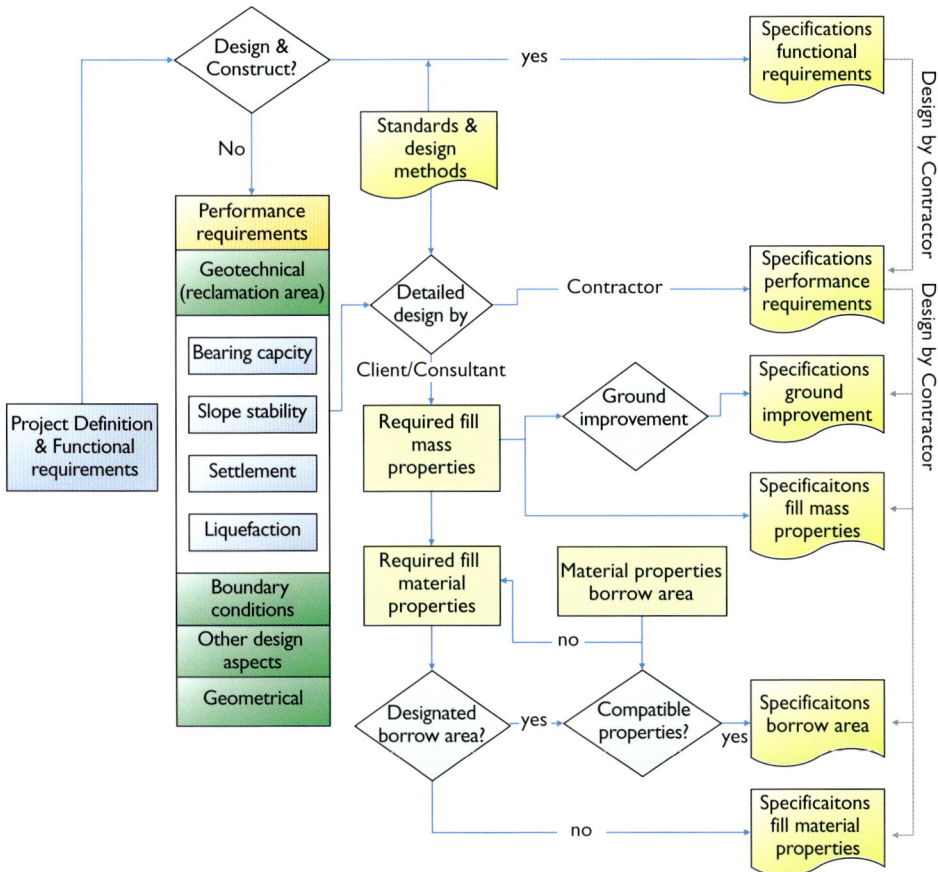

Figure 12.2 *Flow chart design specifications.*

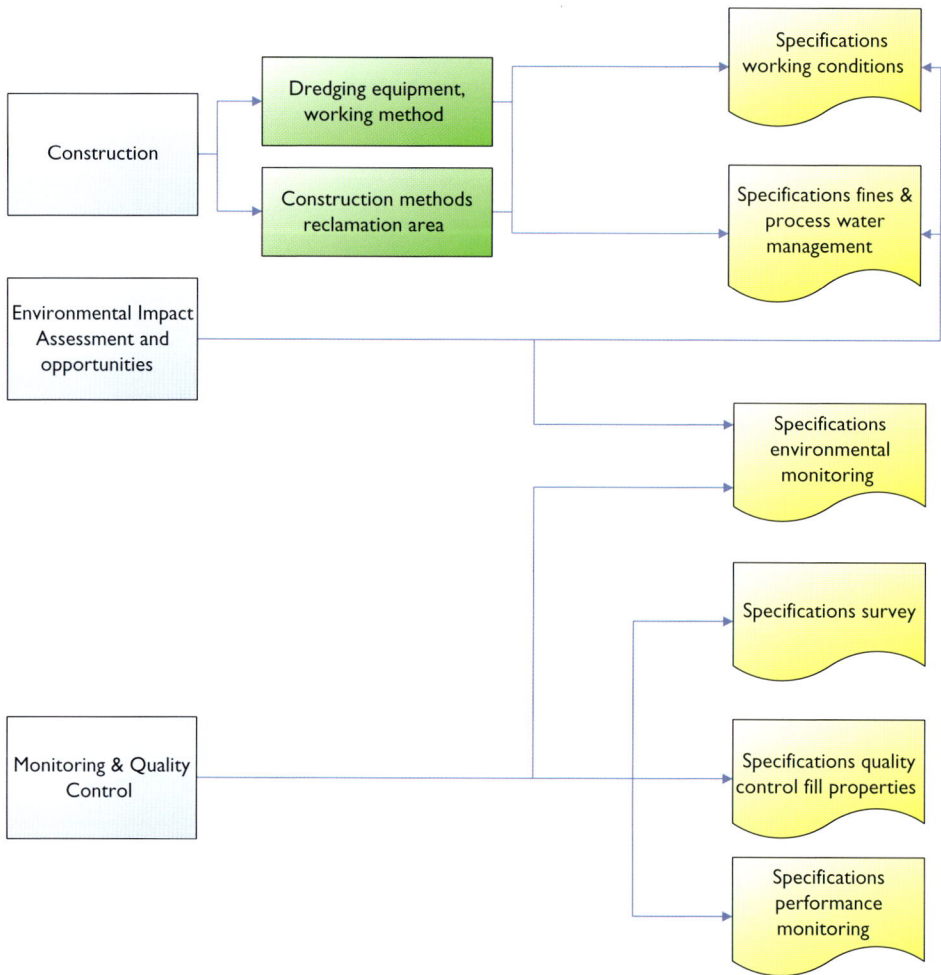

Figure 12.3   *Flow chart construction and monitoring specifications.*

A more detailed overview of items to be considered when writing a set of Specifications, can be found in the checklist presented in Table 12.1.

Irrespective of the party responsible for the project, all topics presented in Table 12.1 should be considered. The Technical Specifications shall in clear terms describe the design, construction and monitoring activities which the Contractor is required to undertake.

## 12.3  Checklist project requirements

Table 12.1  *Checklist project requirements.*

| Chapter/Topic | Scope for functional requirement | Scope for performance requirement | Technical specifications/requirement | Remarks |
|---|---|---|---|---|
| 3. Data collection | collect sufficient information<br><br>bathymetrical, hydraulic, meteorological, morphological, environmental, geotechnical | setup additional soilinvestigation campaign<br><br>setup of a soil model<br>define soil characteritics | type of (soil)investigation<br>Number of boreholes and CPT's, line spacing seismic....<br>required type and number of labtests<br>responsibility (FIDIC or any other form of contract) | should preferably be completed before start of the tender stage. |
| 4. Dredging equipment-planning and working method | complete project in time and achieve the planned milestones<br><br>limit hindrance to shipping<br>dredging method should be compatible with the environmental restrictions | The selected working method must be such to achieve the specified milestones<br><br>vessel flexibility | project deadline, milestones execution period<br><br>self propelled or stationary, use of anchors<br>working hours<br>seasonal restrictions overflow and turbidity | the selection of dredging equipment and work method is usually the contractors responsibility |
| 5. Selection borrow area | define borrow area as material source | borrow area should be physically accessible for the equipment | waterdepth (min., max.) | |

*(continued)*

437

Table 12.1 (*Continued*).

| Chapter/Topic | Scope for functional requirement | Scope for performance requirement | Technical specifications/requirement | Remarks |
|---|---|---|---|---|
| | (a designated borrow area or an area for capital dredging works such as a harbour basin) | reduce percentage of fines by overflowing | realistic specifications for overflow | |
| | | how to deal with poor quality overburden and poor quality intermediate layers | disposal area for poor quality material or storage within project boundaries | |
| | | material should be dredgeable crossing of boundaries allowed (e.g. deeper layer > better quality material?) sufficient volume of fill material within borrow area boundaries | see also dredging equipment define tolerances in case of capital dredging | |
| | | layer thickness and borrow area dimensions should allow for efficient dredging | size of borrowarea | |
| | | available material must coincide with specified fill quality | PSD envelope percentage of fines carbonate content…. angle of internal friction at specified relative density | specification for fill material characteristics must comply with the characteristics of the material in the borrow area |
| 6. Construction methods reclamation area | retain fill material and/or fines within the fill boundaries | construction method must be compatible with the design of intermediate and final project phases | stability of slopes for bunds permeability of bunds mudwaves, flow liquefaction | in case the construction method is prescribed, it should be physically feasible (sufficient waterdepth for dumping….) |
| | | | site clearance (prior to the start of the works) | |
| | limit hindrance | accessiblity of waterways and roads | installation of sinkerlines, floating line in waterways day and/or night work maximum height of equipment define allowable noise and vibration level | |
| | | air traffic noise, vibrations existing infrastructure other Contractors tourism | define Contractor interfaces seasonal limitations | |

| | fines management | fine materials to be stored in project area avoid segregation | size of siltpond, required height of bunds location, available space, restrictions |
| | | volume of fines to be stored in silt pond disposal area's for unsuitable material | waterlevel in bunded area, artificial drainage |
| | | sufficient drainage capacity, discharge design | |
| | process water management | | |
| | safety | | safety procedures |
| | provide access to site (site traffic) | 2 way traffic for specific type of equipment | minimum width/height of bund |
| Environmental impacts, assessmentsEIA and opportunities | compliance with | fines content of discharge water | turbidity limits specify monitoring |
| | | turbidity around dredger noise | campaign compatible working method |
| | | generation impact on fauna and flora | seasonal restrictions for working period |
| 8. Design of the reclamation area | split up the area as per future function and define functional requirements for each sub-area | bearing capacity | angle of internal friction |
| | | slope stability | undrained shearstrenght |
| | | | loading scenarios |
| | | | safety factor/approach |
| | | rate of deformation | degree of consolidation | by correlation with relative density |
| | | | | (specify method and corrections),… |
| | | | | e.g. 90% consolidation after 6 months |
| | | allowable residual settlement | magnitude primary and secondary settlement |
| | | | loading scenarios (future loading conditions, preload) |
| | | allowable differential settlement | distance for differential settlement |
| | | liquefaction resistance | calculation method |
| | | | design earthquake (acceleration, magnitude, return period) |
| | | deformations as a consequence of earthquakes | | generally not practical for land reclamation |
| | geometry: xyz | placement tolerances | define height datum, projection method |

*(continued)*

439

Table 12.1 (Continued).

| Chapter/Topic | Scope for functional requirement | Scope for performance requirement | Technical specifications/requirement | Remarks |
|---|---|---|---|---|
| | design of perimeter bunds | project boundaries | required overheight for settlement compensation | minimum shear strength |
| | | slope stability erosion overtopping | (see above) see the Rock Manual specify applicable standards | |
| 7. Ground improvement | in case performance design requirements are not met then improve the soil characteristics | increase soil strenght increase relative density (fill and/or subsoil) | specify compaction | generally not usefull |
| | | | moisture content = optimum moisture content +/−2 % | this should not be a specifications, but a recommendation! |
| | | | define max. % fines in relation to compaction technique | |
| | | reduce residual settlement | application of surcharge, consolidation period, magnitude of preload installation of vertical drains | linked to allowable residual settlement |
| | | increase consolidation rate (fill and/or subsoil) | | |
| 9. Special fill materials and problematic subsoils | cohesive & fine grained fill materials (verify design and adjust if necessary) | apply surcharge to prevent future settlement and stability problems check performance of special fill material | apply surcharge (>future design load) | |
| | carbonate sand (verify design and adjust if necessary) | | perform zone load test to simulate future loading conditions | |
| | rock (verify design and adjust if necessary) | | define correction methods when adopting correlations for quartz sands | |

| | | | | |
|---|---|---|---|---|
| 10. Other design aspects | sufficient drainage of subsoil prevent pond formation | provide sufficient permeability of fill material or construct drainage system | define percentage of fines | assessment of permeability based on PSD |
| | drainage of reclamation area and hinterland existing infrastructure, utilities and pipelines | existing dewatering channels to be deviated temporary drainage,…. | size and location of drainage canals | |
| | study long term sedimentation and change in currents due to land reclamation | setup 3D model | specify model maximum current in navigation channel morphological changes plant gras, use coarse toplayer compatible placement method | to be verified by outcome computer simulation preferable before tender phase |
| | limit hindrance due to wind erosion | specify area's with potential wind erosion | wave height, frequency of occurrence | |
| | reclamation must resist tsunami's | assess tsunami risk | | |
| | reclamation must resist earthquake's | assess earthquake risk | design earthquake (acceleration, magnitude, return period) | |
| 11. Monitoring and quality control | check quality fill material/ compaction…. | | specify Standards for testing | |
| | monitor settlement | | specify settlement beacons + prediction method grid | |
| | check permeability verify layout | specification survey system specification volume calculation | use approved terrain modelling system and interpolation method (e.g. triangulation) | |
| Maintenance | maintenance requirements | setup maintenance plan for entire life span of the structure | maintenance free period inspection interval | |

## 12.4 Commented examples of technical specifications

### 12.4.1 *Introduction*

This section presents some examples of Technical Specifications extracted from various tender documents and a variety of projects, following the structure of the Manual. It is not the intention to be complete or to provide standard Specifications, but meant to serve as guidance for subjects to be considered in dredging and reclamation projects. Only those subjects relevant within the context of this Manual will be dealt with and back referenced. Special attention is paid to subjects often leading to discussions.

Following subjects are discussed:

- Description of the works;
- Standards;
- Data Collection;
- Dredging equipment and work method;
- Selection borrow area & quality of fill material;
- Construction methods reclamation area;
- Environmental impact;
- Design of a land reclamation:
  o Lines and levels;
  o Bearing capacity and slope stability;
  o Allowable residual settlement;
  o Differential settlements;
  o Rate of deformation;
  o Liquefaction resistance;
  o Design of perimeter bunds.
- Ground improvement:
  o Increase of soil strength by compaction;
  o Reduction of the residual settlement;
  o Increase of the consolidation rate of the fill and/or subsoil.
- Special fill materials:
  o Cohesive or fine grained fill material;
  o Carbonate sand.
- Other design aspects:
  o Drainage;
  o Existing infrastructure and pipelines;
  o Wind erosion and dust creation.
- Monitoring and quality control
  o Geometry
    • Tidal levels;
    • Bathymetrical survey dredging area;
    • Bathymetrical/topographical survey reclamation area.

- ○ Testing fill material properties;
- ○ Testing fill mass properties;
- ○ Settlement monitoring;
- ○ Performance testing;
- ○ Reporting;
- ○ Monitoring and quality control program.

Notes:

- Example Specifications are shown in italic *text*. These examples can be used as a guideline for the setup of specifications.
- Where deemed necessary the example Specifications have been commented.
- Examples of inappropriate Specifications are also shown in *italic text*, but presented in a white textbox. The comments as to why these specifications are to be avoided are added below these textboxes.
- Comments on specific topics are presented in a shaded textbox.

### 12.4.2   *Description of the works*

The Specifications normally start with a detailed description of the Works to be executed under the Contract. The Specification should be arranged in agreement with the Contract drawings. The order of priority of the documents shall be defined in the Contract documents.

*Example:*

> *The Works cover dredging and transportation of dredged material and its deposition in reclamation areas to achieve the lines and levels showed on the Drawings. The reclamation works comprise supply, placing and compaction (if required) of suitable fill material from approved sources.*

The following additional subjects can be identified:

- (a) site clearance, removal and disposal of all unsuitable materials;
- (b) the provision, transport, placing, compaction, monitoring and testing of suitable excavated or dredged materials, to form the permanent reclaimed platform;
- (c) the placement of containment bunds;
- (d) the provision, transport, and placing of suitable materials to form any temporary reclaimed structure required to provide access for the Contractor and others;
- (e) the provision of boundary structure(s) to retain and protect the reclamation platform from erosion;
- (f) the procurement and maintenance of borrow sites in compliance with the provisions of the Contract;
- (g) compliance with all requirements of the relevant Authorities.

### 12.4.3 Standards

*Example:*

> *The following Standards and Codes of Practice are referenced within this Specification. All materials, workmanship and testing shall conform to the requirements of the latest editions of the following Standards and Codes of Practice except as explicitly varied by this Specification.*

### 12.4.4 Data collection (see Chapter 3)

The data collection phase should preferably be completed before the start of the Tender phase. This will allow the Contractor to make an accurate assessment of the working conditions and will also avoid annoying surprises for all Parties involved during the execution of the Works.

The following topics can be identified:

**Bathymetrical and topographical (see Section 3.4.1)**
An accurate in-survey of the borrow area and reclamation site is essential and should be carried out prior to the start of the Tender phase.

**Geological and geotechnical (see Section 3.4.2)**

*Example:*

> *The available geotechnical information that can be made available to the Contractor is included in the tender documents. The Contractor must make his own evaluation of the geotechnical conditions. He shall, at his own cost, carry out additional investigations in the dredging and borrow areas if required. The Contractor shall be fully responsible for the provision of reclamation material of specified quality and sufficient quantity.*
>
> *No obligation on behalf of the Employer is given with this site investigation and the Contractor is advised that he must satisfy himself as to the soil and sub-soil conditions, water table, location of aquifers and other site conditions.*

Comments:

> In case an additional soil investigation is required to verify the soil conditions prior to the Tender date, then sufficient time should be foreseen in the Tender schedule. The potential Contractors should be able to witness the soil investigation campaign and to advise on the completeness of the collected data.

For contractual responsibility of the data, see Section 3.2.

**Hydraulic, meteorological, morphological and environmental (see Section 3.4.3)**
Sufficient data contained in a morphological study should be available prior to the Tender phase. Eventual sedimentation may be resulting from dredging activities or natural processes. Prediction of morphological changes as a result of the realization of the project should be made in an early stage of the design and could be part of the EIA.

For the assessment of the environmental impact caused by dredging and reclamation works, sufficient long term monitoring data of background values should be available (see Section 3.4.3.3).

### 12.4.5 *Dredging equipment and working method (see Chapter 4)*

The selected equipment and working method should be such that all project milestones can be met and that the project will be completed in time taking into specific working conditions (see Section 4.1). Some examples and subjects within this context are presented below:

**Method statements**

Contractor shall provide Method Statements which in detail describe the proposed techniques for carrying out the dredging and reclamation Works. Environmental constraints may dictate that from time to time, dredging plant may not be permitted to operate in an "overflow mode" (if applicable). During such times the Contractor shall be deemed to have made allowances to operate in a "non-overflow mode".

**Work program**
The Contractor shall submit a program for the Works within the time frame and in the form stated in the Appendix to the Contract. The Contractor shall submit revised programs to the Engineer whenever reasonably required (see clause 7.2 FIDIC, first edition 2006).

**Plant**
Contractor shall provide adequate plant for the appropriate and timely execution of the Works required under the Contract. The Contractor has sole responsibility to assess necessary plant and equipment to dredge, transport and place all specified material. The Contractor shall be deemed to have made due allowance for site constraints, including but not limited to the following:

– *compliance with the provisions of the Environmental Management Plans, EIS and any Conditions of Approval;*
  – *water depth ranges;*
  – *wave climate;*
  – *wind climate;*
  – *currents;*
  – *pumping distances and heights.*

**Working hours**

Generally, dredging vessels are working 24 h/day and 7 days/week. Any limitations towards working hours or seasonal restrictions should be stated in the Specifications.

**Positioning**

Contractor shall provide and maintain positioning systems on board of all dredging plant.

**Hindrance to shipping (see Section 4.3.3)**

In case a dredging vessel is working in a busy shipping lane or harbour basin, it might be necessary to interrupt the dredging operations when vessels are passing. This is generally referred to as "shipping delay" and will result in a reduced vessel's efficiency. In some Contracts shipping delays are the Client's responsibility. In order to estimate the impact on the project duration of such shipping delays, detailed information about the frequency of shipping is required.

The dredging operations shall be carried out with minimum interruption to shipping movements.

**Interference**

> *The Contractor shall make due allowance in his programme for any sequence of operations necessary for construction works and co-ordination with other parties and contractors on the Site. The phasing of the works shall be such as to cause minimum interference and disruption to other parties with access to the working area or right of way in the Site or its approaches.*

Comments:

> Possible interfaces with activities on site by other parties and contractors should be defined before the tender phase.

**Use of anchors**

Another important issue is the use of anchors. Cutter suction dredgers may have to deploy an anchor inside the shipping lanes. The following options exist for the Specifications:

– anchoring in the shipping lane or basin is not allowed;
– anchoring in the shipping lane is allowed, but the anchor has to be removed when vessels are passing. This Specification may result in a substantial loss of production capacity which delays the Project;
– anchoring in or across the shipping lane is allowed. The anchor wire has to be slacked when vessels are passing.

**Pipelines**

The use of a floating pipeline may cause hindrance to shipping, especially if the line has to cross the shipping lane. Another option is to use a sinker line. The sinker line can be installed on the seabed of the shipping lane. This will reduce the available water depth by +/- 1 m (depending on the pipe diameter). In case this is not possible, the submerged pipeline can be installed in a dredged trench. This should then be specified.

**Navigational aids**
The Contractor shall provide temporary navigational aids in order to maintain safe navigation throughout the progress of the Works.

**Permits and licences**
The Employer shall obtain those permits, licences or approvals in respect of any planning, zoning or other similar permission required for the Works to proceed, as stated in the Appendix. The Contractor shall obtain all other permits, licences and approvals required for the Works, with the reasonable assistance of the Engineer.

12.4.6   *Selection borrow area—quality fill material (see Chapter 5)*

The main purpose of a borrow area is to provide suitable fill material. Generally the required properties of the material can be split up per area or depth below the surface. High quality fill materials (as may be required for the top layers of a land reclamation) are often referred to as "selected fill". Bulk material is referred to as "general fill".

The following subjects can be identified:
– grading (minimum and maximum grain size, fines content, uniformity);
– plasticity;
– mineralogy;
– particle shape;
– strength.

It should be kept in mind that the feasibility of the requirements for the fill material within the reclamation depends on the characteristics of the material encountered in the borrow area. In order to check the feasibility of the Specifications, sufficient geotechnical data collected in the borrow area are required (see Chapter 3: Data Collection).

*Example:*

> *The ground investigation indicates that the material dredged from the approach channel and berth pocket used for the core will be predominantly a uniform, fine grained sand.*

> *The Contractor shall note that the materials are not receptive to self-weight compaction under and above the water level. It is likely to be in a loose state after placement, particularly when placed underwater. There will be a need to undertake some form of ground improvement to ensure that the performance of the finished causeway does not suffer from adverse total and differential settlement and liquefaction under seismic*

*load. The Contractor should take note of the issues highlighted above when designing the causeway utilising dredged fill material for the core.*

Comment:

> In this example, the Contractor is made aware of the consequences related to the use of specific fill materials. Since the Contractor has the responsibility for realising the performance requirements, these should clearly be defined in the specifications (see 12.4.9).

As discussed in Chapter 5, the properties of the source material can be altered as a result of the dredging and reclamation process. For instance, if a reduction of the fines content by overflowing is envisaged, then overflow in the borrow area should not be prohibited.

**Fines content within fill material**

Specifications generally refer to a maximum percentage of fines in the range of 10%–15%. The background for such a defined maximum however, is not always clear.

There are generally two distinct sets of reasons for limiting the fines content

1. limit the amount of fines for drainage and trafficability reasons. This especially applies to the top layer of the fill and can be addressed by defining a layer of selected fill with a limited allowance for fines (see Section 10.2).
2. limit the amount of fines to allow for a specific compaction method (see Chapter 7).

The effect of the fines content on the fill's geotechnical behaviour should be tested in the laboratory (see Section 9.1.3.4, Table 9.3).

Depending on the governing Standards, fines refer to the fraction smaller than 60 μm, 63 μm or 75 μm (see Section 3.6.1).

**Fines inclusions**

The presence of fines inclusions within the reclamation (e.g. as a result of segregation) does not necessarily mean that its behaviour with respect to bearing capacity or settlement would not satisfy the performance requirements (see Section 8.4 and 8.5). Before taking the decision to remove and replace silt inclusions, Zone Load Tests which simulate the behaviour of future shallow foundations or other loading conditions could be considered. The impact of silt inclusions may be limited, especially if those inclusions are encountered at several meters below the final reclamation level.

Another issue is the detection of the layers of fines. This can be done by drilling boreholes and subsequent laboratory testing of samples that have been visually selected as potentially containing too much fines.

A more practical approach is to use CPT testing. A criterion related to the cone resistance and friction ratio can be used to define a layer of fines. When necessary this criterion can be checked with boreholes.

*Example:*

> *The maximum allowable layer thickness of fine material in the reclamation area shall be or have:*
>
> – *100 mm per layer for all areas;*
> – *a cumulative thickness not greater than 100 mm for Zone A;*
> – *a cumulative thickness not greater than 200 mm for Zone B;*
> – *a cumulative thickness not greater than 300 mm for Zone C.*

*Example:*

> *In case silt inclusions of limited thickness are encountered within the reclamation fill, then the Contractor shall perform a geotechnical analysis to verify that the requirements with respect to bearing capacity and settlement under future loading conditions are satisfied. Remedial measures shall be undertaken if necessary.*

Often, the borrow area is defined in the Specifications. The source of the fill material can be a channel, a harbour basin or a dedicated borrow area. The quality and suitability of the in-situ soil in the borrow area will be reflected in the quality of the material deposited in the reclamation area (see Section 5.2).

The following additional subjects with respect to the borrow area should be addressed:

– Location, coordinates and depth of the borrow area.
– In case an unsuitable soil layer is covering the suitable material, this unsuitable material will have to be removed first. A disposal location for this material should be defined.
– When defining the boundaries, special attention should be paid to the volume and layer thickness of suitable material, the water depth, the orientation and the extent of the borrow area. Failing to do so will have a serious impact on the vessel's production, project cost and hence the feasibility of the project.
– Dredging tolerances.

*Example tolerances:*

> *The vertical tolerance on dredged level is 0.0 m above and 0.5 m below the profile defined on the drawings. The horizontal tolerance on the toe of a dredged batter shall be ±2 m from the line shown on the Drawings.*

Comment:

> The achievable tolerance depends on the type of equipment and environmental circumstances (waves, currents). A vertical tolerance of

0.5 m can be rather strict in certain circumstances, and may lead to reduced production and efficiency.

Is over-dredging allowed in case of the presence of a suitable soil layer underneath the design level of a channel?

*Example unsuitable fill material:*

> *Material that has been deposited in the reclamation which does not comply with the specified requirements shall be removed from site, disposed of by the Contractor and replaced by suitable fill. Unsuitable fill material includes amongst others:*
>
> − *peat, logs, stumps and perishable materials;*
> − *materials which are contaminated or susceptible to spontaneous combustion;*
> − *domestic and industrial wastes;*
> − *soils susceptible to deterioration/change of properties;*
> − *gypsum.*

---

**Low quality fill material**

Economic reasons generally play an important role. It may be cheaper to use low quality fill material from a borrow area located at a close distance from the reclamation site instead of high quality fill material from a borrow area at a long sailing distance. The use of low quality fill material will require a different approach towards the technical specifications. Specifications usually adopted for high quality fill material may not be applicable. In order to comply with the performance requirements, more intensive ground improvement works will be required.

Another option is to reconsider the design of the superstructures and their foundations, resulting in a relaxation of the performance requirements.

The extra costs for the ground improvement and/or modifications of foundation design should be balanced against the extra costs for the longer sailing distance of the high quality fill material.

---

### 12.4.7   *Construction methods reclamation area (see Chapter 6)*

The main requirement with respect to the construction methods of a land reclamation is to retain the fill material within its boundaries. Another important issue is the management of process water and fines.

*Examples:*

> *Preparation of the reclamation*
>
> *Part of the preparation for reclamation is site clearance. Rubbish, polluted soil, vegetation etc. needs to be removed before filling can start. This may be contaminated material which requires special handling, treatment or disposal.*
>
> *Methods*
>
> *Prior to the commencement of the Works, the Contractor shall submit the proposed methods for carrying out the work which shall include the layout and construction of any temporary bunds required to retain the dredged material, methods of working within bunded areas to achieve the required levels, method of limiting the escape of fines, arrangement of layers and compaction, and methods to prevent accumulation of fines.*
>
> *The placement methods shall be compatible with the design of the intermediate and final project phases.*
>
> *Reclamation Plant*
>
> *The Contractor shall provide plant fit for purpose for the proper and timely execution of the Work required under the Contract.*
>
> *Pipelines*
>
> *Pipelines required for deposition of fill material shall be of the shortest length possible and laid along routes and corridors approved by Client's representative. Any road and water crossings of pipelines shall be constructed such as to minimise interference with traffic both during construction and operation of pipelines, and to the approval of Client's representative.*
>
> *Safety*
>
> *Prior to the commencement of the works, the Contractor shall submit a safety plan.*

Process water and fines management

See Section 12.4.8

Access to the project
In case the project area needs to be accessible during construction (by land or by water) the location and design of the access road and channels should be specified.

Slope stability
Fill material shall be deposited in a manner such that all temporary and permanent slopes created by the filling operations are stable at all times. This does not apply for slopes which occur at the fill front during the fill process unless such instabilities may be of consequence for the quality of the fill mass.

Segregation of fines
Segregation of the fill is inherent to the hydraulic deposition method and cannot be avoided. Only measures can be taken to limit segregation as much as possible.

*Example:*

> *The materials shall be discharged at such points over the fill area so as to optimise the homogeneity of filling; in particular, the Contractor shall take measures to avoid producing pockets of fines in the reclamation.*

> *If pockets of fines, being particles of 63 microns and less, occur and the cumulative depth at any location exceeds 150 mm then the Contractor shall remove the fines and replace it with suitable material or mix with acceptable material to the satisfaction of the Engineer.*

*Siltation:*

> *The Contractor shall take all precautions to avoid or prevent siltation of completed areas of the Works and in the vicinity arising from his execution of the Works and shall remove any such siltation to the satisfaction of the Client. This does not include siltation as a result of morphological changes related to the design of the Works.*

*Reclamation tolerances:*

> *Final reclamation levels shall be to the design levels compatible with the construction of the paving, crane beam, rock protection, and other structures and subject to permitted overall tolerance of ±100 mm.*

### 12.4.8    *Environmental impact*

The environmental impact of a dredging and land reclamation project has been discussed in a general matter only in this Manual. Issues involved are fines content of discharge water, turbidity around the dredger, noise generation, impact on fauna and flora. For more information, reference is made to specialized literature such as Bray 2008, Environmental aspects of dredging.

*Examples:*

> *The criteria applicable to dredging and reclamation relate to good dredging practice and any pollution caused by operations shall be minimized. The Contractor shall take all necessary preventive measures to minimize water pollution and mitigate effects on the environment of other vessels at or in vicinity of the Working Area.*

> *The run-off water from the reclamation areas shall be discharged via temporary stilling ponds if necessary, to ensure that it is clean and free*

*from suspended matter. Suspended solids of the final discharge water to the sea shall not exceed xxx mg/l.*

Comments:

The specified turbidity value of the run-off water depends on local environmental requirements and physical conditions such as:

o  the characteristics of the fill material;
o  availability of space for the construction of a settling pond;
o  sufficient residence time of discharge water;
o  background turbidity values;
o  measuring distance from the outlet.

If stilling ponds are specified then a sufficient large area should be available on site in order to guarantee their effectiveness.

**Loss of dredged material during transport**

*Example:*

*Contractor's equipment used for transporting dredged material shall be properly maintained so that losses of material transported between the dredging area and the disposal area are kept to a minimum.*

12.4.9  *Design of a land reclamation (see Chapter 8)*

For design and construct projects, the functional requirements must be specified. It is strongly advised to split-up the reclamation area into various sub-areas based on the future function and use of these areas. This results in a cost-effective design compared with the traditional approach where the entire reclamation should have the same performance requirements. For instance, when designing a reclamation for an airport, the area underneath the runways will have more stringent requirements compared to, for instance, areas reserved as separation zones or car parking.

The following subjects can be identified:

–  lines and levels;
–  bearing capacity and slope stability;
–  allowable residual settlement;
–  differential settlement;
–  rate of deformation;
–  liquefaction resistance;
–  design of perimeter bunds.

**Lines and levels**

The reclamation shall be constructed within the boundaries indicated on the contract Drawings. This includes x-y coordinates, design levels, slopes, etc. Vertical and horizontal tolerances should also be specified.

*Example:*

> *Final reclamation levels shall be as shown on the Drawings subject to permitted tolerances of +/− xxx mm.*

Comment:

> Horizontal tolerances are less common in reclamation specifications, since these generally relate to the boundary structures, such as embankments or quay walls.

**Bearing capacity and slope stability (see Section 8.2.1)**

Bearing capacity and slope stability requirements can be translated into technical requirements by specifying the fill material properties such as required angle of internal friction or shear strength. Since the friction angle is not easy to measure in the field, often a correlation with other parameters such as the Relative Density, degree of compaction or the Cone Resistance is used.

*Example:*

> *Fill for use behind earth-retaining structures shall have a minimum internal friction angle of 33 degrees when compacted. The corresponding degree of compaction of the fill material shall be tested in a laboratory (triaxial or shear box tests).*

See Section 5.3.1 and Section 8.4.2.3.

For design and construct projects, the Contractor must guarantee bearing capacity and slope stability. The applicable Standards, the required safety factors and the design approach should therefore be specified (see 8.4.3.2). For the required bearing capacity, the Specification must describe the design load, the dimensions and the depth of the foundation structure.

*Example:*

> *The bearing capacity of the reclamation area shall be 100 kN/m², applicable to a foundation size of 5 m × 5 m at a depth of 1.0 m below the final level.*

**Allowable residual settlement (see Section 8.2.2)**

Reclamations constructed on soft subsoil will experience settlements to occur as a function of time. Specifications normally set requirements with respect to a maximum allowable residual settlement after construction of the reclamation at time of handover. The magnitude of the residual settlement should depend on the future use of the reclamation. When defining the final reclamation level, this residual settlement should be taken into account. An option is to compensate for the anticipated residual settlement in the design level to be specified. When constructing a reclamation on subsoil consisting of thick layers of cohesive material, considerable secondary settlement may occur. Usually residual settlements of the order of 10 to 30 cm are allowed, also accounting for secondary compression (creep). It should be clearly specified if this residual settlement includes the effect of future loading conditions.

It is noted that specifying a very stringent residual settlement may result in very high additional project cost.

**Differential settlements (see Section 8.2.2 )**
Specifying maximum allowable differential settlements for large reclamation areas generally makes no sense, since this would require a very intensive soil investigation program and a very dense monitoring grid. A differential settlement requirement is more relevant for the design of individual structures such as roads or buildings to be constructed on the reclamation. For these structures a specific soil investigation should preferably be undertaken after the reclamation phase. Specifying a temporary preload may be an option to limit future differential settlements.

Differential settlement problems may occur at "hard spots" such as previous foundations or buried structures. Naturally occurring deposits often vary so gradually that differential settlements are rarely an issue. However, in case the subsoil is highly heterogeneous, differential settlements may occur.

**Rate of deformation (see Section 8.2.2)**
Another way of specifying an allowable residual settlement is to define the required degree of consolidation at time of handover (for instance a consolidation degree of 90%). It is important to realise that the consolidation degree only refers to the primary consolidation and does not include the secondary compression (creep deformation). When specifying the required degree of consolidation, the corresponding load must also be specified (own weight of the fill, preload, future design load).

**Consolidation and secondary compression**

It is often specified that a certain percentage (90%–95%) of the primary consolidation must have occurred at the time of handover of the reclamation. It is noted that a requirement stipulating 100% consolidation is not realistic since settlement is a function of the logarithm of time. In fact the time needed for occurrence of the last 5% consolidation settlement is, on an average, more or less equal to the time period required for occurrence of approximately 95% consolidation, even when vertical drains are installed.

Requiring a higher degree of consolidation (under the weight of the fill) will imply the use and specification of a temporary preload.

Specifications do not always consider or refer to secondary compression (creep). The contribution of the secondary compression to the total settlement can be considerable when thick very soft cohesive layers underlie the reclamation site. Vertical drains do not help to accelerate the secondary compression. Reduction of the secondary compression after handover can only be realized by means of a temporary preload that is in place for a sufficient long time period which is not always available

**Seismic conditions (see Section 10.6)**
For seismic design, following elements should be specified:

– Applicable Standard (e.g. Eurocode 8);
– Performance objective (no local damage versus limited damage allowed);
– Design earthquake: Peak ground acceleration PGA (bedrock level or fill level), earthquake magnitude M. Alternatively the Contractor can be instructed to perform a site specific seismic study;
– Importance class of structures to be built and corresponding importance factor.

See Section 8.6.4

There is generally no need for a seismic design in countries or regions with a very low seismic activity. This is defined by Eurocode 8 as either a reference Peak Ground Acceleration at bedrock level not greater than 0.04 g or where the site specific amplified design ground acceleration is less than 0.05 g.

For stability calculations in earthquake conditions, the design method and related safety factors should be specified.

**Liquefaction resistance (see Section 8.6)**
One of the performance requirements for land reclamation is that it should withstand a certain design earthquake. Specifications related to liquefaction and deformations due to earthquakes should be developed with this requirement in mind. There are three issues in developing a specification: (i) the design earthquake hazard needs stating, for example 2% chance of exceedance in 50 years (and their may be mandatory local codes stating what this requirement should be); (ii) the required minimum material properties (density) will depend on the analysis method used, so the methodology needs stating; and, (iii) the fill will be variable even with the most exhaustive work and that variability needs to be recognized. Finally, the specification must be written in terms of what is to be measured on site so that there is no interpretation of the specified requirements during project execution.

*Example:*

*Contractor shall eliminate the risk of liquefaction in a UBC Zone 2A earthquake, with magnitude not exceeding 6 and a maximum ground acceleration of 0.15 g at bedrock level. Calculations shall be based on recognized methods of liquefaction analysis as defined by Youd et al.; Geotechnical and Geo-environmental Engineering Journal, Oct. 2001. The minimum safety factor against liquefaction shall be 1.25 (see 8.6.5.1). Verification that this target has been met shall be by CPT soundings and 90% of the measured data shall be better than the penetration resistance profile so established.*

*The amount of fines in a fill will affect the penetration resistance, and this can be a source of dispute as to whether low resistance is caused by "high fines" content or by too loose fill. One way this issue has been dealt with is by explicitly including the consideration in the Specification rather than simply providing a required minimum penetration resistance profile, figure 12.4 below illustrating the approach where the 'friction ratio' (F) measured using the CPT is used to directly adjust the required penetration resistance. Also note that this example imposes a requirement for both an average density (the "50% greater than") and a consideration over loose pockets (the "90% greater than"). The y-axis in this figure is the "normalized" penetration resistance $Q = q_t/\sigma_v'$, which removes depth effects on the required penetration resistance.*

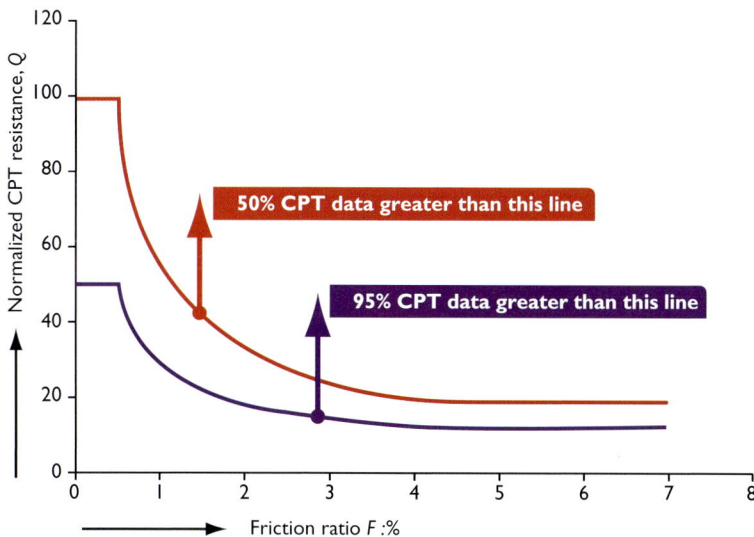

Figure 12.4   *Example of minimum required percentage of CPT values for a sandfill as a function of the friction ratio F.*

**Design of perimeter bunds (see Section 10.4)**
Besides the stability, other topics such as erosion and overtopping are to be taken into account. Reference is made to the Rock Manual (2007).

---

**Over-specification**

When primary specifications are properly defined, many other commonly used material specifications become unnecessary.

*Example:*

1. *Sufficient bearing capacity for buildings with strip foundations of 1 m wide at 0.5 m depth, carrying 100 kN/m².*
2. *Residual settlement <20 mm (under 100 kN/m² on strip foundation)*
3. *No liquefaction of fill material for design earthquake*
4. *Coefficient of uniformity >4*
5. *Percentage of fines <10%*
6. *No silt layers allowed in the fill*

In order to comply with the performance requirements 1 to 3, the Contractor will probably have to compact the fill material. The lower the coefficient of uniformity (poorly graded sand), the more compaction energy will be required to achieve a certain Relative Density or Cone Resistance (req. no. 4). The higher the percentage of fines, the more compaction energy will be required (req. no. 5). In case of thick silt layers, the residual settlement might become too large. It might be necessary to improve or remove this layer (req. no. 6). The requirements 4, 5 and 6 are, in principle, the Contractor's responsibility since he is bound by the specified requirements 1 to 3.

---

### 12.4.10 Ground improvement (see Chapter 7)

The function of ground improvement is to improve the soil characteristics in case the performance requirements cannot be met.

The following subjects can be identified:

– increase of the soil strength by compaction;
– reduction of the residual settlement;
– increase of the consolidation rate of the fill and/or subsoil.

**Increase of the soil strength by compaction**
Well-graded granular material without fines is normally, without additional measures after deposition, suitable in terms of bearing capacity and stiffness. Project sites located in earthquake areas may, however, require additional

compaction of the fill in order to increase its resistance against liquefaction. Sometimes the compaction method is prescribed in the Specifications (for instance: compaction shall be done by vibroflotation). However, allowing for alternative compaction methods can be more practical for technical and economic reasons.

> **Compaction in small layers**
>
> A requirement to compact in small layers (30 to 50 cm) is not compatible with reclamation works were large quantities of fill material are discharged in a short period. Special compaction techniques that allow for compaction of thicker layers are, at present, commonly used in the dredging industry.

*Examples:*

> *During the reclamation process the Contractor shall place approved material within the reclamation area to achieve an in-situ density not less than 90% of the Maximum Dry Density (MDD) throughout the full thickness and lateral extent of the fill. The uppermost zone of the fill shall be placed and treated to ensure that the top 900 mm has an in-situ density not less than 95% MDD.*
>
> *Compaction of the fill material shall be carried out by an agreed ground treatment method in order to achieve the specified minimum Piezocone penetration test Cone Resistance (PCPT) criteria.*

Comment:

> Special attention should be paid to the saturation of the cone when testing fill placed above the water table. Since saturation of the cone can be time consuming, often only standard CPT testing is performed when significant fill layers are present above the water table.

**Timing of performance verification testing**

*Example:*

> *Unless agreed otherwise with the Engineer, a minimum period of two weeks shall be allowed to elapse between completion of any ground densification in an area and the commencement of testing in that area.*

Sometimes compaction is specified both in terms of minimum relative density and minimum cone resistance. This is an over-specification since the two requirements can be correlated to each other.

**Optimum moisture content**

Often moisture content related requirements can be found in Specifications.

*Example:*

> *The water content of the fill material shall be within +/- 2% of the optimum moisture content during compaction.*

Comment:

> This is a compaction related requirement originating from a thin layer compaction scheme (such as for road construction). It is not suitable for reclamation works whereby hydraulic fill is generally placed in thicker layers.

A more appropriate example is:

*Fill placed in the dry shall when necessary be watered by spraying, but the Contractor is expected to arrange his operations so as to make the best use of the natural moisture content of the fill material to obtain the required in situ density.*

**Reduction of the residual settlement (see Section 8.5.3.4 and Section 7.4.1)**
The residual settlement can be reduced by installing a temporary surcharge.

In case the surcharge method (magnitude of the preload and required degree of consolidation under the surcharge load) is specified, then additional specifications for residual settlements after construction may not be compatible.

Reference is made to Section 12.4.9 paragraph "rate of deformation" for some further comments on this topic.

**Increase of the consolidation rate of the fill and/or subsoil (see Section 8.5.3.4 and Section 7.4.2)**
The consolidation rate can be accelerated by installing prefabricated vertical drains. The design of the vertical drain system is usually the responsibility of the Contractor, since it is related to the adopted construction program. Sometimes the drain spacing (triangular or rectangular) and depth are specified in Contract documents. Since the performance of a vertical drain system is not accurately predictable, field trials are recommended. Sufficient time for such trials should then be foreseen in the project schedule or executed before tender stage.

It should be noted that the shorter the construction period, the more ground improvement will be required to achieve the reclamation criteria. Whilst incurring a loss of revenue, allowing an increase in construction time will often reduce the construction cost.

**Preliminary trial for ground improvement**

*Example:*

> *If a ground densification technique is to be adopted to meet the perform-ance criteria detailed herein, for each particular type of ground densi-fication, prior to the main ground densification a preliminary field trial shall be carried out. Each trial area shall be of minimum plan dimen-sions of 10 metres by 10 metres and the location of the preliminary field trial shall be agreed between the Engineer and the Contractor.*

> *The design, sequence and timing of activities and equipment shall be exactly the same as those proposed for the main works. Quality control and performance testing procedures for the trial are to be in general accord-ance with those proposed for the main works. If the Contractor wishes to carry out more than one design layout then the size of the trial area shall be extended, subject to the Engineer's agreement. If more than one trial area is used there shall be a minimum of 10 m between each trial area.*

### 12.4.11 *Special fill materials (see Chapter 9)*

**Cohesive or fine grained fill material (see Section 9.1)**
When using cohesive or fine grained material as fill material, ground improvement techniques may be required to satisfy the performance requirements.

**Carbonate sand (see Section 9.2)**
In literature different correlations are found for the Relative Density and the Cone Resistance as a function of the effective stress. For equivalent Relative Densities, the Cone Resistance measured in calcareous sand is less than in quartz sand because of the compressibility and crushing effects of the particles. Depending on the compressibility of the (calcareous) sand, compared with "reference" sand, a shell factor or cone correction factor has to be used. For clarity, the Specifications should preferably prescribe the following subjects:

- the correlation between relative Density and Cone Resistance applicable for the project;
- the value for the shell- or cone correction factor in case the fill material consist of crushable sand.

In case of lack of experience or doubts with the available fill material then it is recom-mended to perform Calibration Chamber Tests which enables the actual generation of such a correlation. These tests are preferably performed before the Tender phase.

*Example alternative fill material:*

> *At the time of tender, the Contractor may propose the use of fill material which initially might not comply with the Specification but does comply with the performance requirements of the reclamation, possibly to be achieved after ground improvement.*

### 12.4.12   Other design aspects (see Chapter 10)

The following subjects can be identified:

– drainage;
– existing infrastructure and pipelines;
– wind erosion and dust creation.

### Drainage (see Section 10.2)

As a result of hydraulic filling and/or precipitation, an elevated water table is often observed in reclaimed areas, which could affect the accessibility, stability and settlement behaviour. Additional measures to lower the water table (to a pre-defined level) may then be required and must as such be specified:

– placement of a top/bottom layer with sufficient permeability;
– horizontal drainage;
– artificial dewatering (pumps);
– ditches.

*Example:*

> *The Contractor shall be responsible for maintaining water courses including land and/or road drainage within the Site in effective working condition at all times.*

### Existing infrastructure and pipelines (see Section 10.5.4)

*Example:*

> *The Contractor shall carry out operations in filling and discharge of water from the fill area without disturbance to existing works and structures and in accordance with methods approved by the Engineer.*

> *The Contractor shall accurately locate and mark with buoys the existing submarine cables, pipelines, etc. across the reclamation and borrow areas in the working area.*

*No dredging works shall be carried out in vicinity of the existing structures until the Contractor has submitted a method statement to the satisfaction of the Client and ensure that the stability of the structures will not be affected.*

Wind erosion and dust creation (see Section 10.3)

*Example:*

*For fill material on those parts of the reclamation where the delivery height is reached measures shall be taken in case the top layer dries up in order to prevent discomfort caused by dust and sand drift until the time of handover.*

*In Section 10.3, different types of measures to prevent wind erosion and dust creation are discussed. These could be prescribed in the Specification.*

### 12.4.13   *Monitoring and quality control (see Chapter 11)*

This Section presents some examples of specifications related to monitoring and quality control.

The following subjects can be identified:
- Geometry
- Bathymetrical survey dredging area
- Bathymetrical/Topographical survey reclamation area
- Testing fill material properties
- Particle size distribution
- Minimum and maximum density
- mineralogy
- Testing fill mass properties;
- Shear strength
- Stiffness
- Density
- Settlement monitoring
- Performance testing
- Environmental monitoring
- Reporting
- Monitoring and Quality Control Program

For all procedures and testing methods reference must be made to the applicable Standards in order to avoid any confusion (Eurocode, British Standards, ASTM, etc.).

12.4.13.1   Geometry

Tidal Levels

*Example:*

> *For the duration of the Contract the Contractor shall install, calibrate and maintain an approved automatic recording tide gauge. The tide recorder shall be installed at a location as close as possible to the project area and accurately levelled and related to the Reference Level.*

**Bathymetrical survey dredging area (see Appendix B.1)**

*Examples:*

> *Positioning*
>
> *The Contractor shall submit details of his proposed method of position fixing for the approval of the Client before any reclamation or survey work is commenced.*
>
> *Echo-sounder*
>
> *All soundings shall be carried out by means of a recording trace dual frequency echo-sounder with sufficient sensitivity to permit measurement of the bed levels to an accuracy of +/- 100 mm. The echo sounding equipment shall be regularly calibrated by means of a sounding plate.*

Comment:

> Multibeam systems become more and more common practice.

Pre-dredge Survey
Before the commencement of dredging, the whole area is to be surveyed and sounded. Plans and sections are to be drawn accordingly.

Interim Survey
The Contractor shall carry out interim surveys to assess the progress of the works (at least on a monthly basis).

Final Survey
*At the completion of the dredging works, Contractor shall undertake a final survey and prepare final plans and sections deemed to represent the final configuration of the specified areas.*

**Bathymetrical/Topographical survey reclamation area (see Appendix B.1)**

*Examples:*

*Pre-fill Survey*

*Prior to commencement of filling and after the site has been cleared the Contractor shall survey the area to be reclaimed on a grid of 10 m maximum spacing.*

*Interim Surveys*

*Client may order levels and soundings to be taken at any time as necessary for the proper supervision and measurement of the work.*

*Acceptance Surveys*

*Contractor shall demonstrate by means of soundings that the work has been executed to the required lines and levels. Within x working days of the notified date of completion of the filling works the Contractor shall carry out surveys using the same grid as was used for the pre-fill survey to demonstrate such completion.*

*Survey Drawings*

*All soundings shall be related to the Reference Level and the Contractor shall plot the results of the surveys at a scale of 1:1000 on a regular xx m square grid. The drawings shall clearly mention the sounding date, coordinate projection method and reference level. The plot shall be agreed and signed by both the Contractor and the Client and used as a basis for measurement. Subsequent alterations to the drawings shall be made if they are mutually agreed by the Contractor and the Client.*

**Morphological processes**

In practice there is always a time period in between the initial bathymetrical survey performed during the design phase and the actual execution of the works. During this period morphological processes may have altered the bed levels. The differences in bed level between the in-survey at time of execution and the initial survey may be compensated for in the Contract.

Sometimes intermediate surveys are specified in the Contract. These surveys could best coincide with the milestones set for payment.

12.4.13.2    Testing fill material properties (see Section 11.3.2)

Testing shall be carried out by the Contractor in the laboratory to determine whether or not the materials placed comply with the fill material properties described in the Specifications.

The tests described in this section can be carried out on disturbed samples. Samples can be taken from a borehole or as a bag sample.

**Bag samples**
The sampling frequency can be time related (e.g. two samples per day) or volume related (e.g. one sample every 10.000 m³). It is also possible to take samples on board of a trailer hopper suction dredger.

**Boreholes**
Boreholes can be drilled to sample the fill material throughout the entire layer thickness of the fill. Alternatively CPT testing can be used after calibration with the borehole results. The advantage of CPT testing is the speed of execution and the continuous recording of the Cone resistance, the friction ratio and the pore pressure.

**Tests**
Material properties frequently tested are:

– particle size distribution;
– mineralogy;
– minimum and maximum density, moisture content/dry density relationship.

The minimum and maximum density of the fill can be tested on the material collected in bag samples. Since these values strongly relate to the particle size distribution, the frequency of testing should be adjusted to the observed changes in the particle sizes.

It is strongly advised to clearly specify which test method is to be adopted for the determination of the minimum and maximum density. Generally the Modified Proctor test is specified for testing the maximum density. However, in case of crushable materials, the use of the vibration table test is more appropriate. (See Section 9.2.5.3).

**Bulk density, specific gravity**

Sometimes other tests on the bulk samples, such as specific gravity or bulk density are also specified:

– The determination of the specific gravity (particle density) on each sample is not relevant. There is generally very little variation in specific gravity. The determination of the specific gravity on a few selected samples of the generic soil type will provide sufficient information.
– The density of a (disturbed) bulk sample is not relevant.

*Example:*

> *Every day during the progress of filling, the Contractor shall take two bag samples (of 25 kg each) of the materials placed in the reclamation at locations directed by the Engineer. Samples shall be taken at a maximum depth of 0.5 m. The Contractor shall carry out the following tests on each of the bag samples:*
>
> *– sieve analysis*
> *– particle size distribution by hydrometer (when appropriate).*

12.4.13.3   Testing fill mass properties (see Section 11.3.3)

Following fill mass properties are relevant for the design of a reclamation area and should be tested in the field. (See Section 12.4.9)

– (shear) strength;
– stiffness;
– density.

Various direct or indirect tests are available to assess these fill mass properties (see also Appendix B).

**Shear strength**
The shear strength of the fill can be tested directly with a field vane test in a borehole in case of cohesive material.
The shear strength can also be correlated to the results of a CPT (or SPT) test (see Appendix C).

**Stiffness**
The stiffness modulus of the fill mass can be tested by following methods

– plate load test for surface layers;
– pressure meter test for deeper layers;
– by correlation with CPT results.

However, it is more common to test and measure the deformations rather than the stiffness modulus itself, (See 12.4.13.4 and 12.4.13.5.

The seismic cone can be used to indirectly test low strain stiffness, an important parameter for the analysis of the liquefaction potential.

The low strain stiffness of a fill can also be tested with the SASW and CSWC test (See Appendix B).

**Density**
In situ density testing (upper zone of fill)

*Example:*

> *In situ density testing shall be carried out on the upper 900 mm of fill material to determine the degree of compaction achieved in accordance with BS1377: Part 9, section 2.1 or 2.2. The rate of testing shall be one test per xxx m².*

Comment:

- o It should be noted that prior to the in situ density test, the test location should be excavated to the specified depth. It is therefore not practical (or safe) to have in situ density tests performed at depths larger than +/- 2 m.
- o BS1377 Part 9, section 2.1 and 2.2 refer to the sand replacement method. Alternatively, surface nuclear density testing can be adopted.
- o Please note that BS1377 Part 9 has partially been replaced by European standards.

*Example:*

> *The Contractor is allowed to use nuclear methods to determine the in situ density test providing the moisture contents and bulk densities for each type of material have been calibrated to the satisfaction of the Client's or the contract's approving authority.*

In situ density testing deeper layers
The in situ density of deeper layers is generally tested by means of a CPT test. Correlations between CPT testing and relative density are used (see appendix C). It is strongly advised to specify which correlations and corrections are to be used.

*Example:*

> *The Contractor shall perform static Cone Penetration Tests on a square grid of x m for the whole extent and depth of the reclamation that has been filled with dredged material. Where the soundings show Cone Resistance values less than the specified profile, the Contractor shall submit his proposal for improving the density of the fill. Repeat soundings shall be undertaken following any ground improvement. When evaluating the cone resistance values, a rolling average over 1 m depth shall be adopted.*

Comments:

- o The sounding grid x generally varies between 25 and 250 m, depending on the heterogeneity of the fill and subsoil, see Section 11. For large areas the need for a narrow grid must seriously be considered since this may lead to an excessive number of CPTs.

○ When reclaiming and compacting in various lifts, often interim CPT soundings are performed. The function of these sounding is to check the progress and effectiveness of the compaction technique and should not necessarily follow the same dense grid as the final CPT soundings.

○ In case the CPT is used to evaluate the compaction results of vibroflotation compaction, the CPT test locations relative to the vibroflotation grid should be specified. (See Section 7.5.3, Figure 7.10).

**Time interval filling operation/CPT testing**

The Cone Resistance may increase in time as a consequence of "aging". It is therefore recommended to allow for a time interval of 3 to 4 weeks in between the filling operations and the execution of CPTs for checking the conformity with the Cone Resistance profile specified in the Contract.

Another reason for the introduction of such a time interval is to allow for a lowering of the groundwater table within the hydraulically placed fill material which may also have a beneficial effect on the cone resistance.

**Site specific correlation Relative Density—$q_c$ (CPT)**

In view of the difficulties related to the correlation of the Relative Density to the Cone resistance values (see Section 8.4.3.5), rather than using literature correlations, a practical way forward could be to setup a site specific correlation. A tempting approach could be to correlate CPT test results with density measurements on undisturbed samples originated from adjacent boreholes. This is however not possible since undisturbed samples cannot be retrieved from boreholes in non-cohesive soils. Therefore, when setting up a site specific Relative Density – $q_c$ correlation, calibration chamber tests are required. New promising developments with nuclear density probes allow for direct measurement of densities (see Appendix B.2.3.2).

**Alternative methods**

*Example:*

> *The Contractor may wish to submit alternative methods for demonstrating that the fill has achieved the specified density. In that case the Contractor shall submit a detailed method statement for approval.*

12.4.13.4   Settlement monitoring (see Appendix B.5.3)

Generally, the settlement of a reclamation area is monitored by means of settlement beacons. Specifications should describe the position, grid spacing, monitoring interval and period of such beacons.

*Example:*

> *The settlement of the reclamation area shall be monitored by surface settlement beacons. These beacons shall be installed on a grid of x m. The surface settlement beacons shall be installed as soon as the fill is raised above the waterline. Settlement readings shall be taken each xxx days in the first yyy days after placement of the fill, and subsequently each zzz days until time of handover.*

Comment:

> The placement of settlement beacons underwater prior to the start of the fill operations is not practical in view of the difficulty of placement and fixation of the markers.

Another possibility to monitor the settlement of individual soil layers is to install magnetic extensometers. Additionally, piezometers can be installed at various depths to monitor the degree of consolidation.

**Piezometers and vertical drains**

The installation of piezometers within a field of vertical drains is useless since the exact position of the piezometer relative to the drain is unknown. It cannot be guaranteed that the piezometers are centrically positioned in between the drains. The closer the piezometer is positioned to a drain, the lower the measured excess pore water pressure which is not representative for the consolidation of the cohesive subsoil.

It may be more appropriate to accurately monitor the settlements as a function of time. Forecasting techniques (such as the Asaoka or hyperbolic method) can be used to evaluate the actual degree of consolidation.

12.4.13.5   Performance testing

Zone Load Test (see Appendix B.5.1.3)

*Example:*

> *For each acceptance area, after completion of the Cone Penetration Testing, the Contractor shall submit the results of the CPT testing to the Employer*

*and propose locations for Zone Load Test(s) (ZLT). The location for the ZLT should be selected to coincide with areas where the CPT results are lowest within the fill depth zone to be significant stressed by the Zone Load Test. Each acceptance area test shall be tested with at least one ZLT.*

*If there is significant variation in the results of the CPT profiles, the number of ZLT's should be increased accordingly.*

*In the area of the ZLT, 4 CPTs shall be carried out prior to the ZLT being performed (one in each corner of the test zone). The Zone Load Test should generally be carried out and reported in accordance with the ICE Geotechnical Engineering Group Specification for Ground Treatment, 1987 (ISBN 0-7277-0388-9).*

*The Zone Load Test shall load a rigid square or rectangular base of minimum 3 m by 3 m dimension in five equal increments to 100% of the design bearing pressure of xxx kPa. Each increment shall be applied for a minimum of 2 hours with the final increment being left until 90% of the primary consolidation has been achieved, with a minimum period of 24 hours. Settlements shall be recorded at the mid-point of each edge of the rigid base. Upon completion of the test the rigid base shall be unloaded and the recovery recorded. The maximum acceptable final settlement under the ZLT shall be 25 mm at the design bearing pressure of xxx kPa. The final settlement shall be calculated by extrapolating the settlement readings to a point of 100% primary consolidation.*

**Preloading trial (see Appendix B.5.1.4)**

*Example:*

*For preloading trials the Contractor shall surcharge the reclamation area by placing a 20 m sided square 6 m high trial embankment. If preloading is provided, this would be carried out when the reclamation has reached its final elevation. The trial embankment shall remain undisturbed for a period to be agreed with the Engineer.*

**CBR Test (see Appendix B.5.1.1)**

*Example:*

*The fill formation shall have a CBR of 30% within the upper 300 mm of fill and shall thus be adequate for pavement design.*

Comment:

The CBR test is generally used for pavement design. The pavement contractor will usually re-compact the top layer of the fill prior to the installation of the pavement. The influence depth of such a test is very limited, and therefore the CBR test is less relevant for the performance of the fill mass.

## 12.4.13.6    Reporting

The Specification should contain a reporting requirement for the Contractor with respect to monitoring and laboratory results. Generally the reclamation area is split up in acceptance areas. For each area a handover report is required.

Additionally, it is important that any activity or event regarding fill or subsoil is recorded with supporting data.

*Example:*

> *The Contractor shall maintain a daily report for each dredging activity and a copy of such report shall be given to the Client on the following day. These reports shall show the following data in a format agreed with the Client:*
>
> a.  *Date and time;*
> b.  *Weather;*
> c.  *Tidal levels and wave condition;*
> d.  *Dredge name and fleet no;*
> e.  *Dredging plant and equipment;*
> f.  *Schedule of dredging time, delays for position shifts, maintenance, down time due to shipping movements, breakdowns, etc.;*
> g.  *Estimated daily volume and production;*
> h.  *Position fix, original ground level, cut depth and cut width at start of each shift (including graphical plot);*
> i.  *Daily site log;*
> j.  *Lengths and diameter of floating line;*
> k.  *Lengths and diameter of shore line;*
> l.  *Presence or otherwise of booster pump or pumps with pump fleet number and position;*
> m.  *Locations and height of board on weir boxes;*
> n.  *Notes concerning operating conditions, delays, etc;*
> o.  *etc.*

## 12.4.13.7    Monitoring and Quality Control Program (see Section 11.2)

The specifications should contain a clear Monitoring and Quality Control Program, indicating the location and frequency of the required measurements and tests. This is, however, extremely project related and cannot be fitted in a standard template. Reference is made to Chapter 11 for more details about such a program.

Table 12.2 below is only presented in order to provide a very indicative basis for determining the quantity and frequency of some of the main activities related to a monitoring program. One should always realize that a Quality Control Program should be "fit-to-purpose" accounting for the quality of the fill material and the performance criteria of the fill mass and underlying subsoil considering the boundary conditions of the project.

*Example:*

*Maasvlakte 2, Port Extension Rotterdam, The Netherlands*

*The quality control for this project is into a great extend based on process monitoring. The following sampling sequence has been adopted for checking the quality of the hydraulic fill material:*

- *trailing suction hopper dredge: 1 sample per hopper load*
- *cutter suction dredge: 1 sample per 50.0000 m³ fill material in place*

*A specific working procedure (placing hydraulic fill as much as possible by dumping/placing dry fill material in lifts of 0.5 m/etc.) is followed in order to achieve an acceptable density. The following sampling and testing sequence has been adopted:*

*Terminal areas:*

- *1 borehole per 250 m × 250 m down to the original seabed;*
- *1 Cone Penetration Test per 250 m × 250 m;*
- *1 Settlement Plate per 200 m × 200 m.*

*Slopes:*

- *1 Cone Penetration Test per 250 m′ (verification of potential for static liquefaction).*

*Line infrastructure:*

- *Proctor testing every 250 m′;*
- *Shallow boreholes including sampling every 500 m′;*
- *Railways: 1 Settlement Plate every 100 m′;*
- *Roads: 1 Settlement Plate every 200 m′.*

Table 12.2 *Example of typical testing frequency.*

| Description of test | Testing conditions | Density of testing | Number of tests (*) Small project | Large project | Frequency of readout | Remarks |
|---|---|---|---|---|---|---|
| Cone Penetration Tests | – final acceptance testing | $1/2,500 \text{ m}^2 - (50 \times 50 \text{ m})$ (**)<br>or $1/10,000 \text{ m}^2 - (100 \times 100 \text{ m})$ | 400<br>100 | 4,000<br>1,000 | after compaction | use a moving average value over 1 m of the mean Cone Resistance |
| | – interim testing | $1/10,000 \text{ m}^2 - (100 \times 100 \text{ m})$<br>or $1/40,000 \text{ m}^2 - (200 \times 200 \text{ m})$ | 100<br>25 | 1,000<br>250 | | |
| Boreholes | | $1/250,000 \text{ m}^2 - (500 \times 500 \text{ m})$ | 4 | 40 | | |
| Field density | – above the waterline | $1/10,000 \text{ m}^2 - (100 \times 100 \text{ m})$ | 100 | 1,000 | | |
| Moisture content/dry density correlation, minimum and maximum dry density, carbonte content, CBR | | $1 \times 90,000 \text{ m}^2 - (300 \times 300 \text{ m})$ | 11 | 111 | | or if the particle size distribution is clearly different in certain area's |
| Bag samples | – in the hopper | 1/load | | | | laboratory tests: particle size distribution |
| | – on reclamation area | 2/day<br>or $1/20,000 \text{ m}^3$<br>or $1 \times 10,000 \text{ m}^2 - (100 \times 100 \text{ m})$ | 250<br>100 | 2,500<br>1,000 | | |

| Surface settlement beacons | 1/10,000 m² – (100 × 100 m) or 1/62,500 m² –(250 × 250 m) | 100 16 | 1,000 160 | weekly/ fortnightly/ monthly | the monitoring frequency should be increased in the period after placement of subsequent fill lifts |
|---|---|---|---|---|---|
| Piezometers, magn. extensometers water standpipe | | | | | at trial embankments, slopes or at specific instumentation clusters (not in combination with PVD's) |
| Inclinometers | every 250 m | | | | along the perimeter of a reclamation in case of risk of instability (soft subsoil) |
| Zone Load Test | 1 × 250,000 m² – (500 × 500 m) | 4 | 40 | | per acceptance area at the weakest location: selection based on CPT results (or as a replacement for other tests) |

(*) small project: 1,000,000 m²- 5 m layer thickness
large project: 10,000,000 m2- 5 m layer thickness
(**) grid to be reduced in case of non compliance

A   Dredging equipment                                             478
    A.1   Cutter Suction Dredgers (CSD)                            478
          A.1.1   General lay-out                                  478
          A.1.2   Dredging method                                  479
          A.1.3   Possibilities and limitations                    483
          A.1.4   Auxiliary equipment                              485
          A.1.5   Production                                       487
    A.2   Trailing Suction Hopper Dredgers (TSHD)                  487
          A.2.1   General lay-out                                  487
          A.2.2   Dredging method                                  488
          A.2.3   Possibilities and limitations                    490
          A.2.4   Auxiliary equipment                              492
          A.2.5   Production                                       493
    A.3   Backhoe dredgers                                         495
          A.3.1   General lay-out                                  495
          A.3.2   Dredging method                                  495
          A.3.3   Possibilities and limitations                    497
          A.3.4   Auxiliary equipment                              498
          A.3.5   Production                                       498
    A.4   Grab dredgers                                            499
          A.4.1   General lay-out                                  499
          A.4.2   Dredging method                                  500
          A.4.3   Possibilities and limitations                    500
          A.4.4   Auxiliary equipment                              502
          A.4.5   Production                                       502
    A.5   Plain Suction Dredger (SD), dustpan dredger             503
          A.5.1   General lay-out                                  503
          A.5.2   Dredging method                                  504
          A.5.3   Possibilities and limitations                    505
          A.5.4   Auxiliary equipment                              506
          A.5.5   Production                                       506
    A.6   Bucket (ladder) Dredgers (BD)                           506
          A.6.1   General lay-out                                  506
          A.6.2   Dredging method                                  506
          A.6.3   Possibilities and limitations                    508
          A.6.4   Auxiliary equipment                              508
          A.6.5   Production                                       508
    A.7   Water injection dredgers                                509
          A.7.1   General lay-out                                  509

| | A.7.2 | Dredging method | 510 |
| | A.7.3 | Possibilities and limitations | 510 |
| | A.7.4 | Auxiliary equipment | 511 |
| | A.7.5 | Production | 511 |
| A.8 | Auxiliary equipment | | 511 |
| | A.8.1 | Workboats | 511 |
| | A.8.2 | Survey vessels | 513 |
| | A.8.3 | Pipeline transport | 514 |
| | A.8.4 | Barges and barge loading/unloading dredgers | 521 |
| | A.8.5 | Spreader pontoons | 523 |
| | A.8.6 | Booster pump stations | 525 |
| | A.8.7 | Reclamation spread/dry earth moving | 525 |

# A   Dredging equipment

There is a wide variety of dredging equipment and plant. Distinction can be made between dredging equipment and auxiliary equipment. In this Appendix the so-called prime movers are discussed such as cutter suction dredger, trailing suction hopper dredger, plain suction dredger, backhoe dredger etc. and auxiliary equipment like multicats, survey vessels, pipes, equipment required on the fill area, etc.

## A.1   Cutter Suction Dredgers (CSD)

### A.1.1   *General lay-out*

The most essential components of a cutter suction dredger are shown in Figure A.1.

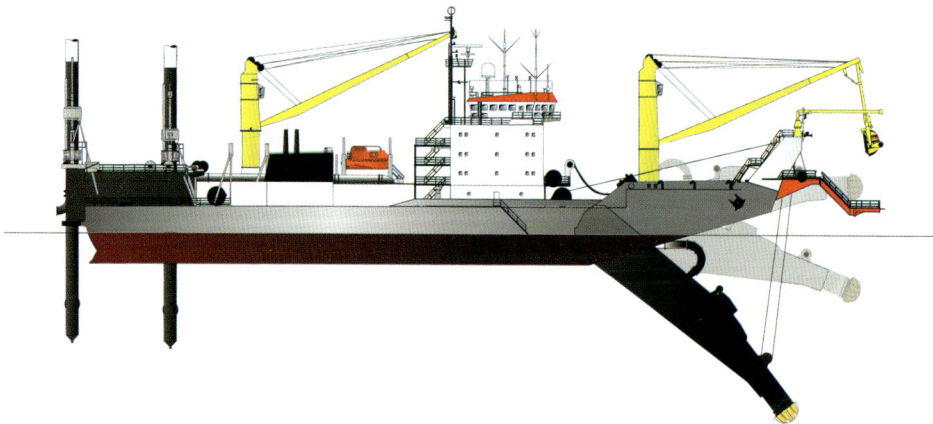

Figure A.1   *Cutter suction dredger.*

The dredger consists of a pontoon hull structure or – occasionally – a ship-form hull that may be self-propelled. At the rear side the dredger is equipped with two spuds. The most common system is the spud carriage system. Less common systems are fixed spuds and walking spuds. At the front-end a ladder is mounted which is provided with a hydraulically or electrically driven cutter head. This ladder can be lowered onto the seabed by a hoisting winch. On each side the CSD has deck (or occasionally ladder) mounted swing winches. The side wires are running to the anchors through sheaves at the lower end of the ladder.

Cutter suction dredgers may – somewhat arbitrarily – be divided into various classes based on their size and installed pumping capacity and cutter power. Table A.1 presents the various classes showing the most important characteristics. In can generally be stated that more cutter power is installed with increasing ladder weight.

Figure A.2 depicts the typical characteristics of a large number of cutter suction dredgers that are currently operational (source "dredgers of the world 2010/2011").

## A.1.2 Dredging method

While fixed at the rear with the working spud firmly imbedded into the ground, the ladder with the cutter head is lowered onto the seabed. The dredger swings from one side to another using swing winches and wires attached to anchors positioned outside the dredging area, see Figure A.3. During such an arc-shaped swing, the rotating cutter head cuts or dislodges the soil or rock and moves it towards a suction intake positioned in the ladder just behind the cutter head. When using a spud carriage system, the CSD is pushed forward at the end of the swing. For the same working spud position the CSD can be moved a number of steps until the carriage has reached the end of its track. At that point the working spud is lifted and the auxiliary spud is lowered to keep the CSD in position. Then the working spud is lifted and the carriage with the working spud is shifted forwards ready to start a new cycle of swings and steps, see Figure A.3. This operation can be repeated until

Table A.1   *Various classes of cutter suction dredgers with typical characteristics.*

| Class | Length pontoon/hull | Width | Draught | Pumping power | Cutter power |
|-------|---------------------|-------|---------|---------------|--------------|
| Small | 62 m | 11.5 m | 2.3 m | 2750 kW | 500 kW |
| Medium | 95 m | 17 m | 3.8 m | 6850 kW | 1500 kW |
| Large | 115 m | 21 m | 4.9 m | 9850 kW | 3250 kW |
| Mega | 130 m | 23 m | 5.5 m | 15.400 kW | 6000 kW |

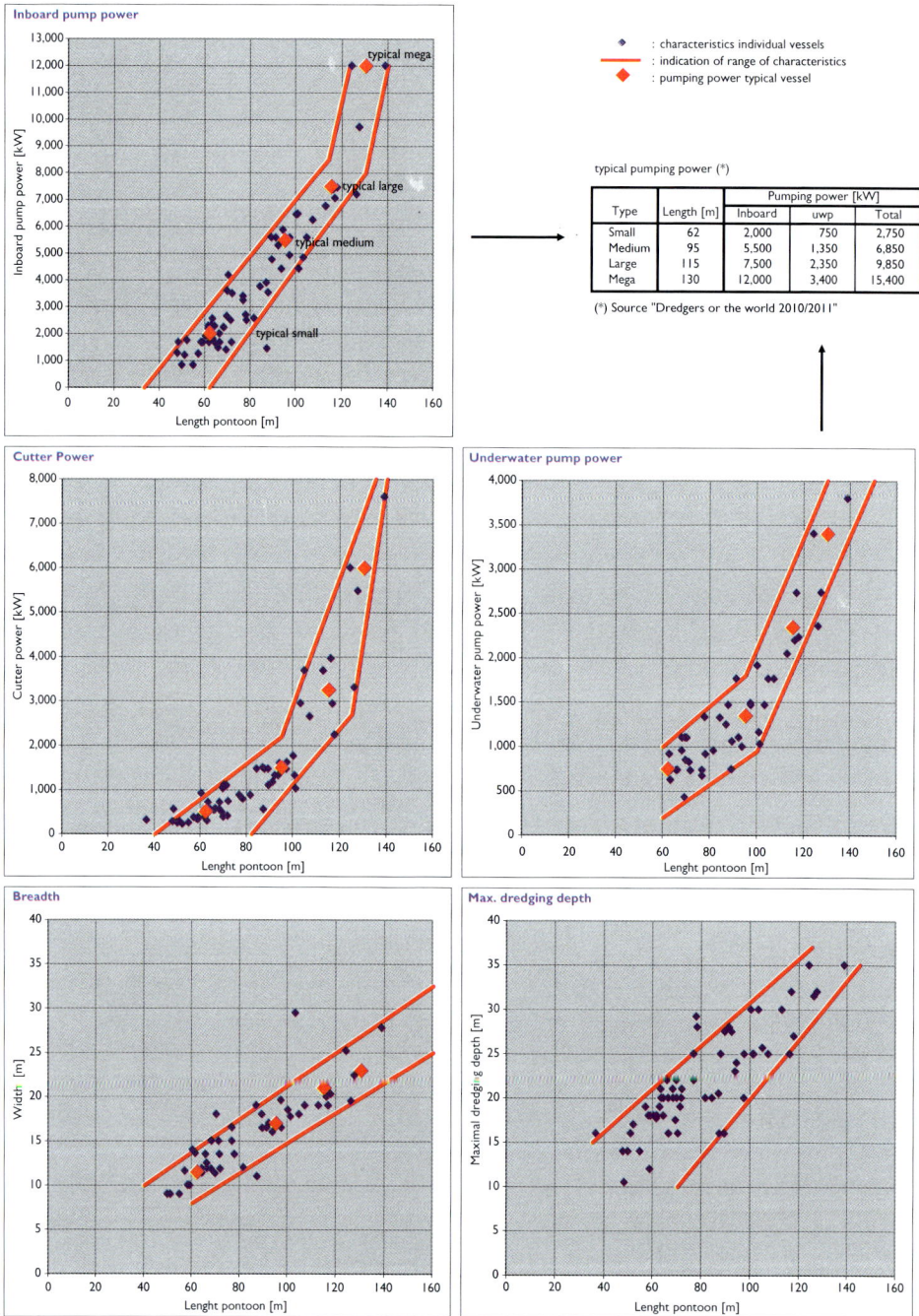

Figure A.2   *Typical characteristics of cutter suction dredgers currently operational.*

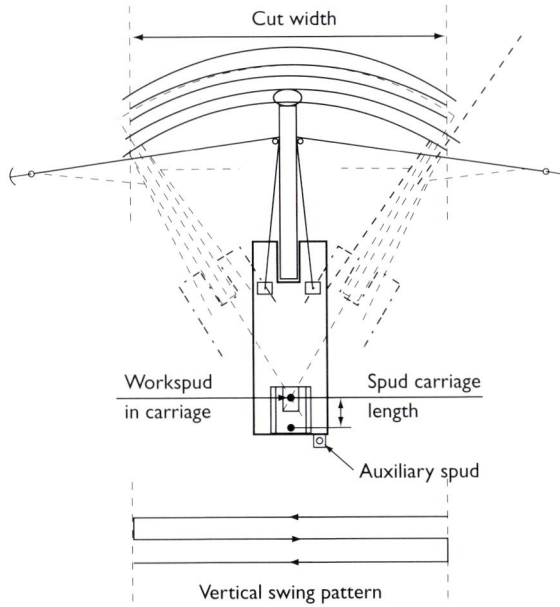

Figure A.3    *Working method cutter suction dredger.*

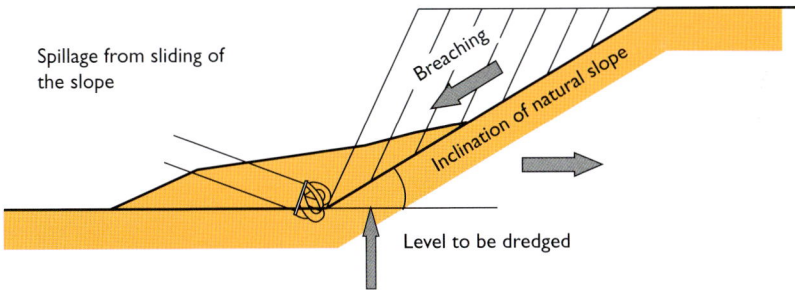

Figure A.4    *Slope formation by breaching and occurrence of spillage down the slope.*

the anchors need to be replaced, the required area has been dredged or further horizontal advancement is not desirable for other operational reasons.

Depending on the thickness and type of soil to be removed, the required depth may be achieved by dredging one or more layers. In less dense non-cohesive material, the cutter can be positioned at the lower end of the profile to be dredged and the material may flow to the suction mouth by the occurrence of a breaching process, see Figure A.4.

Cohesive material and dense sand has to be cut. When cutting more layers this may be done in the same working spud position or in a terrace shaped pattern within one anchor location, see Figure A.5.

Once the full dredging depth has been reached, the spill, if required, is removed by a last sweep and the cutter head is raised. Occasionally, cutter suction dredgers may fix their position by using anchors only. This working method requires a special arrangement of the anchors (so-called Christmas-tree) using three anchors instead of a spud, see Figure A.6. The advantage of this method is the reduced sensitivity of the dredging operation to waves and swell, but it is at the cost of accuracy and production. Moreover, this working method is not feasible in hard materials. This method is therefore rarely adopted.

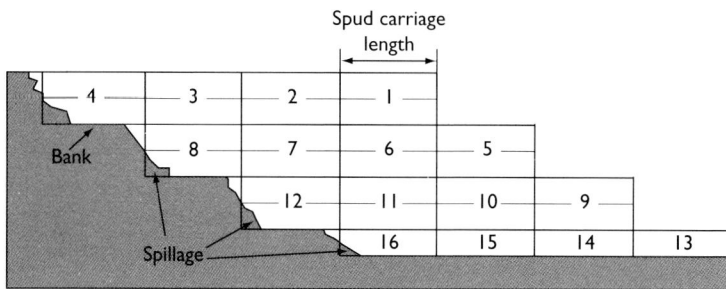

Figure A.5    *Dredging in terrace shaped layers of a number of spud carriage length.*

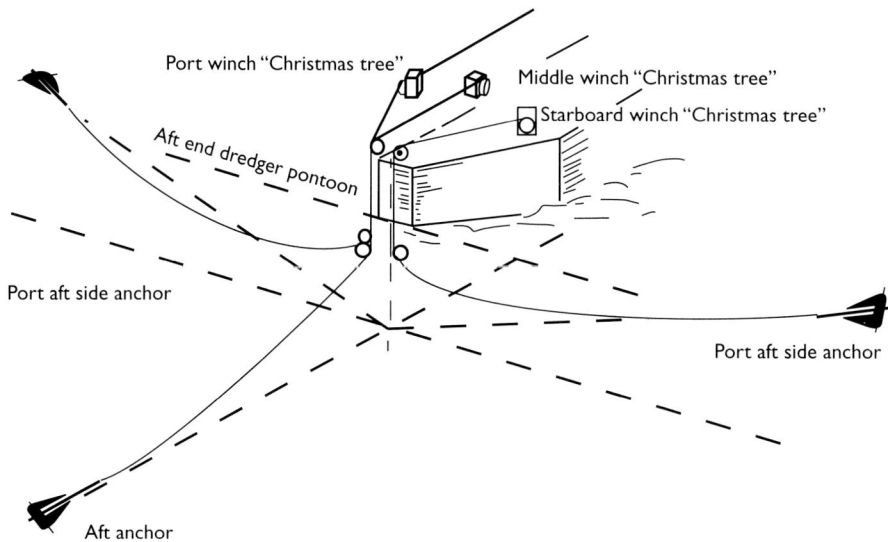

Figure A.6    *Christmas tree anchor system.*

One or more centrifugal dredge pumps transport the soil-water mixture from the suction intake at the lower end of the ladder through inboard pipelines to the stern of the dredger where the inboard pipeline is usually connected to a floating pipeline. This pipeline may in its turn be connected to a sinker line or an onshore pipeline. Alternatively, the CSD may be equipped with a barge loading facility allowing for transport of the dredged materials by barges. Occasionally dredged material can be side casted or placed by using a nozzle (so-called 'rainbowing') or a spreader pontoon at the end of the floating pipeline.

### A.1.3 *Possibilities and limitations*

The arrangement of the CSD makes it a very powerful and versatile tool that can dredge a wide variety of materials ranging from silts and soft clays to rock and transport them without interruption to the required location.

The maximum strength (unconfined compressive strength) of rock that can be dredged not only depends on the installed cutter power, the weight of the ladder, the capacity of the swing winches, the anchor's holding capacity and the dredging depth, but also on natural discontinuities like joints and fractures occurring in the in-situ rock mass.

Should this be required, most modern dredgers can be used to profile slopes to a required angle using a dredging method that may be computer controlled. A more economic operation is dredging of box cuts (see Figure A.7) resulting in faces that will reshape themselves until its slopes have become stable.

The CSD, however, is sensitive to sea conditions and should preferably work in sheltered areas. This implies that when using a CSD it may be important to create sheltered conditions in an early stage of the project. Maximum wave heights for the large type of dredgers are typically 1.5 m, while for smaller vessels these should not exceed 0.6 m.

As indicated in Figure A.8, pumping distances without booster depend heavily on installed pumping capacity, the nature of fill material and the particle size distribution of the fill, but are usually limited to 3–4 km.

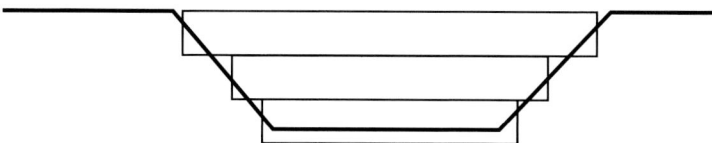

Figure A.7    *Box cut.*

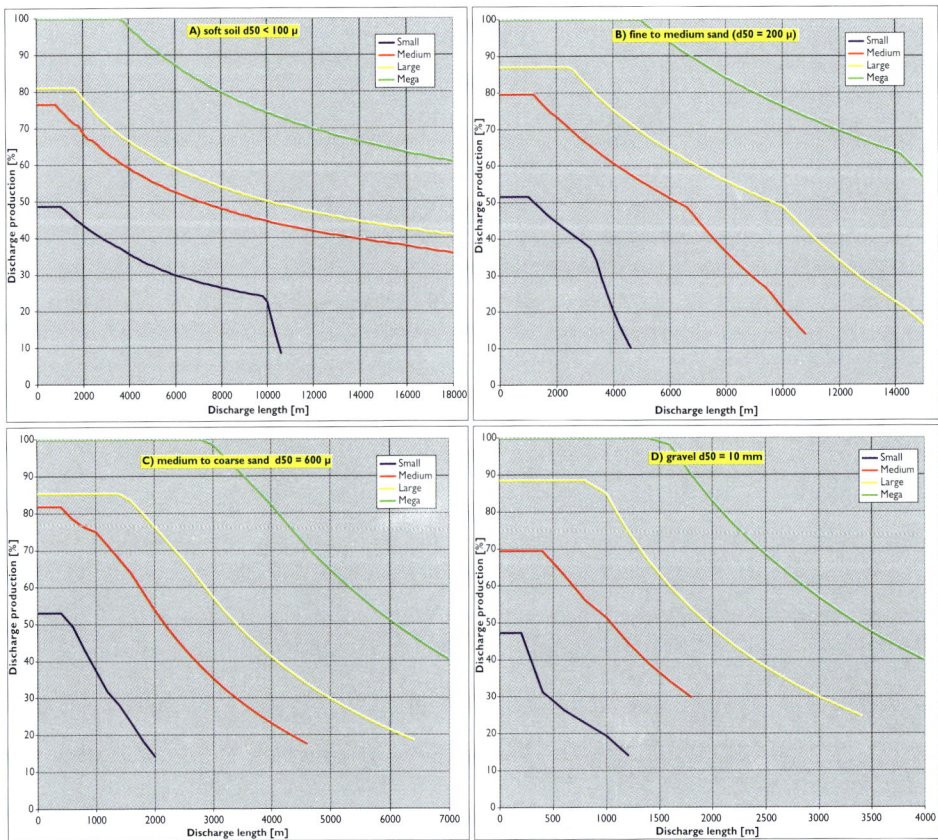

Figure A.8    *Discharge capabilities of typical cutter suction dredgers in various soil classes.*

The maximum water depth in which it can operate depends on the length of the ladder, but is generally limited to 30 m–35 m for the largest CDS's. It can operate in shallow water as it is able to create its own water depth. After dredging, however, the minimum water depth must allow for the draught of the CSD. For the larger CSD, this minimum dredge depth is often 6 to 9 m. Depending on the water depth, the dredging depth and the dimensions of the dredger (width, length and draught) a CSD has a certain minimum dredging width to prevent grounding, see Figure A.9. For the larger CSD, this minimum dredge width is often 50 to 150 m. Placement of the anchors requires often ample space and care has to be taken that the wires between swing winches and anchors are running free. Cutter suction dredgers may be equipped with anchor booms to facilitate placement of the anchors.

In case a CSD is used just to mine fill material then the thickness of the layers should preferably not be less than the diameter of the cutter to ensure an economic operation. Spill as a result of the open structure of the cutter can be significant.

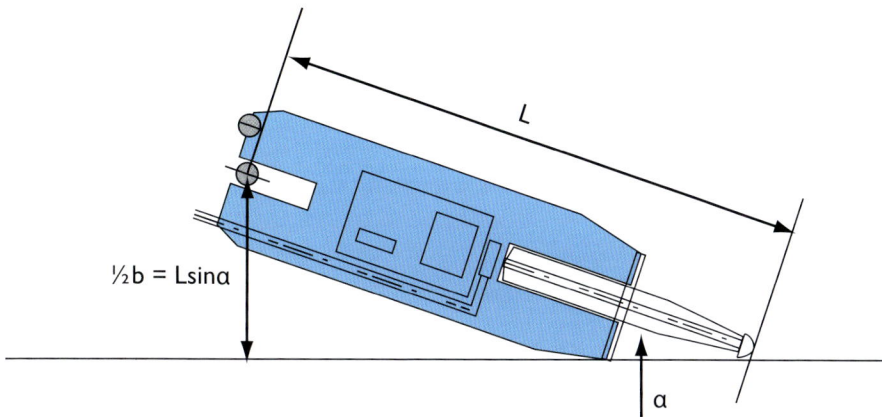

Figure A.9    *Minimum dredging width for cutter suction dredger.*

Wear and tear of the cutter head (in particular the pickpoints placed on the blades of the cutter crown, see Figure A.10), the pipelines and the pumps when dredging abrasive materials like rock may raise the costs of a CSD considerably. The use of a CSD may, however, still be cost effective when compared to other dredging equipment in combination with drilling and blasting. Abrasiveness of the dredged materials may depend on the strength of the rock mass, the mineralogy and the grain shape of the broken fragments/grains.

Mobilization and demobilization of the CSD is generally considered relatively expensive, in particular when the dredger is not self-propelled. Operations of a CSD may hamper other shipping activities in the area as the dredger is fixed by anchors and its spud while it is often connected to a floating pipeline.

### A.1.4    *Auxiliary equipment*

The use of a CSD will typically require the following auxiliary equipment: a multicat, a survey vessel, sufficient pipelines or a barge loading facility and, depending on the scale of operations, a crew boat.

Should the CSD not be self-propelled then a tug must always be stand-by to allow for an immediate evacuation of the dredger in case of an emergency.

The operation of the large cutter suction dredgers is generally a 24 hours/day and 7 days/week, although occasionally noise nuisance or other local restrictions may not allow for such continuous operations. Most large CSDs have crew accommodation on board.

| Rock cutter |  |
| Clay cutter |  |

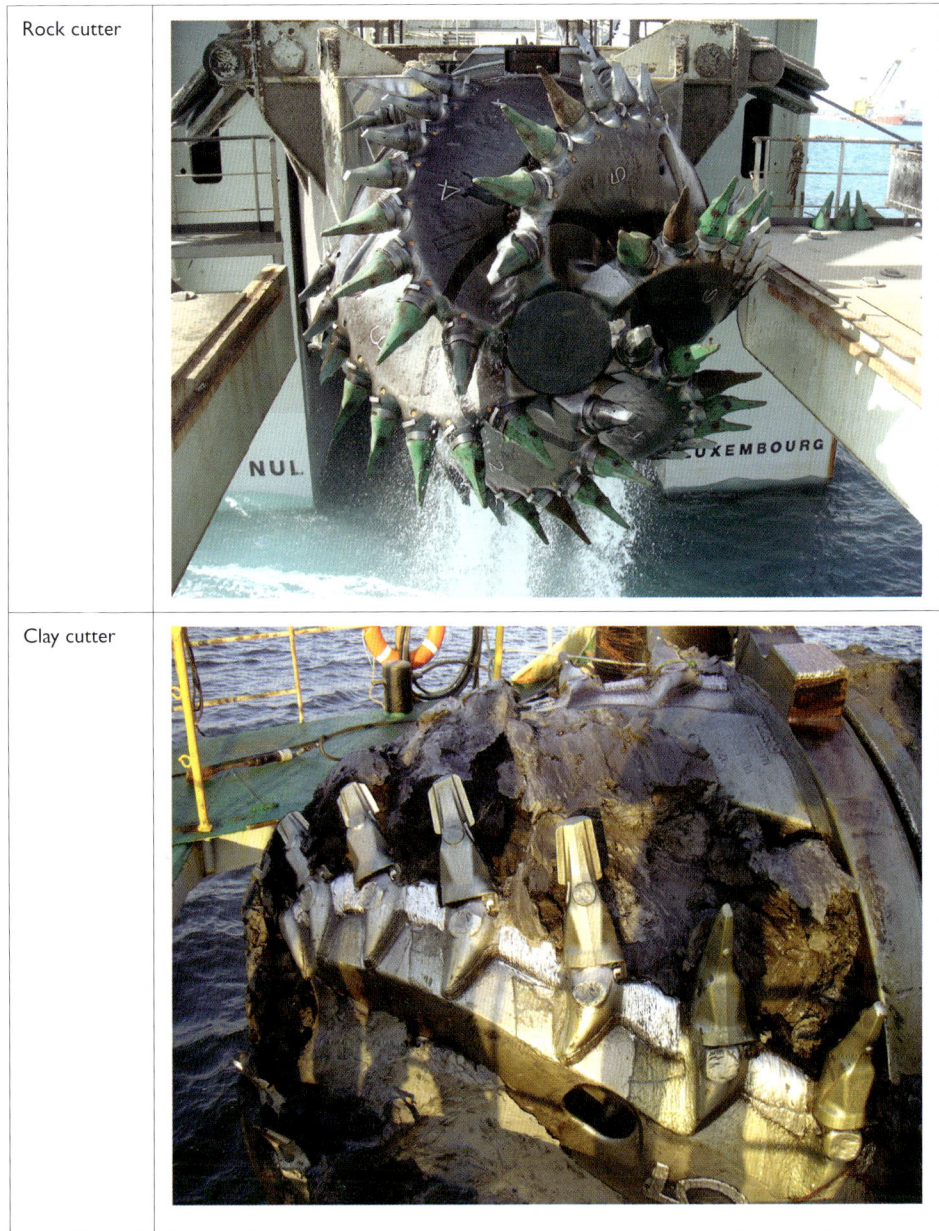

Figure A.10    *Various types of cutter heads.*

## A.1.5  *Production*

The production of a CSD ranges between 100 m³/hour (small sized CSD) to 4500 m³/hour (large sized CSD) depending on the power installed on the cutter head, the pumping capacity of the dredger and boundary conditions like subsoil conditions, geometrical requirements (slopes, depth) of the borrow pit, etc. Also the means of transport of the dredged material (pipeline, barge, sidecasting, etc.) is of influence. Operational hours may strongly be affected by the state of maintenance of the dredger, the wear and tear caused by the materials to be dredged and the wave climate and weather conditions in the working area. Depending on the abrasiveness of the dredged materials replacement of worn pickpoints of the cutter head can interrupt the dredging operations and, hence, influence the production and costs of the CSD significantly.

## A.2  **Trailing Suction Hopper Dredgers (TSHD)**

### A.2.1  *General lay-out*

Figure A.11 presents the general lay-out of a trailing suction hopper dredger (TSHD).

The TSHD is a self-propelled ship – containing a hopper with bottom discharge and often also pump discharge facilities. To load the dredged material into the hopper, the dredger is fitted with one or two suction pipes and centrifugal pumps. A draghead is attached to the lower end of each suction pipe. The centrifugal pumps are either located inboard or mounted in the suction pipe (submerged pumps).

The dragheads (Figure A.12), often equipped with water jets and sometimes fitted with scrapers or ripper teeth, can be lowered onto the seabed using winches and

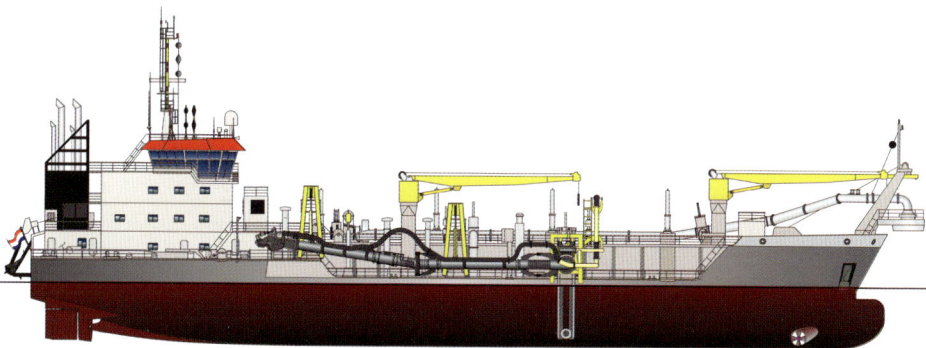

Figure A.11   *A trailing suction hopper dredger.*

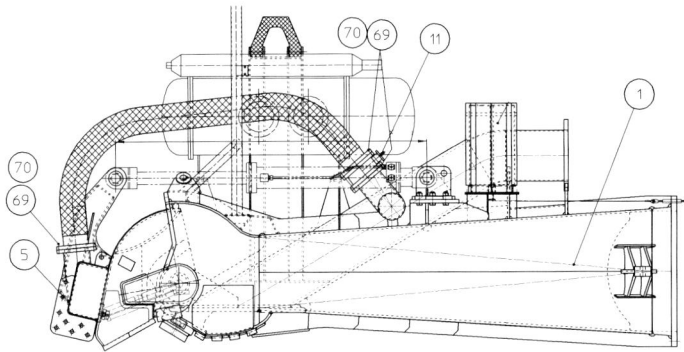

Figure A.12    *Drag head of trailing suction hopper dredger.*

Table A.2    *Typical characteristics of trailing suction hopper dredgers currently operational.*

| Class | Hopper capacity | Length | Width | Draught | Maximum dredging depth | Pumping power |
|---|---|---|---|---|---|---|
| Small | 700 m³ | 60 m | 10 m | 3.5 m | 18 m | 350 kW |
| Medium | 4.500 m³ | 100 m | 18 m | 6.8 m | 30 m | 2.500 kW |
| Large | 10.000 m³ | 130 m | 23 m | 8.8 m | 50 m | 5.500 kW |
| Jumbo | 23.000 m³ | 170 m | 29 m | 11.4 m | 90 m | 10.500 kW |
| Mega | 35.000 m³ | 210 m | 34 m | 13.8 m | 110 m | 16.000 kW |

gantries or A-frames. Various types of dragheads have been developed in order to allow for the variation of soils to be dredged. An example of a draghead is shown in Figure A.12.

There is a wide variety of trailing suction hopper dredgers that can be divided into classes based on the capacity of the hopper. Table A.2 presents a summary of typical characteristics of a large number of trailing suction hopper dredgers (source "dredgers of the world 2010/2011").

### A.2.2    Dredging method

When approaching the dredging area, the ship's speed is decreased and the dragheads are lowered to just above or on the seabed. While moving forward, water jets and teeth loosen the soil. At the same time the centrifugal pumps create a vacuum within the dragheads entraining a soil water mixture into them. Via suction pipe and inboard pipelines this mixture is discharged into the hopper.

At start of dredging the hopper is partly filled with water. After reaching overflow the process water leaves the hopper through the overflow while the solids settle in the hopper. This process of settling and overflowing continues until the hopper is loaded to its mark. The filling of the hopper is basically a sedimentation process implying that coarse material will settle immediately near the inlet while fine material will take much longer to settle and may even be discharged with the water through the overflow system. This loss of fines with the process water reduces the fine content of the material settled in the hopper and overflow is therefore often used to improve the quality of the dredged material, see Figure A.13. The process of removing fines from dredged material can be stimulated by fluidizing the sand in the hopper using a system of nozzles fitted at the bottom of the hopper. This will create an upward flow in the hopper entraining the fines to the surface of the mixture where they come into suspension in the supernatant water that is being discharged.

Once the hopper is full, the suction pipes are hoisted and stowed on deck and the trailer sails to the discharge location.

Depending on the arrangements of the specific TSHD there is a number of methods to discharge the hopper:

– either by dumping through doors or valves fitted in the bottom of the TSHD (note: this requires sufficient water depth below the keel of the vessel);

Figure A.13   *Overflow on board a trailing suction hopper dredger.*

Figure A.14   *Gantry for hoisting and lowering suction pipe on board a trailing suction hopper dredger.*

– by pumping through a (floating) pipeline that is connected to the bow of the trailer;
– by pumping the soil-water mixture through a bow-mounted nozzle (so-called 'rainbowing');
– or by pumping through their suction pipe (so-called 'reverse pumping').

Figure A.15 shows the various discharge methods. Most TSHD's are equipped with a water jet system to fluidize the material in the hopper in order to create a soil-water mixture that can be discharged by pumping. In addition, water jets may be used to clean the hopper and to reduce the load that remains in the hopper after discharging.

### A.2.3   *Possibilities and limitations*

Although recent developments with rippers mounted on the dragheads of the largest class of dredgers have extended the range of materials to be dredged (from firm clay to stiff clay and weak rock), the TSHD is less versatile than the cutter suction dredger with respect to the materials that can be dredged. In particular, when

| | |
|---|---|
| Discharge through valves or bottom doors |  |
| Discharge through floating pipelines |  |
| Discharge by rainbowing |  |

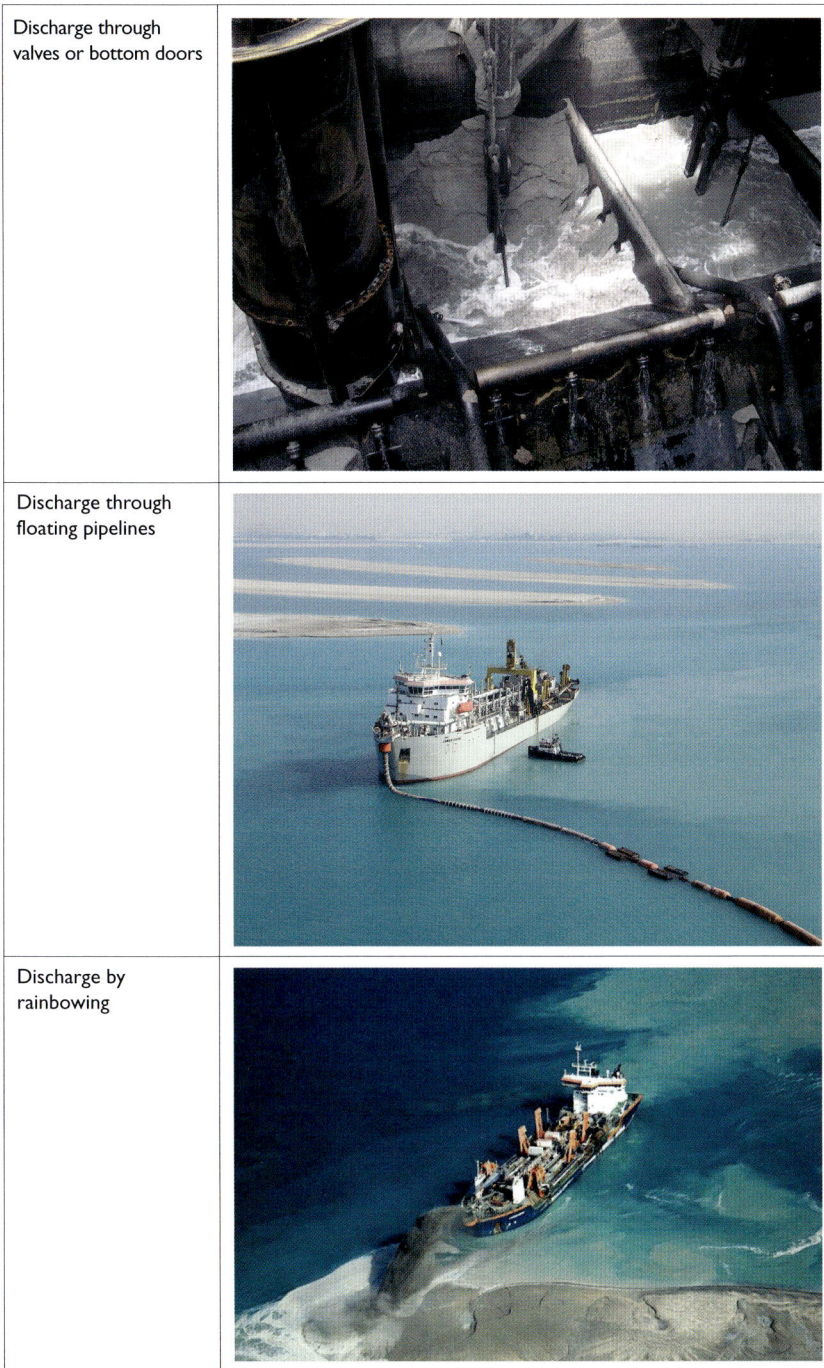

Figure A.15  *Various discharge methods for a trailer suction hopper dredger.*

dredging clays and weak rocks production rates may be low. Highest productions will be achieved in loose, fine to medium grained sands and soft muds/clays.

Using a TSHD it may prove very difficult to produce an even seabed as the drag-head tends to create gullies. Unlike the CSD this type of dredger is not suitable to properly trim slopes to a required angle. The TSHD cannot operate in restricted areas as it is sailing during dredging and needs manoeuvring space and a certain minimum trail speed to cut the soil. Its manoeuvring capabilities, however, reduce the impact on other shipping activities in the area. When discharging, the TSHD may have more effect on other shipping operations as it might have to connect to a floating pipeline.

Depending on its draught in loaded condition the TSHD will require a certain minimum water depth. This may vary from 4 m for the smaller trailers to 17 m for the largest dredgers currently available. The maximum water depth in which the TSHD can operate ranges from 15 m for the smallest type of dredgers to more than 120 m for the largest classes. However, operations in very large water depths may require specially designed suction pipes with submerged pumps. The TSHD is less vulnerable to wave climate and weather conditions than the cutter suction dredger and can, depending on its dimensions, therefore also operate in open seas.

Economical use of the TSHD generally requires relatively long haulage distances. Hopper capacities vary significantly: the smallest class may take no more than 500 m$^3$ while the largest TSHD have hoppers that can transport more than 40.000 m$^3$. The hopper is not always loaded up to its full capacity. The actual load of the vessel depends on the carrying capacity and corresponding maximal draught. The maximal draught is indicated on the hull of the trailer as a collection of marks usually called the loadlines (or Plimsoll marks), see Figure A.16.

Dumping arrangements including the fluidization and water jet system and the type of bottom doors may not be suitable for all soil types. Depending on this arrangement a trailer may be more suitable to operate either in sands and silts or in cohesive materials. As the TSHD is self-propelled, mobilization/demobilization costs are generally less than those of a CSD.

A.2.4  *Auxiliary equipment*

The TSHD needs little auxiliary equipment. When it has to discharge through a floating pipeline it may need assistance of a small vessel to connect to the discharge point. Depending on the project conditions it could further be necessary to regularly undertake a bathymetric survey of the borrow area. This will require a survey vessel. Occasionally a crew boat might be needed.

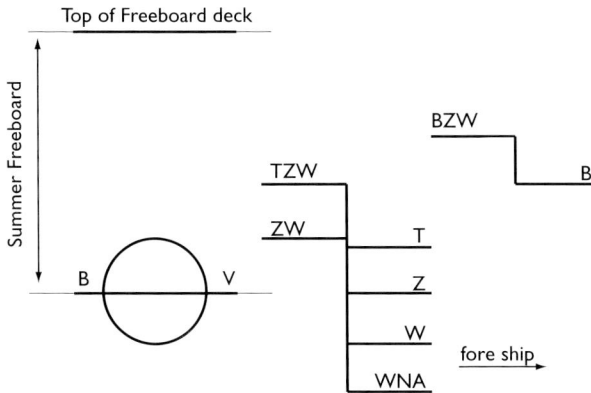

Figure A.16    *Plimsoll mark.*

| | | |
|---|---|---|
| TZW | = | Tropical Fresh Water Freeboard |
| | | -    Tropical Fresh Water Load line    =    TF |
| ZW | = | Summer Fresh Water Freeboard |
| | | -    Fresh Water Load line    =    F |
| T | = | Tropical Freeboard |
| | | -    Tropical Load line    =    T |
| Z | = | Summer Freeboard |
| | | -    Summer Load line    =    S |
| W | = | Winter Freeboard |
| | | -    Winter Load line    =    W |
| WNA | = | North Atlantic Winter Freeboard |
| | | -    Winter North Atlantic Load line    =    WNA |
| BZW | = | Dredge-Fresh Water Freeboard |
| B | = | Dredging Freeboard |

## A.2.5    *Production*

The production of a TSHD is mainly determined by its hopper capacity and the production cycle consisting of time required to load the hopper, to transport the fill to the required location, to discharge the fill into the designated area and to return to the borrow area.

The loading time of the trailer depends strongly on the in-situ characteristics (density, particle size distribution, etc.) of the materials to be dredged, the geometry of the borrow area and any restriction with respect to overflow. Generally loading times in suitable borrow areas may range between 40 and 150 minutes, depending on the size of the vessel.

The sailing speed of a fully loaded trailer may vary from 10 knots (small trailers) to 19 knots (large trailers), but could significantly be affected by weather conditions, currents, water depth, other shipping movements and navigational restrictions. The sailing speed when empty will generally be higher.

The time required to discharge a TSHD mainly depends on the discharge method, the nature of the dredged material and the volume to be discharged. A bottom discharge of sand may be completed in only a few minutes, while mooring, connecting to a floating pipeline and pumping the material into the reclamation may take much more time depending on installed pumping capacity, the volume and nature of the fill material and the required pumping distance.

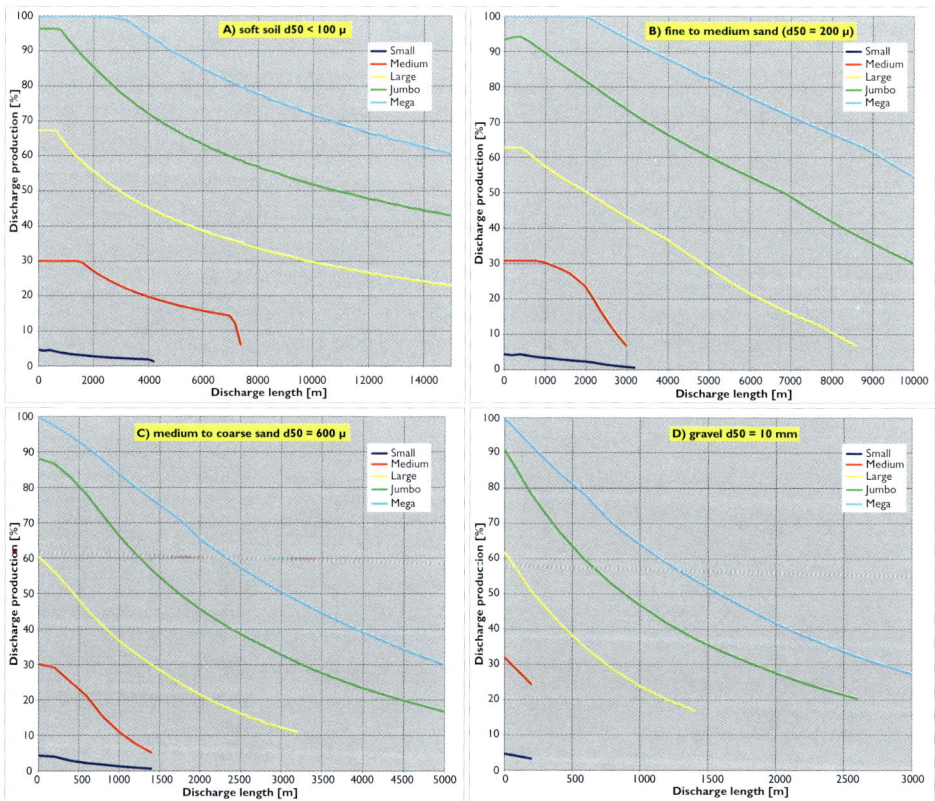

Figure A.17    *Discharge capabilities of typical trailer suction hopper dredgers in various soil classes.*

## A.3 Backhoe dredgers

### A.3.1 General lay-out

A backhoe dredger consists of 2 main elements: a pontoon, generally without pro-pulsion, and a hydraulic backhoe excavator usually mounted on a pedestal located at the front end of the pontoon (see Figure A.18).

The pontoon of the backhoe dredger is usually equipped with 3 spuds that provide the reaction forces for the excavation operations. The stern spud can be tilted or has been mounted on a spud carriage. The backhoe excavator consists of a power unit mounted on a turntable, a boom and stick and a bucket attached to the outer end of the stick, all hydraulically powered. Stick and bucket can (relatively easy) be adapted to local conditions like water depth, strength and type of materials to be dredged.

### A.3.2 Dredging method

While firmly fixed at its position by the spuds and its pontoon partly lifted out of the water, the backhoe excavates the seabed within its reach to the required depth while dumping the dredged material in a hopper barge moored alongside the dredger. Occasionally dredged materials may be sidecasted, but this will generally

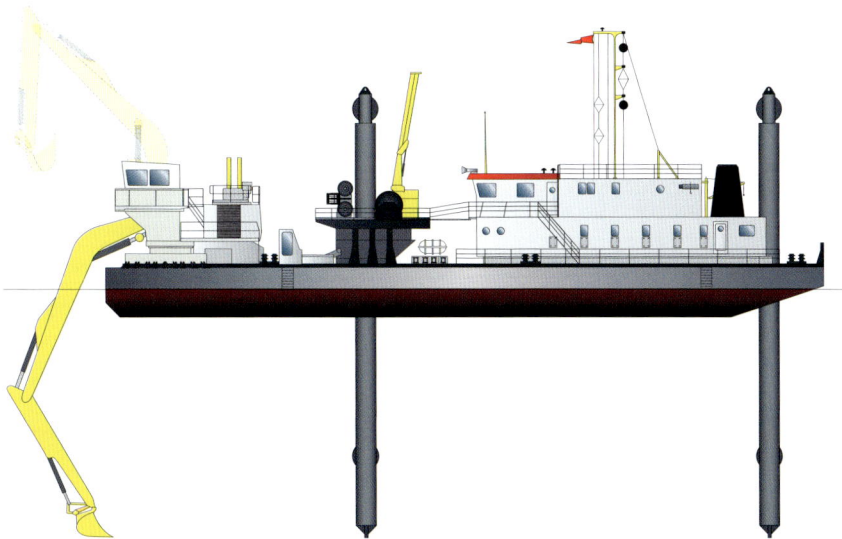

Figure A.18    *Backhoe dredger.*

be restricted to shallow water depths as the outreach of the backhoe is limited and dredged materials may slide back into the excavation.

From its fixed position, the dredger can swing around resulting in an arc-shaped cut. After completion of a cut, the two forward spuds are raised and the dredger is moved into a next position by using the stern spud equipped with a tilting step facility or a spud carriage. Before dredging is resumed, the two forward spuds are lowered to the seabed and the pontoon is partly lifted ensuring a stable position of the dredger. Figure A.19 shows the principles of the dredging method of the backhoe dredger.

To limit spill the backhoe dredger preferably moves backward (i.e. its pontoon is positioned over the area to be dredged), but if required by circumstances it is possible to operate the dredger by moving forward. The traditional backhoe dredger is a hydraulic crane fixed on a pontoon. This may be a crane on tracks or a crane placed on a turning table integrated with the pontoon. A recent development is a further integration of the crane and the pontoon leading to a major increase of

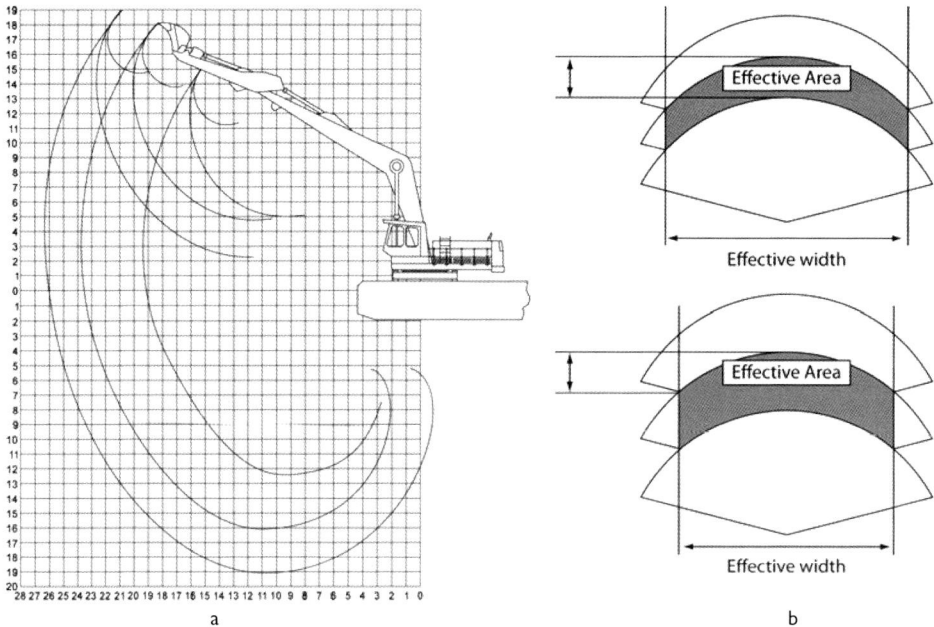

Figure A.19   *Typical reach of backhoe dredger (a) and relation between the size of the step and the effective width of the dredged area (b).*

crane capacity, where the crane drive is not placed in the crane itself but inside the pontoon, the so-called Backacter.

### A.3.3  *Possibilities and limitations*

The backhoe dredger is a versatile tool to dredge a wide variety of soils ranging from soft clays to boulders and weak rocks. In general, the maximum size of the bucket reduces with increasing strength of the materials to be dredged and with increasing dredging depths. The dredger can also remove rubbish and debris depending on its size. The stronger backhoe dredgers can dredge hard clays, dense sand and weak rock.

As a result of an accurate control of position and depth of the bucket and the absence of anchors and anchor wires, the backhoe dredger is able to work in confined areas with complex geometries requiring small dredging tolerances. The size of the pontoon (typical length 30–65 m), however, may limit its application.

The maximum dredging depth of the largest backhoe dredger is approx. 30 m. The dredger can create its own water depth, but the minimum depth after dredging should allow for the draught of the pontoon, generally 3–4 m and the draught of the barge transporting the dredged material.

As a result of its relatively low rate of production, a backhoe dredger may not be the most appropriate dredger to mine fill material at a large scale. It is, however, a suitable tool to construct bunds over shallow water, to trim slopes of borrow areas and bunds to specified slope angles and to dredge trenches. Moreover, because the dredging method allows for a visual inspection of the dredged material before discharging, the backhoe dredger may be used for selective dredging should this be required.

An advantage of the backhoe dredger (and the grab dredger) in comparison with a cutter suction dredger or a trailing suction hopper dredger is the limited disturbance and dilution of the materials being dredged. This will result in higher densities when dumped into a barge. In particular when dredging cohesive materials this may be important as it results in a reduced loss of strength and stiffness of the dredged materials after deposition.

The backhoe dredger is sensitive to waves (wave heights limited to 0.8 m depending on the size of the dredger) and should preferably operate in sheltered conditions.

Mobilization/demobilization costs are relatively high as the backhoe dredger is not self-propelled.

### A.3.4  *Auxiliary equipment*

Dredging with a backhoe dredger will generally require the following auxiliary equipment: barges, a crew boat, a survey vessel and a tug boat to move the dredger and to assist in case of emergency evacuation.

### A.3.5  *Production*

Production rates of backhoe dredgers depend heavily on the type and manufacturer of the excavator. Backhoe dredgers are generally rated according to the maximum bucket size the dredger can handle. The bucket size may range from 1 to 40 m$^3$.

Dredging with a backhoe involves the following production phases: lowering of the bucket to the seabed, filling the bucket, raising the bucket above water, swinging the bucket to the point of discharge, discharging the bucket and swinging back to mark.

The most important factors that define the production of this dredger are the characteristics of the materials to be dredged and the dredging depth. Depending on this, the cycle time may range between 1 and 3 minutes in cohesive and non-cohesive soils. In rock this may take longer. Both factors influence the maximum bucket size that can be used which affects the production directly. In addition, the average percentage of bucket fill depends on the characteristics and the layer thickness of the soils and rocks being dredged, while the time required to lower and raise the bucket to and from the seabed is directly related to the dredging depth. An example of a bucket for dredging in hard clay, very dense sand or weak rock is presented in Figure A.20.

If sidecasting is not applicable then the maximum production of a backhoe dredger can only be achieved if sufficient hopper barges are available to discharge the dredged materials.

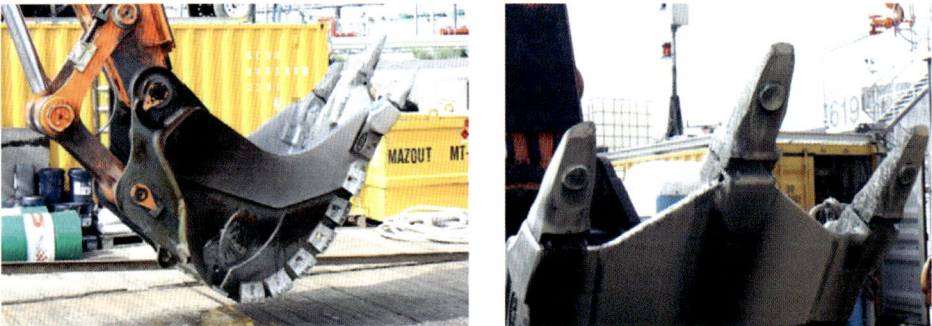

Figure A.20  *Backhoe bucket for dredging in hard material.*

## A.4  Grab dredgers

### A.4.1  *General lay-out*

Two types of grab dredgers can be distinguished:

– A grab pontoon dredger, consisting of a grab crane mounted on a simple pontoon without propulsion. The pontoon may be equipped with spuds and/or anchors and anchor winches for positioning.
– A grab hopper dredger consists of one or more grab cranes mounted on a self-propelled ship equipped with a hopper. The hopper may have a bottom discharge arrangement. The dredger may be fitted with spuds or anchors and anchor winches.

A grab crane consists of a cabin with a power unit and winch drums, a long boom and a grab. The cabin is mounted on a turntable that is fixed to the deck of a pontoon or a hopper dredger. The lower end of the boom is attached to the base of the cabin and can be lowered and raised by a hoisting wire rope. Sheaves at the upper end of the boom allow for the guidance of the wire ropes required to lower, raise and operate the grab.

Figure A.21 shows the typical arrangements of a pontoon based grab dredger and in Figure A.22 a typical arrangement is shown of a self-propelled grab hopper dredger.

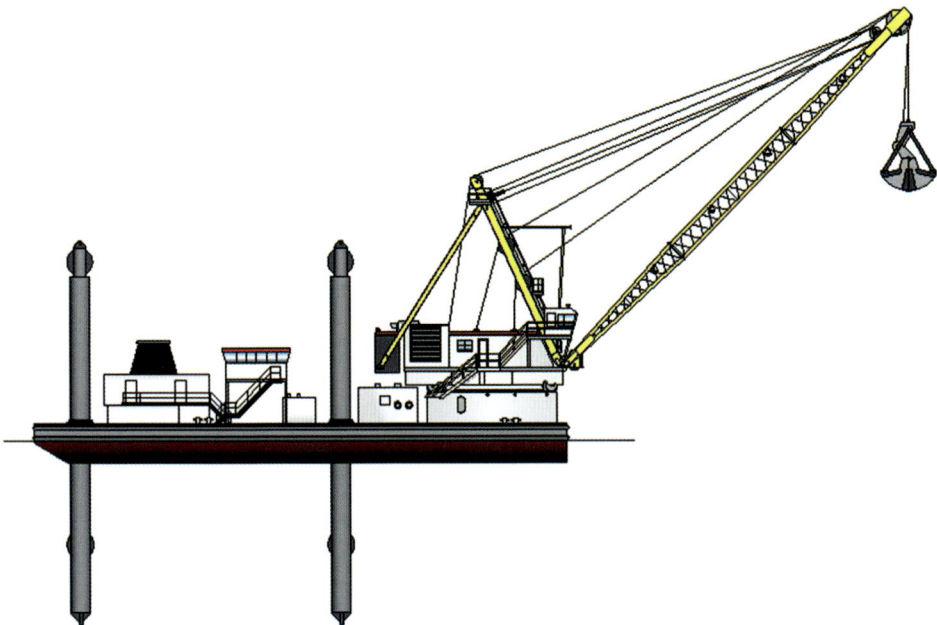

Figure A.21   *Grab pontoon dredger.*

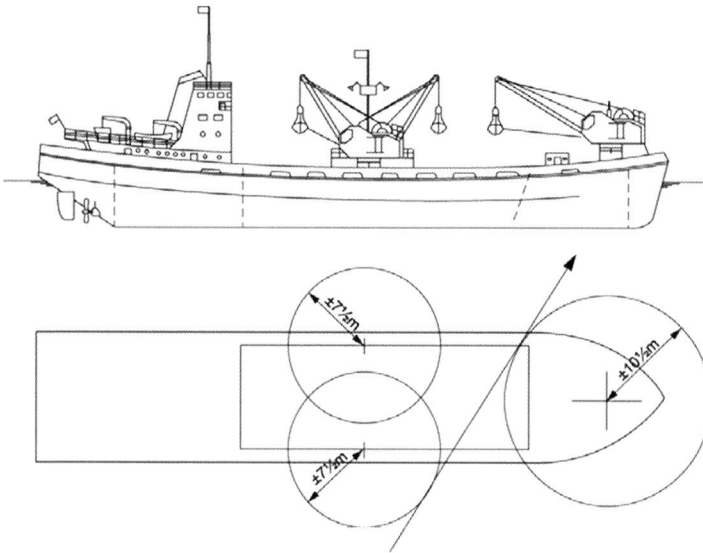

Figure A.22 *Self-propelled grab hopper dredger.*

Various grabs, buckets or clam shells have been developed to suit different soil characteristics: strong materials may require heavy grabs of reduced size equipped with teeth, while muds and loose silts are usually dredged with large buckets with straight cutting edges.

### A.4.2 *Dredging method*

After a fixed position of the dredger is secured by using the anchors and/or the spuds, dredging is carried out in the area within reach of the grabbing crane(s). The dredged materials are either discharged into a barge moored alongside the pontoon, into the hopper of the dredger or side casted. After filling a barge, the grab pontoon dredger can continue immediately with the next barge. This assumes however, that a sufficient number of barges is available. After filling its hopper, a self-propelled grab hopper recovers its anchors (and/or raises its spud) and sails to the discharge area. Once the area is dredged to the required level the grab hopper dredger moves to a next location by adjusting the anchor winches or the spuds.

### A.4.3 *Possibilities and limitations*

The material that can be dredged by a grab dredger ranges from soft cohesive clays and silts to stiff clays and sands. It can also handle cobbles and boulders, rubbish and debris. Since the penetration force into the material to be dredged is determined by the weight of the grab, the grab dredger is less suitable to dredge hard clay and dense sand or very weak rock.

Productions of the grab dredgers are relatively low. This type of dredgers may therefore not be suitable for large scale borrowing operations. However, small scale mining operations on rivers and lakes are often carried out by grab dredgers. In particular when mining loose sands and silts, relatively large volumes can be dredged without the need to move the dredger because these soils tend to flow along the slopes towards the grab.

Grab pontoon dredgers may also be used to construct bunds, to dredge limited areas like (narrow) trenches or to remove debris, wreckages and other obstacles from the seabed prior to filling. When many small spots need to be dredged, a grab pontoon dredger can be very effective.

The grab pontoon dredger has the advantage over the grab hopper dredger that it allows for a more or less continuous borrowing operation provided that the number of hopper barges is sufficient. This results in a higher production per unit.

Because the grab dredger with spuds operates in a stationary position, its boom has a rather wide reach and the wire attached grab can be lowered to great depths, this type of dredger is suitable to be used in confined areas. These areas may have complex geometries or are located immediate adjacent to existing structures like quay walls, jetties, etc. The presence of currents could affect the dredging accuracy of this type of dredgers.

The accuracy of producing a level bottom is limited which often results in the need for considerable overdredging when guaranteeing a certain specified dredge level. In deep water, positioning of the grab may prove to be difficult, which has an adverse effect on the dredging accuracy.

Grab dredgers can, in general, create their own water depth. Most grab pontoon dredgers can operate in shallow water due to the shallow draught of their pontoons. But – if sidecasting is not applicable – the required water depth in the immediate vicinity of the dredger may not be dictated by the draught of the dredger but by the draught of the hopper barge loaded alongside or tug boat required to transport the dredged materials. If not equipped with spuds, laying and recovery of anchors may also require a certain minimum water depth to access the locations of the anchors. The maximum dredging depth can be very large (more than 50 m), especially when operating on anchors only.

Due to the stationary nature of the operation and the possible usage of anchors, grab dredgers may interfere with other shipping movements within the dredging area. Anchor wires may also limit the manoeuvring possibilities of the hopper barges.

Similar to the backhoe dredger, the grab dredger does not cause excessive remoulding and dilution of the materials being dredged. In general, this results in a higher density of the dredged material after dumping in a barge or a hopper than when using a suction dredger. For cohesive materials, this results in a lesser loss of strength and stiffness of the dredged materials.

### A.4.4  *Auxiliary equipment*

The following auxiliary equipment may be required when dredging with a grab pontoon dredger: barges to transport the dredged materials, a crew boat, a multi-cat/tug for laying and recovering the anchors and a tug boat to move the dredger to a next position or a safe location in case of emergency.

The grab hopper dredger may need less assistance since it is self-propelled transporting the dredged materials in its own hopper. Occasionally a crew boat may be required.

When bathymetric surveys are regularly needed a survey vessel may be required.

### A.4.5  *Production*

The dredge cycle of a grab pontoon dredger involves the following successive activities: lowering the grab bucket to the seabed, filling the grab bucket, raising the grab bucket above water, swinging the bucket to the point of discharge, discharging and swinging back to mark.

If sidecasting is not applicable then the economical use of the grab pontoon dredger requires the availability of sufficient barges so that, once a barge is filled, the dredger can continue to discharge into the next barge if moored at the other side of the pontoon. This, however, also depends on the available water depth.

The grab hopper dredger has the same production cycle with the exception that once the hopper is filled the dredger first has to sail to the designated fill area to discharge the dredged material and then return to the borrow area before dredging can be continued.

Similar to the backhoe dredger, production rates of the grab dredger strongly depend on the nature of the materials to be dredged and the water depth at the dredging location. In general, with increasing soil strength, the size of grab bucket decreases while its weight will increase. The sizes of grab range generally between 1 m³ and 30 m³. The larger the water depth, the more time it takes to lower and raise the grab.

The use of anchors for positioning may have a deleterious effect on the production as laying and recovering of the anchors can have a significant impact on the cycle time. Moreover, the presence of anchor wires may also hamper the manoeuvring of the hopper barges.

## A.5 Plain Suction Dredger (SD), dustpan dredger

### A.5.1 *General lay-out*

The plain suction dredger, sometimes called deep suction dredger, is typically used for sand winning purposes. It is a stationary dredger that usually works within a sand winning pit. Figure A.23 shows the general arrangement of a stationary suction dredger.

The plain suction dredger is basically a dredger equipped with a suction pipe that can be lowered to the seabed by a hoisting winch. The suction pipe is connected to a dredge pump that may be located inboard, but could also be submerged, fitted within the suction pipe. The principle of sand winning is based on the erosive action of sand flowing towards the mouth of the suction pipe. No cutter head is attached to this mouth to loosen the material but often jets are installed for this purpose The slope formation while dredging is the result of a breaching and fluidization process which can, more or less, be predicted based on the soil

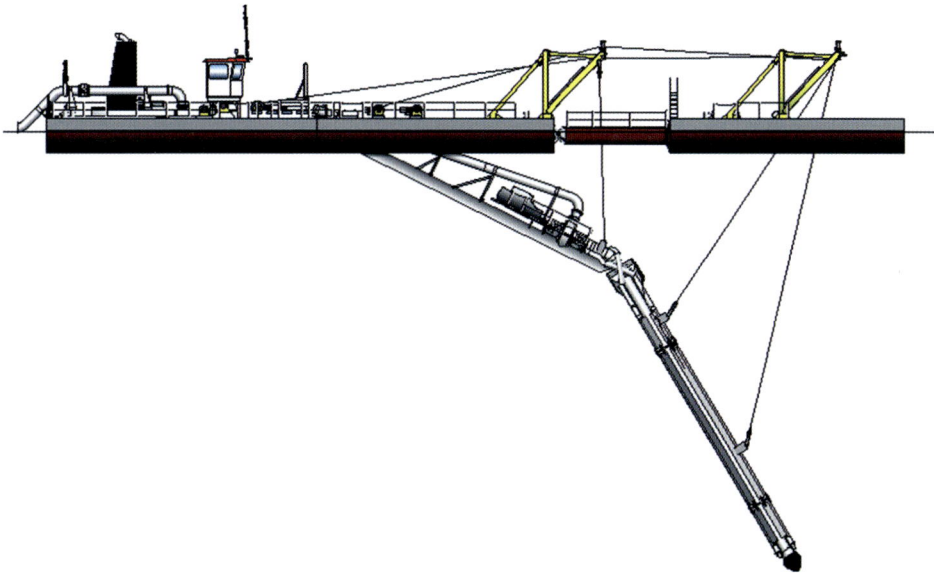

Figure A.23  *Stationary suction dredger.*

characteristics and the dredging process. The nature of this sand winning process entails much less accurate dredging than with the previously described equipment. The dredger may have as many as 6 anchors for positioning during dredging.

The so called dustpan dredger is a stationary suction dredger with a widely flared suction head which may be as wide as the hull of the dredger itself (see Figure A.24). This dredger is developed to achieve a better controlled dredging process than the plain suction dredger. The suction head is usually equipped with water jets to dislodge and fluidize the sediments. This type of dredger may be self-propelled, but dredging is always carried out while being winched.

### A.5.2    *Dredging method*

The dredger operates whilst moored on anchors and is only used to mine granular loose deposits like loose sands and (fine) gravels. From its fixed position the suction pipe is lowered to a position just above seabed. Pumping assisted by water jets will fluidize the sand and entrain it into the suction intake creating a pit with the shape of an inverted cone. During dredging the pipe is gradually lowered into this pit. Production increases with increasing depth of the pit as the sand starts to flow along the

Figure A.24    *Dustpan dredger.*

slopes towards the suction intake. After having reached the target depth, the dredger is slowly winched forward on its bow anchor while dredging continues.

The dustpan dredger operates slightly different: it acts more like a huge vacuum cleaner cutting series of relatively shallow, parallel swaths while pulled forward with hauling winches.

### A.5.3   *Possibilities and limitations*

This type of dredger can – if equipped with a submerged pump – be used in borrow pits with water depths exceeding 50 m. The minimum water depth in which the dredger can operate depends on the draught of the dredger and the length of the suction pipe. For deeper dredging depths the suction pipe could be extended.

Economic use of the stationary suction dredger requires the presence of loose to very loose sand deposits of sufficient thickness (more than approximately 6–8 m). As the equipment is sensitive to vertical movements it is generally only used in inland waters, rivers and sheltered nearshore areas to mine sand or gravel. This dredging method may successfully be used to mine sand from below a soft cohesive layer of limited thickness.

As stated earlier, it is not possible to control the slopes as the dredging method is based on destabilization of the slopes. Depending on the state of the soil and the use of water jets, this may sometimes result in uncontrolled slope failures or flow slides (static liquefaction) during and after dredging. There are reports of large volumes of sand suddenly sliding onto the suction pipe causing capsizing of dredgers. Plain suction dredging creates an uneven seabed with large craters.

Because the dredging method differs from plain suction dredging, the maximum dredging depth of a dustpan dredger is limited to 20–25 m. It produces a more or less level seabed. This type of dredger is suitable to remove relatively thin layers of sediment (minimum thickness approx. 0.25 m) and is generally used for maintenance dredging of inland navigation channels. It is less effective for sand winning in a borrow pit.

Both dredging methods are vulnerable to obstacles like boulders, old tree stumps and other coarse debris present within the dredging area.

Plain suction dredgers and dustpan dredgers both pump the dredged material through pipelines to the reclamation or into barges alongside the dredger which transport the material to the fill area. During barge loading, process water with suspended fine material flows over board which settles on the river- or seabed. This improves the quality of the fill material but will cause local turbidity.

A.5.4  *Auxiliary equipment*

Depending on its design, a stationary suction dredger may require a multicat/tug to move the anchors and a survey boat to carry out regular bathymetric surveys. The multicat may be able to tow the dredger, but for longer distances a tug boat is needed.

A.5.5  *Production*

Depending on the soil conditions and pumping capacity the production of a plain suction dredger may vary between 400 and 2500 m$^3$ per hour. Because of its specific use to dredge in layers the production of a dustpan dredger is often expressed in terms of square meters per hour.

A.6  **Bucket (ladder) Dredgers (BD)**

A.6.1  *General lay-out*

The bucket (ladder) dredger consists of a pontoon with a central well in which a ladder is suspended from a main gantry. The ladder supports an endless chain of buckets that at each end passes over sprocket wheels or tumblers. The ladder can be lowered to the seabed. Just below the top tumbler, the dredger is equipped with chutes that feed the dredged material into barges that are moored alongside the dredger. The pontoon may or may not be self-propelled. When dredging, the bucket dredger is held in position by 6 anchors, see Figure A.25.

Classification of the bucket dredgers is usually based on the size of the buckets that may range between 50–1200 litres.

A.6.2  *Dredging method*

Once the ladder is lowered to the seabed, the bucket chain rotates and the buckets scoop up material from the bottom, raise it to the top of the ladder and overturns at the highest point of the chain. The dredged material is tipped into the chute and slide into barges moored alongside the dredger. While dredging, the pontoon swings around the bow anchor by using the anchor winches (see Figure A.26).

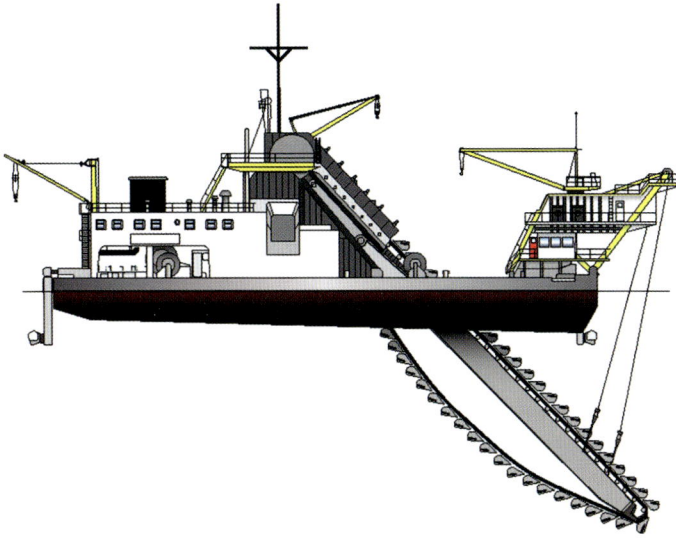

Figure A.25   *Bucket dredger with a raised ladder.*

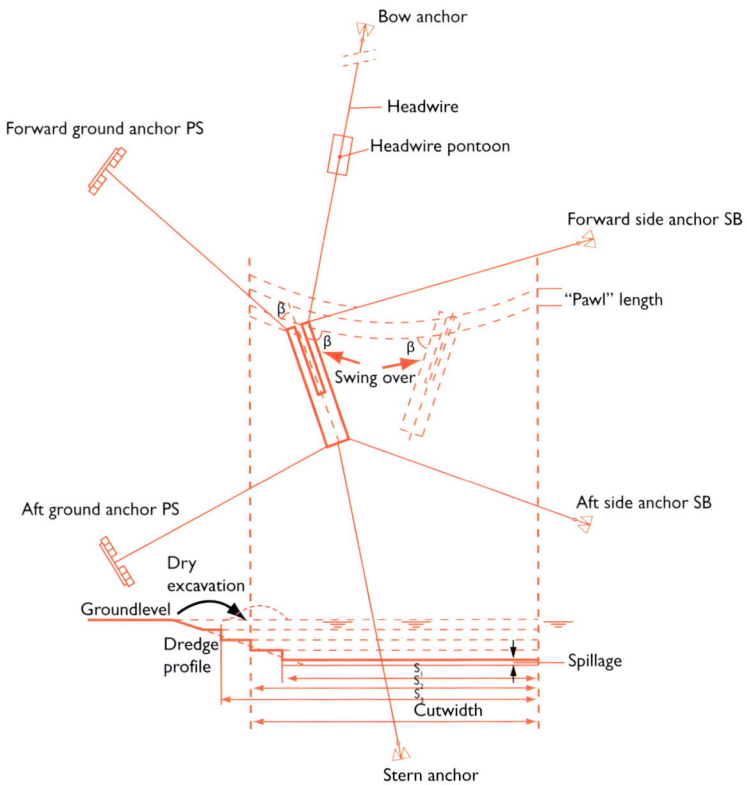

Figure A.26   *Positioning of the bucket dredger and the anchors.*

A.6.3 *Possibilities and limitations*

Today the bucket dredger, one of the oldest types of stationary dredgers, is not frequently used anymore. Its deployment is generally limited to environmental dredging projects and it is occasionally used in mining (gravels, ores) and deepening operations (harbours, waterways, etc.).

Depending on its installed power, the bucket dredger can dredge a wide variety of materials ranging from soft clays to soft rocks. It may also be used to remove blasted rock as it is able to handle a large range of fragment sizes. The bucket sizes generally reduce with an increasing strength of the in-situ materials. Excavation by a bucket dredger does generally not dilute the dredged material which is advantageous for as well production as cost. This limited increase of water content causes relatively little loss of strength of cohesive materials.

The bucket dredger is very sensitive to waves and swell and can therefore only be used in sufficiently sheltered areas. The maximum depth at which the largest bucket dredgers can dredge is approximately 30 m. As a result of the position of the buckets (and the subsequent degree of filling), the bucket dredger is not suitable to dredge at depths less than 4–6 m. However, in shallow water it is capable to dredge its own water depth. The usual face height of a single cut does not exceed 5 m.

As the bucket dredger relies upon transport by barges, it may be used in projects with large transport distances. The presence of the anchors and (often long) anchor wires may interfere with other shipping activities in the area. The use of a bucket dredger in densely populated areas may be restricted as a result of its noise nuisance during operation.

A.6.4 *Auxiliary equipment*

A stationary bucket dredger requires a multicat/tug to move the anchors and a survey boat to carry out regular bathymetric surveys. The multicat may be able to tow the dredger, but for longer distances a tug boat is needed.

A.6.5 *Production*

The production of a bucket dredger depends on the nature and strength of the subsoil, the installed power and the fill degree of the buckets. A rough figure indicates a production ranging between 200 $m^3$ and 750 $m^3$ per hour.

## A.7 Water injection dredgers

Water injection dredging is not suitable for constructing a hydraulic fill but may be used to remove fine unsuitable materials from the seabed in a borrow area (prior to mining) and/or at the footprint of a reclamation area (before placing the fill).

### A.7.1 General lay-out

A water injection dredger consists of a self-propelled pontoon or vessel equipped with a ladder or pipe that can be lowered onto the seabed. At the lower end of the ladder or pipe, a T-bar is attached. A fixed array of water jet nozzles is mounted on this T-bar. Water is pumped through the pipe, T-bar and nozzles and injected into the seabed at relatively low pressure (approximately 150 kPa) and flow rate (2–3 m³/s). The water injection dredger is illustrated in Figure A.27 while Figure A.28 shows the T-bar with the nozzles.

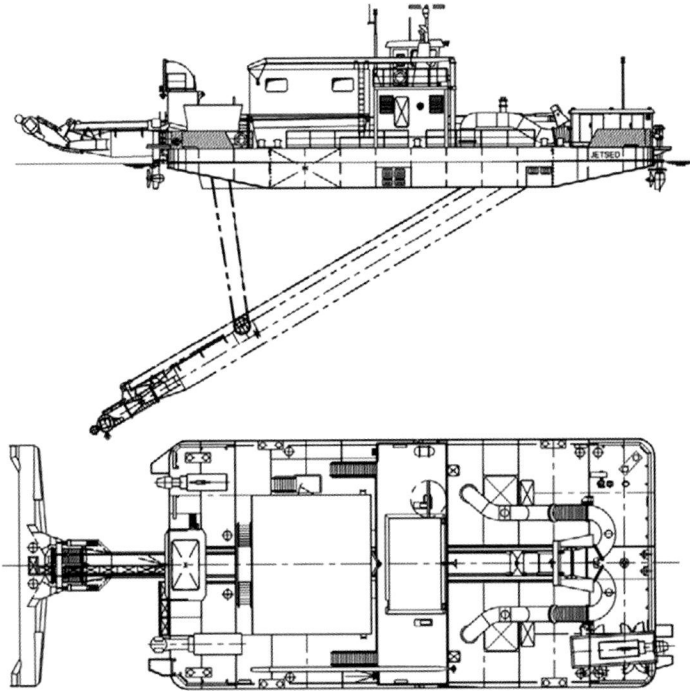

Figure A.27    *Typical lay-out of water injection dredger.*

Figure A.28 *T-bar (with nozzles) at the end of the ladder of a water injection dredger.*

## A.7.2 *Dredging method*

Unlike all other dredgers discussed before, the water injection dredger does not raise the dredged material to the surface nor does it transport the material through a pipeline or by barge to the designated fill area. Instead, it induces a density current at seabed level after fluidizing the in-situ material by injecting water into the seabed. This density current (typical density 1050 kg/m³) will flow under influence of gravity to a lower level and transport the solids until the grains start settling. To create such a density current, the dredger lowers the T-bar to the seabed and injects water while slowly moving in the opposite direction of the induced density flow. The thickness of the fluidized layer may be as much as 2 m, while the velocity of the density current may vary between 0.3–0.5 m/s.

## A.7.3 *Possibilities and limitations*

Water injection dredging is generally restricted to low strength, fine grained sediments (very soft silty clay, clayey silt). Its efficiency depends on aspects such as the penetration depth of the water jets, the viscosity of the fluidized mixture and

510

the settling velocity of the solids. The ability to fluidize the in-situ material and to transport the suspended solids depends on both local and operational conditions.

Important local conditions include:

– the properties of the seabed materials such as in-situ density, shear strength, grain size and grain size distribution, grain shape and particle density;
– the seabed gradient, existing currents and bathymetry (depending on the size of equipment the water depth should typically range between 3 m and 30 m).

Operational factors comprise:

– the characteristics of the dredger such as the type of nozzles, the flow rate of the injected water, the length of the T-bar;
– the working method including sailing speed, distance between nozzles and sea-bed, the dredging geometry.

When fluidizing silts, the transport distance of the density current may be several kilometres while sand may already settle at short distances. After fluidization and deposition the resulting seabed is generally flat. The limiting wave height is typically 0.8 to 1.0 m and cross-currents typically maximum 1.5 knots.

### A.7.4 *Auxiliary equipment*

Auxiliary equipment is limited to a survey vessel. Depending on the type of hull, transport of the dredger over longer distances may require a tug boat.

### A.7.5 *Production*

The theoretical production of a water injection dredger can be calculated by multiplying the penetration depth of the water jets (or the thickness of the layer that can be fluidized), the length of the T-bar and the sailing speed of the dredger. The penetration depth of the water jets depends on the properties of the subsoil and increases with increasing grain size and permeability.

## A.8  Auxiliary equipment

### A.8.1 *Workboats*

Most types of dredgers – and in particular the large stationary units – need attend-ant workboats to assist during dredging operations. The kind of assistance depends

Figure A.29 *Multicat.*

Figure A.30 *Launch type work boat.*

on the type of dredger and the operations to be undertaken and include a wide variety of activities like transport of personnel, consumables and other accessories, handling of anchors and/or floating pipelines, light towing and bunkering. These workboats must therefore have a shallow draught, good manoeuvrability, sufficient power, lifting and towing ability, deck space and bunkering capacity.

Generally two types of workboats can be distinguished: a pontoon-type (multicat) and a launch-type workboat. The hull of a multicat usually consists of a rectangular box with a small cabin/control room providing a robust construction for heavy duty operations with adequate clear deck space. A multicat is often equipped with a hydraulic crane for lifting operations. As a result of its relatively low speed its radius of action is generally limited to the vicinity of the dredgers. Figure A.29 shows a typical multicat.

The launch-type tender is less versatile than the multicat but may be employed for transport of personnel, light towing operations and surveying. As a result of the shape of the hull it can operate at a higher speed making it more suitable for longer distance haulage. An example of the launch-type workboat is presented in Figure A.30.

### A.8.2  *Survey vessels*

Hydrographic surveying of the borrow area and the reclamation site requires a boat with adequate sheltered accommodation to host all electronic equipment and personnel. In addition, it must have a shallow draught, good manoeuvrability and, as the sailing distances may be large, a sufficiently high speed. Usually the survey vessel is a launch-type boat. Figure A.31 shows an example of a survey vessel.

Figure A.31  *Survey vessel.*

### A.8.3   *Pipeline transport*

In case transport distances are not too long, pumps and pipelines are often used to transport the mixture of water and dredged material directly from the dredger to the reclamation.

These pipelines must be:

- transportable both over land and over water;
- strong enough to withstand high internal pressures;
- resistant against abrasion by the dredged materials;
- flexible enough to allow for movements of the dredger and swell and wave actions;
- manageable on the reclamation as they may have to be shifted during the filling operations.

The required diameter of the pipeline is mainly determined by the pumping capacity of the dredger but may also be influenced by the type of material to be transported, the length of the pipeline, the number of branches, etc  If the diameter of the pipeline is too small, friction as a result of high velocities may cause excessive head loss, while a too large diameter may result in deposition of fill material inside the pipeline.

Generally, pipelines can be divided into:

- floating pipeline (flexible pipeline consisting of reinforced rubber or jointed steel pipes);
- sinkerline (welded steel pipeline placed on the seabed);
- onshore pipeline (steel pipeline with bolted flange connections).

As the wear of a pipeline may vary widely it is important to check the wall thickness on a regularly basis. To evenly distribute the wear over their full cross section and, hence, to extend their lifetime, pipelines are regularly rotated.

**Floating pipeline**
Floating pipelines consist of reinforced rubber incorporated in a floating jacket (see Figure A.32) but could also be constructed of steel pipes connected by ball-joints or rubber hoses. These pipes are equipped with a floating jacket or may be mounted on small steel pontoons (see Figure A.33).

Reinforced rubber pipelines provide the greatest flexibility against movements of the dredger or the swell and are therefore used in more exposed, nearshore conditions. Wear and tear and capital cost of such pipelines are generally high. Steel pipelines have a higher resistance against wear and are used in more sheltered

514

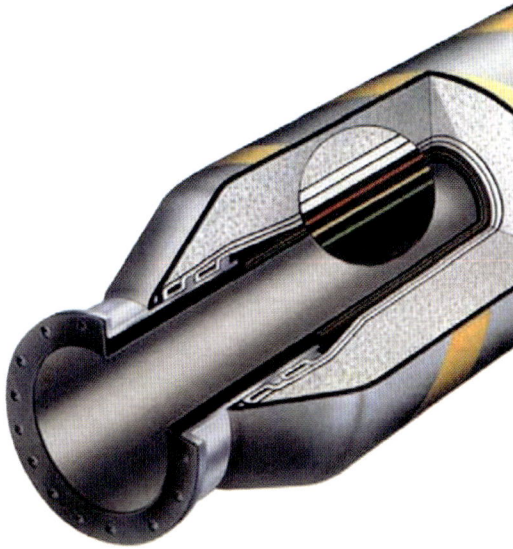

Figure A.32    *Floating pipe line of reinforced rubber.*

Figure A.33    *Floating pipe line on small steel pontoons.*

515

areas or when the dredged material has a high abrasiveness (coarse quartz sand, broken rock). Floating pipelines are often a combination of steel and rubber sections. When transporting highly abrasive materials they may consist of only steel sections coupled by ball joints. The use of pontoons may be preferable above floating jackets when currents affect the dredging area.

Floating pipelines have the disadvantage of obstructing other water-borne traffic and the route of the floating pipeline between the dredger and the shore should therefore be carefully selected. The floating pipeline is connected to either a sinkerline or an onshore pipeline but it may also feed a pontoon mounted nozzle, a spreader pontoon or a diffuser in case the fill is placed below the water line.

### Sinkerline

As an alternative to the floating pipeline, a sinkerline can be used to reduce the interference with other activities in the dredging area and/or the adverse effects of currents, waves and swell. A sinkerline consists of a welded steel pipeline that is sunk onto the seabed. At both ends, a riser pipe is used for transition from the seabed to the surface. Near the dredger the sinkerline is connected to a floating pipeline. At the land side it is connected to either a floating pipe line or an onshore pipeline. Its hydraulic resistance is generally less than that of a floating pipeline as it is usually a straight, welded pipeline with a limited number of balljoints.

Disadvantages of a sinkerline may be the difficulty to install (and recover) it on the seabed in exposed conditions with waves and currents, to inspect the quality of the pipeline (wear, joints, etc) and to clean up a blocked pipeline.

### Onshore pipeline

An onshore pipeline usually consists of steel pipeline sections with bolted flange connections. Onshore pipelines are generally placed on ground level, but they might have to cross roads or other existing infrastructure like bridges, culverts, etc On the reclamation, ramps are often constructed to allow traffic and equipment to cross the pipelines. It is important to plan the route of the onshore pipeline carefully as a relocation of the pipeline including its crossings is time-consuming and costly. If there are significant differences in elevation along the route, snifters may have to be installed to expel trapped air from the pipeline. Figure A.34 shows a snifter and its principle.

For operational reasons, onshore pipelines are equipped with valves and/or may be branched. Note that the use of multiple discharge points of an onshore pipeline may require careful engineering and operating to prevent a too high resistance or a too low velocity of the slurry in the pipeline.

Figure A.34    *Snifter on discharge pipe line.*

High discharge capacities of the latest generation of dredgers may require pipes fitted with a quick coupling system and/or sufficient branches in combination with control valves. A quick coupling system reduces the time required to couple two pipes to 10 seconds, but is more sensitive to wear and may result in more hydraulic resistance within the pipeline.

The end of the pipeline may be equipped with a spray hood to avoid the development of an erosion pit, see Figure A.35.

517

Figure A.35    *Spray hood.*

**Operational aspects**

It is clear that the most economic operation corresponds to a straight alignment of the pipeline running from the dredger to the point of discharge at the reclamation. However, other aspects influencing the pipeline alignment are:

– operational movements of the dredger;
– local bathymetry;
– other water born activities within the dredging area;
– waves, swell and currents;
– the presence of infrastructure on land;
– local regulations;
– construction activities on the reclamation area.

These aspects need to be taken into account when planning the route of the discharge pipeline. Relocation of the pipeline during the execution of the project may not be desirable as it could interrupt the filling operations. The arrangement of the pipeline (including possible control valves, snifters and branches) further strongly depends on the way the filling works are carried out.

**Critical velocity and production**

Hydraulic transport of a sand-water mixture in a pipeline requires the right balance between:

– the pressure provided by the pump(s) of the dredger (or booster pump station);
– the velocity of the soil-water mixture (slurry) in the pipeline;
– the concentration of the solid particles in the slurry.

For a given pipeline diameter, concentration and average grain size of the solid particles in the soil-water mixture, there is a critical velocity below which solids will settle at the bottom of the pipeline. As a result of the settling process, the effective diameter of the pipeline will be reduced increasing the hydraulic resistance of the pipeline and raising the required pressures of the pump(s). At higher mixture velocities, however, pressure losses in the pipeline will also be higher (increased friction losses and presence of bends, control valves, joints, etc.). This will also result in an increased pressure demand. This effect is more pronounced at higher slurry concentrations. Figure A.36 presents the principle of this mechanism.

For a given pipeline arrangement, the critical velocity depends on the size and shape of the grains, the concentration and the specific gravity of the solid particles in the slurry. The highest discharge production will be achieved by the optimum combination of velocity and concentration of the slurry within the capacity range of the dredging pump. Note that the design of the pipeline arrangement must account for the natural variability of the fill material since sudden changes in properties may adversely affect the hydraulic transport operations. This may have a significant impact on the production and may even result in a blocked discharge pipelines which is very troublesome especially in case of floating pipelines or sinker lines.

Generally, the design of a hydraulic transport operation will focus on the most economic supply of fill within given boundary conditions such as the nature of the available material, the bathymetry in the vicinity of the reclamation and the project planning.

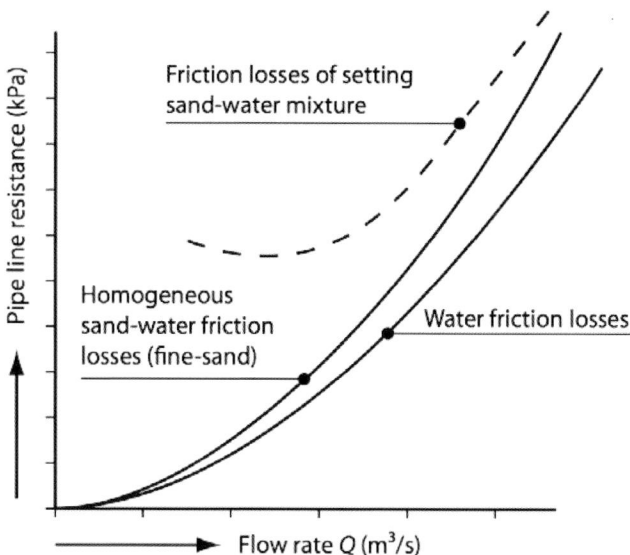

Figure A.36 *Pressure demand and pump characteristic.*

**Wear caused by hydraulic transport**
Regular inspections of the joints and measurements of the thickness of the walls
of the pipeline are important. Bursting of the pipeline at the prevailing high pressures and velocities may result in hazardous situations, accidents and unwanted
interruptions of the filling process.

Table A.3  *Effect of various variables on the wear of a pipeline.*

| Variable | Element | Variation | Wear/m³ fill |
|---|---|---|---|
| Particle size distribution | Solid particles in slurry | More coarse | More |
| Hardness | | Higher | More |
| Angularity | | Higher | More |
| Particle density | | Higher | More |
| Hardness | Construction material | Higher | Less |
| Elasticity | pipeline | Higher | Less |
| Surface roughness | | Higher | More |
| Velocity in pipeline | Slurry | Higher | More |
| Mixture density | | Higher | Less |
| Salt content | | Higher | More |

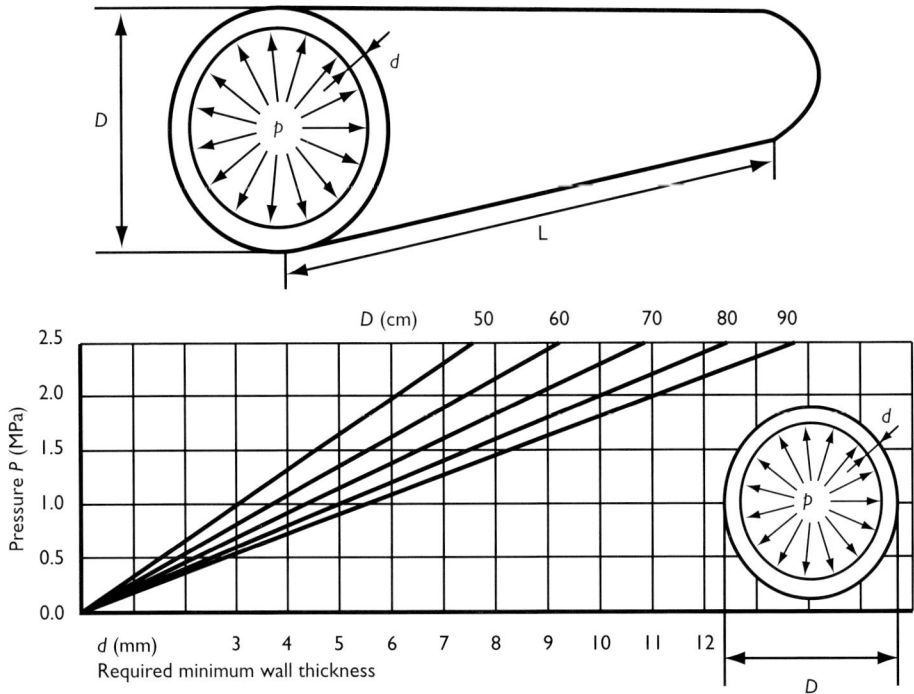

Figure A.37  *Required minimum wall thickness as a function of the pipe line diameter and pressure
within the pipe.*

The mixture is often modelled as slurry but, in practice, coarse and fine solids will segregate in the pipeline. The coarse fraction is rolling along the bottom while the finer fraction is suspended in the water over the full cross section of the pipe. This results in considerably more wear of the bottom of the pipeline. To achieve a more uniformly distributed wear along the full cross section, pipe sections are turned regularly and in this way its life time is extended.

In addition, more than average wear occurs at bends, locations of branching, joints and other irregularities (dents) of the pipeline. Table A.3 presents other variables and their relative effect.

Figure A.37 presents a simplified chart indicating the required minimum wall thickness as a function of the diameter of the pipe and the pressure.

### A.8.4  *Barges and barge loading/unloading dredgers*

Barges (or hopper barges) are used to transport all types of dredged material from – mostly stationary – dredgers to the points of discharge. Capacities may range between 50 m$^3$ to more than 3000 m$^3$. The number of barges required to complete a project in time is usually dictated by the planning of the project, the total fill volume to be transported, the capacity of the barges and the transport cycle time of a barge. The latter is dictated by the time required to load the barge (dependent on the production rate of the dredger), the sailing time loaded (distance and average speed of loaded barge), the time required for discharging and the sailing time unloaded (distance and average speed of unloaded barge).

Barges are usually loaded by cutter suction or stationary suction dredgers using purposely built loading facilities, see figure A.38, or by backhoe, grab or bucket dredgers discharging their buckets straight into the hopper. They may, however, also be filled by trailing suction hopper dredgers. Often, barges are equipped with overflow facilities for removal of fines.

Barges can be self-propelled or may be towed or pushed. They may be sea-going or can be limited to operate in sheltered areas only.

A barge can discharge its load by dumping (through bottom doors or by splitting of the hull) or by pumping (either by an external barge unloading dredger or an internal pumping system), see Figure A.39. Before dumping, attention must be paid to the resulting water depth in relation to the draught of the barge.

The barge unloading dredger is a special stationary suction dredger equipped with a pumping and jetting system. The amount of water available within a barge is, generally, not enough to allow for fluidization. The dredger is therefore equipped

Figure A.38    *Barge loading.*

Figure A.39    *Barge unloading dredger at work.*

Figure A.40   *Jetting water into barge to fluidize dredged material for unloading.*

with a conventional centrifugal dredging pump to unload the sand – water mixture and a special water pump to deliver sufficient process water in the barge (see Figure A.40). The barge unloading dredger is usually moored near the shore and connected to an onshore pipeline.

### A.8.5   *Spreader pontoons*

When the existing subsoil at the reclamation consists of very soft to soft cohesive materials (clayey silt, clay, etc.) then it could be necessary to place the fill in thin layers of uniform thickness in order to avoid instabilities and mud waves in the soft deposits. Over water this may be achieved by using a spreader pontoon. A spreader pontoon consists basically of a wide chute mounted on a barge and connected to a floating pipeline. The pontoon is winched forwards and backwards over the fill area using 4–6 anchors, while the sand-water mixture, supplied by the floating pipeline, flows over the chute and drops into the water (see Figure A.41). Often an underwater diffuser is installed on the spreader pontoon to reduce the turbidity.

A layer with a uniform thickness is placed on the seabed by controlling the speed of the pontoon as a function of the known flow and concentration of the sand-water mixture. The latest generation spreader pontoons are equipped with a computer

Figure A.41    *Spreader pontoon in action.*

controlled winching system. Input is provided by electronic concentration and flow meters located just before the discharge point of the pipeline. Knowing the width of the chute it is possible to calculate the required speed of the pontoon to place a fill layer of specified thickness. The hauling in and paying out speed of the 6 anchor winches are adjusted accordingly.

As a result of the large discharge capacities of modern trailing suction hopper and cutter suction dredgers, the shifting speed of the spreader pontoons must be high (0.5 m/s) to enable placement of layers with a thickness of 0.25 m–0.50 m.

Although this method allows for placing thin fill layers on top of very soft deposits with relatively limited risks of instabilities it also has some disadvantages:

– accessibility of the area for hydrographic surveying is very limited as a result of the presence of the anchor wires and the floating pipeline;
– the area that can be covered without shifting the anchors is limited while shifting of anchors interrupts the operations;

– spreader pontoons usually have a draught of 1.20 m to 2.00 m implying that it is not possible to use this method at smaller water depths. This can be overcome by temporarily setting up the water level within the containment bund of the fill area.

### A.8.6  *Booster pump stations*

Booster pump stations may have to be used in case additional pumping capacity is required. This can be due to the facts that:

– the pumping distance through pipelines becomes too long;
– the difference in elevation between dredger and discharge point of the pipeline is too much;
– the grain size of the fill material is too coarse (gravel, cobbles, clay balls, etc.).

These booster pump stations usually consist of a centrifugal pump (capacities ranging between 0.6 and 4 m³/s) with a diesel or electric drive mounted on a skid or trailer for on land applications or on a pontoon when used over water To avoid the development of a vacuum in the pipeline it is imperative that the position of the booster pump within the pipeline system is such that a positive pressure at the intake side of the booster pump is ensured at all times.

### A.8.7  *Reclamation spread/dry earth moving*

Distribution and levelling of the fill during placement at the reclamation requires equipment. Once the fill has reached a level above the water line, dozers are employed to level the area in front of the pipe and to prevent a built-up of material at the discharge point of the pipeline, see Figure A.42.

In general dozers should not be used to transport fill over a distance of more than 20–30 m. These operations will also provide an initial compaction of the fill mass. In case of a limited bearing capacity of the reclamation, light weight dozers with wide tracks can be used. Dozers may also be used to construct containment bunds before the placement of subsequent lifts. Although in the past a number of 2–3 dozers was generally sufficient to keep up with the production rates of the (small to medium sized) dredgers, high discharge capacities of the latest generation of CSDs and TSHDs may require significantly more spreading and levelling capacity on the reclamation. Other important factors for defining the required capacity of the fill spread include the number of dredgers pumping the fill ashore, the continuity of the fill supply, the number of and distance between the discharge locations (and the distance between these locations), the thickness of the lifts placed, etc.

Figure A.42    *Distribution of fill material by bulldozer.*

Figure A.43    *Large capacity weir box.*

Depending on contract specifications it may be required to limit the suspended solids in the discharge water drained back into the environment by installing one or more adjustable weir boxes in the bunds. Figure A.43 shows a typical example of a weir box. The required number of boxes depends on the specified maximum concentration of suspended solids in the water running through the weir box, the area and volume enclosed by the bunds, the nature and volume of the fines in the fill and the discharge rate of the dredgers.

Transport of individual pipeline sections at the reclamation will be done by front-end loaders equipped with special forks. The coupling capacity needed to provide adequate onshore pipelines depends on the type of coupling, the dimensions of the reclamation and, hence, the length of the pipeline, the thickness of the lifts, the size of the filling operations, etc.

In case significant earthmoving operations are required (for instance to shift or remove surcharge) front-end loaders and dump trucks or scrapers may have to be used. Hydraulic backhoes are often used for excavations. For ground improvement operations like compaction or other stabilization techniques reference is made to section 7.

FIELD AND LABORATORY TESTS

B   Site investigation & monitoring techniques and methods        530
   B.1     Bathymetrical seabed surveys                              530
      B.1.1      Type of echo sounding equipment                 531
      B.1.2      Survey process                                  534
   B.2     Geological and geotechnical investigation                 536
      B.2.1      Geophysical methods                             536
         B.2.1.1      Seismic reflection survey             536
         B.2.1.2      Seismic refraction survey             538
         B.2.1.3      Geo-resistivity                       539
      B.2.2      Sampling methods                                541
         B.2.2.1      Drilling                              541
         B.2.2.2      Vibrocores                            543
         B.2.2.3      Jet probe                             546
         B.2.2.4      Drop (gravity) corer                  546
         B.2.2.5      Box corer                             547
         B.2.2.6      Manually operated piston sampler      547
         B.2.2.7      Grab samples                          547
         B.2.2.8      Test pit                              548
      B.2.3      Testing                                         548
         B.2.3.1      Laboratory testing                    548
         B.2.3.2      In situ tests                         557
   B.3     Hydraulic, morphological and meteorological data          566
      B.3.1      Water levels                                    567
      B.3.2      Currents                                        568
      B.3.3      Waves                                           569
      B.3.4      Water temperature and salinity                  570
      B.3.5      Wind                                            571
      B.3.6      Air pressure                                    573
      B.3.7      Visibility                                      573
      B.3.8      Storm tracks                                    573
      B.3.9      Deposition simulation test                      573
      B.3.10     Turbidity                                       574
      B.3.11     Dissolved oxygen                                576
      B.3.12     Total Suspended Solids (TSS)                    576

| B.4 | Detection of seabed obstruction | | 576 |
| | B.4.1 | Side Scan Sonar (SSS) | 577 |
| | B.4.2 | Multibeam sonar or multibeam echo sounder (MBES) | 578 |
| | B.4.3 | Magnetometry | 579 |
| B.5 | Monitoring | | 581 |
| | B.5.1 | Monitoring bearing capacity | 581 |
| | | B.5.1.1 California bearing ratio | 581 |
| | | B.5.1.2 Plate bearing test | 581 |
| | | B.5.1.3 Zone load test | 581 |
| | | B.5.1.4 Trial embankment | 583 |
| | B.5.2 | Monitoring slope stability | 583 |
| | | B.5.2.1 Electrical piezometer | 583 |
| | | B.5.2.2 Open standpipe | 583 |
| | B.5.3 | Monitoring deformation | 584 |
| | | B.5.3.1 Settlement plate | 585 |
| | | B.5.3.2 Extensometer | 585 |
| | | B.5.3.3 Settlement hose | 586 |
| | | B.5.3.4 Inclinometer | 587 |
| | | B.5.3.5 Trial embankment | 588 |
| | | B.5.3.6 SASW, CSCW | 588 |
| | B.5.4 | Environmental monitoring | 589 |
| | | B.5.4.1 Visual inspection | 589 |
| | | B.5.4.2 Aerial & satellite imagery | 589 |
| | | B.5.4.3 Water sampling | 589 |
| | | B.5.4.4 Point monitoring | 590 |
| | | B.5.4.5 Mobile vessel-based monitoring | 590 |
| | | B.5.4.6 Acoustic Doppler Current Profilers (ADCP's) | 590 |

# B   Site investigation & monitoring techniques and methods

## B.1   Bathymetrical seabed surveys

A bathymetrical survey is a technique that uses acoustic sounding for the measurement of the water depth. The principle is based on emitting a short acoustic signal from a transducer below a ship downwards to the sea floor. This sound wave is reflected by the sea floor and returns to the receiver below the ship. Based on the travel time and velocity of the sound wave, the water depth can be calculated (taking into account corrections for vessel motion and the depth of the transducer).

After correction with the tidal height, the registered water depths can be related to a predefined reference level such as the lowest astronomical tide level (LAT) or mean sea level (MSL). An overview of water levels and heights is presented in Figure B.1.

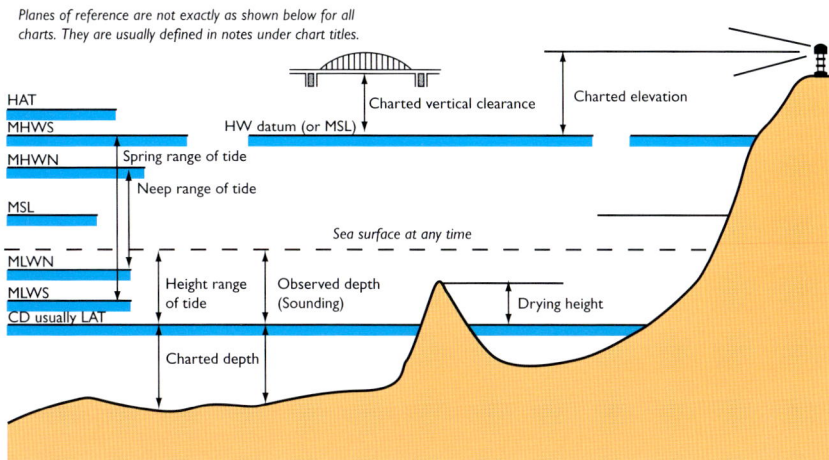

Figure B.1   *Overview of tidal levels (Source chart 5011, UKHO).*

## B.1.1   *Type of echo sounding equipment*

Echo sounding is carried out with acoustic equipment normally mounted underneath a ship. The two kinds of systems widely used in bathymetrical measurements are:

− Single beam echo sounder;
− Multibeam echo sounder.

**Single beam echo sounder**
Single beam systems consist of a transmitter and a receiver. Each time the transmitter is activated one single acoustic pulse is generated and emitted. By sailing parallel survey lines an area can be covered. The seabed depth along these lines is recorded. A bathymetrical map can be made by interpolating the data in between these lines. The correctness of the results of this interpolation depends on the distance between the survey lines and the depth variation in the surface profile. A broad range of echo-sounding systems can be used for shallow and deep-water applications. Various frequencies can be chosen. Two standard frequencies are used: 33 kHz and 210 kHz.

These two frequencies are complementary to each other in the sense of reflection and penetration properties. Although the seabed is usually a good acoustic reflector, in some situations it is not. For example, mud layers close to the sea bottom can disturb the clearness of this reflector. Using the two frequencies simultaneously, the mud layer can be detected: the mud will be penetrated by the 33 kHz and reflected by the 210 kHz. A combination of these two frequencies gives insight in both the depth and the thickness of a mud layer at sea bottom. In some cases, a low density mud might be allowable for navigation (nautical depth). Lower frequencies (in the range of a few kHz) will penetrate the sea floor and are used for seismic investigations. This technique is used for geological or geotechnical investigations. Low frequencies will in general penetrate more than high frequencies, but low frequencies provide less detailed information and are therefore less accurate in terms of depth calculation. A singlebeam echo sounder can achieve an accuracy of 10 cm up to a water depth of 100 m (ref. Lekkerkerk 2006). "Handbook of Offshore Surveying").

A typical non-interpolated result of a single beam survey is displayed in Figure B.2.

**Multibeam echo sounder**
Multibeam echo sounders used in bathymetric surveys measure water depth in a broad track under the ship. These measurements result in high-resolution 3D images of the seabed. A multibeam echo sounder is composed of an array of singlebeam echo sounders. The beams are received under a range of several angles, each beam illuminating a different part of the sea bottom. All beams together are used to create an image of the sea bottom beneath the ship. The swathe width of a multi beam is usually about 2 degrees in direction of the ship's axis and about

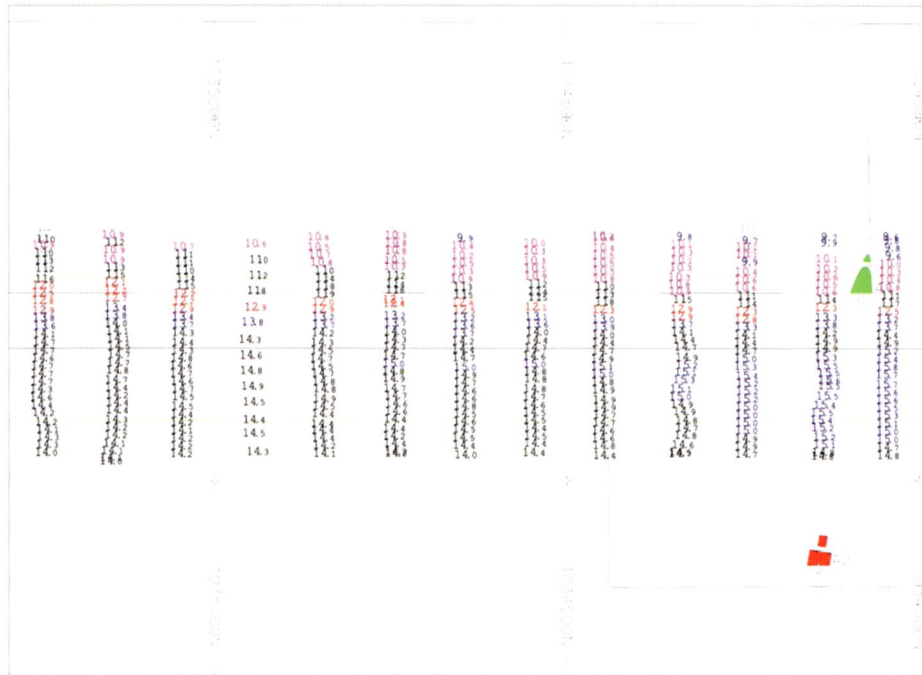

Figure B.2    *Typical results single beam survey.*

150 degrees perpendicular to it. Along the sailing line, a full coverage measurement is obtained. The area covered by the multi beam echo sounder is called the footprint. The dimensions of this area depend on:

- water depth;
- swathe width;
- transmission angle.

The footprint becomes larger with increasing water depth. The width of the footprint is generally about 4 times the water depth. The footprint increases in size with increasing water depth, increasing beam width and transmission angle. The position of the ship, its roll, pitch and heave need to be monitored accurately because they influence the exact area of the sea bottom which is illuminated. An accuracy of 1–15 cm can be achieved depending on the water depth. (ref. Lekkerkerk 2006 "Handbook of Offshore Surveying").

A typical processed result of a multibeam survey is displayed in Figure B.3.

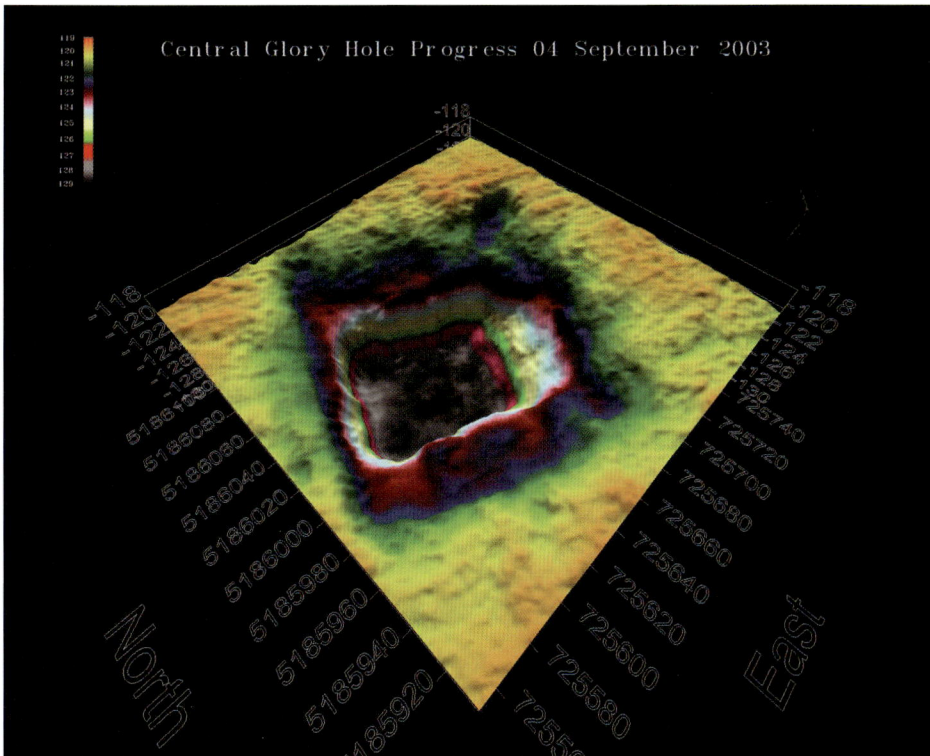

Figure B.3    *Typical result post dredging multibeam survey.*

533

B.1.2 *Survey process*

**Production**

The area covered by a single-beam bathymetric survey depends mainly on the sailing speed of the ship. Under normal conditions, a speed of 5 knots (about 9 km/h) will result in a production of about 80 to 90 km a day (10 hours net production time per day).

– Single beam

  For bathymetry measurements carried out with a single beam echo sounder the total length of the survey lines equals the sailing distance.

– Multibeam

  When surveying with a multibeam system, the total area that is covered depends on the footprint of the multibeam for a specific survey. Assuming a multibeam system with an opening angle of 110 degrees installed at 15 m above the seabed, an area of about 3.4 km$^2$ can be covered in a 10 hour shift. However, often a beam overlap of 100% is required, so the actual production would be 1.7 km$^2$/shift.

**Line density**

Depending on the aim of the bathymetric survey, the equipment and line density is chosen. For an overall bathymetric survey that is carried out for example on the North Sea, a line density of one line every 25 or 50 metres using a single beam echo sounder can be sufficient. On the other hand, in case of monitoring the transport behaviour of so-called sand waves much more detail is needed and hence the use of multibeam echo sounding to obtain a full coverage of the sea bottom is advised.

For dredging works, this choice depends on the available survey equipment and the amount of detail that is desired. Usually a multibeam survey resulting in a 3D image of the sea floor is conducted.

**Data processing**

Processing of the bathymetric data does not only rely on the quality of the acoustic reflections, but also on the following aspects:

– Acoustic wave velocity;
– Position of the ship;
– Motion of the ship;
– Reference level and tide correction.

Calculating the water depth depends on the travel time and the acoustic wave velocities in the water. This speed of sound is roughly 1500 metres/second in water. The main parameters that affect the speed of sound are temperature, salinity and pressure. These parameters can drastically change the way sound travels through water. Changes in sound speed, either vertically in the water column or horizontally along the water surface during sailing, can cause an error in the water depth measurement. Regular checks of the sound velocity profiles over the full water column are therefore required to correct the measured data.

In contrast to land surveys that are mostly static, bathymetric measurements are dynamic as the ship is sailing during the survey. For a bathymetric survey, it is highly recommended that the orientation and the positions of all sensors are well known. If this is not the case, significant errors in depth and position will occur. Sonar systems derive their positional information from external navigation systems. These are typically a combination of GPS units and inertial navigation systems (INS = motion sensor and computer based navigation system).

Motion sensors measure the roll, pitch and heave of the ship. The position of the motion sensor on the ship is especially important for measuring the heave. Because of the movement of the ship, the best position is in the centre of gravity, where all movements except the heave are least. The yaw of the ship can be measured using the gyrocompass.

**Ping rate: detecting shallow and deep water**
In principle, the maximum water depth that can be detected with echo-sounding techniques is unlimited. If a variety of highly efficient transducers are available, single beam measurements can be carried out from extreme shallow water to the deepest trench of 11,000 metres.

An echo sounder is activated by a so-called ping. Each ping is the start of a new measurement. Tuning this ping rate, especially for deep-water measurements, is therefore important. For example, if the sea bottom is about 750 m of depth, an acoustic pulse will travel one second before arriving at the transducer again. Ping rates of 1, 2, 3, 4, etc. pings per second will cause the outgoing ping to coincide with the reflected incoming wave and the measurement will fail. For single beam measurements, a high ping rate is still possible, as long as the reflected wave does not arrive at the transducer at the time a new signal is emitted. This can be achieved by having 0.3 second between each ping, for example. In general, the ping rate is lowered with increasing water depth. A disadvantage of lowering the ping rate is of course the point density that will decrease while the sailing speed remains constant.

The multi beam echo sounder is used in coastal areas where water depths range from about 2 metres below the ship up to about 1000 metres or more. Deep echo sounding with a multi beam system is possible deeper than 1000 metres, but the data coverage will decrease due to the travel time of the acoustic waves. This causes a loss of sea bottom detail (an important aspect of multi beam). For deep echo sounding, a single beam echo sounder is generally preferred because it is easier to use than multi-beam echo sounders.

**Blind depth**

For shallow depths, on the other hand, the pulse length is an important parameter in order to minimize the so-called 'blind depth'. This depth exists as a result of the impossibility of the echo sounder to send and receive simultaneously. Thus, for shallow depths the pulse length may not be too large: the blind depth is equal to the half of the pulse length. Because the velocity of the acoustic wave in the water is roughly 1500 m/s, the pulse length and therefore the blind depth is limited by the repeat frequency of the pulse, which is emitted by the echo sounder. With an increase of the repeat frequency, the pulse length and the depth of the blind zone decrease, if all other parameters remain constant. The acoustic power also affects the depth of the blind zone. By using a separate source and receiver, blind depth as described here can be prevented.

## B.2   Geological and geotechnical investigation

### B.2.1   *Geophysical methods*

#### B.2.1.1   Seismic reflection survey

The seismic reflection method, also called sub bottom profiling, involves the recording of the travel time interval of seismic (acoustic) waves, which are emitted from the surface and reflected by the seabed and underlying soil layers. A seismic source generates a seismic (acoustic) signal with a relative high frequency.

Various systems exist such as:

– pinger (3 kHz–7 kHz);
– boomer (0.5–3.5 kHz);
– sparker (high energy signal);
– chirp (varying frequency 500 Hz–12 kHz).

This high-frequency signal partly penetrates the sea bottom and is reflected on layer discontinuities in the subsoil. The layer discontinuities are horizons with contrasting acoustic impedance, i.e. layers with a different situ density and a different sound wave velocity. The reflected signals (P-waves) are captured by so-called hydrophones and converted to voltages. By making an assumption about the propagating velocities of the acoustic signal in the different layers, the thickness of these layers can be calculated.

The lower the chosen frequency, the deeper the signal will penetrate in the subsoil. However, this also means that the resolution of the resulting layer profile will diminish. Some systems (e.g., the Chirp) emit a signal with variable frequency.

The seismic source is generally towed behind or mounted on a small survey boat (Figure B.4). The repetitive triggering of the seismic source and the recording of the reflected acoustic signal produces "acoustic" cross sections over the surveyed area (Figure B.5). Depending on the used frequency of the seismic source, the vertical resolution of these cross sections is around 0.5 m. Layers with a smaller thickness are not traceable. With a maximum navigation speed of around 5 knots, a total of 110 km of seismic profile can be recorded in 12 hours' time. In that way it is a very quick and relatively cheap method to record the geological build-up of an area of interest such as a future sand borrow area. However, it must be kept in

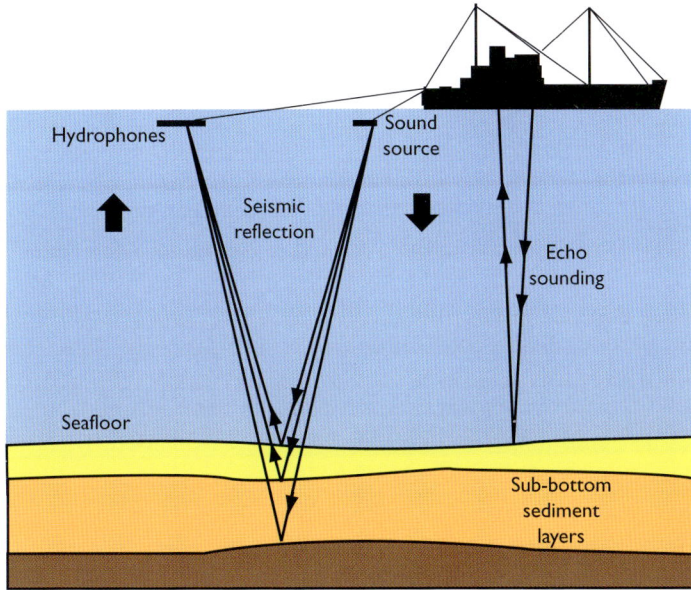

Figure B.4    *Seismic reflection survey.*

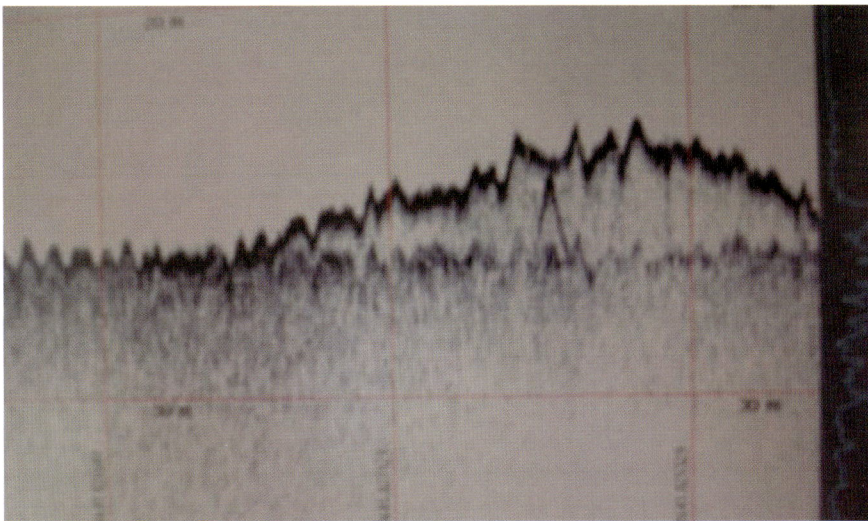

Figure B.5    *Example of a 3.5 kHz Boomer seismic.*

mind that with this method only differences in acoustic impedance are visualized. Calibration of the different acoustic layers on a seismic profile with piston samples, vibrocores, or any other direct sampling method is indispensable.

### B.2.1.2 Seismic refraction survey

While for the seismic reflection method, the reflected signal of an acoustic wave is studied, the seismic refraction method interprets the arrival time of the refracted wave. By putting several hydrophones in a known configuration on the seabed, it is possible to calculate the velocity of an acoustic wave front through different layers of the subsoil. Sediment has sound velocity values, which typically range from:

- 1550–1600 m/s for soft clay and silt;
- 1950 m/s for very dense (cemented) sand and hard clay;
- 2000 m/s–6500 m/s for rock (for example 2000 m/s for very weak mudstone up to 6500 m/s for extremely strong basalt and granite or fossil limestone).

The major advantage of the seismic refraction method is that the derived sound velocity of an individual layer correlates very well to the compactness of the sediment or to the Unconfined Compressive Strength rock strength (UCS). However, whereas the UCS value is a rock strength property; the seismic sound velocity is more a rock mass characteristic that may strongly be affected by joints, fractures and other discontinuities. These properties in combination with the UCS determine the dredgeability of rock.

Where a seismic reflection survey is rather easy to perform – seismic source and receiver are both towed behind the survey vessel (Figure B.4) – the realization of a seismic refraction survey is a challenging operation with a much more complex set-up with several hydrophones and a multitude of data to be interpreted (Figure B.6).

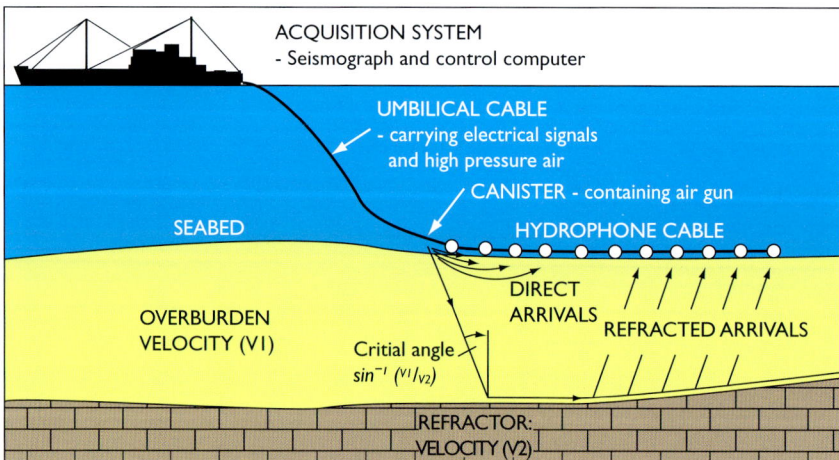

Figure B.6 *Configuration of a seismic refraction survey.*

### *Static seismic refraction survey*

In some cases (for example to assess the sound velocity profile along an offshore pipe or cable route) a so-called static seismic refraction survey is performed. A group of hydrophones (a streamer) is installed "static" on the seabed by divers. A daily progress of 125 to 300 m is the upper limit for such a static seismic refraction survey.

### *Dynamic seismic refraction survey*

For the mapping of a dredge area, the static method is not suitable since the progress is too slow.

The Gambas system (developed by FUGRO, Australia) allows recording sound velocity profiles in a more dynamic way resulting in a daily progress of up to 40 km. Because of the high cost, seismic refraction studies are limited to (big) projects where a considerable amount of hard rock needs to be dredged. A typical plan view of the results of the seismic refraction survey using the Gambas system is presented in Figure B.7.

A disadvantage of the seismic refraction method is that during processing of the data one has to assume that the rock layers show an increasing sound velocity with depth. Therefore, a hard "caprock" layer overlaying a softer rock cannot correctly be detected.

### B.2.1.3   Geo-resistivity

The geo-resistivity method is based on an electrical current, which is injected into the water column & subsurface by means of 2 current electrodes. Two voltage electrodes placed in between the current electrodes are measuring the voltage gradient of the electrical field in the subsurface. Based on the measured values of the current and the voltage the average resistivity of the subsurface is then calculated. The penetration depth of the current depends on the distance between the current electrodes. Larger electrode distances are associated with increasing penetration depths. When measurements of the resistivity are repeated with progressively increasing current electrode distances, information is obtained from progressively deeper geological structures.

Seawater has a much lower resistivity (higher conductivity) than most sediments and rocks. This implies that the resistivity of sediment and rock highly depends on its porosity, water saturation and the degree of water resistivity (salt versus brackish). In sediment and rock, the resistivity is highly variable and covers a range of more than eight orders of magnitude. In general marine sediments have a good porosity, resulting in low resistivity, while rock has a low porosity and shows very high resistivity. As sediments have a remarkably lower resistivity than rocks, a geo-resistivity survey is often executed to map rock outcrops (for example Panama Canal (Figure B.8)). Calibration with any of the direct sample methods is again indispensable.

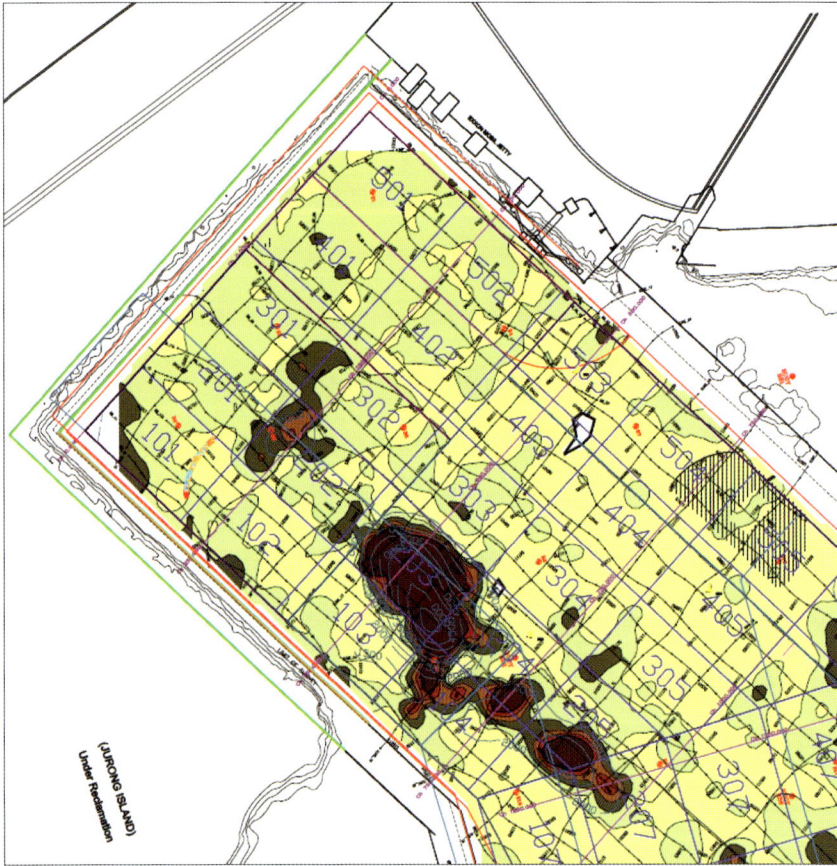

Figure B.7   *Typical plan view with results of a (Gambas) refraction seismic survey, the darker patches indicate rock outcrops. Seismic data from Banyan Basin – Singapore.*

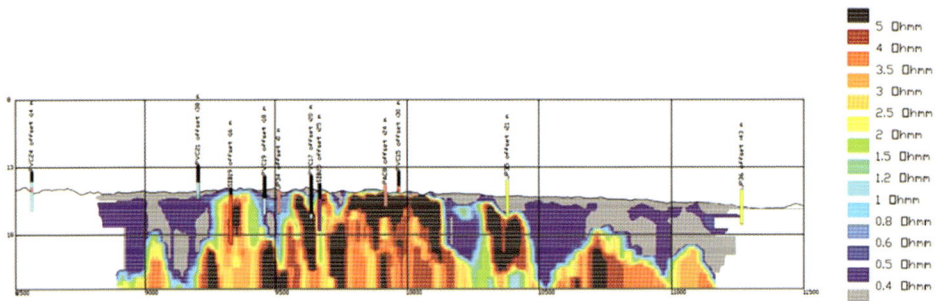

Figure B.8   *Typical cross-sectional view with results of geo-electric survey (data from Panama Canal – Pacific section). A rock head can be clearly identified in the middle of the section. Boreholes were performed to verify the nature of the rock head.*

## B.2.2   Sampling methods

Reference is made to Eurocode 7 Part 2 Ground investigation and testing. In EC7, 3 sampling categories A, B, C are defined. Category A provides the highest quality of samples. Additionally there are 5 quality classes for samples, 1 to 5, where class 1 has the highest quality and lowest disturbance during sampling.

### B.2.2.1   Drilling

For drilling operations, a distinction can be made between the advancing method and the sampling method:

- advancing method: method by which the borehole is advanced in the seabed
  - shell and auger (bailing)
  - chiselling
  - pushing
  - flushing/washing
  - rotary drilling
- sampling method: method by which the soil samples are retrieved
  - Drive samplers
    - Open samplers (e.g. Shelby tube/SPT sampler)
    - Piston samplers
  - Rotary sampling (Core barrel)
  - Block sampling

Depending on the type of soil and required sampling quality, one of the above techniques is chosen. A description of the sampling methods and resulting sample quality class can be found in EN-ISO 22475-1, 2006 (Geotechnical investigation and testing).

### *Drilling in sediment*

When drilling in sediment, a casing is progressively lowered during the drilling process to avoid collapse of the borehole below the underground water table or sea level. If the drilling is performed without casing (rather exceptional), the borehole can also be kept open by using a drilling mud – i.e., a high density fluid such as bentonite. This provides sufficient lateral support to avoid collapse of the borehole.

Table B.1   *EC7 part 2 – sampling categories.*

| Quality class | 1 | 2 | 3 | 4 | 5 |
|---|---|---|---|---|---|
| Sampling category | A | | | | |
| | | | B | | |
| | | | | | C |

Depending on the type of sediment to be drilled, different sampling instruments are deployed in the drill hole:

- a chisel breaks up (very stiff to hard) clay;
- a shell or bailer with a non-return valve collects non cohesive (loose) material;
- a wash pipe provides wash water below, which transports pieces of broken material or sediment upwards;
- a Shelby tube is a thin-walled pipe by which undisturbed samples can be taken by pushing it into the sediment (drill rods are necessary;
- a piston sampler can be used, in which a piston is moved upwards in the sampling tube during the sampling process.

The compactness of undisturbed soil below the drill hole can be measured in situ by means of a probing test (CPT – see B.2.3 "testing") or a standard penetration test (SPT). After successfully performing an SPT-test, the split spoon contains a (disturbed) soil sample, which can be used for the subsequent lab test programme.

**Drilling in rock**

In rock, cylindrical cores are cored by using a circular core barrel that is encrusted with high-strength diamond or tungsten carbide crystals. This method is known as rotary core drilling. There are many different core barrels, which differ in diameter, in type of drill head, in the way the core is held and in the way the core is brought to the surface. The used diameter of the rock core and the way of drilling has a serious impact on the rock core quality. In general, a bigger size core diameter usually improves the rock core quality. To prepare a rock sample for a UCS test a minimum core diameter of 70 mm (HQ size) is preferred. To avoid the breaking-up of the drilled rock it is important to promote a smooth drilling process, without hammering. A more stable drill hole and higher quality samples are achieved by deploying a double tube or even a triple tube drill configuration where respectively two or three tubes rotate in each other.

In principle, there is no limit to the drilling depth that can be achieved, provided that the drilling machine can generate sufficient torque. Generally a drilling depth of 30 to 40 m is not a problem, and is sufficient for most dredging projects.

During the drilling process the so-called RQD% (Rock Quality Designation) is measured per core run. This RQD value stands for the percentage of intact core pieces that are obtained which have a length greater than 100 mm (measured along the core axis). It reflects the natural fractured state of the rock. The Fracture Index is defined as the number of discontinuities per metre run, measured over any length of reasonably uniform character, which is not necessarily the core run length. The SCR% (Solid Core Recovery) gives the percentage of solid rock material (versus sediment) per drilled metre. To estimate the TCR% (Total Core Recovery) recoveries per drilled metre are measured. A typical borehole log is presented in Figure B.9.

**Borehole Field Log**

Borehole: **PGC-02**
Easting (m): 650449
Northing (m): 997901
Penetration (m): 21.95

Project: North Entrance Dredging of the Pacific Access Channel for the New Third Set of Locks Complex Pacific Side

Client: ACP

Country/Area: Panama/Panama City

Equipment: DSB 1 Nordmeyer
BH Diameter (m): 145mm
Core Diameter (mm): 102mm
Fluid Flush: Water
Method: Rotary drilling, wire line double tube
Casing Depth (m): N/A

Water Depth (m): 28.97mPLD
Tide (m): N/A
Corr. Depth (m): 29.02mPLD
Vertical Datum: PLD
Operator: Thijssen Drilling
Logger: BCO

Borehole Details:
Start Date: 30/5/2010
End Date: 31/5/2010
Weather Info:
General: Sunny
Wind: Little wind
Sea State: N/A

| Elev. (m) | Depth | LithoLog | (Elevation m) | Description | Sample No. | Test Data | SPTn | Rock Quality TCR | Rock Quality SCR | Rock Quality RQD | Remarks / Particulars |
|---|---|---|---|---|---|---|---|---|---|---|---|
| 29 | 0 | | | (0.00, 1.07) MADE GROUND, medium to coarse gravel and cobble size clasts of medium strong basalt in matrix of brown, sandy, slightly clayey silt. | | CR-1 | | 88 | 0 | 0 | Top of borehole at 29.02mPLD |
| | | | | | | CR-2 | | 100 | 0 | 0 | |
| | | | | | | CR-3 | | 40 | 0 | 0 | |
| 28 | 1 | | | (1.07, 1.93) Medium strong to strong, dark grey BASALT to ANDESITE, abundant black micro crystals. Fine 1-3mm incipient fractures with black, very fine grained material, sometimes white hard mineral (quartz?), 20°, 45°, 70° with spacing <5cm. - 1,7-1,93m: Coarser grained. Black crystals <1mm, white crystals <0.6mm in dark grey fine grained matrix. | C-1 | CR-4 | | 95 | 71 | 62 | |
| 27 | 2 | | | (1.93, 21.95) Weak to medium strong, bluish dark grey, AGGLOMERATE. Mainly fine to medium gravel sized subrounded & subangular clasts of andesite, basalt (often amygdaloidal), fine grained tuff (mudstone) & some rare clasts of green very weak clayey material in very fine grained dark grey matrix (little matrix). Rare coarse gravel & cobbles of mainly andesite and mudstone (Tuff). - 1,93-3,1m: Joints medium spaced, irregular, rough surface, subhorizontal. | C-2 / C-3 | CR-5 | | 100 | 100 | 100 | Top 8cm of core show clear marks of drill bit |
| 26 | 3 | | | | | | | | | | |

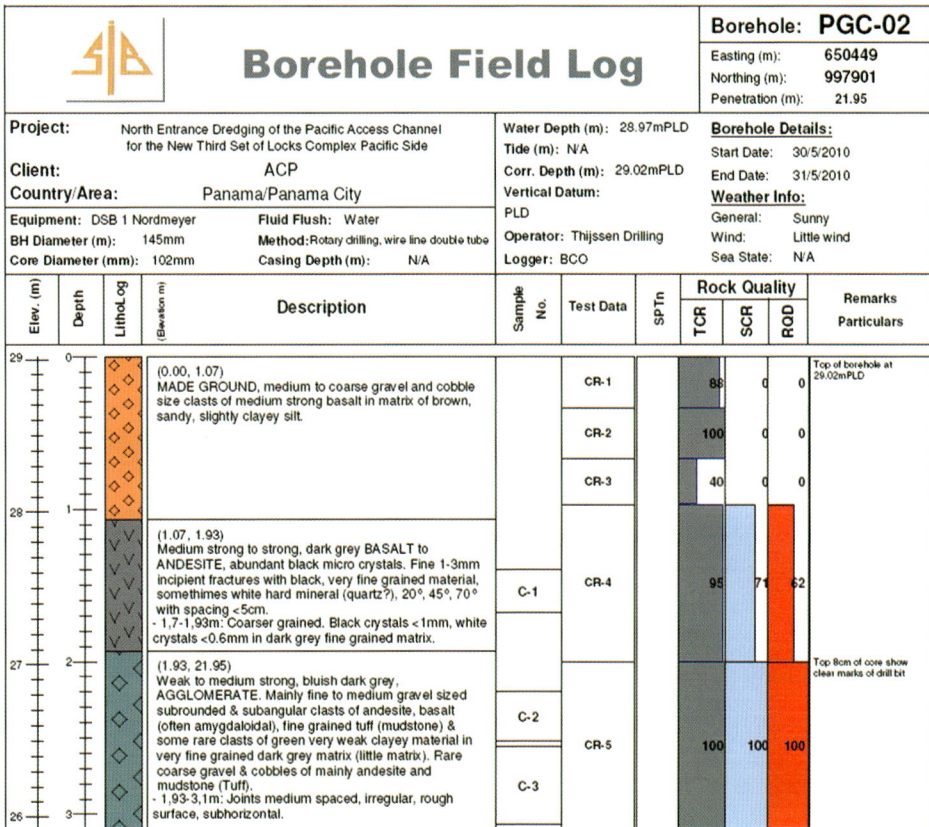

Figure B.9   *Typical borehole log – rotary drilling in rock.*

For shallow water depths, drilling is preferably performed from a jack-up platform. See Figure B.10.

For deeper water depths a dedicated drilling vessel is required. See Figure B.11.

Some coring and drilling techniques exist, which operate from the seabed and can be deployed from smaller vessels. See Figure B.12.

The disadvantages of these systems are the shallow penetrations and often small diameter of samples. However, under certain circumstances (such as unavailability of drilling vessels or jack-ups) this might be the only option to collect seabed samples.

### B.2.2.2   Vibrocores

The vibrocore is a useful tool to sample superficial sand layers. The vibrocore is generally not suited to sample soft cohesive sediments.

Figure B.10   *Drilling from a jack-up platform.*

Figure B.11   *Geotechnical drill vessel.*

Figure B.12    *Bottom corer (168 kg).*

A vibrocorer consists of a heavy (up to 2 tons) vibrating block, electrically powered, which is able to push a steel barrel with a length of 3 to 8 m into the soil (Figure B.13). Inside the steel barrel, a PVC liner (with a diameter of 10 cm) stores a stacked soil sample. After sampling – the vibration process itself generally takes less than 1 minute – the PVC liner is taken out of the barrel and is cut into 2 halves exposing the soil sample taken. A core catcher at the lower end of the barrel prevents the soil sample from escaping out of the liner. A continuous generally disturbed sample can be retrieved from the barrel.

With a weight up to 3000 kg and a height of 8 m for a powerful vibrocorer this kind of site investigation requires the deployment of an appropriate vessel with a spacious aft deck (6 × 10 m). A crane with a hoisting capacity of 10 tons (for working in stiff clay) is necessary to put the vibrocorer into the water. Quite often, a supplier type of vessel is suitable.

When a significant amount of sand is required, an offshore sand search making use of a vibrocorer is a frequently adopted method, see section 3.5. With a favourable sea state and calm wind conditions, up to 20 vibrocores can be taken in a time period of only 12 hours. A swell of 1 m is generally the upper limit for vibrocore sampling. When planning a sand search, attention should be given to the seasonal sea and weather conditions.

Figure B.13   *Deployment of a vibrocorer.*

The penetration depth of a vibrocorer depends on the compactness of the subsoil and is generally limited in dense sand and firm clay. Stiff clay will reduce the penetration depth significantly, while a cemented sand layer (for example cap rock) will prevent deeper sampling.

### B.2.2.3   Jet probe

A metal tube is connected with a hose to a pump (5–12 HP). In order to create a powerful jet, both water jetting and air jetting are possible. In loose/soft sediment the metal tube easily penetrates the soil up to > 6 m. Jet probing is often used to check if a hard layer (a hard gravel layer, a coral body) is present above dredging level. Operated by divers jet probing is a relative low cost method, which allows finishing up to 10 sample points per day depending on the water depth and the number of divers. The main disadvantage is that no representative soil sample can be collected because the fines are preferentially washed out during the process of jetting.

### B.2.2.4   Drop (gravity) corer

A drop or gravity corer is a useful tool to sample superficial clayey sediments.

Gravity corers are deployed on a rope, and dropped in free-fall mode to the seabed. Core recovery is typically in the range of 1 m, depending on the soil characteristics.

Figure B.14   *Drop (gravity) corer.*

A valve in the core head closes when the deployment rope becomes slack, and provides suction during recovery.

### B.2.2.5   Box corer
A box corer is a large mechanical coring device (Figure B.15) containing a box that takes a large, relatively undisturbed sample when lowered to the seafloor. When the deployment line is retracted, a rotating spade closes off the bottom of the box before the box corer is lifted.

### B.2.2.6   Manually operated piston sampler
A manually operated piston sampler (Figure B.16) consists of a metal tube (length 2 m) in which a piston with rubber sealing perfectly fits (diameter 50 mm). Because of combined action of creating a vacuum while raising the piston and simple man-power, a hollow metal tube can be pushed into the subsoil over a distance of 2.0 m in loose/soft sediment. A disturbed sample with a penetration depth of 2.0 metres is retrieved. In a marine environment, (2) divers operate the piston sampler. It is often used for small sand searches (<1,000,000 m³) in shallow water depths (<25 m).

### B.2.2.7   Grab samples
A grab sampler, often called a Van Veen grab, is used to take surface samples of the seabed or surface samples of dredged material in the hopper bin. The size of the taken soil sample varies between 0.5 kg and 5 kg as function of the size of the two metal grab shells. A grab sampler is often used to verify the soil type of the seabed as given on the Admiralty charts. Penetration depth is maximum 10–25 cm (Figure B.17).

Figure B.15   *Box corer.*

### B.2.2.8   Test pit

If an excavator is available on site, digging a test pit might often give a quick visual indication of the geology of the top layers. See Figure B.18.

### B.2.3   *Testing*

### B.2.3.1   Laboratory testing

A detailed description of the available lab tests is outside the scope of this manual. Some of the following text and tables are thankfully copied from "Site investigation requirements for dredging" (PIANC report of working group 23, 2000). A brief overview of possible tests is presented in Table B.2, Table B.3 and Table B.4.

A variety of national and international standards apply to these laboratory tests and the most widely used standards are included in the reference list.

Figure B.16    *Piston sampler.*

Since density tests are very common in reclamation works, these tests are described more in detail in this section. For a description of the other laboratory tests, reference is made to the corresponding standards and textbooks.

**Minimum dry density**
The minimum dry density ($\rho_{d\ min}$) is easy to determine by drying the soil and pouring it through a funnel into a calibrated volume.

**Determination of dry density/moisture content relationship using 2.5 or 4.5 kg hammer (Proctor tests)**
This laboratory test was developed to deliver a standard amount of mechanical energy (compaction effort) to determine the maximum dry density of a soil. The test was initially intended for more or less cohesive soils, whereby the water content is greatly affecting the dry density after compaction.

In the standard Proctor test a dry soil specimen is mixed with water and compacted in a cylindrical mould of 1 litre (standard mould) by repeated blows from the mass of a hammer of 2.5 kg falling freely from a height of 300 mm (British Standard sizes). The soil is

Figure B.17    *Grab sampler.*

Figure B.18    *Excavation of a test pit in hydraulic fill.*

Table B.2  *Summary of laboratory test requirements for cohesive soils.*

| Material characteristics | Type of test | Remarks | Required quality of samples (EC7) |
|---|---|---|---|
| Particle size distribution | Sieve analysis | Only if coarse fraction present | 1, 2, 3, 4 |
| | Sedimentation test, Hydrometer test | | |
| Strength | Laboratory vane | In very soft clays only | 1 |
| | Unconfined compressive strength test | | |
| | Triaxial test | Usually only for slope design | |
| | Direct shear test | | |
| Stiffness | Oedometer test | | 1 |
| | Triaxial test | | |
| Consolidation | Oedometer test, Constant Rate of Strain test | | 1 |
| Permeability | Falling head test | | 1, 2 |
| Plasticity | Atterberg limits tests | | 1, 2, 3, 4 |
| Density/water content | Natural bulk density tests | | 1, 2, (3) |
| | Dry density test | | |
| Particle density, density of solids | Water Pycnometer | | 1, 2, 3, 4 |
| Mineralogy | Visual examination | | 1, 2, 3, 4 |
| Rheological properties | Viscosity test | Very soft/semi-fluid soils only | 1, 2, 3, 4 |
| Organic content | Loss on ignition/hydrogen peroxide digestion | | 1, 2, 3, 4 |

**Triaxial test**

The triaxial test is a widely used test method to determine the shear strength and stress strain behaviour of soils.

Following test methods can be adopted:

- UC (unconfined compression test): quick determination of undrained shear strength of saturated clays.

- CD (consolidation drained test): method used to determine the drained shear strength parameters to analyse long term stability of a soil mass.

- CU (consolidation undrained test): method used to determine the drained and undrained shear strength parameters. The excess pore water pressure should be measured to determine the drained characteristics.

> – UU (unconsolidated undrained test): determination of undrained shear strength of a saturated soil.
>
> – Cyclic triaxial testing: determination of liquefaction and breaching potential.

Table B.3  *Summary of laboratory test requirements for non-cohesive soils.*

| Material characteristics | Types of test | Remarks | Required quality of samples (EC7) |
|---|---|---|---|
| Particle size distribution | Sieving | | 1, 2, 3, 4 |
| | Sedimentation | Only if significant fine fraction present | |
| Density | Natural bulk density | Difficult for sands | 1, 2, (3) |
| | Dry bulk density | | |
| Strength | Triaxial test (CD) | | 1 |
| | Direct shear test | | |
| Stiffness | CBR test | | 1, 2, 3, 4 |
| Compaction characteristics | Minimum and maximum density tests | | 1, 2, 3, 4 |
| Particle specific gravity | Water Pycnometer | | 1, 2, 3, 4 |
| Angularity/roundness | Visual examination | | 1, 2, 3, 4 |
| Permeability | Laboratory permeability | More usually established by field tests | 1, 2 |
| Organic content | Loss on ignition/ hydrogen peroxide digestion | Priority 1 if organic content greater than 5% | 1, 2, 3, 4 |
| Mineralogy | Visual examination | Petrographic analysis may be necessary | 1, 2, 3, 4 |
| | Carbonate content | | |

compacted in three layers, each of which is subjected to 27 blows. It is noted the ASTM uses slightly different sizes (0.94 litre mould, falling height of 305 mm and 25 blows).

A modified Proctor test was initially developed for compaction of an airfield to support heavy aircraft loads. In this test, a hammer with a mass of 4.5 kg falls freely from a height of 450 mm and the soil is compacted in 5 layers with 27 blows per layer in the standard 1 litre mould (BS sizes). The difference between the two Proctor tests is schematised in Figure B.20.

Table B.4  *Summary of laboratory test requirements for rock.*

| Material characteristics | Types of test | Remarks | Required quality of samples (EC7) |
|---|---|---|---|
| Strength | Unconfined compressive strength | Most widely used measure of strength | 1 |
| | Triaxial test | Unusual, weak rock only | |
| | Indirect tensile strength test (such as Brazilian splitting test) | | |
| | Point load test (*Is50*) | Must be supported by compressive strengths | |
| Elasticity | Youngs modulus | | 1 |
| Density | Natural bulk density | | 1, 2 |
| Particle specific gravity | Immersion test (ASTM C127-04) | | 1, 2, 3, 4 |
| Mineralogy | Visual examination | Petrographic analysis may be necessary | 1, 2, 3, 4 |
| | Carbonate content | | |

A larger (CBR) mould of 3.24 litres will be used in the standard and modified Proctor test when coarse gravel sized particles are present. However, when more

**Shape and angularity**

The shape and angularity of the grains of the fill material are important properties as they define to a certain extent the shear strength and the stiffness of a fill mass. These material properties may depend on the depositional environment of the borrow area in the geological past: If the material has been transported by wind or water over large distances before deposition at its current location it may often be rounded to sub-rounded, but if the fill is an in-situ weathering product of the parent rock or is composed of broken skeletal remains, shell fragments, etc. without being transported it may be very angular.

Angularity may affect wear and tear etc. See Appendix A.

Angularity or roundness is difficult to quantify and is generally determined on a visual basis by comparing the grain shapes to roundness scales as proposed by Powers (see Figure B.19) or the ASTM D 2488.

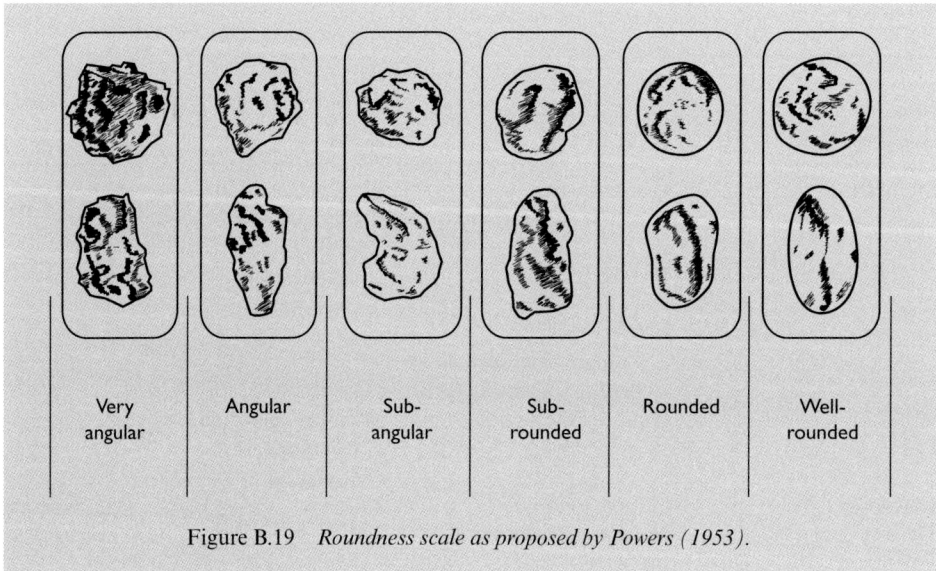

Figure B.19    *Roundness scale as proposed by Powers (1953).*

than 10% of the particles are larger than 37.5 mm the test is not applicable according the BS.

Four or more tests are conducted on the soil using different water contents. The last test is identified when additional water causes the bulk unit weight of the soil to decrease. The results are plotted as dry unit weight (ordinate) versus water content in Figure B.21. A line connects the measurements in this graph and the top of the curve indicates the Maximum Dry Density ($\rho_{d\,max}$). The water content at which this maximum dry unit weight is achieved is called the optimum water content.

In cases where the test is performed on stiff cohesive soils (i.e., clay), which need to be chopped in lumps, the result of the test depends on the size of the resulting pieces. Furthermore, the densities obtained in the test will not necessarily be directly related to densities in situ.

As Proctor Maximum Dry Density is frequently used as reference to achieve a certain compaction on the field, it is of the utmost importance that the type of test (standard or modified) be specified and agreed upon.

**Maximum density by vibrating hammer**
This maximum density test covers the determination of the dry density of the soil, which may contain some particles up to coarse gravel size, when it is compacted by vibration in a specified manner over a range of moisture contents. The range

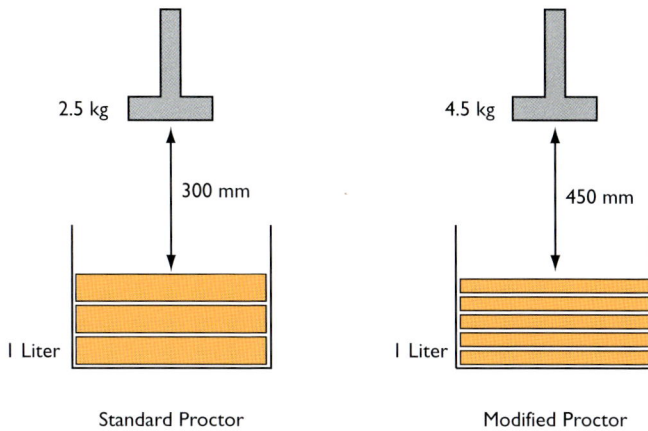

Figure B.20 *Standard and modified Proctor test for granular soils containing no coarse gravel (BS 1377-4, ASTM D1557–02).*

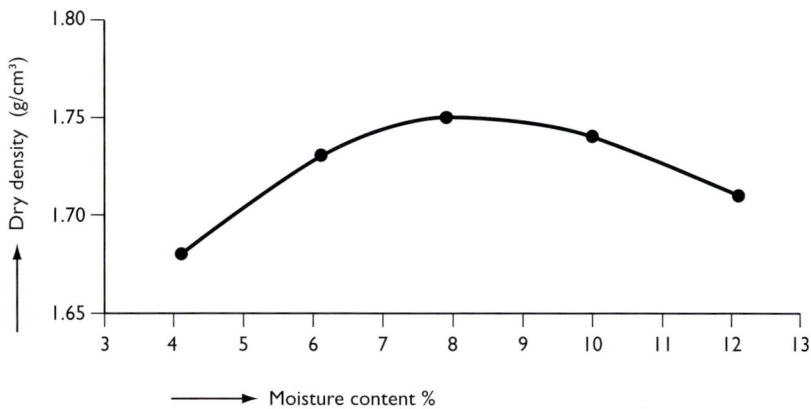

Figure B.21 *Example of Proctor test result.*

includes the optimum moisture content at which the maximum dry density for the specified degree of compaction is obtained. In this test the soil is compacted into a CBR mould (3.24 litre) using an electrically operated vibrating hammer.

The test is suitable for certain soils containing not more than 30% by mass of material retained on the 20 mm test sieve, which may include some particles retained on the 37.5 mm sieve. It is not generally suitable for cohesive soils.

**Maximum density by vibrating table (ASTM D4253)**
This test method is used to determine the maximum density of a free draining soil using a vibratory table.

**Remarks**

An important note has to be made with regard to crushable soil. Owing to the impact of the hammer, soil might be crushed resulting in higher densities. This is especially the case in soils consisting of carbonate sand (with shell origin) The ASTM also stresses this feature.

"ASTM1557-02: Degradation – Soils containing particles that degrade during compaction are a problem, especially when more degradation occurs during laboratory compaction than field compaction, the typical case. Degradation typically occurs during the compaction of granular-residual soil or aggregate. When degradation occurs, the maximum dry-unit weight increases so that the resulting laboratory maximum value is not representative of field conditions. Often, in these cases, the maximum dry unit weight is impossible to achieve in the field."

For samples susceptible to crushing, no reuse of samples for tests at different water contents is allowed.

The Proctor tests used for the determination of the optimum moisture content is only useful for non-free-draining soils.

"BS1377 part 4 For some highly permeable soils such as clean gravels, uniform graded and coarse clean sands, the results of the laboratory compaction test may provide only a poor guide for specifications on field compaction. The laboratory test might indicate meaningless values of moisture content in these free-draining materials and the maximum dry density is often lower than the state of compaction which can be readily obtained in the field. For these soils one of the maximum dry density tests described in clause 4 would be more appropriate".

It should also be kept in mind that the Proctor test was intended for surface compaction in road works and airport runways, where high degrees of compaction are required. In these circumstances it might be important to perform the compaction at a water content close to the optimum water content. Otherwise the required degree of compaction might not be achievable. For large-scale land reclamation projects, however, the required degree of compaction is in general significantly lower compared to the road works standards. Therefore the required degree of compaction can often be achieved at a water content far from optimal. Moreover, in hydraulic reclamation projects, the fill will be applied in a larger layer thickness than what is customary in road building. (The applied layer thickness is generally larger than 1 m). It is often practically impossible to achieve a uniform water content over the total extent of this layer.

When compacting a hydraulic fill below the waterline, it is obvious that the water content cannot be adjusted.

This test should be preferred over the Proctor test when dealing with crushable soils.

### B.2.3.2   In situ tests

The main types of in situ tests, which are appropriate to dredging investigation, are summarised in Table B.5.

In situ tests are valuable because, when properly carried out, they reflect the characteristics of the overall material mass without any disturbance caused by sampling.

### Field vane shear test

The field vane shear test can be undertaken in boreholes formed by cable tool and wash boring methods. It is used to measure the shear strength of soft clays and clayey silts by means of a small vane, which is pushed into the soil. An increasing torque is applied to the vane until the cylinder of soil fails at which point the applied torque is noted. It is often the only way in which the shear strength of very soft soils can be measured reliably because of the difficulty of obtaining and testing truly undisturbed samples in such material.

### Hand vane and pocket penetrometer

The hand vane and pocket penetrometer tests are also commonly used for a quick determination of the shear strength of a soil.

### *Quasi-static cone penetration test*

The quasi-static cone penetration test CPT(U) has many variants most of which involve the measurement of the force required to push a rod with a conical tip into

Table B.5   *In situ tests appropriate for soil investigations (modified PIANC "Site investigation requirements for dredging works", 2000).*

| Test | Material type | Measured properties or characteristics | Remarks |
|---|---|---|---|
| Field vane shear test | Soft to firm clay, clayey silts | Undrained shear strength, remoulded shear strength | Carried out in boreholes |
| Hand vane test | Soft to firm clay | Undrained shear strength, remoulded shear strength | |
| Pocket penetrometer | Soft to stiff clay | Undrained shear strength | |
| Quasi-static cone penetration test CPT(U) | Most soils except coarse gravels, cobbles and boulders | Relative density of granular soils, shear strength of cohesive soils, permeability | |

*(continued)*

Table B.5  (*Continued*).

| Test | Material type | Measured properties or characteristics | Remarks |
|------|---------------|----------------------------------------|---------|
| Standard penetration test (SPT) | Most soils except cobbles and boulders, weak rocks | Relative density of granular soils, indicative shear strength of cohesive soils, indicative strength of cohesive soils | Carried out in boreholes |
| Dynamic cone penetration test (DCPT) | Sands and gravels | Qualitative evaluation of compactness/relative density, qualitative evaluation of sub-soil stratification | |
| Pressure meter test | | Stress strain relationship | |
| Dilatometer test | Fine grained soils | Strength and deformation properties | |
| Plate loading test, CBR test | All soils | Stiffness | Carried out a the surface of a reclamation area |
| In situ density test Sand replacement Rubber balloon Nuclear methods surface | Granular soils | In situ density | Carried out above the water table |
| In situ density Nuclear method deep measurement | All soil types | In situ density | In combination with CPT test |
| Reference density test Proctor test Minimum density test Vibrating table or hammer | Granular soils | Minimum and maximum density | |
| Permeability test | Granular soils | Mass permeability | Carried out in boreholes |
| Trial dredging | Soils and rocks | Dredgebility, Production rate, wear and tear | |

the ground at a constant speed. Above the cone, a friction sleeve can be mounted. A picture of such a CPT cone is shown in Figure B.23.

The forces on the cone end and on the friction sleeve are continuously measured. When a so-called piezometric cone is used (CPTU), the pore water pressure is recorded just above the probing cone. The CPT is sometimes referred to as the 'Dutch' cone penetration test, since it was initially developed at the Dutch Laboratory for Soil Mechanics in Delft. The test can be used to make an interpretative identification of the soil type and to derive a variety of engineering parameters including soil strength and relative density.

The friction ratio (i.e., sleeve friction/cone resistance × 100%) can give an indication of sandy versus clayey nature of the penetrated soil. Starting from these two parameters it is possible to determine the soil type. Measurement of the pore pressure provides additional info about the encountered soil type.

When adding a seismic sensor (usually a geophone) to the CPT cone, it is possible to measure the shear wave velocity in the soil. This shear wave is generate at the surface. The shear wave velocity can be correlated to the low strain stiffness of a soil. It is a useful tool to evaluate the liquefaction potential of the soil. For further information on this technique reference is made to the ISSMGE publication "seismic cone downhole procedure to measure shear wave velocity".

Some soil investigation companies provide cone penetration tools, which are mounted on a rigid frame. The frame is then lowered onto the seabed. Especially for deep areas this might be a good solution for collecting geotechnical data. e.g. Seacalf system Fugro (see Figure B.24), Roson CPT system.

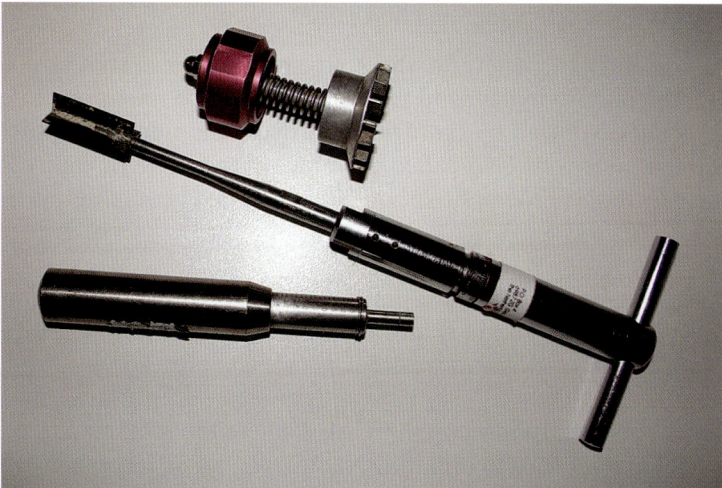

Figure B.22   *Field equipment hand vane and pocket penetrometer.*

Figure B.23    *CPT cone.*

Special care should be taken to the interpretation of these CPT data. Following corrections might be applicable:

– shallow depth correction (Appendix C.4.5);
– correction in case a non-standard cone size is used.

### *Standard penetration test (SPT)*

The SPT is the most widely used of all in situ tests and is carried out in bore-holes formed by cable tool percussion methods or by washboring. It involves the hammering of a standard split spoon into the ground and measuring the number of blows required to do this, using a standard hammer weight and drop height.

The number of hits to penetrate the split spoon over a standard length of $3 \times 15$ cm in the underlying subsoil is counted. Because the upper 15 cm of the subsoil are often disturbed by the drilling process the number of hits from 15 to 45 cm contributes to the so-called N-value of the SPT. In case of gravel, a closed cone can replace the split spoon because gravel can clog up the hollow cone.

The SPT test provides quantitative data on the degree of compaction (relative density) of granular soils and may also be used, at an indicative level, to measure the strength of cohesive soils and weak rocks (see Chapter 3.6.1).

The test is also useful in rotary core boreholes where the rock quality is such that good recoveries are not easy to obtain. In such cases, an SPT is usually carried out after each 'core run' at interval of 1.5–3.0 metres.

Figure B.24    *Deployment of Roson system, Fugro.*

Remarks:

– The SPT test is not a continuous test;
– The accuracy of the SPT values is limited; especially in very loose sand or soft clay it might not give an accurate indication of the soil properties. (E.g. SPT = 0 in very soft clay);
– Since no friction is measured, it is difficult to distinguish clayey soils from sandy soils. It is therefore advised to perform CPT testing as much as possible;
– It is important to know which type of SPT apparatus was used and which corrections have been applied in the reported SPT-value.

Possible corrections are:

– N60-value: correction for energy efficiency level of SPT device. N60 represent the SPT value that would be measured with an SPT device with 60% energy efficiency (EN-ISO 22476-3: 2005);
– correction for fines;
– correction for shallow SPT's readings.

### Dynamic cone penetration test (DCPT)

The dynamic cone penetration test, like the quasi-static cone penetration test, comes in a variety of forms. It is undertaken in a manner similar to the standard cone penetration test but executed continuously to gather a full profile of penetration resistance, rather than at discrete intervals in a borehole. It is a useful test but is not very widely used in the dredging industry, which prefers the quasi-static test because the registration of the sleeve friction is not possible for a DCPT test.

### Determination of in situ density

Most of the available methods to determine the in-situ density of material depend on the removal of a representative soil sample from the site and then determining the mass and volume it occupied before being removed. Mass determination is common to all methods and straightforward. The variation lies in the several procedures used for measuring the volume. Both ASTM and BS recommend the following methods, normally restricted to soil within 2 m of the surface and above groundwater level only:

- the sand replacement or sand cone method (BS1377, ASTM D1556 and AASHTO T191);
- the core cutter method or drive cylinder method (BS 1377, ASTM D2937 and AASHTO T204);
- the weight in water and water displacement method (BS 1377 and AASHTO T233);
- the rubber balloon method (ATSM D2167 and AASHTO T205).

### Sand replacement method

This method is widely used and accepted all around the world. It consists of a plastic jar with a funnel connected to the neck of the jar (Figure B.25). First a hole is excavated at the test location. The removed soil is collected and sent to the laboratory to determine the water content and the dry weight. The hole is filled with the standard sand by inverting the sand cone apparatus over the hole and opening the valve. The remaining weight of the sand in the jar is measured. The volume of the filled hole can be calculated this way.

Depending on the particle size of the soil, different sizes of jars and funnels are used following the Standards. This test requires quite some effort as well in the field as in the lab. Therefore, one team can only make about 10 measurements a day.

### Rubber balloon method

In this method a hole is dug into the soil and the weight of the excavated soil is determined. After that a metal plate is placed over the hole and a water-filled cylinder is placed on top of the plate (open in the middle). At the end of the cylinder a balloon is placed, which can be filled with water from the cylinder. This is done by putting pressure on a handle of the cylinder. The balloon filled with water

Figure B.25    *Performing the Sand Cone Test.*

takes the same volume as the dug hole. A scale on the side of the cylinder shows the volume of water that went in the balloon and thus the volume of the hole. See Figure B.26.

### Nuclear density gauge – surface measurement

The nuclear density apparatus is a handy device to rapidly obtain the unit weight and water content of the soil. Soil particles cause radiation to scatter to a detector tube and the amount of scatter is counted. The scatter count rate is inversely proportional to the unit weight of the soil (Figure B.27 left). If water is present in the soil, the hydrogen in water scatters the neutrons and the amount of scatter is proportional to the water content (Figure B.27 right). The radiation source is either radium or radioactive isotopes of caesium and americium.

In the field a hole is made to the depth of the source (maximum 30 cm) to apply the apparatus.

The nuclear density apparatus is first calibrated using the manufacturer's reference books. This calibration serves as a reference to determine the unit weight and water content of a soil at a particular site.

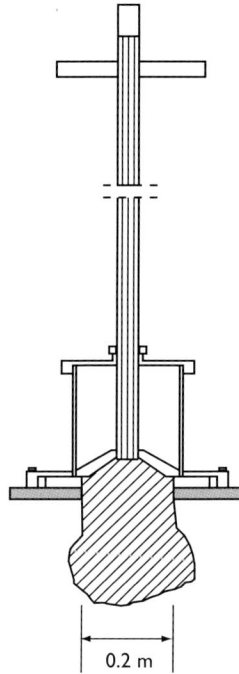

Figure B.26 *Rubber balloon method.*

In the field the apparatus is calibrated before each series of test to measure the backscatter of the area. Figure B.28 and Figure B.29 show the preparation and measurement with a nuclear probe next to the hole of a sand cone test for calibration purposes.

A big advantage of this apparatus is the great number of tests that can be performed in a limited time. The only limiting factor is the battery that needs to be recharged, but one can assume that easily 40 measurements can be made a day as the measurement itself takes less than 1 minute. The result is displayed immediately on the read out unit.

Additionally, no laboratory tests are required once the device is calibrated.

**Nuclear density cone – deep measurement**
A variety of probes has been developed that contain a radioactive source and a detector, which can be used in combination with a Cone Penetration Test. An example of such a cone is presented in Figure B.30.

The in-situ density is continuously recorded during the penetration of the cone. Several types of nuclear density probes are available on the market. For a more

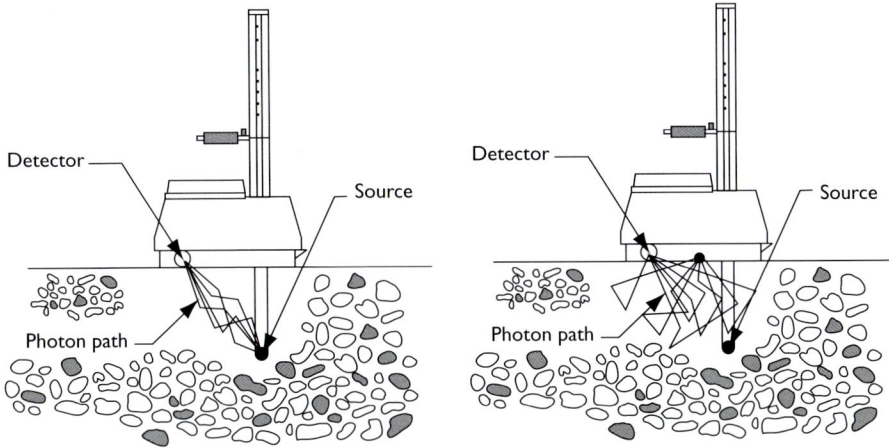

Figure B.27   *Nuclear density measurement.*

Figure B.28   *Making hole for nuclear measurement next to hole of sand cone test.*

Figure B.29   *Nuclear density measurement next to hole from sand cone test.*

detailed description of the instrumentation and its capabilities reference is made to the latest brochures of geotechnical instrumentation companies. For nuclear methods reference is made to ASTM D2922 and AASHTO T238.

## Permeability test

Permeability tests can be undertaken in boreholes and are usually done using the 'falling head' method. Despite the importance of permeability in the dredging process, particularly in respect of granular soils, these tests are rarely carried out during dredging ground investigations.

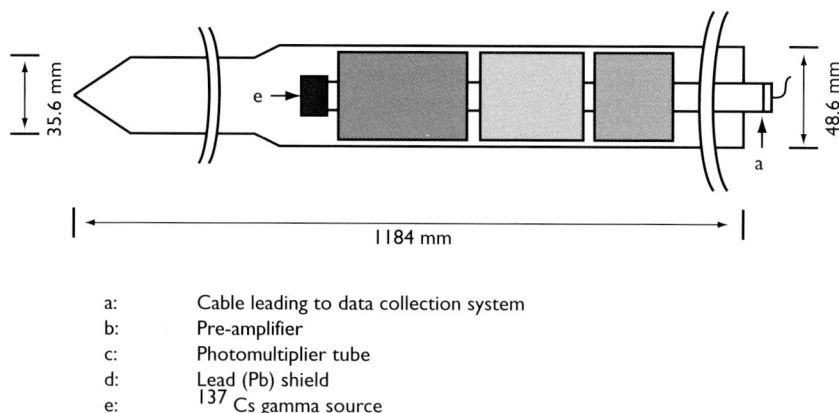

a:     Cable leading to data collection system
b:     Pre-amplifier
c:     Photomultiplier tube
d:     Lead (Pb) shield
e:     $^{137}$Cs gamma source

Figure B.30   *Example and principle Nuclear Density Cone (Ref. Development and calibration of radio-isotope cone penetrometers. A.K. Shrivastava).*

## Trial dredging

There are some projects where the complexity of the geological conditions or other special circumstances calls for trial dredging in order to confirm whether or not direct dredging is possible. If the trial shows that dredging is not possible, other methods such as drilling and blasting will have to be adopted.

Trial dredging has the advantage that it will also provide information on the rate of production or frequency of delays caused by boulders or other obstructions. However, trial dredging is expensive and is usually only justifiable in the case of large projects and/or where mobilisation costs are small. It is important that dredging trials are planned in such a way as to be readily interpreted and the results applied to the type of equipment, which is likely to be used for the main project. Trial dredging areas should be representative for the range of ground conditions which will be encountered and care must be taken to report all of the characteristics of the plant used, in addition to the results of the trials and the ground conditions at the trial locations.

## Trial dredging pitfalls

When trial dredging is done with a trailer hopper suction dredger, only a superficial top layer is sampled. Often this top layer will have very different characteristics compared to the deeper layers. It might therefore be important to focus the trial dredging on a small area, trying to reach the deeper soil layers, rather than cover a large area removing only the top layer.

## B.3   Hydraulic, morphological and meteorological data

In the following sections the hydraulic, morphological and meteorological data collection is discussed. Relevant data are:

Hydraulic data:

- Water levels
- Currents
- Waves
- Water temperature
- Water salinity

Meteorological data:

- Wind
- Air pressure
- Precipitation
- Air temperature, visibility, storm tracks

Morphological and environmental data

- Deposition simulation test
- Turbidity (background levels)
- Dissolved oxygen
- TSS

### B.3.1 *Water levels*

Water level data can be derived from various sources and in various formats: measured or hind casted time-series, sets of tidal constituents or tables with average or maximum tide or river levels. The main sources and formats are:

- Tide tables (e.g. Admiralty Tide Tables (ATT) from the United Kingdom Hydrographic Office (UKHO, www.ukho.gov.uk));
- Databases with tidal constituents (e.g. International Hydrographic Organisation (IHO www.iho.int, Topex/Poseidon));
- Literature and project reports with sets of tidal constituents from previous projects or surveys in the area of interest;
- Hind cast time-series from scientific or commercial databases (e.g. Ocean weather Inc.);
- Digital databases with measured water level time-series (e.g. University of Hawaii Sea Level Center (UHSLC, http://uhslc.soest.hawaii.edu) or the National Oceanographic Data Center (NODC, www.nodc.noaa.gov));
- Water level measurements at nearby stations or measurements dedicated to the project.

In coastal areas and estuaries, diurnal and higher order tidal constituents can be derived from water level measurements of one month as a minimum by means of harmonic analysis. Longer measurements (of one or more years) enable the assessment of seasonal constituents and extreme water levels by means of extreme value analysis.

Validation of two – or three-dimensional numerical models (like Delft3D, Mike21, Telemac) in the coastal area can be based on relatively short duration water level measurements of one month. Validation can be done by comparing observed time-series of water level with computed time-series, or in the frequency domain by comparing amplitudes and phases of the main tidal constituents.

The tidal component of water level measurements can be separated from the non-tidal component by means of harmonic analysis. The tidal component is translated in sets of tidal constituents, of which the most important are: M2, S2, O1 and K1. The non-tidal component, or residual, may include tidal noise, wave set-up and positive or negative water level surge heights as a result of meteorological (wind and barometric) variations.

In rivers, water level measurements of at least one year, but preferably more, are required to assess short-term and seasonal water level variations, as well as extreme water levels.

Water levels are usually recorded with pressure sensors (tide gauge) that are fixed to jetties or quay walls. An important aspect is the vertical reference level of the sensor relative to the local datum and mean sea level. The water level readings are to be corrected for the atmospheric pressure variations.

The Intergovernmental Panel on Climate Change (IPCC) (www.ipcc.ch) is the leading platform for information on global climate change issues. In their Fourth Assessment Report (Working Group 1 Report, Table 10.7) the present average lower and upper estimates for future sea level rise are given: respectively as 0.2 to 0.45 m over a period of 100 years. Regional differences relative to the global estimates are given in Figure 10.32 in the same IPCC report.


B.3.2   *Currents*

In tidal areas current monitoring is typically conducted for a period of at least two weeks (covering one neap-spring cycle). In coastal or estuarine areas with distinct seasonal variability (e.g. wet and dry seasons), it is advised to perform measurements during both seasons.

Currents can be measured at single-points in the water column by a propeller-type current meter (with a vane to also allow measuring directions) or by electromagnetic flow meter (so-called ECM's or DRCM's). The instruments can also be combined with CTD's to measure conductivity, temperature and water depth (see further below).

Acoustic Doppler Current Profilers (ADCP) measure currents velocities and directions at various depth ranges. ADCP's can be mounted looking sideward (in rivers), looking upwards (on seabed) or looking downwards (moored to a ship). An ADCP measures the shift in frequency of backscattered sound pulses ('Doppler effect') in so-called bins, typically at bin heights of 1 m. Discharges in rivers or arbitrary cross-sections are derived by integrating the sideward or downward ADCP recordings. For the latter, a ship will need to sail a pre-defined zigzag pattern.

Same as for the analysis of water level measurements, the current measurements can be analysed in the time domain or in the frequency domain. Because of the more turbulent nature of currents, the level of noise following the harmonic analysis will be higher than the noise following the harmonic analysis of water levels.

### B.3.3   *Waves*

Usually wave measurements will not be available at the reclamation site or borrow areas. As short duration measurements are of limited use, it is common practice to transform offshore wave data (e.g., from scientific or commercial databases or offshore wave buoy measurements) to the site by means of numerical wave modelling.

Some company offer world-wide wave statistics based on satellite measurements (ARGOSS: http://www.waveclimate.com, Oceanor: http://www.oceanor.no).

Waves are normally recorded by wave buoys in offshore waters (Figure B.32). In shallow water, waves can also be recorded by an ADCP (by translating wave orbital velocities to wave parameters).

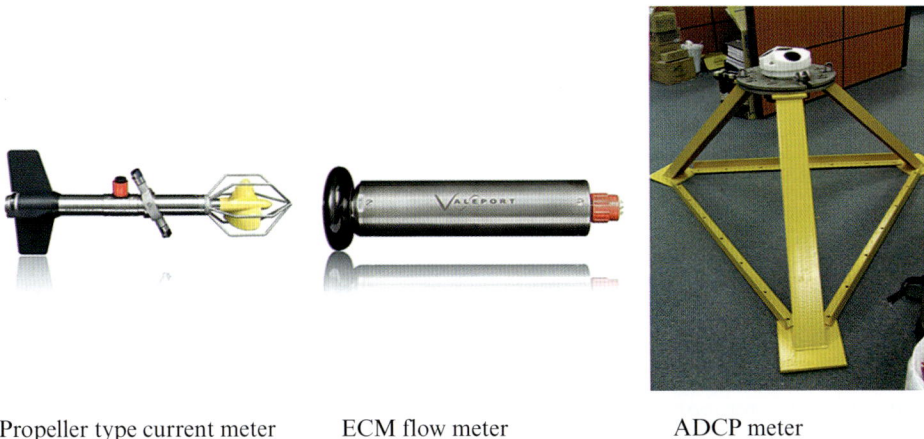

Propeller type current meter       ECM flow meter                    ADCP meter

Figure B.31   *Various types of current meters.*

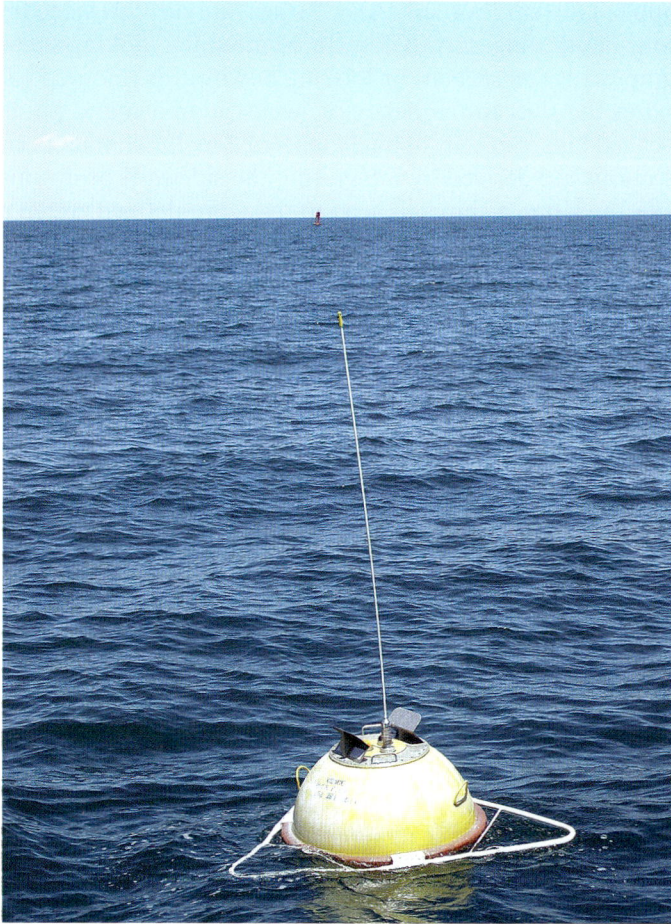

Figure B.32    *Directional wave rider buoy (Datawell).*

For reclamations in areas sheltered from sea and swell waves (like estuaries, lakes or rivers) it is relevant to know the potential impact of ship-induced waves. Ship-waves can be derived from in-situ measurements, from general relations (e.g., PIANC rules) or from modelling.

### B.3.4    *Water temperature and salinity*

Data on water temperature can be derived from in-situ measurements with thermometers, CTD's or ADCP's. The salinity of (sea) water is usually determined by measuring the conductivity of the water by means of a CTD (Conductivity – Temperature – Depth). The conductivity is usually expressed in μS/cm (micro-Siemens/centimetre) and will require proper calibration before use.

B.3.5   *Wind*

Wind data is available in various formats and can be collected from various sources. A general overview of wind data sources and types is given in Table B.6.

Table B.6   *General overview of wind data sources and type.*

| | Climate tables and roses | Time-series | Time- and spatial varying wind fields | Storm tracks |
|---|---|---|---|---|
| Weather handbooks, Pilots | ▨ | | | |
| Various internet sources | ▨ | ▨ | | ▨ |
| Met-office, airport or government authorities | ▨ | ▨ | | |
| Scientific or commercial databases | | ▨ | ▨ | |

Operational wind climates and extreme wind conditions can be derived from the literature or from analyses of multi-year time-series of wind speed and direction, see Table B.7. For the latter, a minimum data record of one year is required, but preferably a time-series of 10 years or more.

It is important to know the following details of wind measurements:

– location (geographical co-ordinates) and height of anemometer;
– if the data has been corrected to the standard level (10 m above ground level);
– averaging time of readings (usually 10-minute or hourly averages);
– presence of surrounding obstacles influencing the wind records (e.g., buildings, mountains);
– location shifts or changes in measurement devices or settings.

The wind time-series are typically analysed as follows:

– extreme wind speeds by means of extreme value analysis, if relevant per directional sector, see Figure B.33.

571

Table B.7   *Example of joint-occurrence table of wind speed versus wind direction (in % of time).*

| Wind | Direction [°N] | | | | | | | | | | | | |
|---|---|---|---|---|---|---|---|---|---|---|---|---|---|
| Speed | −15 | 15 | 45 | 75 | 105 | 135 | 165 | 195 | 225 | 255 | 285 | 315 | |
| [m/s] | 15 | 45 | 75 | 105 | 135 | 165 | 195 | 225 | 255 | 285 | 315 | 345 | Total |
| <2.0 | 1.2 | 1.3 | 1.1 | 1.6 | 1.6 | 1.6 | 1.6 | 1.6 | 1.5 | 1.6 | 1.4 | 1.3 | 17.4 |
| 2.0:4.0 | 1.9 | 2.1 | 3.2 | 3.4 | 3.0 | 2.5 | 2.0 | 2.3 | 2.8 | 3.8 | 4.1 | 2.9 | 33.7 |
| 4.0:6.0 | 0.4 | 0.6 | 3.2 | 4.8 | 2.0 | 1.0 | 0.3 | 0.4 | 1.4 | 3.8 | 4.4 | 1.5 | 23.9 |
| 6.0:8.0 | 0.1 | 0.1 | 2.7 | 4.1 | 0.9 | 0.2 | 0.0 | 0.1 | 0.3 | 3.0 | 2.7 | 0.4 | 14.5 |
| 8.0:10.0 | 0.0 | 0.0 | 1.2 | 3.0 | 0.2 | – | – | – | 0.1 | 1.6 | 1.4 | 0.0 | 7.5 |
| 10.0:12.0 | – | – | 0.3 | 1.0 | – | – | – | – | – | 0.9 | 0.3 | – | 2.6 |
| 12.0:14.0 | – | – | – | 0.1 | – | – | – | – | – | 0.2 | 0.1 | – | 0.4 |
| >14.0 | – | – | – | – | – | – | – | – | – | – | – | – | – |
| Total | 3.6 | 4.1 | 11.7 | 18.0 | 7.7 | 5.2 | 3.9 | 4.3 | 6.2 | 14.9 | 14.4 | 6.2 | 100.0 |

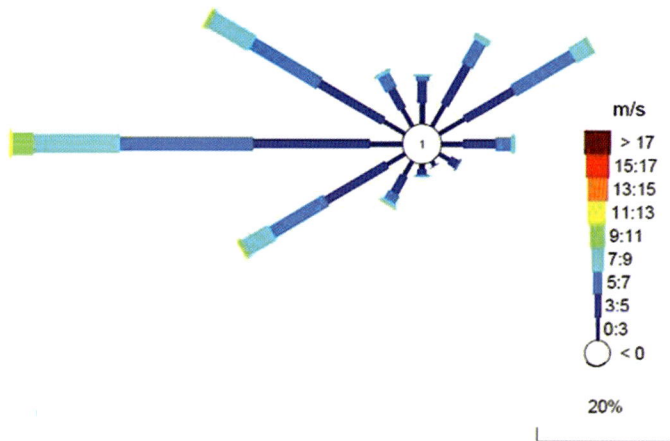

Figure B.33   *Example of wind speed rose.*

Numerical modelling of extreme storm events is usually applied in areas where extreme storms determine the extreme wave and water level conditions. Input wind fields (and air pressure fields) can be derived from scientific or commercial hind cast databases of which some well-known are: ECMWF (www.ecmwf.int), Ocean weather (www.oceanweather.com), KNMI (www.knmi.nl), BMO (www.metoffice.gov.uk).

When using hind cast data, the user should verify if the temporal – and spatial resolution of the data is sufficient and if the data is statistically representative for the regional conditions. Figure B.33 shows a typical example of a wind speed rose.

### B.3.6   *Air pressure*

Usually the air pressure data forms part of the wind data purchased from scientific or commercial databases.

### B.3.7   *Visibility*

Historical information about visibility can be obtained at the meteorological offices.

### B.3.8   *Storm tracks*

Various websites offer information about the path and intensity of previous storms in the region. Examples of such web sites are:

–  http://weather.unisys.com/hurricane/
–  http://www.csc.noaa.gov/hurricanes

Typical storm/hurricane tracks are shown in Figure B.34.

### B.3.9   *Deposition simulation test*

The deposition simulation test is a simple test used to determine the sedimentation rate of a soil-water mixture. The soil-water mixture is poured into a plexiglass

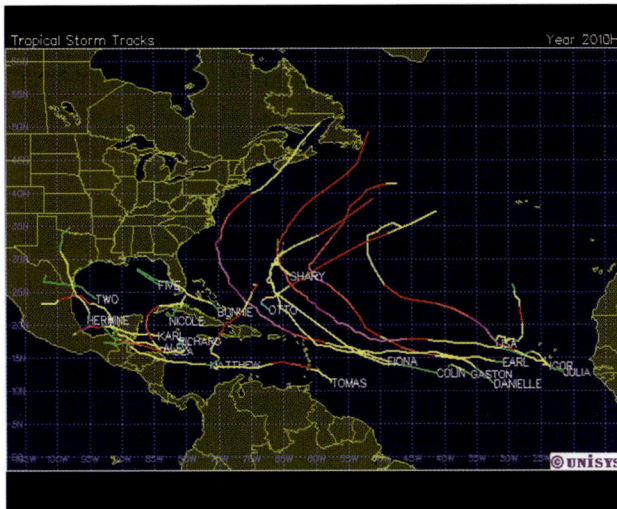

Figure B.34   *Typical storm/hurricane tracks for the Atlantic ocean in 2010 (source http://weather.unisys.com/hurricane).*

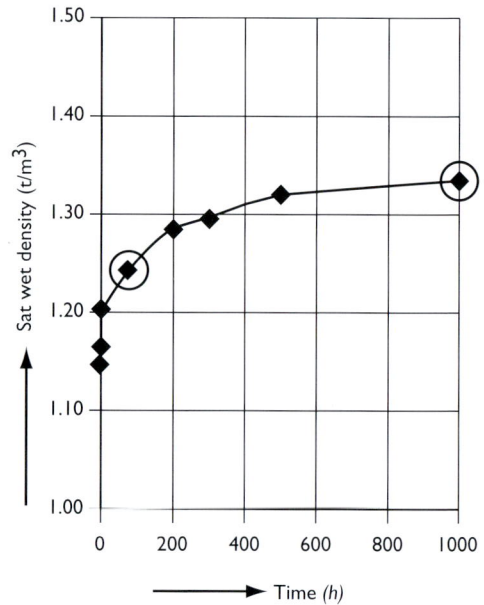

Figure B.35   *Deposition simulation test.*

cylinder. The interface between the clear water and sedimented sample is recorder at regular time intervals. Based on these readings, the density of the sedimented part of the sample can be calculated, see figure B.35.

In case the behaviour of the silt/clay fraction, resulting from the segregation process during the reclamation is to be studied, the sand & gravel fraction of the soil samples has to be sieved off, prior to the test. The resulting fine fraction has to be mixed with water (seawater were appropriate!). It is advised create a sample with a mixture wet density of approximately 1.1 t/m³.

### B.3.10   *Turbidity*

A turbidity meter measures the intensity of light scattered by the suspended particles in a water body, expressed in Nephelometric Turbidity Units (NTU). Because the amount of suspended particles corresponds to the intensity of scattered light, the measurement of turbidity is proportional to the amount of suspended solids (TSS) (EN-ISO 7027: 1999, ASTM D 6698: 2007, ASTM D, 6855: 2010). The intensity of reflection depends on the particle's nature and therefore determination of the conversion coefficient between NTU and TSS concentration is required.

Figure B.36 *Example Secchi disk (ref www.geoscientific.com) and turbidity meter.*

Alternatively, the transparency of a water body can be estimated by means of a Secchi disk (Figure B.36). This black and white disk is lowered into the water column until the disk disappears from sight. The distance over which the disk is visible gives an indication with respect to the transparency of the water body.

The inorganic solids, or suspended sediments, are usually measured as Total Suspended Solids (TSS): the dry-weight mass of sedimentary material that is suspended in the water per unit volume of water (Bray, 2008), typically in mg/l.

TSS may also be related to the acoustic backscatter from an Acoustic Doppler Current Profiler (ADCP). Using the ADCP technique information about the complete water column is obtained but the problem is the complexity of the system calibration. While the optical signal responds almost linear to the size of the particles the acoustic signal changes exponentially. Also the accuracy achieved using the acoustic backscatter technique will be much lower when the concentration over the succeeding cells decreases from high to low than for increasing concentrations. Hence, this technique will always require calibration against in-situ water samples or simultaneous measurements conducted with optical sensors.

The ADCP can be fixed on the seabed or buoy, or mounted to a survey vessel to, for instance, monitor a dredging plume.

The TSS from in-situ water samples are measured by weighing the amount of solids (e.g. collected by running the water sample through a filter) included in a known volume of water.

### B.3.11 *Dissolved oxygen*

Various sensors are available which make use of two fundamental techniques for the measurement of dissolved oxygen (DO): galvanic and polarographic method. Both methods use an electrode system whereby the DO reacts with the cathode to produce a current. Dissolved oxygen is usually expressed in mg/l or ppm.

### B.3.12 *Total Suspended Solids (TSS)*

The TSS of a water sample is determined by pouring a measured volume of water (typically one litre) through a pre-weighed filter, followed by drying the residue in an oven. The weight f the residue is a measure of the dry weight of the Total Suspended Sediments (TSS) present in the sample, which is typically expressed as milligrams per litre or mg/l (NEN-EN 872, 2005, APHA 2540 D, 2005, International Standard EN-ISO 11923:1997).

## B.4 **Detection of seabed obstruction**

Over the past ten years the high-resolution marine geophysical survey field has witnessed significant advances in survey investigation equipment. New equipment is based on optical, acoustic, electrical, and magnetic sensors. Some of these techniques can be used to detect and map pipelines, cables, wrecks and other seabed obstructions. Common used techniques include:

- Acoustic (high frequency sound pulses)
  o Side Scan Sonar (SSS)
  o Multibeam sonar or Multibeam Echo Sounding (MBES)
- Acoustic (low frequency sound pulses)
  o Sub-bottom profiler
- Magnetic (magnetic fields)

Acoustic methods are ideal for deep or turbid waters (unclear because of stirred-up sediment). Unlike optical methods, such as a laser or underwater camera, water characteristics such as turbidity and light penetration do not affect the acoustic sonar sensor. When a seabed obstruction is not completely covered by sand, it can easily be detected by sonar. However when a seabed obstruction is completely covered, standard sonar-based surveying techniques are not sufficient. In these cases, sub-bottom profiling and magneto-gradiometry are used to detect and localize the obstructions. Another problem arises since although these techniques can 'see' beneath the seabed, they are not as accurate as the side scan or multibeam sonar.

The following paragraphs give a more detailed description of techniques that can be used for the detection of seabed obstruction. Table B.8 gives a summary of some important characteristics.

### B.4.1  Side Scan Sonar (SSS)

The side scan sonar emits fan-shaped acoustic pulses down towards the seabed across a wide angle perpendicular to the path of the ship and is used to efficiently create an image of large areas of the sea bottom. It consists of a series of transducers, fitted in a fish-shaped vehicle (Figure B.37) that is towed behind or attached to the hull of a ship a few metres above the seabed.

Objects and obstacles located on the seabed give a sharp reflection of these pulses and create an acoustic shadow on their lee side. Thus, an image is obtained similar to an aerial photo at sunset, where shadows are accentuated, so that protrusions can be detected and an estimation of their height can be made. An absolute height measurement cannot be performed with the side scan sonar, for this echo sounding or multi-beam techniques are required.

The side scan sonar gives a clear image of irregularities in a wide swath on both sides of the ships path. However, there is always a trade-off between resolution of the image and coverage (swath), which can be 25 to 300 m. The side scan with a

Figure B.37    *Retrieving a side scan sonar device (Photo: NOAA).*

wide swath is convenient for a first detection of objects. The type and size of those objects can then be determined with measurements that are more precise.

The side scan sonar images can also be used to get an indication of the nature of the superficial soil layers, see Figure B.38 and Figure B.39.

### B.4.2   *Multibeam sonar or multibeam echo sounder (MBES)*

The multibeam is sometimes referred to as Swath echo sounder. The multibeam is in principle equal to the side scan sonar and the sub-bottom profiler, except that

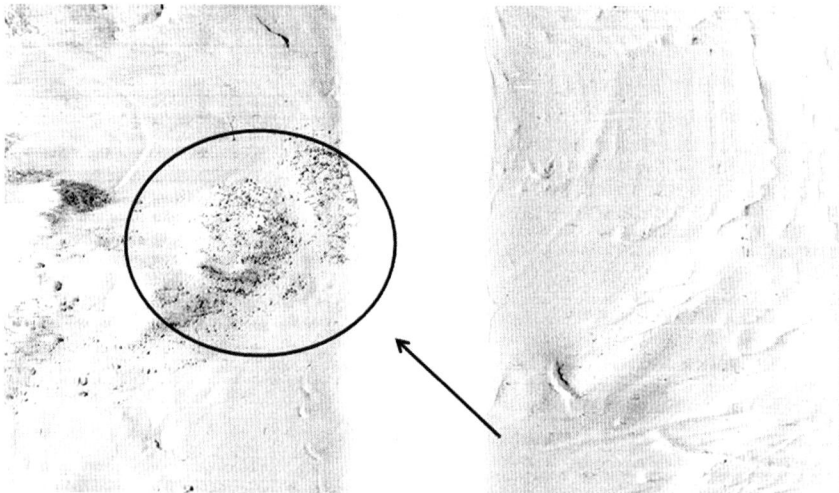

Figure B.38    *Side scan sonar image indicating the presence of gravel and rock.*

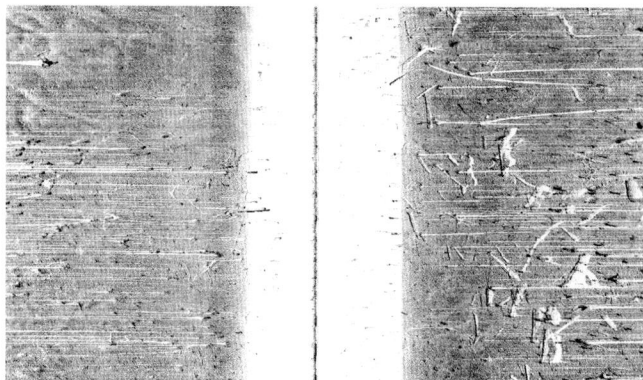

Figure B.39    *Side scan sonar image indicating the presence of tree trunks.*

the transducers are not moved mechanically over the area of interest. An acoustic pulse is spread at once over the area, so that an acoustic image of the entire area is obtained. The width of the area depends on the angle of aperture, which can be for instance 90, 120 or even 180 degrees. Figure B.40 presents a typical image of a shipwreck detected with a multibeam echo sounder.

Using multibeam sonars and GPS it is possible to locate obstacles with an accuracy of 0.3 to 0.4 m. However, because many of the obstacles are buried beneath the seabed and the multibeam uses high frequencies (200 kHz or higher) it is also necessary to use other techniques.

An advantage of the multibeam sonar compared to the side scan sonar is that the multibeam gives depth information (accuracy 0.05 to 0.10 m in shallow water) over the entire scanned area instead of only the area directly under the side scan sonar 'fish'. However, the range of the multibeam is smaller than that of a side scan and the acoustic image is much less detailed.

### B.4.3    *Magnetometry*

The magneto-gradiometry or magnetometer detection technique measures anomalies in the earth's magnetic field. Anomalies in the magnetic field are caused by the presence of ferromagnetic objects. The size of the anomalies depends on the mass of the iron object and its depth relative to the measurement level. Figure B.41 shows a magnetometer image possibly indicating a wreck with a large amount of debris scattered around.

Magnetometry techniques such as these can determine the contours of obstacles with an accuracy of about 1 to 2 m. Unfortunately, the sub-bottom depth of an obstacle cannot always be determined using this technique. Still, magnetometry is very useful for a final survey of a site to make sure everything that should be removed, has been removed. If anything is left behind, the mass of the

Figure B.40    *Left: Multibeam sonar on survey vessel (Image: ONHO). Right: Multibeam image of a shipwreck, displayed as three-dimensional grid).*

-500 -75  -30  -15 -7.5 -2.5 -1.5  0  1.5 2.5 7.5  15  30  75 500

**nT/m**

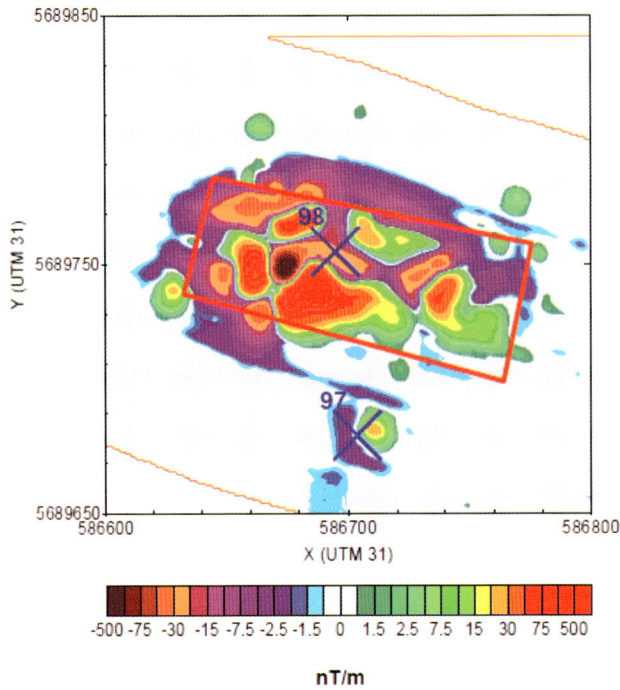

Figure B.41     *Magnetometry imaging shows a cluster of magnetic anomalies spread over a zone of 125 m × 50 m, possibly indicating a wreck with a large amount of debris scattered around.*

Table B.8    *Characteristics of commonly used survey techniques.*

|  | Common freq. ranges | Coverage area or footprint | Horizontal resolution | Vertical resolution | Bottom penetration | Complexity and costs |
|---|---|---|---|---|---|---|
| Side scan sonar | 100–500 kHz | Wide, 5–15 × water depth | 0.10 m | 1 m | No | Moderate |
| Sub-bottom profiler | 2–20 kHz | Narrow | 0.10–0.75 m | 0.10 m, locally | Yes | Moderate |
| Multibeam sonar | 200–450 kHz | Medium 2–7 × water depth | 0.30–0.40 m | 0.05–0.10 m | No | High |
| Magnetometry | – | 10 m | 1–2 m | n/a | Yes | High |

remaining parts is estimated from the measurements taken with the magnetic field sensor.

The technique is not affected by water depth or by bottom type, but the magnetic anomaly is strongly influenced by the orientation of the object, and expert knowledge is required to interpret the results. The complexity and costs for acquisition and processing are high.

Table B-8 presents an overview of frequency ranges, coverage and resolution of the discussed survey techniques used for the detection of seabed obstructions.

## B.5   Monitoring

### B.5.1   *Monitoring bearing capacity*

#### B.5.1.1   California bearing ratio

The California Bearing Ratio (CBR) of a soil is tested in situ by causing a cylindrical plunger to penetrate the soil at a given rate and comparing the relationship between force and penetration into the soil to that for a standard material. The test is normally used to evaluate the bearing capacity of soils directly underneath roads, airports, etc., especially for the design of the pavement. The method is not applicable for testing the bearing capacity of shallow foundations. Testing procedures for the in situ determination of the California Bearing Ratio are presented in BS1377, Part 9 or ASTM D4429–09a.

#### B.5.1.2   Plate bearing test

The Plate Bearing Test is a direct method to determine a bearing capacity and load-deflection relationship, but the results are only valid for the geometry and dimensions of the test plate used and the embedment depth of the plate. The test consists of loading a circular plate whilst measuring the settlement (deflection) of that plate as a function of the load. The plate can be loaded by means of deadweight or hydraulically by means of a jack. The larger the diameter of the plate, the deeper the influence zone of the load into the subsoil. The influence depth is approximate twice the diameter of the plate. Normally a plate diameter of 600 mm is used. Testing procedures for the Plate Bearing Test are presented in BS1377, Part 9 or ASTM D1196. Given the ultimate bearing capacity as determined by the plate bearing test it may be possible to calculate the shear strength of (the influenced upper zone of) the subsoil using one of the general bearing capacity formulae as proposed by Terzaghi, Brinch Hansen or Meyerhof, corrected for the circular geometry of the plate and the embedment depth.

#### B.5.1.3   Zone load test

The Zone Load Test is also a direct method for checking the bearing capacity of the reclaimed land. The Zone Load Test is, in principle, a large-scale Plate Bearing Test and, hence, also a direct method to determine the bearing capacity of a shallow foundation placed on or in the fill mass. This test is executed with a rigid square or rectangular plate of dimensions that may be equal to those of the proposed foundations (practical maximum size approximately $3 \times 3$ m), which implies that the fill is tested over an influence depth comparable to that of the future foundation (usually a considerably larger depth than when subjected to the Plate Bearing Test as shown in Figure B.42). The Zone Load Test is therefore considered to be a more meaningful test for verification of the bearing capacity of shallow foundations on a reclamation project.

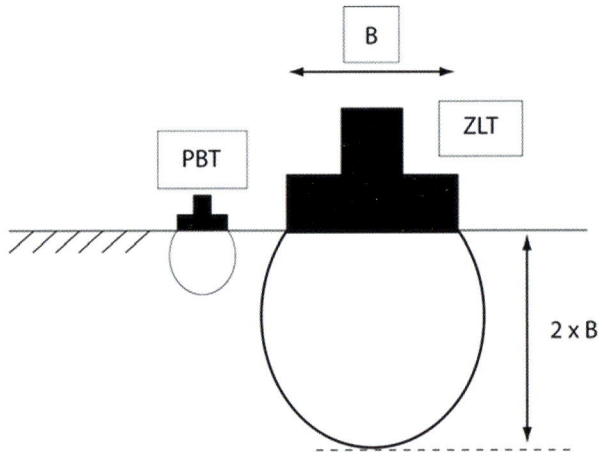

Figure B.42    *Influence depth of Plate Loading Test (PBT) and Zone Load Test (ZLT).*

Useful results are obtained if the test load significantly exceeds the future design load, which is imposed by the planned structure. In principle, a number of 5 to 6 equal load increments to a certain percentage (125%–150%) of the required allowable bearing pressure are applied. Each increment is maintained for a minimum of 2 hours. It is recommended to maintain the required working load for a minimum of 24 to 48 hours. Settlements should be recorded during the entire duration of the test at prescribed intervals. Figure B.43 presents an example of the arrangement needed to execute a Zone Load Test. Further reference is made to "Specification for Ground Treatment", The Institution of Civil Engineers, 1987.

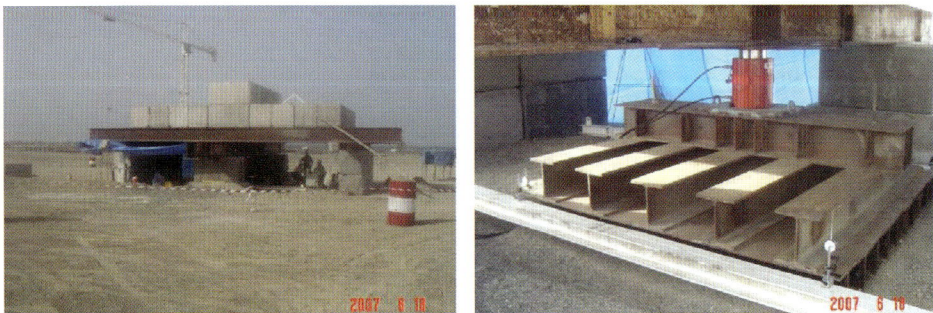

Figure B.43    *Arrangement for a Zone Load Test.*

582

### B.5.1.4 Trial embankment

The advantage of performing a Trial Embankment is that in an early phase of the construction information is obtained about the actual behaviour of the reclamation. The dimensions of a Trial Embankment (e.g. $10 \times 10$ m or $20 \times 20$ m) should be representative for the future loading conditions on the reclamation. Instruments can be installed such that the behaviour of the structure is monitored with respect to all relevant aspects. Corrective measures may be taken in case of non-compliance and optimisations can be considered in case the reclamation behaves better than anticipated.

### B.5.2  *Monitoring slope stability*

The following tools can be used to monitor the geotechnical stability of slopes:

### B.5.2.1 Electrical piezometer

As a result of the low permeability of fine-grained soils pore water pressures in such materials (both fill and subsoil) tend to increase when loaded. These excess pore pressures have a negative effect on the shear strength and, hence, on the stability. Because of the low permeability of cohesive soils, consolidation (i.e., dissipation of excess pore pressure after loading) takes time. Generally a decrease of the excess pore pressure is associated with settlements due to volume reduction in fine-grained materials which, on its turn, simultaneously results in an increase of the shear strength. Therefore monitoring the pore pressures during subsequent filling phases is essential.

Electrical piezometers (John Dunnicliff, 1993 and www.slopeindicator.com) are normally used to monitor pore water pressures on reclamation projects. The measured pore water pressures are corrected with records of the phreatic level and atmospheric pressure. Piezometers only need a very small volume of water entering the sensor to detect and measure a change of pore water pressure. They therefore display a more or less immediate response to changing conditions, even in fine-grained soils of low permeability. They are consequently very suitable when strongly varying pore water pressures are expected. The principle of a vibrating wire piezometer is presented in Figure B.44.

Different types of electrical piezometers are available on the market. For a more detailed description of the instrumentation and its capabilities reference is made to the websites and the latest brochures of geotechnical instrumentation companies.

### B.5.2.2 Open standpipe

An open standpipe piezometer consists of a vertical pipe, which is perforated over the bottom part such that ground water can flow into the pipe. This monitoring

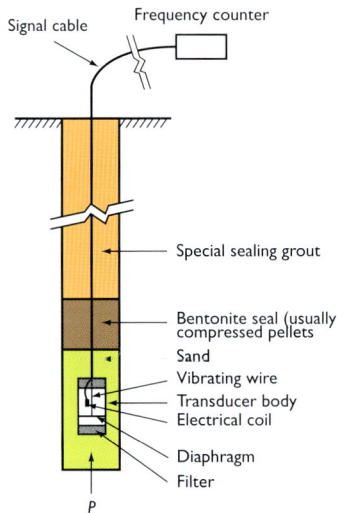

Figure B.44  *Example and principle Vibrating Wire Piezometer.*

method is less suitable for landfill projects than the electrical piezometers because the relatively large volume of water required for a change of the water level within the standpipe will result in a slower response to varying pore water pressures. This is especially true in soft cohesive material with a low permeability.

### B.5.3  *Monitoring deformation*

The following tools and devices are available for monitoring the deformations of a landfill:

Figure B.45  *Settlement plates.*

### B.5.3.1  Settlement plate

The settlement plate is the most simple and efficient tool to monitor vertical deformations (settlements) in and underneath a reclamation. The plates are installed at a certain level after which the vertical position of the pipe (fixed to the plate) is frequently measured. The pipes or rods can be extended during subsequent reclamation phases. An example of such a device is presented in Figure B.45.

Each individual measurement must include:

– the ground level at the location of the settlement plate;
– the length of the extension rods;
– the level of the top of the extension rod.

### B.5.3.2  Extensometer

An example of an extensometer is the spider magnet type. The extensometers are installed around a hollow plastic pipe at the desired distances and prepared for anchoring in the soil with the provided release system. This release system consists of a small chain secured by a pin that tightens the steel "springers" to the PVC pipe (Figure B.46).

The fully prepared PVC pipe is then installed in an open borehole (Figure B.47). Once the tube is in place, the casing is retrieved and the extensometers are launched by pulling the locking pins out of the release system; the extensometers are now anchored in the surrounding soil. In the end the entire borehole is grouted with a well-defined mixture of cement, bentonite and water. The idea is that the extensometers move with the surrounding soil during consolidation. The depth of their

Figure B.46    *Extensometer before release.*

Figure B.47    *Installation of extensometers in borehole.*

position in the subsoil is measured through the hollow pipe with another magnet-based instrument that indicates when the extensometer is reached.

Extensometers have the same purpose as settlement plates and are often used in combination with them. Their accuracy, however, is less than that of the plates and it is noted that some extensometers may not function well due to faulty grouting or a failed release system.

### B.5.3.3   Settlement hose

The settlement hose is a flexible hose, which is installed underneath the landfill at the initial ground or bed level. The hose, which is filled with water, follows the deformation of the subsoil under the weight of the landfill. A pressure meter is inserted into the hose, which measures the water pressure at fixed intervals. The water pressure gives accurate information on the vertical depth of the probe relative to the water level at both ends of the flexible hose. The measurements are used to establish a longitudinal depth profile.

## B.5.3.4   Inclinometer

The construction of landfills on soft subsoil initiates horizontal ground deformations especially at discontinuities along the edges (side slopes) of the reclamation. Lateral deformations are measured with inclinometers. For this purpose, a vertical flexible pipe is installed in the ground after which a probe equipped with an inclinometer is inserted. The angle of the probe is measured at regular intervals, for instance, every 0.5 m. Comparison of subsequent measurements provides information on the development of horizontal deformations for pipe and thus subsoil in time. The principle of the inclinometer is presented in Figure B.48.

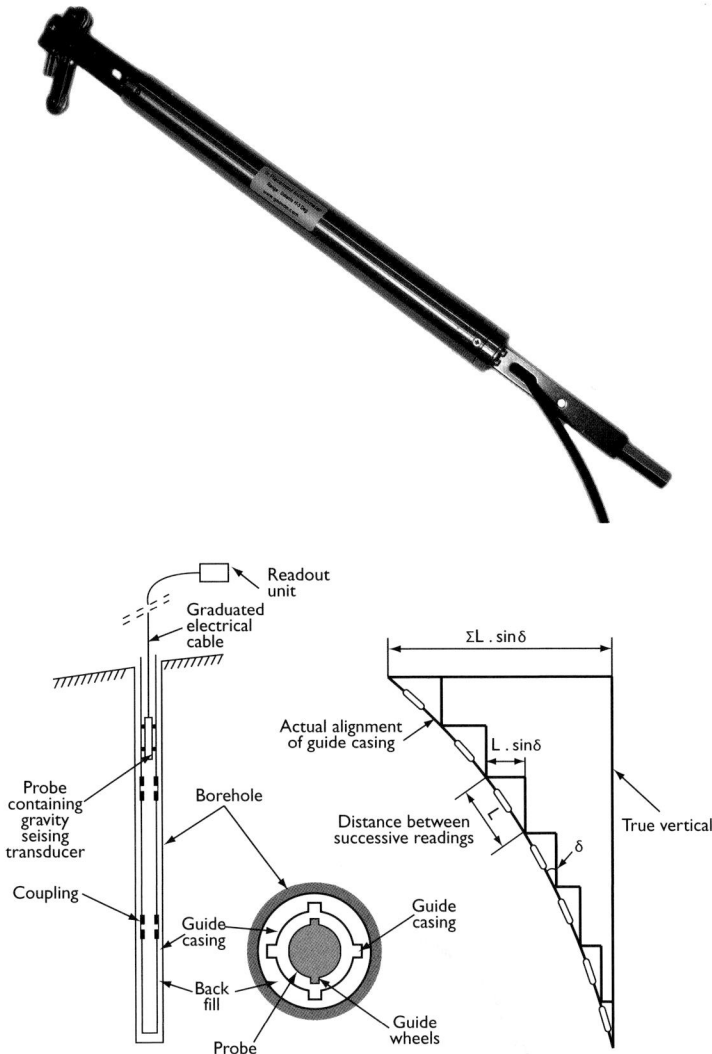

Figure B.48   *Example and principle inclinometer.*

There are several types of inclinometers (John Dunnicliff, 1993) on the market. For more detailed descriptions of the instrumentation and their capabilities, reference is made to the websites and latest brochures of geotechnical instrumentation companies.

### B.5.3.5 Trial embankment

During a Trial Embankment relevant information is obtained about the actual stiffness (deformations) of the reclamation under future loading conditions. The principle is already described in section 8.5.3.2 and section 11.3.3.2.

### B.5.3.6 SASW, CSCW

The SASW test (Spectral Analysis of Surface) is a non-intrusive test method for testing the small strain stiffness of the soil by analysing the velocity of Rayleigh surface waves. The Rayleigh wave is generated by a hammer impact. The waves are recorded by a pair geophones placed at variable distance from the source.

In CSWS tests (Continuous Surface Wave System), the Raleigh wave is generated by a computer controlled vibrator, see Figure B.49. The waves are recorded by multiple geophones placed at fixed interval distance from the source.

SASW and CSWS test can reach a depth penetration of 10–15 m in soft soils and up to 30 m in stiff soils.

The CSWS test is regarded as more accurate and effective than the SASW method. (Molnar 2007)

A relate technique measuring the shear wave velocity, is the seismic cone downhole method (see B.2.3.2).

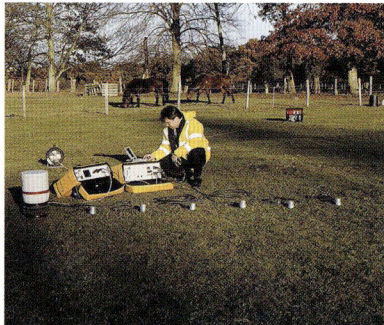

Figure B.49    *CSWS test.*

### B.5.4  *Environmental monitoring*

#### B.5.4.1  Visual inspection

Visual inspection is the first step to identifying whether a significant sediment plume is created. When supported by photographs it can be a useful tool, but it does not provide any information on the TSS concentrations. Visual inspections are limited to daylight hours and can be significantly affected by sun angle, sea state and light conditions.

#### B.5.4.2  Aerial & satellite imagery

During favourable weather and sea conditions, aerial photography or sophisticated remote sensing techniques are very useful for demonstrating the visual extent of the plume at a certain moment. A sequence of photographs can further provide information on the temporal and spatial dispersion of the plume during varying sea states and tidal conditions. The only drawback of this monitoring method is that it does not provide information on the extent of the sediment plume below the penetration extent of the satellite image.

#### B.5.4.3  Water sampling

Water sampling, followed by testing in a certified laboratory, is a clear and effective tool to record the state of the water quality. Depending on the analysis, it can provide relevant information on any physical parameter or chemical component (International Standard EN-ISO 5667: 2006, Techniques of Water-Resources Investigations, 2008). For dredging and reclamation projects, water samples are mainly used to determine the amount of TSS in a water body. In addition, water sampling is often performed simultaneous with turbidity measurement used to determine the proportional relation between turbidity and the amount of TSS contained in the water column. Niskin water samplers (Figure B.50) are specifically designed for the collection of a water sample. As the device is opened and lowered to the desired depth, water is trapped inside the container by remotely activating a tripping mechanism.

Figure B.50    *Example Niskin water sampler ( Ref. www.kc-denmark.dk).*

### B.5.4.4 Point monitoring

Point or static monitoring provides information about the water quality and/ or hydrodynamic parameters of any single location. Single point monitoring is commonly applied at series of predefined stations, which are normally located along the project boundaries or at locations with identified sensitive receptors. By mounting the equipment to a buoy, pole or frame placed on the seabed, the state of the water column can be continuously monitored. Monitoring data can be transmitted by means of GPRS or a radio signal to an onshore station, where data can be assessed and evaluated.

At reclamation sites, information about the presence of TSS in the decant water can be obtained by collecting water samples from the outflow location or by installation of a turbidity sensor in the weir box.

### B.5.4.5 Mobile vessel-based monitoring

When environmental information is required from numerous stations or when the installation of offshore buoys, poles or frames is not feasible, environmental data can be collected using a monitoring vessel. Commonly, mobile monitoring is performed on board a survey vessel in a non-continuous way. This flexible method allows the environmental surveyor to observe the water quality at any relevant location.

### B.5.4.6 Acoustic Doppler Current Profilers (ADCP's)

ADCP's are more frequently used to monitor sediment plumes, which may occur during dredging, and reclamation works. "Backscatter" information from the

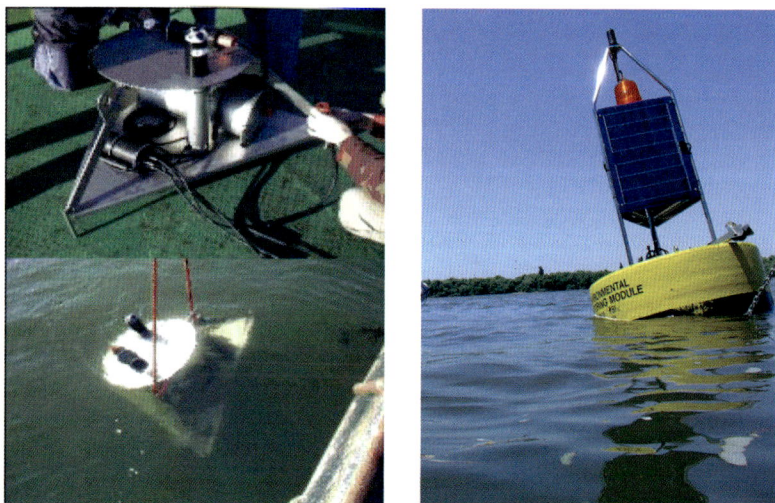

Figure B.51    *Seabed installed frame and monitoring buoy [Ref. http://www.osil.co.uk, http://www.ysi.com].*

ADCP in combination with turbidity sensors enables determination of the dispersion of sediment plumes in water. The instrument can be mounted on a buoy or installed on the seabed, but it can also be applied in a dynamic mode on board a survey vessel. Information is obtained about the temporal and spatial extent of a sediment plume.

# Appendix C

## CORRELATIONS AND CORRECTION METHODS

| | | | |
|---|---|---|---|
| C | Correlations and correction methods | | 593 |
| | C.1 | Density | 594 |
| | | C.1.1 $e_{max}$, $e_{min}$ – $C_u$ (uniformity), angularity (Youd) | 594 |
| | | C.1.2 $\rho_{d\,max}$ (Maximum proctor density) – $C_u$ (uniformity) | 594 |
| | | C.1.3 Relative density $R_e$ – $q_c$ (Baldi) | 595 |
| | | C.1.4 Relative density $R_e$ – SPT (Gibbs and Holtz) | 595 |
| | C.2 | Strength | 595 |
| | | C.2.1 Effective friction angle $\phi'$ – $q_c$ | 595 |
| | | C.2.2 Effective friction angle $\phi'$ – $q_c$, $K_o$ | 598 |
| | | C.2.3 Effective friction angle $\phi'$ – SPT | 599 |
| | | C.2.4 $q_c$, – SPT | 599 |
| | | C.2.5 $c_u$ – SPT | 600 |
| | C.3 | Stiffness and deformation | 601 |
| | | C.3.1 Young modulus – $q_c$ | 601 |
| | | C.3.2 Undrained modulus – $c_u$, $I_P$, OCR | 602 |
| | C.4 | Relative density – correction methods | 603 |
| | | C.4.1 $q_c$ relative density correlations for compressible sands | 603 |
| | | C.4.2 $q_c$ relative density correlations for calcareous and carbonate sands | 603 |
| | | C.4.3 Correction for saturated sand deposits | 606 |
| | | C.4.4 Correction for fine content | 606 |
| | | C.4.5 Correction for shallow readings | 608 |

## C   Correlations and correction methods

Appendix C presents some correlations between various soil properties. The correlations which are referred to in the main text of this Manual are gathered in this appendix and are sorted per subject, following the structure of the Manual. It is not the intention of this appendix to prescribe which correlation should be used, nor is the intention to give a complete overview of all available correlations. The subject correlations can only be used within the specific conditions as described in the corresponding literature. In view of the particular importance of the relative density as a soil property in a hydraulic fill, some correction methods are proposed for particular soils or circumstances.

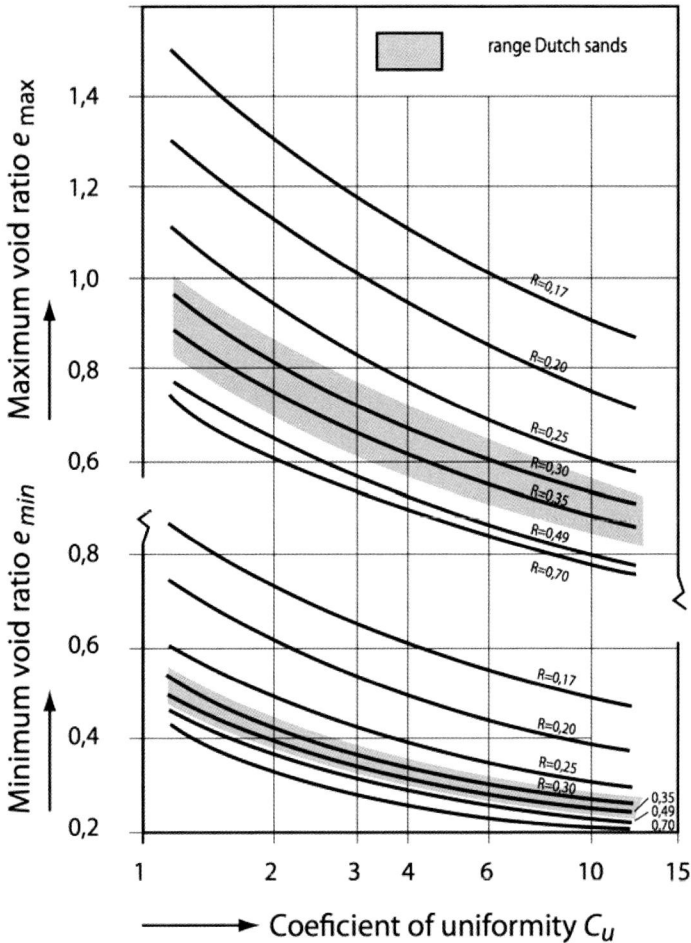

Figure C.1   *Minimum and maximum void ratio as a function of angularity and uniformity (Youd 1973).*

## C.1   Density

### C.1.1   $e_{max}$, $e_{min}$ – $C_u$ *(uniformity), angularity (Youd)*

Given the angularity and coefficient of uniformity, an estimate of the minimum and maximum void ratio of a sand can be made based on Figure C.1. (Youd 1973).

### C.1.2   $\rho_{d\,max}$ *(Maximum proctor density) – $C_u$ (uniformity)*

Various researchers have investigated the relationship between the uniformity coefficient and the maximum Proctor density. (Figure C.2).

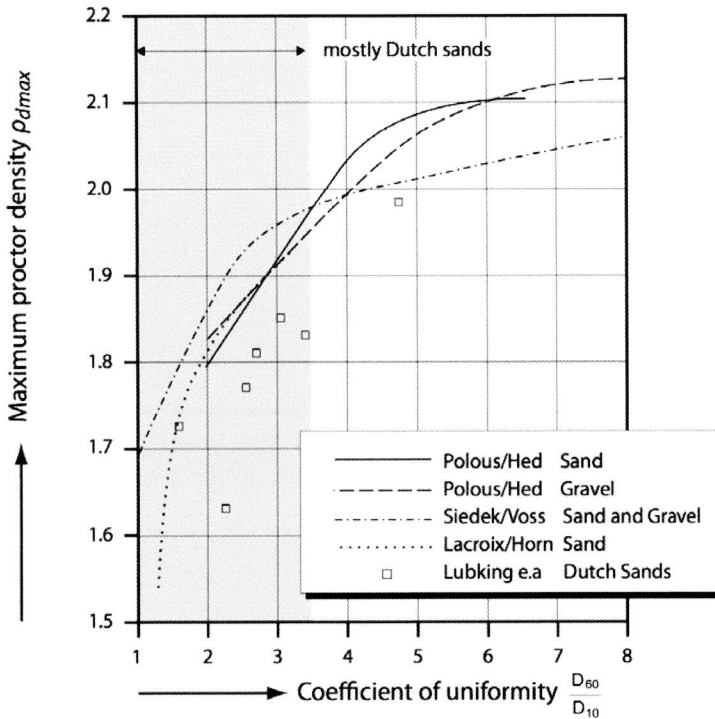

Figure C.2    *Relationship between $C_u$ and $\rho_{d\,max}$ (modified Proctor).*

## C.1.3    *Relative density $R_e - q_c$ ( Baldi)*

Baldi (1986) conducted extensive calibration chamber studies on moderate compressible Ticino quartz sand and obtained the following relationship to evaluate $D_r$ (Lunne *et al.*, 1997 after Baldi *et al.*, 1986).

## C.1.4    *Relative density $R_e - SPT$ ( Gibbs and Holtz)*

Gibbs and Holtz (1957) presented a method to estimate the relative density of a soil based on the SPT value and effective vertical stress.

## C.2    **Strength**

## C.2.1    *Effective friction angle $\phi' - q_c$*

Robertson *et al.*, 1983 presented a method to determine the effective friction angle $\phi'$ for a silica sand based on the cone resistance and effective vertical stress.

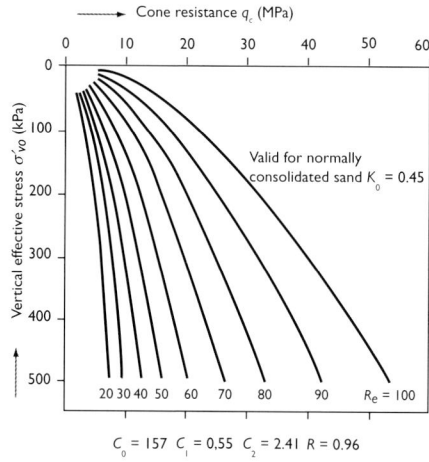

Figure C.3   $q_c$ – $R_e$ relationship for normally consolidated Ticino sands (after Baldi et al., 1986).

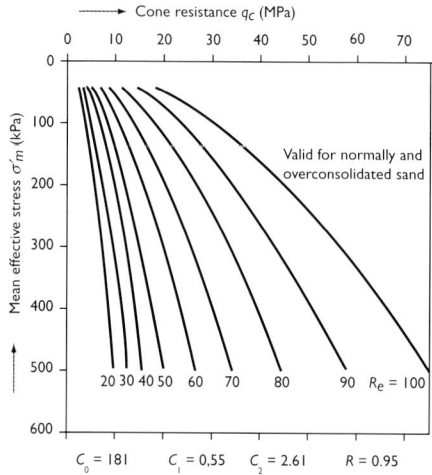

Figure C.4   $q_c$ – $R_e$ relationship for over-consolidated Ticino sands (after Baldi et al., 1986).

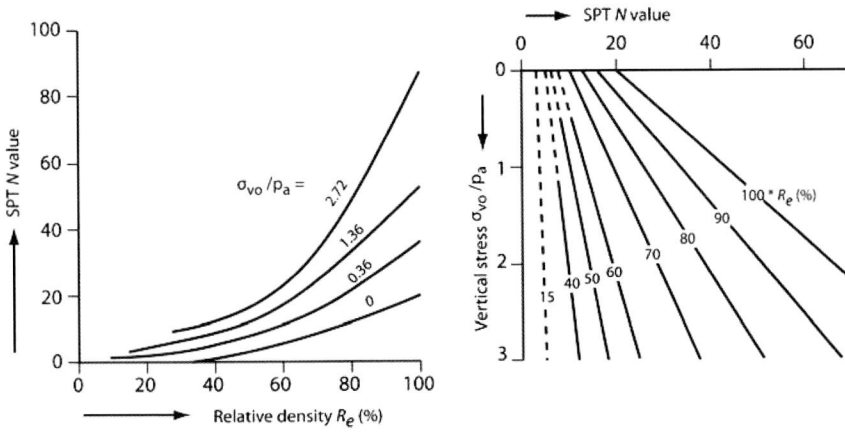

Figure C.5   *Gibbs and Holtz 1957 correlation $R_e$ – SPT.*

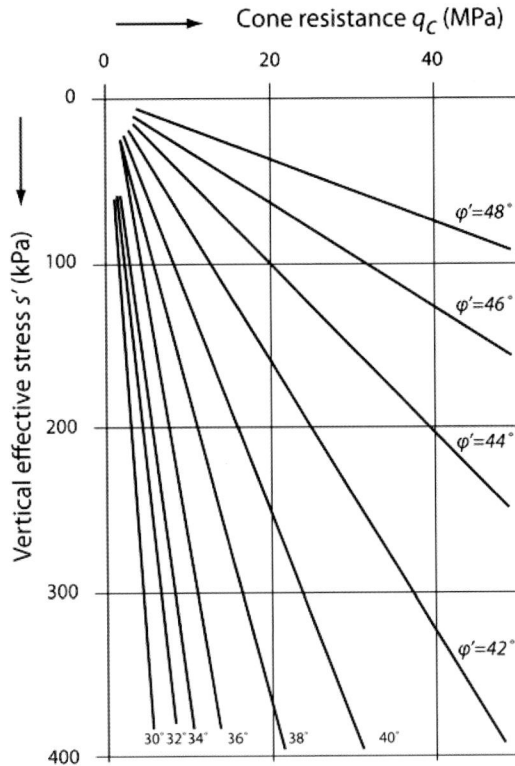

Figure C.6   *Effective friction angle from CPT, for silica sands (ref Robertson et al., 1983).*

## C.2.2  Effective friction angle $\phi' - q_c$, $K_o$

Dorgunoglu and Mitchell, 1975 presented a method to determine the effective friction angle $\phi'$ for a silica sand based on the cone resistance, the effective vertical stress and the coefficient of active earth pressure at rest $K_o$.

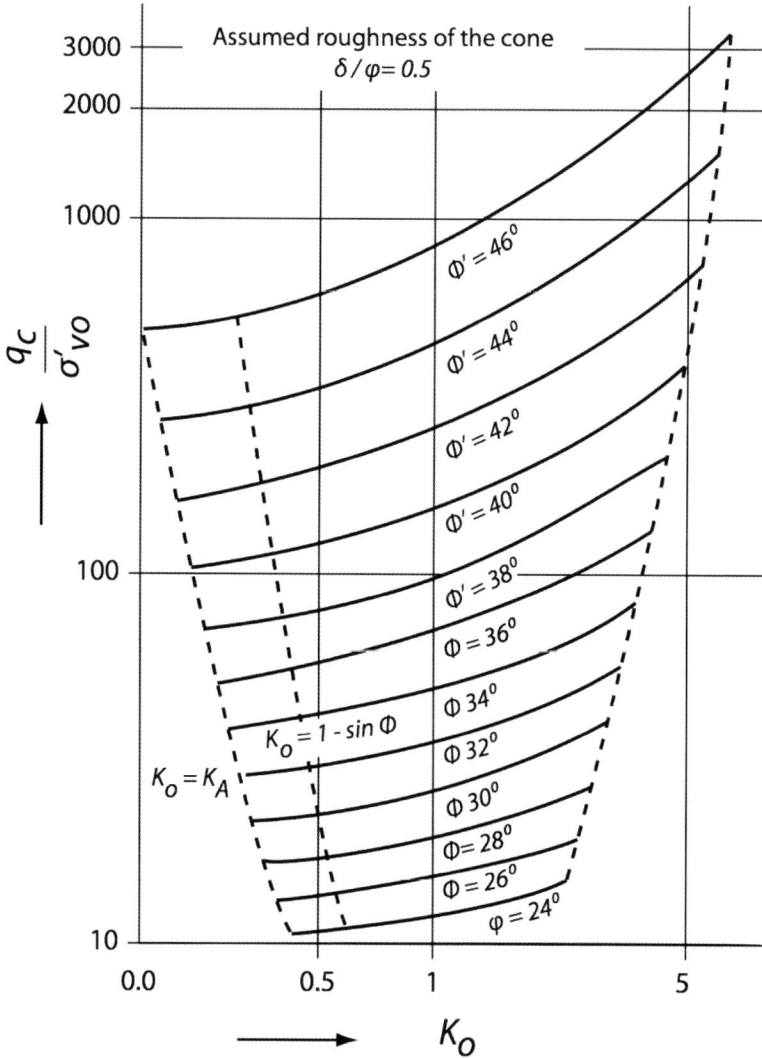

Figure C.7  *Effective friction angle from CPT for silica sands (Dorgunoglu and Mitchell, 1975).*

### C.2.3  *Effective friction angle φ′ – SPT*

Peck *et al.*, 1974 presented a correlation between the effective friction angle $\phi'$ and the SPT value.

### C.2.4  $q_c$ *– SPT*

Robertson *et al.*, 1983 presented a correlation between the SPT value and cone resistance values $q_c$, based on the mean particle size of the soil.

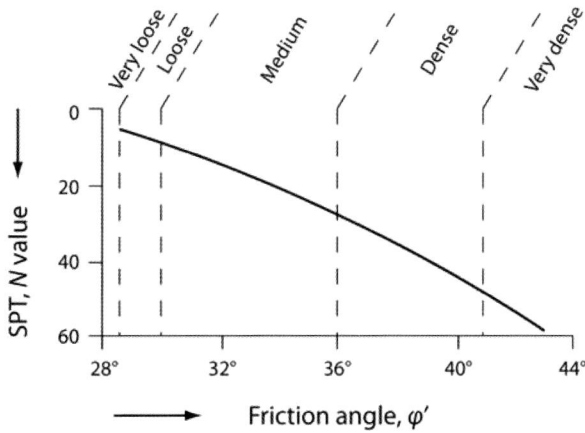

Figure C.8   *SPT-φ′ correlation (Peck et al., 1974).*

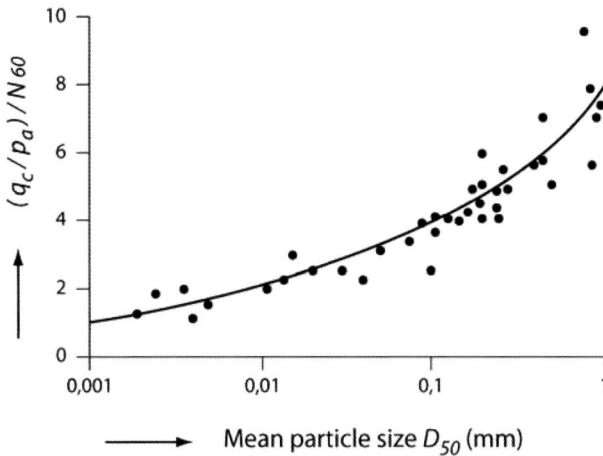

Figure C.9   *Correlation between CPT and SPT-$N_{60}$ based on soil type characterised by $D_{50}$ (after Robertson et al., 1983).*

## C.2.5 $c_u - SPT$

Various researchers presented correlations between the undrained cohesion $c_u$ and SPT value in fine grained soils.

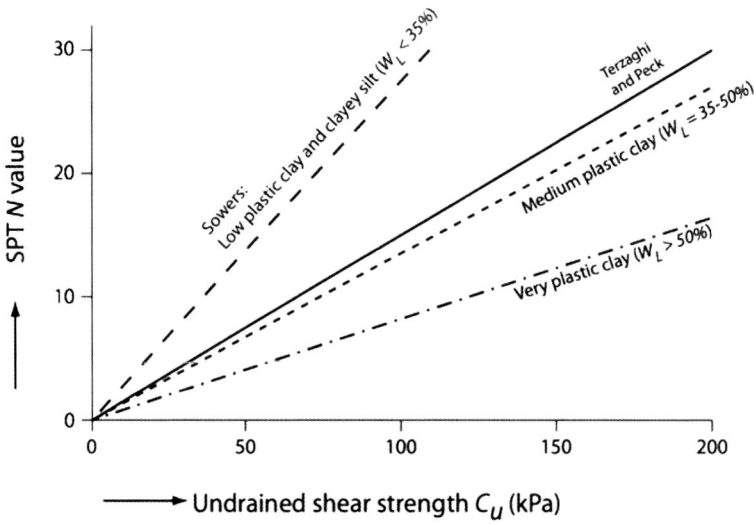

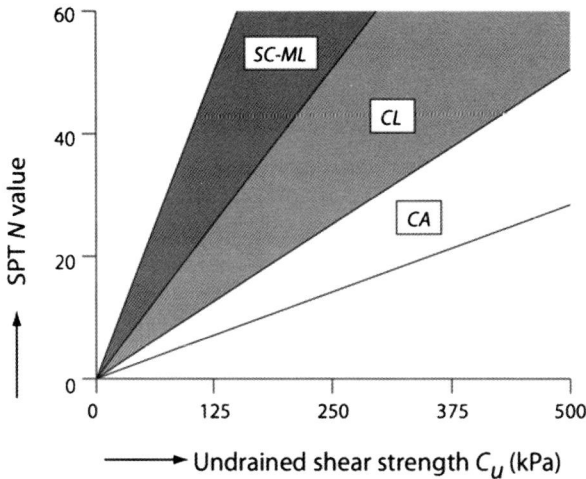

Figure C.10 *Correlation between SPT-N and $c_u$ according to different authors and taking into account the plasticity of the clay.*

## C.3 Stiffness and deformation

### C.3.1 *Young modulus – $q_c$*

The Young's modulus can be correlated to the cone resistance by means of Figure C.11. and Figure C.12.

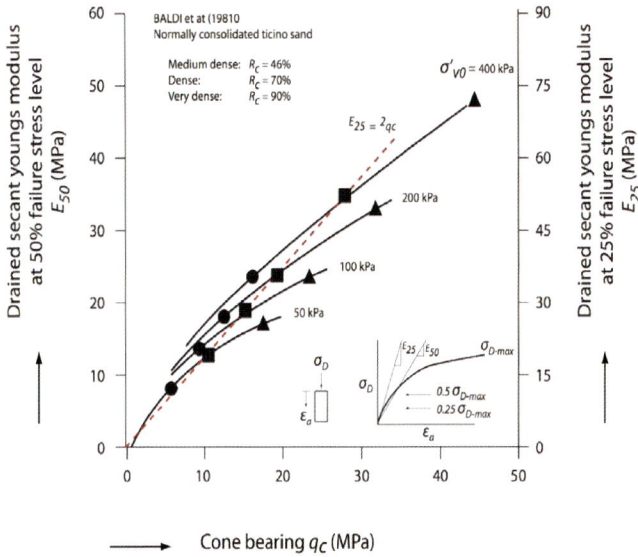

Figure C.11 *Young's modulus $E_{25}$ and $E_{50}$ from CPT testing.*

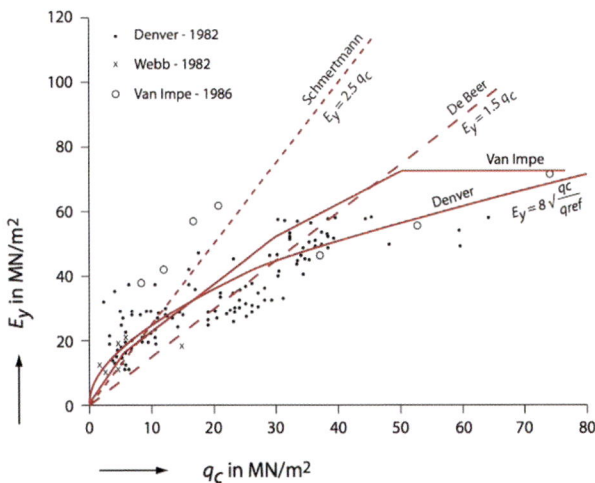

Figure C.12 *Young's modulus from CPT testing.*

Table C.1  *Correlation between cone resistance $q_c$ and stiffness for fine-grained soils (Sanglerat, 1972).*

| Soil type | Cone resistance | $\alpha$ | Other characteristics |
|---|---|---|---|
| $E_s = \alpha \cdot q_c$ | | | |
| Clay of low plasticity (CL) | $q_c < 0.7$ MPa<br>$0.7 < q_c < 2.0$ MPa<br>$q_c > 2.0$ MPa | $3 < \alpha < 8$<br>$2 < \alpha < 5$<br>$1 < \alpha < 2.5$ | |
| Silts of low plasticity (ML) | $q_c > 2.0$ MPa<br>$q_c < 2.0$ MPa | $3 < \alpha < 6$<br>$1 < \alpha < 3$ | |
| High plastic silts and clays (MH, CH) | $q_c < 2.0$ MPa | $2 < \alpha < 6$ | |
| Organic silts (OL) | $q_c < 1.2$ MPa | $2 < \alpha < 8$ | |
| Peat and organic clay ($P_t$, OH) | $q_c < 0.7$ MPa | $1.5 < \alpha < 4$<br>$1 < \alpha < 1.5$<br>$0.4 < \alpha < 1$ | $50 < w < 100$<br>$100 < w < 200$<br>$w > 200$ |

Table C.2  *Coarse-grained soils (Lunne and Christoferson, 1983).*

| $q_c$-values | $E_s$ |
|---|---|
| Normally consolidated sands<br>$q_c < 10$ MPa<br>$10$ MPa $< q_c < 50$ MPa<br>$q_c > 50$ MPa | <br>$4\,q_c$<br>$2\,q_c + 20$ (MPa)<br>120 MPa |
| Overconsolidated sands<br>$q_c < 50$ MPa<br>$q_c > 50$ MPa | <br>$5\,q_c$<br>250 MPa |

Other correlations exist between $q_c$ and $E_s$, e.g. according to Sanglerat (1972) for fine-grained soils (Table C.1) or Lunne and Christopherson (1983) for coarse-grained soils (Table C.2).

### C.3.2  *Undrained modulus – $c_u$, $I_P$, OCR*

The undrained stiffness of a cohesive soil can be linked to the undrained shear strength as shown in Figure C.13.

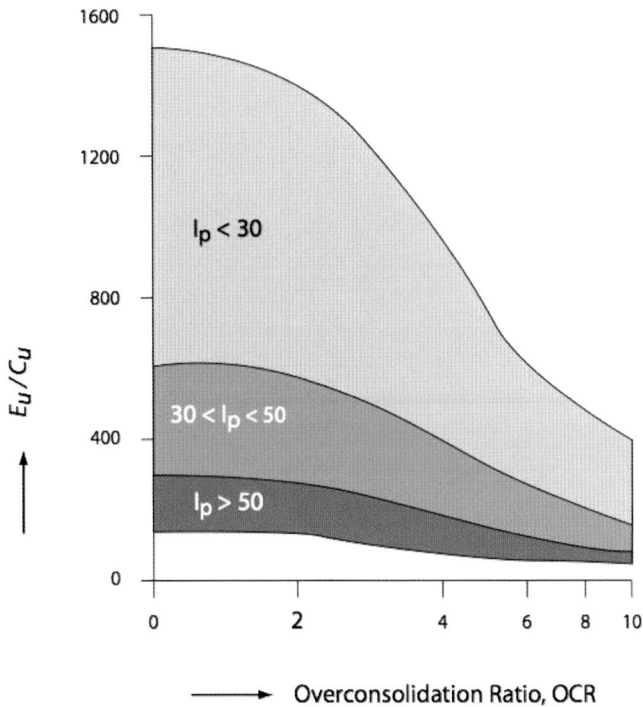

Figure C.13   $E_u$ as a function of $c_u$ (Duncan and Buchignani, 1976).

## C.4   Relative density – correction methods

### C.4.1   $q_c$ relative density correlations for compressible sands

The Baldi correlations as presented in Figure C.3 and in Figure C.5 were setup for moderately compressible Ticino sand. Comparing the Baldi correlations to correlations setup by other researchers (for other types of sand) clearly demonstrates the influence of the compressibility on the $q_c$/relative density relationship. In Figure C.14, the Baldi (1982) correlation is compared to the Schmertmann correlation (1976) and the Villet & Mitchell correlation (1981).

As can be seen on Figure C.15, a $q_c$ reading can correspond to a large range of relative densities depending on the compressibility of the sand.

### C.4.2   $q_c$ relative density correlations for calcareous and carbonate sands

Reference is made to section 9.2 for a detailed description of carbonate sands.

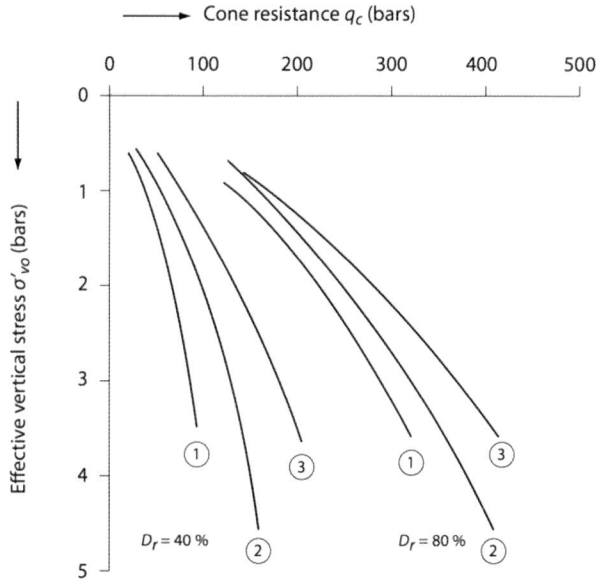

Figure C.14 *Effect of compressibility on relative density relationships (after Robertson & Campanella, 1983).*

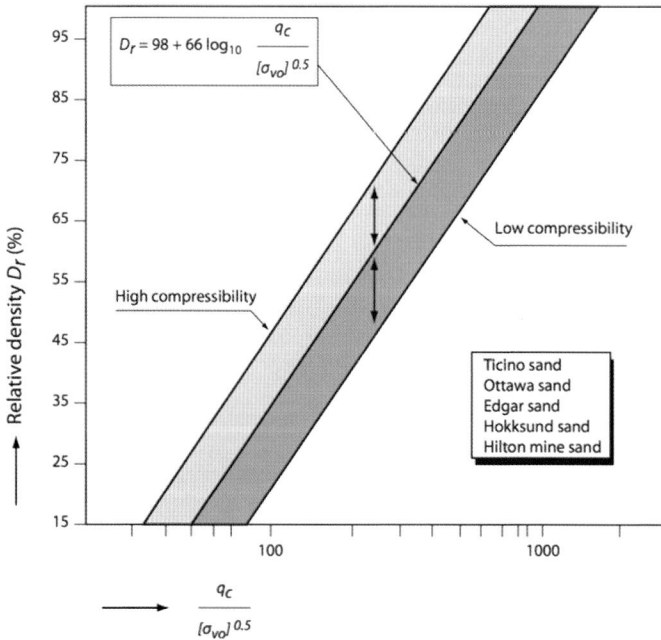

Figure C.15 *Influence of compressibility on NC uncemented, unaged predominantly quartz sands (Jamiolkowski et al., 1985).*

Calcareous or carbonate sands are often compressible sands. Additionally, these sands are subject to some degree of crushing during any dynamic testing (for example CPT). This crushing effect might not be problematic for the use as fill material since the stress levels causing the crushing during a CPT test might be much higher than the stress levels during the lifetime of the fill. In order to correctly assess the relative density based on CPT testing of such carbonate or calcareous sands, some corrections have to be applied to the standard correlation formulas setup for siliceous sand (e.g., the Baldi correlation for Ticino sand).

Some authors have compared the cone resistance in carbonate and siliceous sands at the same relative density during calibration chamber testing. Carbonate sands, such as Dogs Bay sand or Quiou sand have been studied. As can be seen from the possible range of resulting correction factors in table Table C.3, the factor to be applied is strongly related to the type of sand involved.

The listed factor is the multiplication factor, which has to be applied on the cone resistance reading in the carbonate sands. After applying this factor, the standard relative density correlations for quartz sand can be adopted.

The only correct way to find a good correlation in calcareous sand (different from the published reference sands) is to setup a calibration chamber test. Unfortunately this is a very time consuming and costly effort and is therefore most often not a realistic approach.

In case there is no time for a calibration chamber test, and there are serious doubts about the correlation to be adopted, it is advised not to use CPT-testing as a means to determine the relative density of the sand. In such cases, other testing meth-

Table C.3  *Multiplication factor cone resistance for carbonate sands.*

| Author | Description | Multiplication factor to be applied on the cone resistance |
|---|---|---|
| Vesic 1965 | comparison quartz sand/quartz sand with 10% shells | 2.3 |
| Belloti/ Jamiolkowski 1991 M. Almeida *et al.*, 1992 | comparison Quiou carbonate sand/Ticino sand | 1.3 to 2.2 ($R_e$) |
| Van Impe *et al.*, 2001 | comparison Quiou carbonate sand/Mol sand (data adapted from Belotti *et al.*, 1991) | 2 to 4.8 ($R_e$) |
| Van Impe *et al.*, 2001 | comparison Dogs Bay carbonate sand/Mol sand (data adapted from Yasufuku *et al.*, 1995) | 3.4 to 9.9 ($R_e$) |
| Gudehus, Cudmani 2004 Wehr 2005 | comparison of Dubai Sand to quartz sand (after removal of large shell fragments) | 1.36 to 1.82 ($R_e$) $\sim 0.0046 \cdot R_e + 1.363$ |

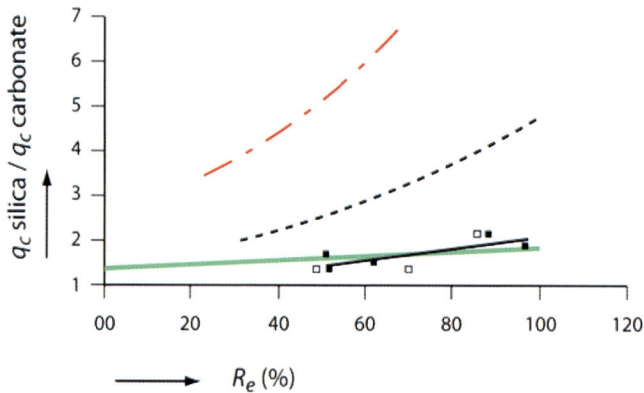

Figure C.16    *Correction factors carbonate sands in function of the relative density.*

ods such as the zone load test, may prove to be more practical for checking the geotechnical properties of the fill.

### C.4.3    *Correction for saturated sand deposits*

Most relative density/cone resistance correlations in literature were setup for dry sands. Research has shown that for saturated sand deposits the standard $q_c - R_e$ correlations lead to an underestimation of the actual relative density and need to be corrected with the following formula: (Jamiolkowski 2001):

$$\frac{R_e(saturated) - R_e(dry)}{R_e(dry)}100 = -1.87 + 2.32\ln\frac{q_c}{(\sigma'_{vo}P_a)^{0.5}}$$

This correction can be applied for any correlation which has been determined by testing on dry sands.

### C.4.4    *Correction for fine content*

- In case the fine content of the tested sand is less than 5% the standard $q_c - R_e$ correlations can be adopted;
- In case the fine content is between 5% and 15%, the correlations can be used after applying some corrections (as explained further in this paragraph);
- In case the fine content is more than 15%, the corrected correlations should not be used (except for preliminary purposes).

One way to consider the required correction for a sand with fine fraction of 5% to 15% is to use the corrections developed for the evaluation of the liquefaction potential (Figure C.17 and Figure C.18).

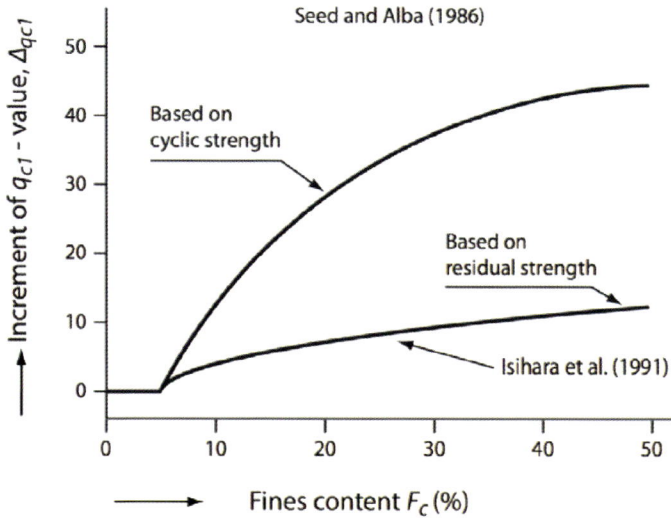

Figure C.17    *Increment D$_{qc1}$ as a function of fines content (Ishihara 1993).*

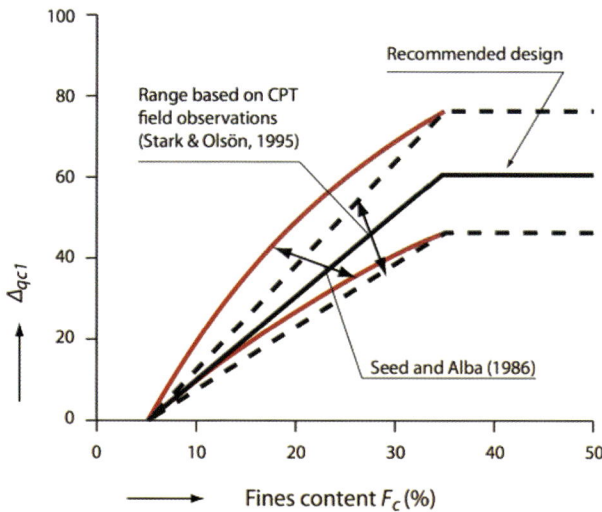

Figure C.18    *Correction for fines content for evaluation of liquefaction potential (after Lunne et al., 1997).*

Note that $qc_1$ is the normalised cone penetration resistance.

In view of the uncertainty involved in this approach the use of a correction range as represented by the lower and upper curves in Figure C.17 and Figure C.18 is recommended.

Another alternative is to directly measure the in situ density using special probes.

## C.4.5  *Correction for shallow readings*

It is generally accepted that the published relationships $q_c$/CPT are not valid for shallow CPT readings. Puech and Foray (2002) published a practical method for the interpretation of shallow depth CPT readings. Based on the bearing capacity models for shallow foundations, a relationship between $q_c$, depth D and effective friction angle $\varphi'$ was setup. By curve fitting, a suitable value for $\varphi'$ can be found. This $\varphi'$ can be correlated to a relative density (Figure C.19).

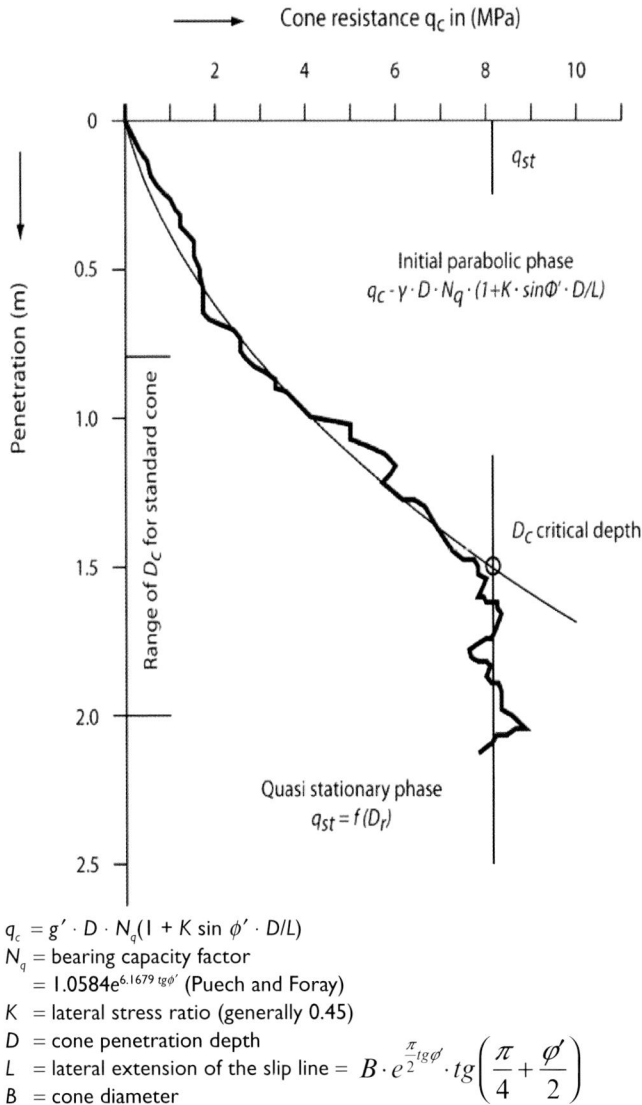

Figure C.19   *Concept for analysis of shallow depth CPT readings (Puech and Foray 2002).*

GEOTECHNICAL PRINCIPLES

| D | Appendix geotechnical principles | | 609 |
|---|---|---|---|
| | D.1 | Density | 610 |
| | D.2 | Strength | 613 |
| | | D.2.1 Bearing capacity calculation – Meyerhof (1951, 1953, 1963) | 613 |
| | | D.2.2 Squeezing – Matar-Salençon calculation rule | 615 |
| | D.3 | Stiffness and deformation | 616 |
| | | D.3.1 Final primary settlement – Terzaghi | 616 |
| | | D.3.2 Vertical consolidation (time-dependent primary settlement) – Terzaghi | 617 |
| | | D.3.3 Horizontal consolidation (use of vertical drains) – Barron | 621 |
| | | D.3.4 Horizontal consolidation – Smear effects: Hansbo | 623 |
| | | D.3.5 Combination of horizontal consolidation and vertical consolidation – Carillo | 625 |
| | | D.3.6 Secondary compression | 626 |

## D   Appendix geotechnical principles

In this appendix, some fundamental geotechnical theories and formulas are briefly presented. For more detailed information, reference is made to geotechnical handbooks such as:

- Das B.M., Principles of geotechnical engineering, 2009;
- Lambe T.W., Whitman R.V. Soil Mechanics, 1969;
- Terzaghi K., Peck R.B., Mesri G. "Soil Mechanics in Engineering Practice", Wiley and Sons, 1996;
- Budhu M., Soil Mechanics & Foundations, 2010;
- ICE Manual of geotechnical engineering edited by John Burland, Tim Chapman, Hilary Skinner and Michal Brown Volume 1 and 2, 2012.

## D.1  Density

Table D.1  *Density relationships for saturated soils.*

| | $w_{sat}$ | $n$ | $e$ | $\rho_s$ | $\rho_d$ | $\rho_{sat}$ |
|---|---|---|---|---|---|---|
| $\rho_s, \rho_d$ | $\dfrac{\rho_w}{\rho_d} - \dfrac{\rho_w}{\rho_s}$ | $\dfrac{\rho_s - \rho_d}{\rho_s}$ | $\dfrac{\rho_s - \rho_d}{\rho_d}$ | | | $\dfrac{\rho_s \cdot \rho_d + \rho_w \cdot (\rho_s - \rho_d)}{\rho_s}$ |
| $\rho_s, \rho_{sat}$ | $\dfrac{(\rho_s - \rho_{sat}) \cdot \rho_w}{(\rho_{sat} - \rho_w) \cdot \rho_s}$ | $\dfrac{\rho_s - \rho_{sat}}{\rho_s - \rho_w}$ | $\dfrac{\rho_s - \rho_{sat}}{\rho_{sat} - \rho_w}$ | | $\rho_s \cdot \dfrac{\rho_{sat} - \rho_w}{\rho_s - \rho_w}$ | |
| $\rho_s, w_{sat}$ | | $\dfrac{\rho_s \cdot w_{sat}}{\rho_s \cdot w_{sat} + \rho_w}$ | $\dfrac{\rho_s}{\rho_w} \cdot w_{sat}$ | | $\dfrac{\rho_s \cdot \rho_w}{\rho_s \cdot w_{sat} + \rho_w}$ | $\dfrac{(1 + w_{sat}) \cdot \rho_s \cdot \rho_w}{\rho_s \cdot w_{sat} + \rho_w}$ |
| $\rho_s, n$ | $\dfrac{\rho_w \cdot n}{\rho_s \cdot (1-n)}$ | | $\dfrac{n}{(1-n)}$ | | $\rho_s \cdot (1-n)$ | $\rho_s \cdot (1-n) + \rho_w \cdot n$ |
| $\rho_s, e$ | $\dfrac{\rho_w \cdot e}{\rho_s}$ | $\dfrac{e}{(1+e)}$ | | | $\dfrac{\rho_s}{1+e}$ | $\dfrac{\rho_s + \rho_w \cdot e}{1+e}$ |
| $\rho_d, \rho$ | $\dfrac{\rho - \rho_d}{\rho_d}$ | $\dfrac{\rho - \rho_d}{\rho_w}$ | $\dfrac{\rho - \rho_d}{\rho_w + \rho_d - \rho}$ | $\dfrac{\rho_d \cdot \rho_w}{\rho_w + \rho_d - \rho}$ | | |

| | | | | | |
|---|---|---|---|---|---|
| $\rho_d,\ w_{sat}$ | $\dfrac{\rho_d \cdot w_{sat}}{\rho_w}$ | $\dfrac{\rho_d - w_{sat}}{\rho_w - \rho_d \cdot w_{sat}}$ | $\dfrac{\rho_d \cdot \rho_w}{\rho_w - \rho_d \cdot w_{sat}}$ | | $\rho_d \cdot (1 + w_{sat})$ |
| $\rho_d,\ n$ | $\dfrac{\rho_w \cdot n}{\rho_d}$ | $\dfrac{n}{(1-n)}$ | $\dfrac{\rho_d}{(1-n)}$ | | $\rho_d + \rho_w \cdot n$ |
| $\rho_d,\ e$ | $\dfrac{\rho_w \cdot e}{\rho_d \cdot (1+e)}$ | $\dfrac{e}{(1+e)}$ | $\rho_d \cdot (1+e)$ | | $\rho_d + \rho_w \cdot \dfrac{e}{(1+e)}$ |
| $\rho_{sat},\ w_{sat}$ | $\dfrac{\rho_{sat} \cdot w_{sat}}{\rho_w \cdot (1+w_{sat})}$ | $\dfrac{\rho_{sat} \cdot w_{sat}}{\rho_w \cdot (1+w_{sat}) - \rho_{sat} \cdot w_{sat}}$ | $\dfrac{\rho_{sat} \cdot \rho_w}{\rho_w \cdot (1+w_{sat}) - \rho_{sat} \cdot w_{sat}}$ | $\dfrac{\rho_{sat}}{(1+w_{sat})}$ | |
| $\rho_{sat},\ n$ | $\dfrac{\rho_w \cdot n}{\rho_{sat} - \rho_w \cdot n}$ | $\dfrac{n}{(1-n)}$ | $\dfrac{\rho_{sat} - \rho_w \cdot n}{(1-n)}$ | $\rho_{sat} - \rho_w \cdot n$ | |
| $\rho_{sat},\ e$ | $\dfrac{\rho_w \cdot e}{\rho_w \cdot (1+e) - \rho_w \cdot e}$ | $\dfrac{e}{(1+e)}$ | $\rho_{sat} \cdot (1+e) - \rho_w \cdot e$ | $\rho_{sat} - \rho_w \cdot \dfrac{e}{(1+e)}$ | |
| $w_{sat},\ n$ | | $\dfrac{n}{(1-n)}$ | $\dfrac{\rho_w \cdot n}{w_{sat} \cdot (1-n)}$ | $\dfrac{\rho_w \cdot n}{w_{sat}}$ | $\dfrac{\rho_w \cdot n \cdot (1+w_{sat})}{w_{sat}}$ |
| $w_{sat},\ t,\ e$ | $\dfrac{e}{(1+e)}$ | | $\dfrac{\rho_w \cdot e}{w_{sat}}$ | $\dfrac{\rho_w \cdot e}{w_{sat} \cdot (1+e)}$ | $\dfrac{\rho_w \cdot e \cdot (1+w_{sat})}{w_{sat} \cdot (1+e)}$ |

Table D.2 *Density relationships for partly saturated soils.*

| | $Sr = 1,\ n_s = 0$ | $w,\ S$ | $w,\ n_a$ | $n,\ n_w$ | $e,\ e_w$ | $\rho_{sat}$ | $\rho,\ w$ | $\rho_d,\ w$ |
|---|---|---|---|---|---|---|---|---|
| $w$ (saturated) | | $\dfrac{w}{S}$ | $\dfrac{w\cdot\rho_s+n_a\cdot\rho_w}{\rho_s\cdot(1-n_a)}$ | $\dfrac{n\cdot\rho_w}{\rho_s\cdot(1-n)}$ | $\dfrac{e\cdot\rho_w}{\rho_s}$ | $\dfrac{(\rho_s-\rho_{sat})\cdot\rho_w}{(\rho_{sat}-\rho_w)\cdot S}$ | $\dfrac{(\rho_s-\rho)\cdot\rho_w}{(\rho-S\cdot\rho_w)\cdot\rho_s}$ | $\dfrac{(\rho_s-\rho_d)\cdot\rho_w}{\rho_d\cdot\rho_s}$ |
| $w$ (partly saturated) | $w$ | $w$ | $w$ | $\dfrac{n_w\cdot\rho_w}{\rho_s\cdot(1-n)}$ | $\dfrac{e_w\cdot\rho_w}{\rho_s}$ | | $\dfrac{(\rho_s-\rho)\cdot S\cdot\rho_w}{(\rho-S\cdot\rho_w)\cdot\rho_s}$ | $\dfrac{S\cdot(\rho_s-\rho_d)\cdot\rho_w}{\rho_d\cdot\rho_s}$ |
| $n$ | $\dfrac{w\cdot\rho_s}{w\cdot\rho_s+\rho_w}$ | $\dfrac{w\cdot\rho_s}{w\cdot\rho_s+S\cdot\rho_w}$ | $\dfrac{w\cdot\rho_s+n_a\cdot\rho_w}{w\cdot\rho_s+\rho_w}$ | $n$ | $\dfrac{e}{(1+e)}$ | $\dfrac{\rho_s-\rho_{sat}}{\rho_s-\rho_w}$ | $1-\dfrac{\rho}{(1+w)\cdot\rho_s}$ | $1-\dfrac{\rho_d}{\rho_s}$ |
| $e$ | $\dfrac{w\cdot\rho_s}{\rho_w}$ | $\dfrac{w\cdot\rho_s}{S\cdot\rho_w}$ | $\dfrac{w\cdot\rho_s+n_a\cdot\rho_w}{\rho_w\cdot(1-n_a)}$ | $\dfrac{n}{(1-n)}$ | $e$ | $\dfrac{\rho_s-\rho_{sat}}{\rho_{sat}-\rho_w}$ | $\dfrac{(1+w)\cdot\rho_s}{\rho}-1$ | $\dfrac{\rho_s}{\rho_d}-1$ |
| $\rho_{sat}$ | $\dfrac{(1+w)\cdot\rho_s\cdot\rho_w}{w\cdot\rho_s+\rho_w}$ | | | $\rho_s\cdot(1-n)+n\cdot\rho_w$ | $\dfrac{\rho_s+e\cdot\rho_w}{1+e}$ | $\rho_{sat}$ | $\dfrac{(\rho_s-\rho_w)\rho}{(1+w)\cdot\rho_s}+\rho_w$ | $\dfrac{\rho_s\cdot\rho_d+\rho_w\cdot(\rho_s-\rho_d)}{\rho_s}$ |
| $\rho_s$ | | $\dfrac{S\cdot\rho_s\cdot\rho_w\cdot(1+w)}{w\cdot\rho_s+S\cdot\rho_w}$ | $\dfrac{\rho_s\cdot\rho_w\cdot(1+w)(1-n_a)}{w\rho_s+\rho_w}$ | $\rho_s\cdot(1-n)+n_w\cdot\rho_w$ | $\dfrac{\rho_s+e_w\cdot\rho_w}{1+e}$ | | $\dfrac{(\rho_s-\rho)\cdot\rho_w}{(\rho-S_r\cdot\rho_w)\cdot\rho_s}$ | $\rho_d\cdot(1+w)$ |
| $\rho_d$ | $\dfrac{\rho_s\cdot\rho_w}{w\cdot\rho_s+\rho_w}$ | $\dfrac{S\cdot\rho_s\cdot\rho_w}{w\cdot\rho_s+S\cdot\rho_w}$ | $\dfrac{\rho_s\cdot\rho_w\cdot(1-n_a)}{w\rho_s+\rho_w}$ | $\rho_s\cdot(1-n)$ | $\dfrac{\rho_s}{1+e}$ | $\rho_s\dfrac{\rho_{sat}-\rho_w}{\rho_s-\rho_w}$ | | $\rho_d$ |
| $S_r$ | $1$ | $\dfrac{w}{\rho_{sat}}$ | $\dfrac{w\cdot\rho_s\cdot(1-n_a)}{w\rho_s+n_a\cdot\rho_w}$ | $\dfrac{n_w}{n}$ | $\dfrac{e_w}{e}$ | | $\dfrac{(\rho_s-\rho)\cdot\rho_w}{(\rho-S\cdot\rho_w)\cdot\rho_s}$ | $\dfrac{w\cdot\rho_d\cdot\rho_s}{\rho_w\cdot(\rho_s-\rho_d)}$ |

## D.2 Strength

### D.2.1 *Bearing capacity calculation – Meyerhof (1951, 1953, 1963)*

The bearing capacity of shallow foundations can be calculated by the Meyerhof theory.

$$q_u = i_q d_q s_q N_q \sigma_v' + i_c d_c s_c N_c \, c' + 0.5 i_\gamma s_\gamma B \, N_\gamma \gamma_k \tag{D.1}$$

with:

| | |
|---|---|
| $q_u$ | Ultimate bearing capacity |
| $d_q, d_c$ | Depth factors |
| $s_q, s_c, s_\gamma$ | Shape factors |
| $i_q, i_c, i_\gamma$ | Inclination factors |
| $N_q, N_c, N_\gamma$ | Bearing capacity factors which are related to the effective friction angle $\varphi'$; in Figure D.1 these factors are shown as a function of $\varphi'$ |
| $B$ | Foundation width |
| $\sigma_v'$ | Effective vertical stress at foundation level next to the foundation |
| $\gamma_k$ | Effective weight of the soil underneath the foundation level |

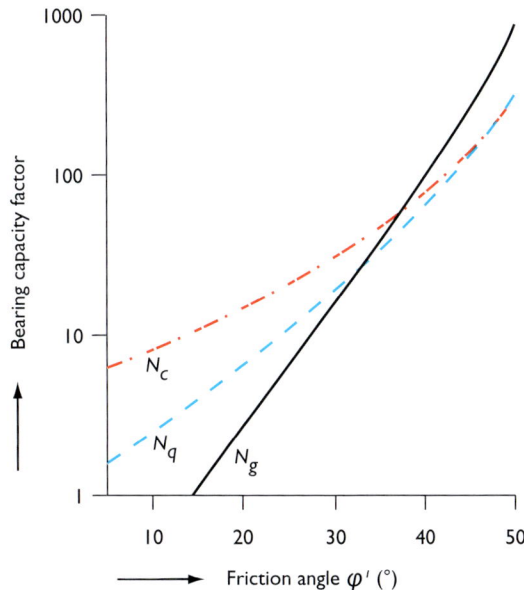

Figure D.1　*Bearing capacity factors according to Meyerhof (1963).*

This formula is based on a certain failure mode of the soil underneath the foundation as shown in Figure 8.14. The literature presents different failure modes and formulas for the above-mentioned factors. Reference is made to Bowles (1997) or Eurocode 7, Part 1, Annex D (2004) for specific formulas. Special consideration is required in case of eccentric and inclined loading conditions. These specific situations will not be discussed in this manual. Reference is made to relevant geotechnical textbooks.

**Calculation example bearing capacity**

The Meyerhof formula for a square footing (width 3 m) on sand with $\varphi' = 32°$ and $c' = 0$ kPa, with a vertical and centric loading while neglecting the depth factors can be written as:

$$q_u = s_q N_q \sigma'_v + 0.5 \, s_\gamma \, B \, N_\gamma \gamma_k$$

This footing can be placed on the ground surface or at 1 m depth. The water table should be taken into account as well. In Table D.3 several options are worked out with:

$s_q = s_\gamma$ = 1.33
$d_q = d_\gamma$ = 1.06
$N_q$ = 23.2
$N_\gamma$ = 22.0
$B$ = 3 m
$d$ = 0 m or 1 m, which is the depth of the foundation level below ground surface
$\gamma_k$ = 16 kN/m³ for dry sand
$\gamma_k$ = 10 kN/m³ when the water table coincides with the foundation level
$\sigma'_v$ = d × 16 for dry sand above the foundation level

Table D.3    *Ultimate bearing capacity of a footing 3 m × 3 m (in kPa).*

| Footing depth | No groundwater | GW level at foundation level |
| --- | --- | --- |
| 0 m | 701 | 438 |
| 1 m | 1264 | 985 |

This example demonstrates that a unique bearing capacity does not exist. The boundary conditions, which need to be clearly specified, finally define the actual bearing capacity.

### D.2.2 Squeezing – Matar-Salençon calculation rule

One of the analysis methods regarding squeezing has been defined by Matar-Salençon (CUR 162, 1996), who presented the following formula to calculate the allowable fill height for a sloping embankment on soft subsoil:

$$\Delta p = \frac{1}{FS} \cdot \left( 4.14 c_u^b + \frac{c_u^o}{h} \cdot x \right) \qquad (D.2)$$

where:

$\Delta p$ = allowable load (kPa)

$FS$ = factor of safety

$c_u^b$ = undrained shear strength at the upper side of the soft layer (kPa)

$c_u^o$ = undrained strength at the lower side of the soft layer (kPa)

$h$ = thickness of the soft layer (m)

$x$ = distance measured from the toe of the slope (m)

The formula gives the "envelope" of the maximum load on the ground, as shown in Figure D.2. The fill's load (and corresponding fill height) should, on average, remains below this envelope. In Figure D.2, the hatched zone indicates the soft layer. The dashed line refers the allowable load as per formula D.2. The area under the allowable filling line, measured from the toe to the crest of the slope, has to be larger than the area under the actual fill load line in order to guarantee stability.

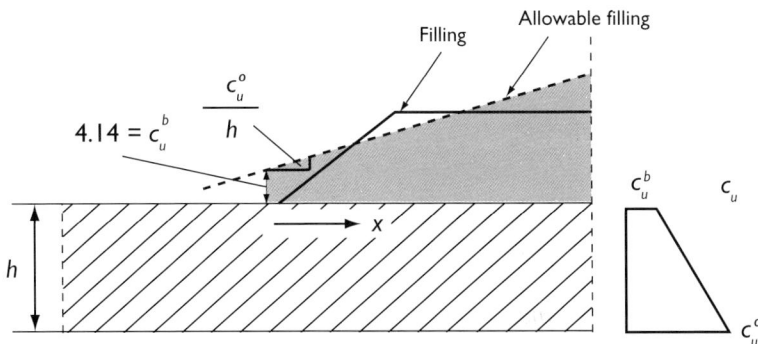

Figure D.2  *Squeezing analysis of a sloping fill surface according to Matar-Salençon (CUR 162, 1996).*

The main advantage of this relatively simple calculation method is that only information about the undrained shear strength of the foundation material is needed. For a soft layer with a uniform undrained shear strength and $x$ equal to $h$, this formula reduces to

$$\Delta p = \frac{5.14 c_u}{FS}. \qquad (D.3)$$

which is derived from the bearing capacity formula (8–2 section 8.4.3.4) for cohesive fine grained soils.

**Example application of the Matar–Salençon method**

A bund has to be constructed for the deposition of hydraulic fill on a 5 m thick layer of soft soil. The water table coincides with the top of the soft layer. The undrained shear strength is expressed as a function of the depth: $c_u = 5 + 1.5\ z$.

The fill for the bund is humid and has a weight of 18 kN/m³. The bund height is 3 m and a slope of $1_V$ over $3_H$ is planned to be realised. The required factor of safety is 1.15.

Question: Is this stable according to the method of Matar-Salençon?

Calculation:

The slope length is 9 m and the load under the slope is $3 \times 9 \times 18/2 =$ 243 kN/m.

Maximum unit load at toe of the slope is ($x = 0$ m): $4.14 \times 5/1.15 = 18$ kPa.

Maximum unit load at the crest of the slope is ($x = 9$ m): $18 + [(5 + 1.5 \times 5) \times 9/5]/1.15 = 37.6$ kPa.

Total allowable load over the slope area is $(18 + 37.6) \times 9/2 = 250$ kN/m

Conclusion: The effective loading is smaller than the allowable load: 243 kPa < 250 kPa: the embankment will not cause squeezing.

## D.3   Stiffness and deformation

### D.3.1   *Final primary settlement – Terzaghi*

Primary settlement calculations are based on Terzaghi's formula, assuming one dimensional deformation:

$$\frac{\Delta h}{h_0} = \frac{\Delta e}{1+e_0} = \frac{C_r}{1+e_0} \times \log\left(\frac{\sigma'_p + \sigma'_{ref}}{\sigma'_0 + \sigma'_{ref}}\right) + \frac{C_c}{1+e_0} \times \log\left(\frac{\sigma'_p + (\Delta\sigma' - (\sigma'_p - \sigma'_0)) + \sigma'_{ref}}{\sigma'_p + \sigma'_{ref}}\right)$$

where:

$h_0$ = initial thickness of the considered layer [m]

$e_0$ = initial void ratio [–]

$\Delta e$ = change in void ratio [–]

$\Delta h$ = change in layer thickness [m]

$\sigma'_0$ = initial effective stress in the middle of the considered layer [kPa]

$\sigma'_p$ = pre-consolidation stress in the middle of the considered layer [kPa]

$\Delta\sigma'$ = effective stress increment in the middle of the considered layer [kPa]

$\sigma'_{ref}$ = small reference stress (usually taken equal to 1 kPa); this is a term to make the formula theoretically correct, but it is often omitted [kPa]

$C_c$ = compression index [–]

$C_r$ = recompression index [–]

$\Delta h$ is normally calculated for (small) sub layers and all $\Delta h$-values are summed to obtain the total primary settlement.

### D.3.2 *Vertical consolidation (time-dependent primary settlement) – Terzaghi*

When only vertical consolidation occurs (vertical flow of water), Terzaghi's one-dimensional consolidation theory is used to calculate the degree of consolidation.

The settlement at time $t$ is defined as:

$$S(t) = S_{total} U_v(t) = S_{total}\left(1 - \sum_{m=0}^{\infty} \frac{8}{\pi^2} \frac{1}{(2m+1)^2} e^{-\frac{\pi^2}{4}(2m+1)^2 T_v}\right)$$

where:

$U_v$ = average degree of consolidation (percentage)

$T_v$ = time factor: $T_v = \frac{c_v}{(aH)^2}t$ (dimensionless)

$S_{total}$ = total consolidation (or primary) settlement

$c_v$ = vertical coefficient of consolidation [m²/y]

$t$ = time [y]

$H$ = layer thickness [m]

$a$ = parameter depending on the drainage conditions of the soft layer: ($a = 1$ for unilateral drainage $a = 0.5$ for bilateral drainage)

In Figure D.3 this relation is given graphically. In the literature approximate formulas can be found for the relation between the degree of consolidation and the time factor. Reference is made to Bo *et al.* (2003).

For practical use the following simple formula represents the relation between the time factor $T_v$ and the average degree of consolidation $U_v$ usually with sufficient accuracy:

$$U_v = \sqrt[6]{\frac{T_v^3}{T_v^3 + 0.5}}$$

It should be noted that the degree of consolidation in a soil layer is an average value. At the top and/or bottom of the subject layer, the excess pore water pressures will have dissipated earlier compared to the centre of the layer. This should also be kept in mind when interpreting the results of piezometer readings.

Using the above formulas, which define the degree of vertical consolidation, the loading nor the magnitude of the settlement have an influence on the consolidation rate.

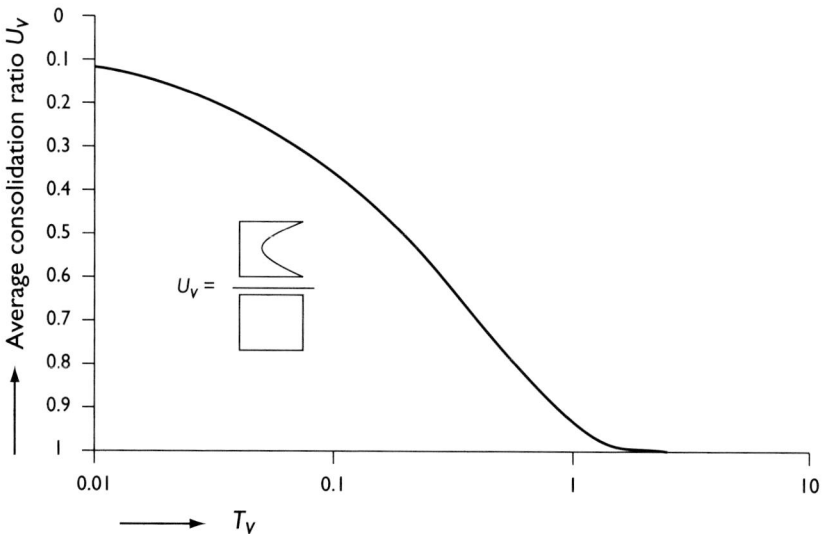

Figure D.3   *Graphical presentation of the relation $U_v$ versus $T_v$.*

**Example of a settlement calculation (without vertical drains)**

A large area has to be raised by 2 m of sand fill. The original ground level is 1.0 m RL, groundwater is encountered at −1.0 m RL. The existing subsoil consists of 2 m soft clay covering a layer of firm clay with a thickness of 1 m. These clay strata are overlying bedrock.

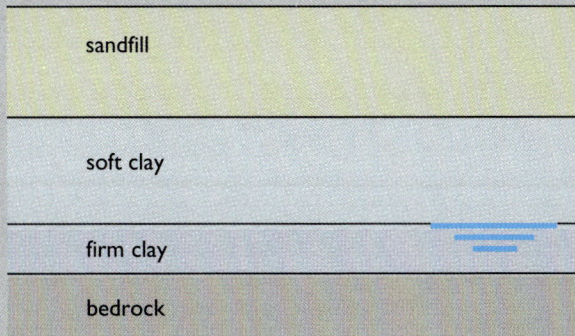

Figure D.4   *Geotechnical profile.*

Results of laboratory testing on samples of the clay layers and the sand fill indicate:

soft clay:  $C_r = 0.19$
$C_c = 0.96$
$c_v = 10^{-8}$ m²/s
$\sigma_p' = 14$ kPa
$e_0 = 2.85$
$\gamma_{sat} = 14$ kN/m³

firm clay:  $C_r = 0.05$
$C_c = 0.39$
$c_v = 2 * 10^{-8}$ m²/s
$\sigma_p' = 50$ kPa
$e_0 = 1.61$
$\gamma_{sat} = 16$ kN/m³

sand fill:  $\gamma_{dry} = 18$ kN/m³

**Calculation of total primary settlement**

The initial effective stress in the middle of the soft clay layer $\sigma_0' = 1 \times 14 = 14$ kPa.

The initial effective stress in the middle of the firm clay layer $\sigma_0' = 2 \times 14 + 0.5 \times (16 - 10) = 31$ kPa.

Comparing these figures with the pre-consolidation stresses as determined in the laboratory suggests that the soft clay is normally consolidated, while the firm clay is slightly over-consolidated.

The effective stress increment in the middle of both the soft and the firm clay layer $\Delta\sigma' = 2 \times 18 = 36$ kPa. Note that as a result of the large area over which the sand fill is placed this effective stress increment is uniform over the full height of the subsoil layers (no stress distribution with depth).

Primary settlement:

soft clay: $s_p = 2 \times [0.96/(1 + 2.85)] \times \log [(14 + 28)/14] = 0.24$ m

firm clay: $s_p = 1 \times [0.05/(1 + 1.61)] \times \log (50/31) + 1 \times [0.39/(1 + 1.61)] \log [(50 + (36 - (50 - 31)))/50] = 0.02$ m

Total primary settlement $s_{p\cdot total} = 0.24 + 0.02 = 0.26$ m

Note that the actual settlement will be slightly less because the lower part of the soft clay will sink below the water table as a result of the settlements occurring in the firm clay layer.

**Calculation of time dependent primary settlement**

What is the total settlement after 16 weeks?

The total settlement at time $t$ is the sum of the settlements originating from both the soft and the firm clay layer at time $t$ (in seconds).

Total primary settlement at end of consolidation:

- soft clay: 0.24 m
- firm clay: 0.02 m

Both layers have one-sided drainage.

Time $t = 16 * 7 * 24 * 60 * 60 = 9{,}676{,}800$ seconds

soft clay: $T_v(t) = [10^{-8}/(1 * 2)^2] * 9{,}676{,}800 = 0.0242$

$$U_v = [(0.0242)^3/(0.0242 + 0.5)^3]^{1/6} = 0.21$$

$$S_{16\,weeks} = 0.21 * 0.24 = 0.05 \text{ m}$$

firm clay: $T_v(t) = [2 * 10^{-8}/(1 * 1)^2] * 9{,}676{,}800 = 0.1935$

$$U_v = [(0.1935)^3/(0.1935 + 0.5)^3]^{1/6} = 0.53$$

$$S_{16\,weeks} = 0.53 * 0.02 = 0.01 \text{ m}$$

Total settlement after 16 weeks will approximately be $0.05 + 0.01 = 0.06$ m.

For $T_v = 2.5$ the degree of consolidation $U_v = 0.95$. In this example $T_v = 2.5$ will be reached after:

- soft clay: $t = 2.5 * (1 * 2)^2/10^{-8} = 10^9$ seconds = 1653 weeks = 31 years
- firm clay: $t = 2.5 * (1 * 1)^2/(2 * 10^{-8}) = 1.25 * 108$ seconds = 207 weeks = 4 years

### D.3.3 *Horizontal consolidation (use of vertical drains) – Barron*

In soil layers where Prefabricated Vertical Drains (PVDs) are installed, part of the consolidation can be attributed to the horizontal water flow towards the drains. The formulas to calculate the consolidation $U_h$ as a result of horizontal drainage, are based on Barron's theory. Reference is made to the literature about this subject (Holtz *et al.*, 1991). In Figure D.5 the basic formulas are given for the design of PVDs. However, alternative formulas exist in which smear effects caused by penetration of the drain in cohesive soil is taken into account as well.

Around each PVD an equivalent cylinder is assumed with a radius of 1.05 times or 1.13 times the distance between the PVDs for respectively a triangular or a square spacing (see Figure D.6).

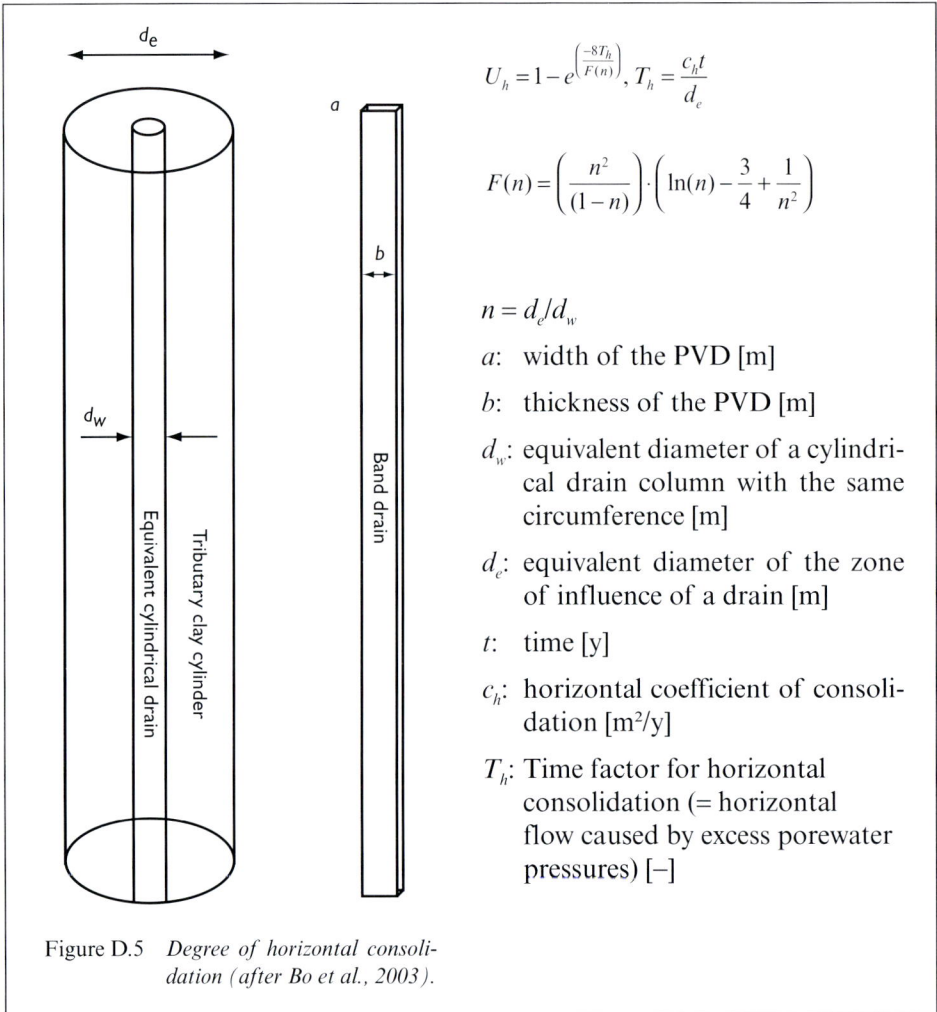

$$U_h = 1 - e^{\left(\frac{-8T_h}{F(n)}\right)}, \; T_h = \frac{c_h t}{d_e}$$

$$F(n) = \left(\frac{n^2}{(1-n)}\right) \cdot \left(\ln(n) - \frac{3}{4} + \frac{1}{n^2}\right)$$

$n = d_e/d_w$

$a$:   width of the PVD [m]

$b$:   thickness of the PVD [m]

$d_w$: equivalent diameter of a cylindrical drain column with the same circumference [m]

$d_e$: equivalent diameter of the zone of influence of a drain [m]

$t$:   time [y]

$c_h$: horizontal coefficient of consolidation [m²/y]

$T_h$: Time factor for horizontal consolidation (= horizontal flow caused by excess porewater pressures) [–]

Figure D.5   *Degree of horizontal consolidation (after Bo et al., 2003).*

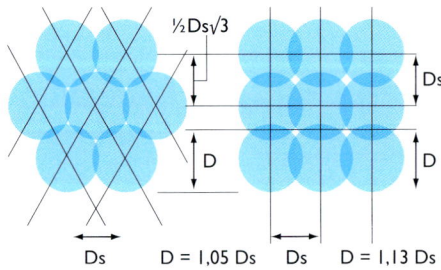

Figure D.6   *PVD installation and equivalent cylinders.*

In natural sediments the horizontal permeability can be up to three times larger than the vertical permeability. The same ratio may apply for the horizontal and vertical coefficient of consolidation. However, when using PVDs caution must be taken in using a larger horizontal coefficient of consolidation: The closer PVDs are installed to each other, the more the effect of smear will become important and the positive effect of a larger horizontal coefficient of consolidation might be diminished. Imanshi 2000 even reported that in case of close PVD spacings, the horizontal permeability became equal or smaller than the vertical permeability.

### D.3.4 *Horizontal consolidation – Smear effects: Hansbo*

It is recommended to use the formulas taking into account smear effects and to use the correct ratio between the horizontal and vertical coefficient of consolidation. This means that the horizontal permeability needs to be defined by means of laboratory or field testing.

The beneficial effect of larger horizontal permeability is partly destroyed during the installation of a PVD. The zone surrounding the PVD in which this effect takes place is called the smeared zone.

The extend of the smeared zone is largely determined by the size of the mandrel which is used to install the PVD's.

Hansbo proposed following correction to the Barron formula:

$$U_h = 1 - e^{\left(\frac{-8T_h}{F}\right)}, \quad \text{where}$$

$$F = F(n) + Fs$$

$Fs = \left(\dfrac{k_h}{k_s} - 1\right) \ln\left(\dfrac{d_s}{d_w}\right)$ (represents the additional factor taking into account the smear effects)

where:

$k_h$ = horizontal permeability
$k_s$ = horizontal permeability in the smeared zone surrounding the drain. (often $k_s$ is set equal to the vertical permeability $k_v$)
$d_s$ = diameter of disturbed zone = $\pm 2d_m$,
$d_m$ = diameter of circle with area equivalent to the cross-sectional area of the mandrel

A typical mandrel size and PVD size is presented in Figure D.7

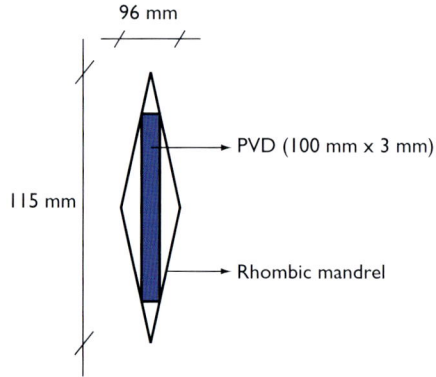

Figure D.7    *Typical mandrel and PVC size.*

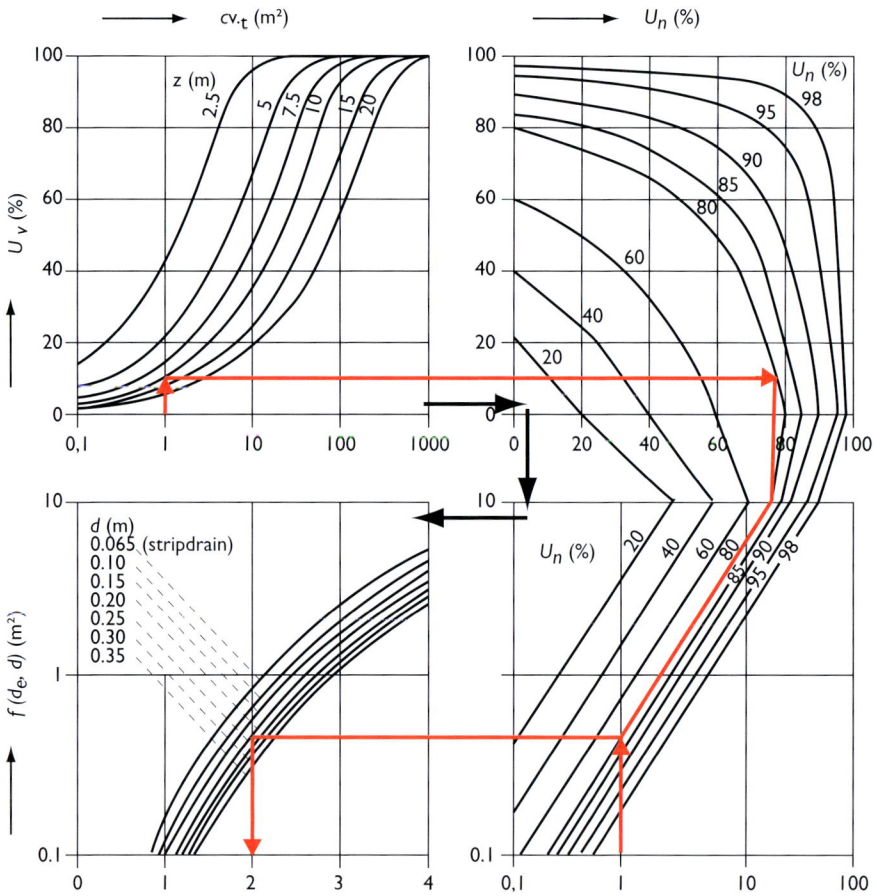

Figure D.8    *Graphical PVD design method in which both horizontal and vertical consolidation are assumed (CUR 162, 1996).*

### D.3.5 *Combination of horizontal consolidation and vertical consolidation – Carillo*

When vertical drains are used and vertical and horizontal consolidation occurs, the average degree of consolidation $U_{vh}$ can be calculated using Carillo's equation:

$$U_{vh} = U_v + U_h - U_v \cdot U_h$$

Figure D.7 represents a nomogram wich can be used to design PVD's, taking into account both horizontal and vertical consolidation. In this monogram smear effects are not taken into account. For the vertical consolidation, unilateral drainage has been assumed. and one way vertical drainage is valid.

**Example on the use of design chart Figure D.8.**

Input data:

required degree of consolidation = 80%

available consolidation period = 6 months

layer thickness soft soil layer = 10 m

$c_v = 2$ m²/y, $c_h = 2$ m²/y.

Drains with an equivalent width d = 0.065 mm will be used.

Calculation:

$c_v \cdot$ t $= 2 \cdot 0.5 = 1$ m²

In the upper left figure $U_v$ is (approximately) 11%.

From the upper right figure (for $U_{vh} = 80\%$) is found (approximately) $U_h = 77\%$.

$c_h \cdot t = 2 \cdot 0.5 = 1$ m²

The intersection of $U_h$ with the vertical line through $c_h \cdot t = 1$ is transferred horizontally to the lower left figure and the intersection with the curve for $d = 0.065$ gives the equivalent soil column diameter $d_e = 1.7$ m.

Result:

The drain spacing for a triangular grid is 1.7/1.05 = 1.6 m.

### D.3.6 *Secondary compression*

There are different theories about secondary consolidation. One approach suggests that secondary consolidation only starts at the end of primary consolidation, the other states that secondary consolidation is a different process to primary consolidation and also starts when the load is applied. The designer should be aware of these different theories, since this should also be reflected in way the consolidation parameters are derived from the consolidation tests.

The secondary compression can be calculated by means of the following formula (Mesri)

$$\frac{\Delta h_s}{h_p} = \frac{C_\alpha}{(1+e_p)} \times \log\left(\frac{t_f}{t_p}\right)$$

where:

$\Delta h_s$ = secondary compression of layer with thickness $h_p$ [m]
$h_p$  = thickness of the considered layer after occurrence of the primary settlement [m]
$C_\alpha$ = secondary compression index [–]
$e_p$  = void ratio of layer with thickness $h_p$ (after primary settlement) [–]
$t_p$  = time at end of primary settlement (= full consolidation) [year]
$t_f$  = time at which the secondary compression has to be calculated [year]

The Mesri approach assumes that the secondary compression only starts when the primary consolidation is completed. To calculate the secondary settlements, the "end of primary settlement" is assumed being the time when the primary consolidation reaches approximately 95%. $h_p$ and $e_p$ are calculated from the primary consolidation results.

The secondary compression is not defined by the magnitude of the loading. However, theoretically, $C_\alpha$ needs to be derived from compression testing in the relevant loading range.

When the secondary compression index is unknown, $C_\alpha$ can be defined from the ratio $C_\alpha/C_c$. Some correlations for several soil types can be found in the literature. (for instance Mitchell, 1993):

$C_\alpha/C_c = 0.04 \pm 0.01$ for inorganic clays and silts
$C_\alpha/C_c = 0.05 \pm 0.01$ for organic clays and silts
$C_\alpha/C_c = 0.075 \pm 0.01$ for peats
$C_\alpha/C_c = 0.015$ to $0.03$ for clean sands

# REFERENCES

Abduljauwad, S.N. and Al-Amoudi, O.S.B. Geotechnical behaviour of saline sabkha soils, *Géotechnique*, 45(3): 425–445, 1995.

Al-Amoudi, O.S.B., Abduljauwad, S.N., El-Naggar, Z.R. and Safar, M.M. Geotechnical considerations on field and laboratory testing of sabkha, Proceedings, Symposium on recent advances in geotechnical engineering III, Singapore, 1–6, 1991.

Almeida, M.S.S., Jamiolkowksi, M. and Peterson, R.W. Preliminary results of CPT tests in calceareous Quiou Sand, Proceedings of the first international symposium on calibration chamber testing, Potsdam, New York, 28–29, June 1991.

Arulanandan, K., and Scott, R.F. Verification of numerical procedures for the analysis of soil liquefaction problems. Proceedings of the International Conference on the Verification of Numerical Procedures for the Analysis of Soil Liquefaction Problems, Vols. 1 and 2, A. A. Balkema, Rotterdam, the Netherlands, 1993.

Asaoka, A. Observational procedure of settlement prediction, Soils and Foundations, 18(4): 87–101, 1978.

ASTM International Standard D 1889–00 Standard Test Method for Turbidity of Water, Section 11 Water and Environmental Technology, Volume 11.01 Water. ASTM Stock Number: S110106, 2006.

Bagnold, R.A. The Physics of Blown Sand and Desert Dunes, Chapman and Hall, London, 194.1.

Bahill, A.T. and Gissing, B. Re-evaluating systems engineering concepts using systems thinking, IEEE Transaction on Systems, Man and Cybernetics, Part C: Applications and Reviews, 28(4): 516–527, 1998.

Baldi, G., Bellotti, R., Ghionna, V., Jamiołkowski, M. and Pasqualini, E. Interpretation of CPTs and CPTUs; 2nd part: drained penetration of sands. IV International Geotechnical Seminar, Singapore, 143–156, 1986.

Been, K. and Jefferies, M.G. A state parameter for sands, *Géotechnique*, 35(2): 99–112, 1985.

Been, K. and Jefferies, M.G. Towards Systematic CPT Interpretation. Proceedings Wroth Memorial Symposium, Thomas Telford, London, 121–134, 1992.

Belotti and Jamiolkowksi. Evaluation of CPT and DMt in crushable and silty sands, third interim report ENEL C.R.I.S., Milan, Italy, 1991.

Beuth, L., Benz, T., Vermeer, P.A. and Wieckowski, Z. Large deformation analysis using a quasi-static material point method. Journal of Theoretical and Applied Mechanics, Sofia, 38(1–2): 45–60, 2008.

Bjerrum, L. The 3rd Terzaghi lecture: Progressive failure in slopes in over-consolidated plastic clay and clay shales. J. Soil Mech. Found. Div. 93(SM5), 1–49, 1967.

Bjerrum, L. Embankments on Soft Ground. Proceedings, Conference on Performance of Earth and Earth-Supported Structures, ASCE, Vol. II, 1–54, 1972.

Borja, R.B., Duvernay, B.G. and Lin, C-H. Ground Response in Lotung: Total Stress Analyses and Parametric Studies. *Geotechnical and Geoenvironmental Engineering,* ASCE, 128(1): 54–63, 2002.

Boulanger, R.W. High overburden stress effects in liquefaction analyses, Journal of Geotechnical and Geo-environmental Engineering, ASCE, 129(12): 1071–1082, 2003.

Boulanger, R.W. and Idriss, I.M. State normalization of penetration resistances and the effect of overburden stress on liquefaction resistance, in Proceedings of the 11th International Conference on Soil Dynamics and Earthquake Engineering, and 3rd International Conference on Earthquake Geotechnical Engineering, D. Doolin et al., eds., Stallion Press, Vol. 2, 484–491, 2004.

Boulanger, R.W. and Idriss, I.M. Cyclic Failure and Liquefaction: Current Issues. Proc. 5th Int. Conf. on Earthquake Geotech Eng., 2011. http://www.5icege.cl/images/stories/congress_sessions/lectures/state_of_the_art/4-Boulanger-SOA-5ICEGE.pdf

Bowles, J.E. Foundation analysis and design, 5th edition, 1997.

Bray, J.D. and Travasarou, T. Simplified Procedure for Estimating Earthquake-Induced Deviatoric Slope Displacements, Journal of Geotechnical and Geoenvironmental Engineering, ASCE, 133(4): 381–392, April 2007.

Bray, R.N. Environmental Aspects of Dredging, Taylor & Francis, 2008.

Bray, R.N., Bates, A.D. and Land, J.M. Dredging, a handbook for engineers, second edition, Butterworth-Heinemann, Oxford 1997.

Brown, R.E. and Glenn, A.J. Vibroflotation and terraprobe comparison, Journal of the Geotechnical Engineering Division, ASCE, Vol. 102, 1976.

Buchan, S. and Smith, D.T. Deep-sea sediment compression curves: some controlling factors, spurious overconsolidation, predictions, and geophysical reproduction, Marine Georesources and Geotechnology, 17(1), 1990.

Budhu, M. Soil Mechanics & Foundations, Wiley, 2010.

Burnaman, M.D., Withkers, K.D. and Wolf, D.J. Integration of well and seismic data using geostatistics. In: Chambers, R.L. et al. (eds), Stochastic Modeling and Geostatistics, Principles, methods and case studies. Oklahoma, 177–200, 1994.

Carman, P.C. Flow of gases through porous media, Butterworths Scientific Publications, London 1956.

Carrier, W.D. and Bromwell, L.G. Logan airport revisited. Sedimentation Consolidation Models, Proc. Symp. Geotechnical Eng. Div. ASCE, 344–355, 1984.

Casagrande, A. The determination of the preconsolidation pressure and its significance, IICSMFE Cambridge, Vol. 3, 1936.

Casagrande, A. Soil mechanics in the design and construction of the Logan airport, Journal of the Boston Society of Civil Engineers, 36(2): 192–205, 1949.

Cetin, K.O., Seed, R.B., Kiureghian, A.D., Tokimatsu, K., Harder, L.F., Kayen, R.E. and Moss, R.E.S. Standard penetration test-based probabilistic and deterministic assessment of seismic soil liquefaction potential, Journal of Geotechnical and Geo-environmental Engineering, ASCE, 130(12): 1314–1340, 2004.

Clark, A.R. and Walker, B.F. A proposed scheme for the classification and nomenclature for use in the engineering description of Middle East sedimentary rocks, *Géotechnique*, Vol. 27, 1977.

Chiles, J.-P. and Delfiner, P. Geostatistics, Modeling Spatial uncertainty, Wiley Series in Probability and statistics. New York: Wiley, 1999.

Chinese Code of Practice for Installation of Prefabricated Drains and the Quality Inspection Standard for Prefabricated Drains (JTJ/T256-96 1996, JTJ/T257-96), 1996.

Chow, V.T., Maidment, D.R. and Mays, L.W. Applied Hydrology, McGraw Hill, 1988.

Chu, J., Leroueil, S. and Leong, W.K. Unstable behaviour of sand and its implication for slope instability, Canadian Geotechnical Journal, 40(5): 873–885, 2003.

Chu, J., Varaksin, S., Klotz, U. and Mengé, P. Construction processes, State-of-the-Art Report (TC 17, ISSMGE), Proc. 17th Int. Conference on Soil Mechanics and Geotechnical Engineering, Alexandria (Egypt), 2009.

Chu, J., Leong, W.K., Loke, W.L. and Wanatowski, D. Instability of loose sand under drained conditions. Geotechnical and Geoenvironmental Engineering, ASCE, 138(2): 207–216, 2012.

CIRIA, CUR, CETMEF. The use of rock in hydraulic engineering (2nd edition). The Rock Manual, CIRIA, London, 2007.

Coop, M.R. The mechanics of uncemented carbonate sands, *Géotechnique*, 40(4), 1990.

Coop, M.R. and Airey, D.W. Carbonate sands, Characterization and Engineering Properties of Natural Soils, Tan et al. (eds), 2003.

Coop, M.R., Sorensen, K.K., Bodas Freitas, T. and Georgoutsos, G. *Particle breakage during shearing of a carbonate sand. Géotechnique*, 54, 2004.

CUR/CIRIA/CETMEF, Rock Manual, C683, 2007.

CUR Recommendation 113 Slope stability at sand borrow pits (in Dutch), CUR, Gouda (Netherlands), 2008.

CUR report 130, Manual on Artificial Beach nourishment, Gouda, 1987.

CROW publication 281, Materialen in (constructieve) ophoging en aanvulling—Richtlijn ter beoordeling van alternatieven voor zand, (publication only available in Dutch) "Materials for (constructive) fill—Guidelines for the assessments of alternatives materials replacing sand", 2009.

Das, B.M. Principles of foundation engineering, fourth edition, California State University, Sacramento, 2009.

Davis, J.C. Statistics and data analysis in geology. Second edition. Canada: John Wiley & Sons, 1986.

Day, Geotechnical Earthquake Engineering Handbook, ISBN 0-07-137782-4, 2009.

Delay and Disruption Protocol, Society of Construction Law, October 2002, ISBN 0-9543831-1-7

De Groot, M.B. Liquefaction, flow and settlement of sand in dredging, hydraulic fill and flow slides, International Conference on Civil and Environmental Engineering, University of Hiroshima, 2004.

De Groot, M.B., Bolton, M.D., Foray, P., Meijers, P., Palmer, A.C., Sandven, R., Sawicki, A. and Teh, T.C. Physics of liquefaction phenomena around marine structures, Journal of waterway, port, coastal and ocean engineering, ASCE, 132(4): 227–243, 2006.

De Groot, M.B., Van der Ruyt, M., Mastbergen, D.R. and Van den Ham, G.A. Bresvloeiing in zand, Geotechniek (in Dutch), 34–39, July 2009.

Delay and disruption Protocol. Society of Construction Law, Oxfordshire, England, October 2002.

Den Haan, E.J. and Sellmeyer, H.J.B. Calculation of Soft Ground Settlement with an Isotache Model, ASCE, 2000.

Deutsch, C.V. and Journel, A.G. GSLIB Geostatistical Software Library and User's Guide. New York: Oxford university Press, 1992.

Dorgunoglu and Mitchell. *Static penetration resistance of soils* I-II. Proceedings of the ASCE Specialty Conference on in situ measurement of soil properties, Raleigh, North Carolina, 1, 151–189, 1975.

Douglas, B.J. and Olsen, R.S. *Soil classification using electric cone penetrometer. Cone Penetration Testing and Experience.* Proceedings of the ASCE National Convention, St. Louis, 209–227, ASCE, 1981.

Dredging News Online, Wreck removal demands sophisticated surveys, http://www.sandandgravel.com/news/article.asp?v1=7631, April 2000.

Drnevich, V., Evans, A.C. and Prochaska, A. A study of effective soil compaction control of granular soils, Joint Transportation Research Program, Project No. C-36-3600, School of Civil Engineering, Purdue University, 2007.

Duncan, J.M. and Buchignani, A.L. *An engineering manual for slope stability studies* University of california, Department of civil engineering, 1975.

Duncan, J.M. and Wright, S.G. Soil Strength and Slope Stability, Wiley, 2005.

Dunnicliff, J. Geotechnical instrumentation for monitoring field performance, John Wiley & Sons, 1993.

DWW-Rijkswaterstaat, Dienst weg- en waterbouwkunde—Richtlijn ophogen met klei uit baggerspecie, (publication only available in Dutch "Guidelines for Filling with dredged clay"), 2005.

Fahey, M. Shear modulus of cohesionless soil: variation with stress and strain level, Canadian Geotechnical Journal, 29, 157–161, 1992.

Fardis, et al. Designers' guide to Eurocode 8: Design of structures for earthquake resistance, Thomas Telford, ISBN: 978-0-7277-3348-1, 2005.

FIDIC Blue book, Form of contract for dredging and reclamation works, FIDIC, ISBN 2-88432-045-6, 2006.

FIDIC Red book, Conditions of contract for Construction for building and engineering works designed by the Employer. ISBN 2-88432-022-9, FIDIC, 1999.

FIDIC Yellow book, Conditions of contract for plant and design-build for electrical and mechanical plant, and for building and engineering works, designed by the Contractor, ISBN 2-88432-023-7, FIDIC, 1999.

FIDIC Silver book, Conditions of contract for EPC/Turnkey projects, FIDIC 1999.

Floss, R. et al. Vergleich der Verdichtungs- und Verformungseigenschaften grobkörniger bindiger Mischböden, Bundesanstalt für Strassenwesen, Wissenschaftliche Bericht der Bundesanstalt für Strassenwesen, Heft 6, 1968.

Fookes, P.G. The geology of carbonate soils and rocks and their engineering characterisation and description, Engineering for calcareous sediments, R.J. Jewell & M.S. Khorshid, eds, Balkema Rotterdam, Vol. 2, 1988.

Fookes, P.G., French, W.J. and Rice, M.M. The influence of ground and groundwater chemistry on construction in the Middle East, Quarterly Journal of Engineering Geology, Vol. 18, 1985.

Galloway, D., Jones, D.R. and Ingebritsen, S.E. Land Subsidence in the United States, U.S. Geological Survey Circular 1182, 2005.

Geo-Slope, SLOPW/W manual 2012, www.geo-slope.com/products/slopew.aspx

Gohl, W.B., Jefferies, M.G., Howie, J.A. and Diggle, D. Explosive compaction: design, implementation and effectiveness. *Géotechnique*, 50(6): 657–665, 2000.

Gibbs, H.J. Properties which divide loose and dense uncemented soils, Earth laboratory report, No. EM-608, United States Department of the Interior, Bureau of Reclamation, Denver Colorado, 1961.

Gibbs, H.J. and Holtz, W.G. Research on determining the density of sands by spoon penetration testing, Proceedings 4th international conference on soil mechanics and foundation engineering, Vol. 1, 35–39, London, 1957.

Golightly, C.R. 1989, *Engineering Properties of Carbonate Sands.* PhD. Thesis, University of Bradford.

Goovaerts, P. Geostatistics for natural resources evaluation. New York: Oxford University Press, 1997.

Gorelick, S.M. and Kolterman, C.E. Heterogeneity in sedimentary deposits: a review of structure-imitating, process-imitating, and descriptive approaches'. Water resources research, 32(9): 2617–2658, 1996.

Graaff, J. Van de. Coastal Morphology and coastal protection, draft, September 2006.

Gudehus, G. and Cudmani, R.O. Experimental investigation of the cone penetration resistance of Dubai sand in a large calibration chamber, Consulting report for Keller Grundbau GmbH, unpublished, 2004.

Hamada, M. "Similitude law for liquefied-ground flow", Proceedings of the 7th U.S.-Japan Workshop on Earthquake Resistant design of lifeline facilities and countermeasures against soil liquefaction, 191–205, 1999.

Hamidi, B., Nikraz, H., Yee, K., Varaksin, S. and Wong, L.T. Ground improvement in deep waters using dynamic replacement. Proc. 20th Int. Offshore and Polar Engineering Conf., Beijing, China, 2010.

Handboek Bodemsaneringstechnieken, deel 4, Sdu Publishers, December 2000.

Handboek Zandboek, CROW, Ede, April 2004.

Hardin, B.O. Crushing of soil particles, Journal of Geotechnical Engineering, Proc. ASCE, 111(10), 1985.

Hazen, A. Discussion of 'Dams on sand foundations by A.C. Koenig, Trans. Am. Soc. Civ. Eng., 73, 1911.

Herbich, J.B. Handbook of dredging engineering, 1992.

Hicks, M.A. and Onisiphorou, C. Stochastic evaluation of static liquefaction in a predominantly dilative sand. *Géotechnique*, 55(2): 123–133, 2005.

Holtz, W.G. and Gibbs, H.J. Discussion of SPT and relative density in coarse sand, Journal of the geotechnical engineering division, ASCE, 105(GT3): 439–441, 1979.

Hosgit, E. and Royle, A.G. Local estimation of sand and gravel reserves by geostatistical methods. Transactions of the Institute of Mining and Metallurgy, Section A, 83, A53–A62, 1974.

IADC. Meeting the challenge of making more land for more people, Terra et Aqua, No. 100, September 2005.

ICE Manual of geotechnical engineering, Edited by John Burland, Tim Chapman, Hilary Skinner and Michal Brown. Volume 1 and 2, 2012.

Idriss, I.M. An update to the Seed-Idriss simplified procedure for evaluating liquefaction potential, in Proceedings of the TRB Workshop on New Approaches to Liquefaction, Publication No. FHWA-RD-99-165, Federal Highway Administration, January 1999.

Idriss, I.M. and Boulanger, R.W. Semi-empirical procedures for evaluating liquefaction potential during earthquakes, in Proceedings of the 11th International Conference on Soil Dynamics and Earthquake Engineering, and 3rd International Conference on Earthquake Geotechnical Engineering, D. Doolin et al. eds., Stallion Press, Vol. 1, 32–56, 2004.

Idriss and Boulanger. Soil Liquefaction during Earthquakes, EERI Monologue, 2008.

Idriss, I.M. and Boulanger, R.W. Soil liquefaction during earthquakes, EERI publication No. MNO-12, Earthquake Engineering Research Institute (EERI), Oakland, USA, 2008.

Imanishi. Design of prefabricated vertical drain method on reclaimed marine clay, 2000.

Institution of Civil Engineers, Specification for ground treatment, London, 1987.

Ishihara, K. Liquefaction and flow failure during earthquake. *Géotechnique*, 43(3): 349–416, 1993.

Ishihara, K., Verdugo, R. and Acacio, A.A. Characterization of cyclic behaviour of sand and post-seismic stability analyses, in Proceedings of the 9th Asian Regional Conference on Soil Mechanics and Foundation Engineering, Bangkok, Thailand, Vol. 2, 45–70, 1991.

Ishihara, K. and Yoshimine, M. Evaluation of settlements in sand deposits following liquefaction during earthquakes. Soils and Foundations, 32(1): 173–188, March 1992.

ISO 7027: Water quality—Determination of Turbidity, 1999.

ISSMGE. Geotechnical & geophysical investigations for offshore and nearshore developments: (Ref. www.issmge.org), 2005.

Jaky, J. The coefficient of earth pressure at rest, Journal of Hungarian Archit. Engrs, 22, 1944.

Jamiolkowski, M. et al. ICSMGE New developments in field and laboratory testing of soils, 1985.

Jamiolkowski, D., Lo Presti and Manassero, M. Evaluation of relative density and shear strength of sands from CPT and DMT. ASCE—Geotechnical special publication No. 119 Soil behavior and soft ground construction, 2001.

Jefferies, M. and Been, K. Static liquefaction: a critical state approach, First Edition, Taylor & Francis, United Kingdom, 2006.

Jefferies, M. and Shuttle, D.A. Understanding liquefaction through applied mechanics. 5th International Conference on Earthquake Geotechnical Engineering, Santiago, Chile, 10–13 January 2011.

Jewell, R.J. and Andrews, D.C. (eds). Proc. Int. Conf. on Calcareous Sediments, Perth, Australia, 1988.

Karlsson, R. and Hansbo, S. Soil classification and identification. Byggforskningsrådet, Document D8, Stockholm, 1989.

Karthikeyan, M. Ganeswara, R.D. and Tan, T.S. In situ characterization of land reclaimed, using big clay lumps. Canadian Geotechnical Journal. No. 41, 242–256, 24(2): 145–156, 2004.

Kempfert, H.G. Ground improvement methods with special emphasis on column-type techniques, Int. Workshop on Geotechnics of Soft Soils-Theory and Practice, Vermeer, Schweiger, Karstunen and Cudny (eds.), 2003.

Kitazume, M. The sand compaction pile method, Taylor & Francis, 2005.

Kramer, S.L. Geotechnical earthquake engineering, Prentice-Hall, New Jersey, USA, 1st edition, 653, 1996.

Kwag, J.M., Ochiai, H. and Yasufuku, N. Yielding stress characteristics of carbonate sand in relation to individual particle fragmentation strength, Engineering for Calcareous Sediments, Bahrain, Al-Shafei (ed.), Balkema, 1999.

Lade, P.V. and Yamamuro, J.A. Evaluation of static liquefaction potential of silty sand slopes. Canadian Geotechnical Journal, 48(2): 247–264, 2011.

Lambe, T.W. and Whitman R.V. *Soil Mechanics*, Wiley, 1969.

Lekkerkerk, H.J. Handbook of Offshore Surveying. Clarkson Research Services Ltd., 2006.

Leroueil, S., Kabbaj, M., Tavenas, F. and Bouchard, R. Stress–strain–strain rate relation for the compressibility of sensitive natural clays. *Géotechnique*, 35(2): 159–180, 1985.

Leshchinsky, D. and Richter, S.D. Dredging research program/US army corps of engineers, Hydraulically transported clay balls, 1994.

Leung, C.F., Wong, J.C., Manivanann, R. and Tan, S.A. Experimental evaluation of consolidation behaviour of stiff clay lumps in reclamation fill. Journal of Geotechnical Testing, 2001.

Leung, C.F., Wong, J.C. and Tan, S.A. Assessment of lumpy fill profile using miniature conepenetrometer. Proceeds of the Fourteenth International Conference on Soil Mechanics and Foundation Engineering. Hamburg, Vol. 1, 527–530, 1997.

l-Homoud, A.S. and Wehr, W. Experience of vibrocompaction in calcareous sand of UAE, Geotechnical and Geological Engineering, No. 24, 2006.

Life06 EMV/FIN/00195, Stable, Technical final report, Controlled treatment of TBT-contaminated dredged sediments for the beneficial use in infrastructure applications. Case: Aurajoki—Turku, Finland (http://projektit.ramboll.fi/life/stable/), 2009.

Li, X-S., Dafalias, Y.F. and Wang, Z-L. State dependent dilatancy in critical state constitutive modelling of sand. Canadian Geotechnical Journal 36, 599–611, 1999.

Lunne, T., Robertson, P.K. and Powell, J.J.M. Cone Penetration Testing in Geotechnical Practice, E & FN Spon, 1997.

Lunne, T. *Guidelines for use and interpretation of CPT in hydraulically constructed fill*. Report No. 20041367-3, NGI, 2006.

Makdisi, F.I. and Seed, H.B. Simplified Procedure for Estimating Dam and Embankment Earthquake Induced Deformations, Journal of Geotechnical Engineering Division, ASCE, 107(GT7): 849–867, 1978.

Manivannan, R., Leung, C.F. and Tan, S.A. Analysis of lumpy fill. Coastal Geotechnical, Engineering in Practice. Nakase and Tsuchida (eds). Balkema, Rotterdam, ISBN 90 5809 1511, 2000.

Manzari, M.T. and Dafalias, Y.F. A critical state two-surface plasticity model for sands. *Géotechnique*, 47, 255–272, 1997.

Marcuson, W.F.H., Hynes, M.E. and Franklin, A.G. Evaluation and use of residual strength in seismic safety analysis of embankments. *Earthquake Spectra*, 6(3): 529–572, 1990.

Massarsch, K.R. Deep Soil Compaction using Vibratory probes, ASTM symposium on design, construction and testing of deep foundation improvement: stone columns and related techniques, ASTM STP, R.C. Bachus Edt, 297–319, 1991.

Massarsch, K.R. Effects of vibratory compaction, Vibratory Pile Driving and Deep Soil Compaction—TRANSVIB2002, Holeyman, Vanden Berghe and Charue Edts., Swets and Zeitlinger, 33–42, 2002.

Massarsch, K.R. and Fellenius, B.H. Vibratory compaction of coarse-grained soils, Canadian geotechnical Journal, 39(3), 2002.

Mathijssen, F.A.J.M., Boylan, N., Molenkamp, F. Leroueil, S. and Long, M. Material behaviour and constitutive modelling of organic soils. *CEDA Dredging days*. Dredging tools for the future. Ahoy Rotterdam, the Netherlands. CEDA, 2009.

Massarsch, K.R. and Fellenius, B.H. Deep vibratory compaction of granular soils, Ground Improvement-Case Histories, Elsevier publishers, B. Indranatna and C. Jian, Edt, 2005.

Massarsch, K. Rainer, and Bengt, H. Fellenius Vibratory compaction of coarse-grained soils. Canadian Geotechnical Journal, No. 39, 695–709, 2002.

Mays, L.W. Water Resources Engineering, Edition, John Wiley & Sons, Inc, 2005.

Mecsi, J., Gölkalp, A. and Düzceer, R. Densification of hydraulic fills by Vibroflotation technique, Proc. 16th Int. Conf. Soil Mechanics and Geotechnical Engineering, Millpress, Rotterdam, 2005.

Meigh, A.M. Cone penetration testing methods and interpretation. CIRIA Ground Engineering Report B2: In-situ Testing. CIRIA, 1987.

Menge, P. Surface compaction of hydraulic fill of limited thickness, Presentation made at TC 17 Workshop at 14ECSMGE, Madrid, TC 17 website: www.bbri.be/go/tc17, 2007.

Meyerhof, G.G. *The ultimate bearing capacity of foundations*, *Géotechnique*, 2(4): 301–331, 1951.

Meyerhof G.G. *The bearing capacity of foundations under eccentric and inclined loads*, 3rd International conference on soil mechanics and foundation engineering, Zurich, Switzerland, 440–445, 1953.

Meyerhof, G.G. *Some recent research on the bearing capacity of foundations,* Canadian Geotechnical Journal, 1(1): 16–26, 1963.

Miedema, S.A. and Ramsdell, R.C. Hydraulic transport of sand/shell mixtures in relation with the critical velocity, Terra et Aqua, No. 122, 2011.

Mitchell, J.M. and Jardine, F.M. A guide to ground treatment, CIRIA publication C573, CIRIA, 2002.

Molnar, S., Cassidy, J.F., Monahan, P.A. and Dosso, S.E. Comparison of geophysical shear-wave velocity methods, Ninth Canadian Conference on Earthquake Engineering, Ottawa, June 2007.

Moseley, M.P. and Kirsch, K. Ground Improvement, Second Edition. Spon Press, New York, 2004.

Nagase, H. and Ishihara, K. Liquefaction-induced compaction and settlement of sand during earthquakes. Soils and Foundations, JSSMFE, 28(1): 65–76, 1988.

Naifeng, H., In Fookes, P.G. and Perry, RHG. (eds.) Engineering characteristics of arid soils, Balkema, Rotterdam, 29–34, 1994.

Newmark. Effects of Earthquakes on Dams and Embankments, London, *Géotechnique*, 15(2): 139–160, 1965.

Ngan-Tillard, D., Haan, J., Laughton, D., Mulder, A. and Nooy van der Kolff, A. Index test for the degradation potential of carbonate sands during hydraulic transport, Engineering Geology 108, 2009.

Nooy van der Kolff, A.H. Engineering geology of (hyper)arid regions, Proceedings Symposium An Overview of Engineering Geology in The Netherlands, Delft University of Technology, 1993.

Nooy van der Kolff, A.H., Lesemann, D. and Petereit, K. The use of dredged sludge as a fill in the Osthafen, Bremerhaven, Germany, Proc. PIANC MMX Congress Liverpool, UK, 10–14 May 2010.

Nutt, N.R.F. and Houlsby, G.T. Calibration tests on the cone pressuremeter in carbonate sand, Calibration chamber testing (ed. A.B. Huang), Elsevier Science Publishing Company, 1991.

Nutt, N.R.F. Development of the cone pressuremeter. PhD thesis, University of Oxford, 1993.

Olgun, C.G., Martin, J.R. and LaVielle, T.H. Liquefaction susceptibility of calcareous sediments along the coastal plains of Puerto Rico, Department of Civil and Environmental Engineering, Virginia Polytechnic Institute and Stae University, Virginia, USA, 2009.

Olsen, S.M. and Stark, T.D. Yield strength ratio and liquefaction analysis of slopes and embankments, Journal of Geotechnical and Geo-environmental Engineering, 129(8): 727–737, 2003.

Olson, S.M. Liquefaction analysis of level and sloping ground using field case histories and penetration resistance, Ph.D. Thesis, University of Illinois-Urbana-Campaign, Urbana, IL, USA, 2001.

Olson, S.M. and Stark, T.D. Liquefied strength ratio from liquefaction flow failure case histories. *Canadian Geotechnical Journal*, 39, 629–647, 2002.

Peck, Hanson W.E. and Thornburn, T.H. *Foundation engineering handbook*, Wiley, 1974.

PIANC Report of Working Group No. 23, Site investigation requirements for dredging works, Brussels, 2000.

PIANC, Seismic design guidelines for port structures, PIANC Report of working group 34, 2001.

Plewes, H.D., Davies, M.P. and Jefferies, M.G. CPT based screening procedure for evaluating liquefaction susceptibility. Proc 45th Canadian Geotechnical Conference, Toronto, 1992.

Potts, D.M. and Zdravkovic, L. Finite Element Analysis in Geotechnical Engineering: Theory, Telford, 1999.

Poulos, S.J., Castro, G. and France, J.W. Liquefaction evaluation procedure, Journal of Geotechnical Engineering, ASCE, 111(6): 772–792, 1985.

Powers, M.C. A new roundness scale for sedimentary particles, Journal of Sedimentary Research, 23(2), 1953.

Price, D.G. Introduction Engineering Geology, q38. Part I. Delft University of Technology, 135–159. Delft, 1986.

Priebe, H.J. The design of vibro replacement, Reprint from Ground Engineering, Technical paper GT 07-13 E, December 1995.

Puech, A. and Foray, P. Refined Model for interpreting shallow penetration CPTs in sands. Offshore technology conference, Houston, Texas, 6–9 May 2002.

Raithel, M., Kempfert, H.G. and Kirchner, A. Geotextile-Encased Columns (GEC) for Foundation of a Dike on Very Soft Soils, Würzburg, 2002.

Robertson, P.K. Soil classification using the cone penetration test, Canadian Geotechnical Journal 27(1): 151–158, 1990.

Robertson, P.K. Suggested terminology for liquefaction, Proc. 47th Canadian Geotechnical Conference, Halifax: conference preprints, Canadian Geotechnical Society, 277–286, 1994.

Robertson, P.K. Discussion of Liquefaction potential of silts from the CPTu. *Canadian Geotechnical Journal*, 45(1): 140–141, 2008.

Robertson, P.K. CPT interpretation—a unified approach, *Canadian Geotechnical Journal*, 46: 1–19, 2009.

Robertson, P.K. Estimating in situ state parameter and friction angle in sandy soils from CPT. *Proc. 2nd intern. symp. on cone penetration testing*, Huntington Beach, 9–11 May 2010.

Robertson, P.K. Evaluation of flow liquefaction and liquefied strength using the cone penetration test. J. Geotech. and Geoenvir. Engrg. 136(6): 842–853, 2010.

Robertson, P.K. and Campanella, R.G. Interpretation of cone penetration tests, Canadian geotechnical journal, 20(4), 1983.

Robertson, P.K., Campanella, R.G., Gillespie, D. and Grieg, J. Use of piezometer cone data. Proceedings, In Situ '86, ASCE Specialty Conference on use of in situ tests in geotechnical engineering, Geotechnical special publication 6, 1263–1280, Blacksburg, VA, ASCE, 1986.

Robertson and Wride. Evaluating cyclic liquefaction potential using the cone penetration test, Can. Geotechn. J., Ottawa, 35(3): 442–459, 1998.

Robinson, R.G., Tan, T.S., Dasari, G.R., Leung, C.F. and Vijayakumar, A. Experimental Study of the Behaviour of a Lumpy Fill of Soft Clay. International Journal of Geomechanics. ASCE, 5(2): 125–137, 2005.

Rock Manual. The use of rock in coastal engineering, CIRIA 683, London 2007.

Rogers, B.T., Graham, C.A. and Jefferies, M.G. Compaction of hydraulic fill sand in Molikpaq core. Proc. 43rd Canadian Geotechnical Conference, Canadian Geotechnical Society 2, 567–575, 1990.

Sandilands, M. and Swan, A.R.H. Introduction to geological data analysis. Oxford: Blackwell Science, 1995.

Sadiq, A.M. and Nasir, S.J. Middle Pleistocene karst evolution in the State of Qatar, Arabian Gulf, Journal of Cave and Karst Studies, 64(2), 2002.

Samson, L. and La Rochelle, P. Design and performance of an expressway constructed over peat by preloading. Can. Geotech. J. 9(4): 447–466, 1972.

Sanglerat, G. *The penetrometer and soil exploration*. Elsevier, Amsterdam, 1972.

Sanin, M.V. and Wijewickreme, D. Cyclic shear response of channel-fill FraserRiver Delta silt. *Soil Dynamics and Earthquake Engineering,* 26(9): 854–869, 2006.

Schellmann, W. An introduction in Laterite, http://www.laterite.de, August 2007.

Schiereck, G.J. Introduction to bed, bank and shore protection, Delft University Press, 2001.

Schmertmann, J.H. Guidelines for Cone Penetration test, Performance and design, Report no FHWA-TS-78-209 U.S, 145, Department of Transportation, Washington, DC, 1978.

Schnabel, P.B., Lysmer, J. and Seed, H.B. *SHAKE—A Computer Program for Earthquake Response Analysis of Horizontally Layered Site,* Report No. EERC 72-12, Earthquake Engineering Research Center, University of California, Berkeley, 1972.

Seed and De Alba. Use of SPT and CPT test for evaluating the liquefaction resistance of soils, Proc. Insitu, ASCE, 1986.

Seed, R.B., Cetin, K.O., Moss, R.E.S., Kammerer, A.M., Wu, J., Pestana, J.M., Riemer, M.F., Sancio, R.B., Bray, J.D., Kayen, R.E. and Faris, A. Recent advances in soil liquefaction engineering: A unified and consistent framework. *EERC-2003–06*, Earthquake Engineering Research Institute, Berkeley, Calif., 2003.

Seed and Harder. SPT-based Analysis of Cyclic Pore Pressure Generation and Undrained Residual Strength, Proc., H.B. Seed Memorial Symp., Vol. 2, BiTech Publishing, Vancouver, B.C., Canada, 351–376, 1990.

Seed, H.B., Martin and Lysmer. The generation and dissipation of pore water pressures during soil liquefaction, 1975.

Seed, H.B. and Idriss, I.M. Simplified procedure for evaluating soil liquefaction potential, Journal of Soil Mechanics and Foundations Division, ASCE, 97(SM9): 1249–1273, 1971.

Seed, H.B., Tokimatsu, K., Harder, L.F. Jr. and Chung, R. Influence of SPT procedures in soil liquefaction resistance evaluations, Journal of Geotechnical Engineering, ASCE, 111(12): 1425–1445, 1985.

Sego, D.C., Robertson, P.K., Sasitharan, S., Kilpatrick, B.L. and Pillai, V.S. Ground freezing and sampling of foundation soils at Duncan Dam. *Canadian Geotechnical Journal*, 31(6): 939–950, 1994.

Shrivastava, A.K. Nuclear Instruments and Methods in Physics Research Section A: Accelerators, Spectrometers, Detectors and Associated Equipment, 539(1–2): 421–426, February 2005.

Shuttle, D.A. and Jefferies, M.G. Dimensionless and unbiased CPT interpretation in sand. International Journal of Numerical and Analytical Methods in Geomechanics, 22, 351–391, 1998.

Shuttle, D.A. and Cunning, J. Liquefaction Potential of Silts from CPTu, Canadian Geotechnical Journal, 44, 1–19, 2007.

Shuttle, D.A. and Cunning, J. Reply to the discussion by Robertson on 'Liquefaction Potential of Silts from CPTu', Canadian Geotechnical Journal, 45, 142–145, 2008.

Skempton, A.W. and Northey, R.D. The sensitivity of clays, *Géotechnique*, 3(1), 1953.

Skempton, A.W. The consolidation of clays by gravitational compaction, Q. J. Geol. Soc. 125, 1970.

Skempton, A.W. and Northey, R.D. The Sensitivity of Clays. *Géotechnique*, 3(1): 30–53, 1952.

Soil Dynamics and Earthquake Engineering, and 3rd International Conference on Earthquake Geotechnical Engineering, D. Doolin et al. (eds), Stallion Press, Vol. 2, 484–491.

Srivastava, R.M. An overview of stochastic methods for reservoir characterization. In: Chambers, R.L. et al. (eds), Stochastic Modeling and Geostatistics, Principles, methods and case studies. Oklahoma: AAPG, 3–16, 1994.

Stapelfeldt, T. Preloading and vertical drains, Helsinki University of Technology, Helsinki, 2006.

Stark and Olson. Liquefaction resistance using CPT and field case histories, Journal of Geotechnical Engineering, ASCE, 121(12): 856–869, 1995.

Stark, Olson, Kramer and Youd. Shear strength of liquefied soils. Proc. Geotechnical Earthquake Engineering and Soil Dynamics Specialty Conf., ASCE Geo-Institute Geotechnical Special Publication No. 75, August 3–6, Seattle, WA, Vol. 1, 313–324, 1998.

Stark and Mesri. Undrained shear strength of liquefied sands for stability analysis, Journal of Geotechnical Engineering, ASCE, 118(11): 1727–1747, 1992.

Stewart, H.R. and Hodge, W.E. Molikpaq core densification with explosives at Amauligak F-24. OTC 5684, Proc. 20th Annual Offshore Technology Conference, Houston, 1988.

Stokoe II, K.H., Wright, G.W., James, A.B. and Jose, M.R. Characterization of geotechnical sites by SASW method, in Geophysical characterization of sites, ISSMFE Technical Committee #10, edited by R.D. Woods, Oxford Publishers, New Delhi, 1994.

Stout, J.E. and Arimoto, R. Threshold wind velocities for sand movement in the Mescalero Sands of southeastern New Mexico, Journal of Arid Environments 74, Elsevier 2010.

Suzuki, Koyamada and Tokimatsu. Prediction of liquefaction resistance based on CPT tip resistance and sleeve friction, Proc. XIV Intl. Conf. on Soil Mech. and Foundation Engrg., Hamburg, Germany, 603–606, 1997.

Swedish Geotechnical Institute, Quick clay in Sweden, Report 65, 2004.

Tan, S.A. and Chew, S.H. Comparison of the hyperbolic and Asaoka observational method of monitoring consolidation with vertical drains. Soils and Foundations, 36(3): 31–42, 1996.

Tavenas, F., Leroueil, S., La Rochelle, P. and Roy, M. Creep behaviour of an undisturbed lightly overconsolidated clay. Can. Geotech. J. 15(3): 402–423, 1976.

Taylor, D.W. Fundamentals of Soil Mechanics. John Wiley, New York, 1948.

Techniques of Water-Resources Investigations, Book 9 Handbooks for Water Resources Investigations National Field Manual for the Collection of Water quality data, U.S. Geological Survey, 2008.

Terzaghi, K., Peck, R.B. and Mesri, G. Soil Mechanics in Engineering Practice, John Wiley and Sons, 1996.

Thiam-Soon Tan, Toshiyuki Inoue and Seng-Lip Lee. Hyperbolic Method for Consolidation Analysis, Journal of Geotechnical Engineering, 117(11): 1723–1737, 1991.

Tokimatsu and Seed. Evaluation of settlements in sand due to earthquake shaking, J. of Geotechnical Engineering, ASCE, 113(8): 861–878, 1987.

Tokimatsu and Yoshimi. Empirical correlation of soil liquefaction based on SPT N-value and fines content, Soils and Foundations, Vol. 23(4), December 1983.

Tsuchida, T. and Gomyo, M. Unified model of e-log p relationship with the consolidation of the effect of initial void ratio, Proc. Int. Symp. On Compression and Consolidation of Clayey Soils—IS—Hiroshima, Japan, 1995.

UBC. Uniform Building Code—Chapter 16 Structural Design Requirements, 1997.

U.S. Department of Agriculture, Natural Resources Conservation Service, Urban Hydrology for Small Watersheds, TR-55, 1986.

U.S. Army, Field Manual 5-410: Military Soils Engineering. June 1997.

Valsamis, A., Bouckovalas, G. and Dimitriadi, V. Numerical Evaluation of Lateral Spreading Displacements in Layered Soils. Proc. 4th International Conference on Earthquake Geotechnical Engineering, Thessaloniki: Paper No. 1644, 2007.

Van Impe, W.F. Soil Improvement Techniques and their Evolution, A.A. Balkema, 125, 1989.

Van Impe, W.F. et al. Cone resistance: quartz vs calcareous sand, 2001.

Van Impe, W.F., De Cock, F., Massarch, K.R. and Mengé, P. Recent experiences and developments of the resonance vibrocompaction technique, Proc. of the 13th Int. Conf. on Soil Mechanics and Foundation Engineering, 3, 1151–1156, 1994.

Vesic, A.S. Ultimate loads and settlement of deep foundations in sand. Proceedings of the symposium on bearing capacity and settlement of foundations in sand, Duke University, Durham, North Carolina, USA. 53–68, 1965 (cited by Wehr, 2005).

Vlasblom, W.J. Lecture notes: Designing dredging equipment, February 2004.

Waltham, A.C. and Fookes, P.G. Engineering classification of karst ground conditions, Quarterly Journal Engineering Geology Hydrogeology, 2003.

Waltham, A.C., Bell, F.G. and Culshaw, M.G. Sinkholes and subsidence, karst in cavernous rocks in engineering and construction, Springer, 2005.

Wasti, Y. and Bezirci, M.H. Determination of consistancy limits of soils by the fallcone test, Canadian Geotechnical Journal, Vol. 23, 1986.

Wehr, W.J. Influence of the carbonate content of sand on vibro compaction, 6th Conf. on Ground Improvement techniques, Coimbra, Portugal, 2005.

Weingart, W., Körner, M. and Saal, S. Investigation into the compaction in depth effect of a vibratory roller with a polygonal drum, in German only, Strasse und Autobahn, Kirschbaum verlag GmbH, Bonn, Germany, 2003.

White, D.J. *An investigation into the behaviour of pressed-in piles.* Ph.D dissertation, University of Cambridge, UK, Cambridge, 2002.

Wiegel, R.L. Oceanographic Engineering, Prentice Hall, Englewood Cliffs, N.J. 1964.

Winterwerp, J.C. and Van Kesteren, W.G.M. Introduction to the physics of cohesive sediment in the marine environment, 1996.

Wride, C.E., Robertson, P.K., Biggar, K.W., Campanella, R.G., Hofmann, B.A., Hughes, J.M.O., Küpper, A. and Woeller, D.J. In-Situ testing program at the CANLEX test sites. Canadian Geotechnical Journal, 37(3): 505–529, June 2000.

Yan, S.W., Chu, J., Fan, Q.J. and Yan, Y. Building a breakwater with prefabricated caissons on soft clay, Proc. of ICE, Geotechnical Engineering, 162(1), 2009.

Yang, L.A., Tan, T.S., Tan, S.A. and Leung, C.F. One-dimensional self-weight consolidation of a lumpy clay fill. *Géotechnique*, 52(10): 713–725, 2002.

Yang, L.A. and Tan, T.S. One-dimensional self-weight consolidation of a lumpy clay with non-linear properties, *Géotechnique*, 55(3): 227–235, 2005.

Youd, T.L. Factor Controlling maximum and minimum densities of sands, Evaluation of relative density and its role in geotechnical projects involving cohesionless soils, ASTM STP 523, American society for testing and materials, 98–112, 1973.

Youd, T.L. A Look Inside the Debate Over EERI Monograph MNO 12. Civil & Env. Eng, Brigham Young University, 2011. http://cgea.org/downloads/calgeo-ac-2011.pdf

Youd and Idriss, I.M. eds. Proceedings of the NCEER Workshop on Evaluation of Liquefaction resistance of Soils, Technical Report NCEER-97-022, 1997.

Youd, T.L., Idriss, I.M., Andrus, R.D., Arango, I., Castro, G., Christian, J.T., Dobry, R., Finn, W.D.L., Harder, L.F., Hynes, M.E., Ishihara, K., Koester, J.P., Liao, S.S.C., Marcuson, W.F., Martin, G.R., Mitchell, J.K., Moriwaki, Y., Power, M.S., Robertson, P.K., Seed, R.B. and Stokoe, K.H. Liquefaction resistance of soils: summary report from the 1996 NCEER and 1998 NCEER/NSF Workshops on evaluation of liquefaction resistance of soils. *Journal of Geotechnical and Geoenvironmental Engineering ASCE*, 127(10): 817–833, 2001.

Zhang, Robertson and Brachman, Estimating liquefaction-induced ground settlements from CPT for level ground, Canadian Geotechnical Journal 39: 1168–1180, 2002.

## British standards

– BS 5930 Code of practice for site investigations

– BS 6349-5:1991 Maritime Structures – Part 5: Code of practice for dredging and land reclamation

– BS 1377-1:1990 Methods of test for soils for civil engineering purposes – Part 1: General requirements and sample preparation

– British Standard BS 1377-2, Soils for civil engineering purposes – Part 2: Classification tests, 1990

- BS 1377-4:1990 Methods of test for soils for civil engineering purposes – Part 4: Compaction-related tests

- BS 1377-9:1990 Methods of test for soils for civil engineering purposes – Part 9: In-situ tests

## European standards

- CEN/TC 341 Geotechnical investigation and testing

- CEN/TC 345 Characterization of soils

- CEN/TC 396 Earthworks

- CEN European Standard EN 14731: Execution of special geotechnical works – Ground treatment by deep vibration, European Committee for Standardization: 2005

- CEN European Standard PrEN 15237: Execution of special geotechnical works – Vertical drainage, European Committee for Standardization, Ref. No. prEN 15237: 2005

- CEN ISO/TS 22476 Geotechnical investigation and testing – Field testing

- CEN ISO/TS 17892 Geotechnical investigation and testing – Laboratory testing of soil

- EC8-1, Eurocode 8: Design of structures for earthquake resistance – Part 1: General rules, seismic actions and rules for buildings, CEN, ref EN 1998-1:2004:E

- EC8-5, Eurocode 8: Design of structures for earthquake resistance – Part 5: Foundations, retaining structures and geotechnical aspects, CEN, ref EN 1998-5, 2004:E

- EN ISO 14688-1: 2002 Geotechnical investigation and testing – Identification and classification of soil - Part 1: Identification and description

- EN ISO 14688-2: 2004 Geotechnical investigation and testing – Identification and classification of soil - Part 2: Principles for a classification

- EN ISO 14689-1: 2003 Geotechnical investigation and testing – Identification and classification of rock – Part 1: Identification and description

- EN ISO 22475-1: 2006 Geotechnical investigation and testing – Sampling methods and groundwater measurements – Part 1: Technical principles for execution

- Eurocode 7: Geotechnical design – Part 1: General rules

- Eurocode 7: Geotechnical design – Part 2: Ground investigation and testing

- NEN-EN 872:2005 (nl) Water quality – Determination of suspended solids – Method by filtration through glass fibre filters. Nederlands Normalisatie Instituut

- NEN 6606:2009 (nl) Water - Determination of perspicacity by means of a disk according to Secchi. ICS-number 13.060.60. Nederlands Normalisatie Instituut

- ISO 11923:1997 – Determination of suspended solids by filtration through glass-fibre filters. International Standard Organisation

- ISO 7027:1999 Water quality – Determination of Turbidity. International Standard Organisation

- ISO 5667-1:2006 Water quality – Sampling – Part 1: Guidance on the design of sampling programmes and sampling techniques. ICS-number 13.060.45. International Standard Organisation

## US standards

– ASTM D2487-11: Standard Practice for Classification of Soils for Engineering Purposes (Unified Soil Classification System)

– ASTM D422-63: Standard Test Method for Particle-Size Analysis of Soils, 2007

– ASTM D1196-93: Standard Test Method for Nonrepetitive Static Plate Load Tests of Soils and Flexible Pavement Components, for Use in Evaluation and Design of Airport and Highway Pavements, 2004

– ASTM D1556-07: Standard Test Method for Density and Unit Weight of Soil in Place by the Sand-Cone Method, 2007

– ASTM D1557-12: Standard Test Methods for laboratory compaction characteristics of soil using modified effort, 2009

– ASTM D2167-08: Standard Test Method for Density and Unit Weight of Soil in Place by the Rubber Balloon Method, 2008

– ASTM D2937-10: Standard Test Method for Density of Soil in Place by the Drive-Cylinder Method, 2010

– ASTM D4253-00: Standard Test Methods for Maximum Index Density and Unit Weight of Soils Using a Vibratory Table, 2006

– ASTM D4429-09a: Standard Test Method for CBR (California Bearing Ratio) of Soils in Place, 2009

– ASTM D6698-07: Standard Test Method for On-Line Measurement of Turbidity Below 5 NTU in Water, 2007

– ASTM D6855-10: Standard Test Method for Determination of Turbidity Below 5 NTU in Static Mode, 2010

– AASHTO T191: Standard Method of Test for Density of Soil In-Place by the Sand-Cone Method

– AASHTO T204: Standard Method of Test for Density of Soil In-Place by the Drive Cylinder Method, 1990

– AASHTO T233: Standard Method of Test for Density of Soil In-Place by Block, Chunk, or Core Sampling, 2002

– AASHTO T205 - Standard Method of Test for Density of Soil In-Place by the Rubber-Balloon Method, 1986

– APHA 2540 D - 2005 Gravimetric Method Total Suspended Solids. APHA 1998, Standard Methods for the Examination of Water and Wastewater, 20th Edition, American Public Health Association, 2005.

## Web sites

http://www.alway-associates.co.uk/legal-update/article.asp?id=14

www.betterground.com

www.boskalis.com

http://cellar.org/showthread.php?t=13414

www.cofra.com

www.deme.be

www.dgi-menard.com

www.fao.org

http://geography.about.com/od/physicalgeography/a/karst.htm

http://geology.about.com/library/bl/maps/blworldindex.htm

http://geology.about.com/od/quakemags/a/European-Macroseismic-Scale-EMS98.htm

www.groundimprovement.ch

www.iadc-dredging.com

www.isrm.net (International society for rock mechanics)

http://www.jagansindia.co.in/2011/03/japan-tsunami-devasting-live-footage.html

www.jandenul.com

www.kc-denmark.dk

www.keller.co.uk

www.landpac.co.za

http://www.nerc-bas.ac.uk/tsunami-risks

http://www.osil.co.uk

http://www.planningplanet.com

www.vanoord.com

www.weru.ksu.edu

www.weru.ksu.edu/weps.html

www.slopeindicator.com

http://www.ysi.com

http://en.wikipedia.org/wiki/Karst#Morphology

www.ukho.gov.uk UKHO (United Kingdom Hydrographic Office)